Ground Support Technology for Highly Stressed Excavations

The performance of ground support as a scheme is essential to constrain failures occurring at the rock surfaces of deep or highly stressed excavations. This book covers laboratory and theoretical developments coupled with field experiments and observations with the implementation of the methodology at mines. It explains the energy dissipation capabilities of reinforcement and support systems leading to the design of complete ground support schemes that can maintain integrity following the dynamic ejection of a mass of rock from an excavation boundary.

The key features of the book are as follows

- It explores the mechanics, demand and capacity of ground support technology.
- It covers the whole gamut of theories, laboratory and field test results and case studies related to ground support technology.
- It includes a comprehensive database of mesh, rock bolts, cable bolts and shotcrete capacity.
- It examines ground support scheme testing and explanation.
- It discusses comprehensive case studies, including de-stress blasting.

This book is aimed at professionals in mining engineering, including civil engineering, geological engineering and geotechnical engineering, and related advanced postgraduate studies.

Ground Support Technology for Highly Stressed Excavations

Integrated Theoretical, Laboratory, and Field Research

Ernesto Villaescusa, Alan G Thompson,
Christopher R Windsor, and John R Player

CRC Press
Taylor & Francis Group
Boca Raton London New York

CRC Press is an imprint of the
Taylor & Francis Group, an **informa** business

First edition published 2023
by CRC Press
6000 Broken Sound Parkway NW, Suite 300, Boca Raton, FL 33487–2742

and by CRC Press
4 Park Square, Milton Park, Abingdon, Oxon, OX14 4RN

CRC Press is an imprint of Taylor & Francis Group, LLC

ISBN: 978-1-032-39972-0 (hbk)
ISBN: 978-1-032-41360-0 (pbk)
ISBN: 978-1-003-35771-1 (ebk)

DOI: 10.1201/9781003357711

Typeset in Times
by Apex CoVantage, LLC

Contents

About the Authors xi
Foreword xiii
Preface xv
Acknowledgements xvii

Chapter 1   Introduction 1
     1.1   Introduction 1
     1.2   The Process of Support and Reinforcement 3
     1.3   Brief History of Ground Support Technology 9
     1.4   Scope and Contents of This Book 11

Chapter 2   Terminology 13
     2.1   Introduction 13
     2.2   Reinforcement System Response 13
     2.3   Continuous Mechanically Coupled 13
     2.4   Continuous Frictionally Coupled 16
     2.5   Discrete Mechanically or Frictionally Coupled 16
     2.6   The Load Transfer Concept 16
     2.7   The Embedment Length Concept 18
     2.8   Materials Behaviour Terminology 18
          2.8.1   Elastic 18
          2.8.2   Plastic 18
          2.8.3   Brittle 18
          2.8.4   Ductile 19
          2.8.5   Resilience 19
          2.8.6   Toughness 19
          2.8.7   Yield 19
          2.8.8   Stiffness 19
     2.9   Performance Indicators of Capacity 19
     2.10 Surface Support System Response 21
          2.10.1 Type of Support Systems 21
                2.10.1.1 Point Support Systems 21
                2.10.1.2 Strip Support Systems 22
                2.10.1.3 Areal Support Systems 22
          2.10.2 Loading Mechanisms 22
          2.10.3 Load Transfer Concepts 23
                2.10.3.1 Point Support Systems 23
                2.10.3.2 Strip and Areal Support Systems 25

Chapter 3   Probabilistic Simulation of Rock Mass Demand 27
     3.1   Introduction 27
     3.2   Rock Mass Characterization 27
          3.2.1   Geotechnical Mapping of Underground Exposures 28
     3.3   Rock Mass Model 30
     3.4   Deterministic Assessment of Rock Mass Instabilities 31

3.5   Probabilistic Assessment of Rock Mass Instabilities 33
        3.5.1   Interpretation of Results..37

Chapter 4   Ground Support Mechanics...43
        4.1   Introduction 43
        4.2   Reinforcement Response 43
                4.2.1   Load Transfer within the Collar Region....................................43
                4.2.2   Load Transfer along the Element and at the Toe Anchor Region 46
        4.3   Surface Support Response 48
        4.4   Failure Geometry..50
        4.5   Surface Layer Toughness..52
        4.6   Ground Support Scheme..56

Chapter 5   Ground Support Design...61
        5.1   Introduction 61
        5.2   Rock Mass Characterization 61
                5.2.1   Structure 61
                5.2.2   Strength 62
        5.3   Tunnel Instability..62
                5.2.1   Spalling Failure 63
                5.2.2   Structurally Controlled Failure 65
                5.2.3   Damage and Deformation Prior to Violent Failure 66
        5.3   Ejection Velocity...67
        5.4   Ground Support Demand..73
                5.4.1   Shallow Depth of Failure (Spalling) 75
                5.4.2   Structurally Controlled Depth of Failure 75
        5.5   Ground Support Capacity 80
                5.5.1   Ground Support Design for Extremely High
                         Demand Conditions..82

Chapter 6   Laboratory Testing...87
        6.1   Introduction 87
        6.2   Testing Documentation...87
        6.3   Component Testing..87
        6.4   Element Testing 90
                6.4.1   Prestressing Strand 90
                6.4.2   Solid Threaded Bar 90
        6.5   Internal Fixture Testing 93
                6.5.1   Slot and Wedge Anchors 93
                6.5.2   Expansion Shell Anchors 94
                6.5.3   Cement Grouts 97
                6.5.4   Resin Grouts..97
        6.6   External Fixture Testing 97
                6.6.1   Split Tube Rings 98
                6.6.2   Barrel and Wedge Strand Anchors 98
                6.6.3   Plates 100
        6.7   Systems Testing 100
                6.7.1   Cable Bolt Pull Tests..104
                6.7.2   Resin-Encapsulated Reinforcement 105
        6.8   WASM Dynamic Test Facility...108
                6.8.1   Force Transfer and Displacement..111

		6.8.2	Simulated Boreholes and Sample Installation	111
		6.8.3	Dynamic Test Results	113
		6.8.4	Analysis of Dynamic Test Results	116
		6.8.5	Dynamic Testing of Mesh and Shotcrete Layers	119

Chapter 7 Energy Dissipation of Rock Bolts 121

7.1 Introduction 121
7.2 Momentum Transfer 121
7.3 Reinforcement System Load Transfer 121
7.4 Continuously Frictionally Coupled 123
    7.4.1 Friction Rock Stabilizer 123
        7.4.1.1 Static Testing 124
        7.4.1.2 Dynamic Testing 124
    7.4.2 Expanded Tube Bolts 126
        7.4.2.1 Static Testing 127
        7.4.2.2 Dynamic Testing 129
    7.4.3 Hybrid Point Anchored Bar and Split Tube Bolts ... 132
        7.4.3.1 Borehole Simulation and Installation ... 132
        7.4.3.2 Dynamic Testing 134
7.5 Continuously Mechanically Coupled 140
    7.5.1 Cement-Encapsulated Threaded Bar 140
        7.5.1.1 Australian Threaded Bar 142
        7.5.1.2 Chilean Threaded Bar 142
    7.5.2 Resin-Encapsulated Threaded Bar 144
7.6 Discretely Mechanically or Frictionally Coupled 147
    7.6.1 Cement-Encapsulated Decoupled Threaded Bar 148
    7.6.2 Resin-Encapsulated Decoupled Posimix 149
    7.6.3 Cement- and Resin-Encapsulated D Bolt 156
    7.6.4 Cement- and Resin-Encapsulated Cone Bolt 157
    7.6.5 Cement- and Resin-Encapsulated Garford Dynamic Bolt ... 161
    7.6.6 Cement-Encapsulated Durabar Bolt 165
    7.6.7 Cement-Encapsulated Yield-Lok Bolt 166
    7.6.8 Self-Drilling Anchor Bolt 170

Chapter 8 Energy Dissipation of Cable Bolts 173

8.1 Introduction 173
8.2 Cement Grout 175
    8.2.1 Physical and Mechanical Properties 175
        8.2.1.1 Fresh Cement Paste 175
        8.2.1.2 Hardened Cement 178
    8.2.2 Grouting Reinforcement Boreholes 185
        8.2.2.1 Toe-to-Collar Grouting 186
        8.2.2.2 Mechanized Grouting 188
8.3 Cable Bolt Types 189
    8.3.1 Modified Strand Cable Bolts 189
    8.3.2 Plain Strand—15.2 mm Diameter 190
    8.3.3 Plain Strand—17.8 mm Diameter 196
    8.3.4 Decoupled Strand 201
    8.3.5 Multiple Dynamic Impact Testing 204
8.4 Cable Bolt Plates 206
    8.4.1 Barrel and Wedge Anchors 207

		8.4.1.1	Anchor Installation	207
		8.4.1.2	Anchor Mechanism and Performance	211
	8.4.2	Dynamic Testing Results		213

Chapter 9 Energy Dissipation of Mesh Support 215
    9.1   Introduction 215
    9.2   Mesh Load Transfer 215
    9.3   Mesh Testing 218
        9.3.1   Boundary Conditions 220
        9.3.2   Loading Method 220
    9.4   Mesh Force and Displacement 220
        9.4.1   Static Results for Welded Wire Mesh 223
        9.4.2   Static Results for Woven Mesh 224
        9.4.3   Dissipated Static Energy 227
        9.4.4   Dynamic Results for Welded Wire Mesh 228
        9.4.5   Dynamic Results for Woven Mesh 228
        9.4.6   Dissipated Dynamic Energy 231

Chapter 10 Energy Dissipation of Shotcrete Support 235
    10.1   Introduction 235
    10.2   Shotcrete Mix Design 235
    10.3   Material Properties 236
        10.3.1   Cement 236
        10.3.2   Mechanism of Hydration of Cement 236
        10.3.3   Supplementary Cementing Materials 237
            10.3.3.1   Silica Fume 237
            10.3.3.2   Fly Ash 238
            10.3.3.3   Slag Cement 238
        10.3.4   Mixing Water 239
        10.3.5   Aggregate 239
        10.3.6   Fibres 240
            10.3.6.1   Steel Fibres 242
            10.3.6.2   Synthetic Fibres 242
        10.3.7   Admixtures 243
            10.3.7.1   Accelerator 243
            10.3.7.2   Superplasticizer 243
            10.3.7.3   Air-Entraining 243
            10.3.7.4   Hydration Stabilizer 244
    10.4   Shotcrete Support System 244
        10.4.1   Rock Surface and Shotcrete Profiles 245
        10.4.2   Deformation Mechanisms 247
    10.5   Static Performance of Freshly Sprayed Shotcrete 248
        10.5.1   Review of Shotcrete Early Strength 249
        10.5.2   Shear Strength of Freshly Sprayed Shotcrete 251
            10.5.2.1   Development of Shear Strength of Shotcrete Paste with and without the Influence of a Chemical Admixture 252
            10.5.2.2   Development of Shear Strength of Shotcrete Paste with Various Combinations of Mixed Components 253

10.5.3   Structural Requirements for a Freshly Sprayed Shotcrete Layer....253
    10.5.3.1   Requirements for Self-Support.......................................253
    10.5.3.2   Required Shotcrete Shear Strength for Self-Weight
      and Block Support 256
10.5.4   Safe Re-entry Time 257
10.6   Static Performance of Cured Shotcrete 259
10.6.1   Uniaxial Compressive Strength (UCS) 259
10.6.2   Tensile Strength..261
10.6.3   Tensile Bond Strength of Rock—Shotcrete Interface 261
10.6.4   Shear Strength 263
10.6.5   Toughness...264
10.7   Shotcrete Failure Mechanisms..265
10.7.1   Shotcrete Load Transfer 266
10.8   Large Scale Static Testing 267
10.9   Large Scale Dynamic Testing 270
10.9.1   Shotcrete Test Set-Up...270
10.9.2   Shotcrete Failure Mechanism...273
10.9.3   Shotcrete Energy Dissipation 274
    10.9.3.1   Fibre-Reinforced Shotcrete 274
    10.9.3.2   Mesh-Reinforced Shotcrete..280

Chapter 11 Dynamic Performance of Ground Support Schemes 283
11.1   Introduction 283
11.2   Load Transfer..283
11.3   Free Body Diagrams...285
11.4   Combined Reinforcement and Mesh Schemes......................................286
11.4.1   Sample Preparation and Testing..287
11.4.2   Data Analysis 288
11.4.3   Data 289
    11.4.3.1   Cement-Encapsulated Rebar and G80/4 Mesh...............290
    11.4.3.2   Cement-Encapsulated Rebar and Welded
      Wire Mesh 295
    11.4.3.3   Decoupled Posimix and Welded Wire Mesh 298
    11.4.3.4   Decoupled Posimix and Woven Mesh............................306
11.4.4   Summary of Energy Dissipation 321
11.5   Large-Scale Testing of Full-Scale Schemes 323

Chapter 12 Reinforced Block Analysis..327
12.1   Introduction 327
12.2   Description of the Problem..327
12.3   Reinforcement Response at a Block Face 327
12.4   Reinforcement Databases 328
12.4.1   A Generic Reinforcement System..332
12.4.2   Measurement of Reinforcement System Responses 333
12.4.3   Reinforcement System Simulations..333
12.4.4   Reinforcement System Responses 333
    12.4.4.1   Component Properties..333
    12.4.4.2   Interface Properties...335

12.4.4.3 Variations of Axial Force-Displacement Responses with Encapsulation Length 335

12.4.4.4 Axial Force-Displacement Responses with External Fixture and Encapsulation Lengths 335

12.4.5 Application of the Reinforcement Databases in Design for a Reinforced Block 337

12.4.5.1 Description of the Analysis 337

12.4.5.2 Estimation of Force-Displacement Responses 337

12.5 Static Analysis of a Reinforced Arbitrarily-Shaped Block 338

12.5.1 The Design Problem 340

12.5.2 Description of the Analysis Method 340

12.6 Dynamic Block Loading 341

12.6.1 Newtonian Mechanics-Based Analysis 342

12.6.2 Momentum-Based Analysis 342

12.6.3 Energy-Based Analysis 343

12.6.4 Summary of Dynamic Analysis Methods 343

12.7 Displacement Controlled Dynamic Analysis Methodology 343

12.8 Example of Implemented Theory 344

12.8.1 Description of the Analysis 344

12.8.2 Imposed Loading and Results 345

Chapter 13 Construction and Monitoring 351

13.1 Introduction 351

13.2 Induced Stress 351

13.3 Construction 354

13.3.1 Excavation Shape 355

13.3.2 De-Stress Blasting 357

13.3.2.1 Mechanics 359

13.3.2.2 De-stress Blasting Patterns 361

13.3.2.3 Explosive Energy 362

13.3.2.4 Micro Seismic Activity 364

13.3.2.5 Geotechnical Concerns 365

13.3.2.6 Case Study—Kanowna Belle Mine, WA, Australia 366

13.3.3 Construction of a High Energy Dissipation Ground Support Scheme 379

13.3.3.1 Clearing of Temporary Face Support 380

13.3.3.2 Mechanical Scaling 380

13.3.3.3 Structural Geological Mapping with Photogrammetry .. 381

13.3.3.4 Shotcrete Application 381

13.3.3.5 Primary Reinforcement Mark-Up 383

13.3.3.6 Installation of Primary Reinforcement and Mesh 384

13.3.3.7 Primary to Secondary Support Installation Sequence 391

13.3.3.8 Final Ground Support Scheme Arrangement 396

13.4 Monitoring 397

13.4.1 Drilling and Blasting 398

13.4.2 Deformation 401

References 405

Index 417

About the Authors

Ernesto Villaescusa is a mining engineer with over 35 years of experience and specialization in underground mining methods and ground support of underground excavations. Over the last 25 years, Ernesto has conceptualized and undertaken several research projects in ground support technology. These projects were undertaken at the Western Australian School of Mines (WASM) in conjunction with many sponsoring companies. They ranged from static and dynamic laboratory testing of support and reinforcement elements to *in situ* assessment of ground support corrosivity. One of the main objectives was to understand the energy dissipation capabilities of reinforcement and support systems leading to the design of complete ground support schemes that can maintain integrity following a dynamic ejection at an excavation boundary. Ernesto has also worked extensively in underground excavation design. In 2014, he published the textbook 'Geotechnical Design for Sublevel Open Stoping'. In May 1997, he established the Mining Rock Mechanics Research Group at WASM in Kalgoorlie. Since then, the WASM team has graduated under his direct supervision a total of 13 PhD students. In addition, more than 50 master's candidates have completed a postgraduate mining geomechanics program at Curtin University. A major benefit of the WASM research is the rapid technology transfer of mining rock mechanics research into industry practice. Also noteworthy is that, over the past 25 years, the research group has received more than 30 million AUD in funding from several mining industry companies.

Alan G Thompson was formally trained in civil and geotechnical engineering and has over 40 years of industrial experience, mainly in the area of ground support for underground rock excavations. Alan has collaborated and worked with mining companies, ground control hardware manufacturers and contractors, consultants and government regulatory authorities located across all states of Australia. He has also worked internationally in Argentina, Brazil, Canada, Chile, Papua New Guinea, South Korea, South Africa and Thailand. These work assignments involved applied research, consulting, independent ground support component and system testing and assessment for development, ground support technology transfer and training. Alan and his colleagues in the early 1980s developed the double embedment length test for evaluating cement grouted reinforcement systems in the laboratory. This test configuration became the standard accepted and used by various organizations throughout the world. The double embedment test has been used in the WASM Dynamic Test Facility for the last two decades. Several results from this test are presented throughout this book. To complement laboratory testing, Alan developed software to simulate load transfer in all types of reinforcement systems. This software can replicate the response of reinforcement systems to static and dynamic loading and provides invaluable insight into the expected *in situ* performance of reinforcement systems. Alan has also been involved in the development of software which integrates various modules ranging from analysis of geological mapping data to the stability assessment of reinforced blocks of rock subjected to either static or dynamic loading. The outcomes from the innovative work have resulted in more than 50 technical papers that have been presented at international conferences in countries such as Australia, Canada, Chile, Hong Kong and the USA. Alan is currently enjoying semi-retirement in the remote seaside town of Esperance in the southeast of Western Australia.

Christopher R Windsor has over 40 years of experience in rock mechanics research and development. He obtained his first degree in civil engineering, followed by an MSc and a DIC (Imperial College of Science Technology and Medicine, London). He spent 16 years at the CSIRO as a senior principal research scientist and manager of the Rock Reinforcement Group, five years as Director of Rock Technology and, finally, 11 years as Principal Investigator of CRC Mining Projects and Associate Professor at WASM. During this period, he managed and participated in five back-to-back, three-year research projects conducted by CSIRO and later by Rock Technology, all sponsored by the Australian and International Mining Industry.

In his initial projects, he concentrated on developing instruments and methods to measure deformation in structured rock and reinforcement, many of which were commercialized. He then worked on developing mapping procedures and computer programs for statistically analysing and characterizing rock structure geometry. These were supplemented with a suite of deterministic and probabilistic block theory procedures and programs for reinforcement design in structured rock. His later projects involved analysing and reconciling the stress in Earth's crust using a stress database comprising most of the rock stress tensor measurements that have been conducted.

Chris has conducted research for over 50 companies on four continents; written over 50 computer programs, over 50 scientific papers and over 50 technical reports; given advanced training to over 60 postgraduates at eight universities; organized and assisted in over ten conferences; given over 50 industrial training courses; and was awarded two patents. He was a joint founder of the Australian Shotcrete Association. In addition, he spent ten years on the national committee for the Australian Underground Construction and Tunnelling Association (AUCTA), ten years as AUCTA editor, eight years as a national correspondent for the International Journal of Tunnelling and Underground Space Technology and editor for the International Journal of Rock Mechanics and Mining Sciences. He gave the Bicentennial Engineering Lecture on Australian tunnelling at ICE in London. In 1996, he was awarded the Schlumberger Prize for Excellence in Rock Mechanics.

John R Player is trained in mining and geotechnical engineering and has 30 years of industrial and research experience. His work experience covers a broad and applied knowledge base in areas of mine engineering (production and planning), mine design (method selection and layouts) and geotechnical assessments (rock mass classification and variability) for mine excavations (entry and non-entry). John has worked as a site-based mining and geotechnical engineer, researcher and consultant. He is the owner and Principal Engineer of the consulting company MineGeoTech. John undertook his PhD at WASM and was responsible for the design, construction and testing at the WASM Dynamic Test Facility from 2001 to 2011.

Foreword

The continued and growing worldwide demand for minerals means that mining operations are being conducted at ever-increasing depths. Safe and efficient mineral extraction at depth is accompanied by significant engineering problems that require the modification of existing practices and the development and implementation of new technologies. This book is timely in that it attempts to address a particularly difficult problem—maintaining the stability of production and infrastructure excavations under changing static and dynamic conditions.

Mineral deposits are geological anomalies situated within a host rock mass, commonly structured by a network of mutually intersecting discontinuities. As depth increases, *in situ* stress increases along with the chance of two types of instabilities which are mechanisms of stress relief and manifest at the free surfaces of excavations. Firstly, there is the possibility of fracture through otherwise intact rock at excavation surfaces. Secondly, there is the possibility of violent detachment of rock blocks formed by discontinuities that intersect the free surfaces. In low-stress, quasi-static environments, instability has been successfully managed by installing arrangements of artificial reinforcement and support. However, at depth, the current design and implementation procedures for these treatments produce inadequate results. The geometry of the rock structure is stochastic and, together with high stress produces a very difficult dynamic problem.

The research programs used to define the dynamic demand of the rock and the dynamic capacity of artificial reinforcement and support provide the framework of this book and demonstrate the use of the scientific method to solve a non-trivial engineering problem. The work involved a multi-decade research program comprising a considered sequence of field observations, *in situ* testing programs, laboratory testing programs and theoretical developments that reconcile with the results, all leading to a computer design methodology with the final proof of validity demonstrated by extensive, full-scale field trials. Each topic has been researched with sufficient and necessary detail to allow firm recommendations to be made on the basis of systematic investigations, observations and formal conclusions.

Clarity is achieved by adopting an internally consistent formal terminology based on the principles of mechanics. Mechanisms of static and dynamic instability and the various artificial devices employed to manage instability are treated as mechanical schemes comprising nested systems of mechanical components. This approach has abandoned the standard design practice of considering the behaviour of mechanisms in terms of forces alone with the attendant assumptions that result in the so-called 'factor of safety'. This is replaced with the formal requirement that any mechanism comprising nested systems must simultaneously achieve force equilibrium and displacement compatibility for all system components in order to be considered stable. It allows arbitrary treatment of the geometry, morphology and spatiality of all system components that comprise the mechanism. Finally, the approach produces results that allow selection and arrangement of artificial components with suitable force-displacement capacities to satisfy the force-displacement demands required to stabilize an unstable mechanism.

This approach defined the requirements for experimental testing and theoretical development. It produces the required engineering design rigour and implementation advantages not necessarily available elsewhere. The extensive and detailed results from static and dynamic, *in situ* and laboratory investigations, preserved in carefully structured databases, are of considerable value. The theoretical investigations and developments have been carefully reconciled with experimental results culminating in computer-based design methodologies, which contribute to the state-of-the-art and can be expected to stand the ultimate test of time.

The book is presented in a logical structure of chapters and written in a style with undergraduate and postgraduate students, mine engineers and consultants in mind for self-learning and reference.

The research reported herein has been conducted by a close, long-term group of dedicated researchers with complementary expertise. The writer is privileged to have had a close association with the group and with the authors of this book throughout the period when the reported work was carried out. During this period, the writer enjoyed several stimulating visits to Kalgoorlie as the group's guest.

The following summarizes his links with the individual authors of this outstanding book:

- The group's leader, Professor Ernesto Villaescusa, was his first PhD student when he returned to Australia from London to take up a position at the University of Queensland by the end of 1987. He has remained Ernesto's friend and mentor ever since.
- He shares with Dr Alan Thompson the distinction of being a civil engineering graduate of the University of Melbourne, albeit ten years earlier than Alan. Probably because of this background, he particularly appreciates Alan's approach to solving engineering problems. He was familiar with the excellent work that Alan and Chris Windsor did at CSIRO, Australia, from 1980 to 1995.
- He taught Chris Windsor when Chris did the master's in engineering rock mechanics at Imperial College, London, during 1983–1984, and has attempted to encourage Chris in his subsequent career.
- He had the honour of observing and reviewing Dr John Player's experimental work during several visits to Kalgoorlie, including acting as an examiner of John's PhD thesis in 2012.

He offers his congratulations to the authors on their manifest contributions to mining rock mechanics, not least through the publication of this book.

Edwin T. Brown, AC
Emeritus Professor, University of Queensland, Brisbane, Queensland, Australia
(formerly Senior Deputy Vice-Chancellor)
President, International Society for Rock Mechanics, 1983–1987
Senior Consultant, Golder Associates, 2001–2018
Brisbane, Australia
25 April 2022

Preface

Underground excavations reach depths below the ground surface, where high energy failure events frequently occur at the excavation boundaries of many tunnels and underground mines worldwide. The observations indicate that stress-driven, violent failures first occur as shallow, strain bursts. However, if the induced stress increases, it can be followed by the mobilization of geological discontinuities, potentially causing larger, structurally-controlled block ejection. The failure depth, area and mass can result in energy demand capable of exceeding the installed ground support capacity.

The performance of ground support as a scheme is essential to constrain failures occurring at the rock surfaces of deep or highly stressed excavations. As the incipient instabilities develop into significant failures, they are likely to first mobilize several reinforcement elements prior to loading the surface support layers. As the deformation increases, several reinforcement elements may fail either through slow, time-dependent deformation or violently. In such situations, the load is transferred to the surface support located immediately adjacent to any failed reinforcement elements. The surface support layers require sufficient toughness to displace and transfer the load to other reinforcement elements located in the periphery of the failure geometry. The surface support needs to be completely connected to all the reinforcement elements within and outside the failure for the load transfer to occur efficiently and contain the failed material.

Over the last 20 years, the Western Australian School of Mines (WASM) has undertaken research to quantify the dynamic energy dissipation capabilities of many commercially available rock bolts, cable bolts and mesh support systems. The Mining Rock Mechanics group at the WASM secured sponsorship from the mining industry and the Western Australian government to develop static and dynamic loading simulation test facilities that use unique methodologies. The laboratory testing results have been implemented at several deep underground mine sites, including the Northern Star Resources Kanowna Belle Mine in Australia and the New Mining Level Mine, which is part of the huge El Teniente Mine of CODELCO, Chile.

The work presented here is largely unpublished and draws heavily from the results of a large number of master's and PhD theses that were completed at the WASM over the last two decades or so. The research work comprised laboratory and theoretical developments coupled with field experiments and observations related to the implementation of the methodology at several working mines that had previously experienced widespread failures. In all cases, the conventional ground support design and implementation approach was replaced by a High Energy Dissipation (HED) ground support design aimed at simultaneously achieving force equilibrium and displacement compatibility for all the system components within the installed ground support schemes.

The authors would like to acknowledge the financial support from several mining companies, including the sponsors of the Mining Development at Great Depth Research Project, comprising Northern Star Resources, CODELCO, Newcrest, BHP, Geobrugg and DSI. The research was undertaken through CRC Mining and later through Mining3. The early financial support of the Western Australian government through the Minerals and Energy Research Institute of Western Australia (MERIWA) was also important and greatly appreciated.

Ernesto Villaescusa
Alan G Thompson
Christopher R Windsor
John R Player
Western Australian School of Mines

Acknowledgements

This book was written with underground rock mechanics engineers, consultants and postgraduate students in mind. It draws heavily on the knowledge, laboratory experiments and mining experience gained during our years of employment at the WASM in Kalgoorlie (1997–2022). It also benefits from early theoretical work and course notes developed by some of the authors at the Rock Reinforcement Group, the CSIRO Division of Geomechanics and the Rock Technology company in Perth. Also, it must be noted that the work presented here would not have been possible without the significant contribution from our research students and technicians at the WASM. In particular, we wish to acknowledge the following key personnel whose contributions eventually helped shape the contents of this book:

- Professor Edwin T. Brown kindly agreed to write the Foreword and provided encouragement from the early stages of conceptualization of this book. His friendship and technical advice continue to be invaluable to this day.
- Ushan de Zoysa, as part of his PhD studies, tirelessly undertook dynamic testing for over a decade at WASM. His friendship, disposition for hard work and commitment to excellence have been an asset during difficult sample preparation, dynamic testing and data analysis.
- Dr Nixon Saw, a close friend, offered his excellence in technical support spreading over many areas. Nixon has undertaken a PhD in Early Strength of Shotcrete at WASM and has contributed significantly to the earlier versions of Chapter 10.
- Dr Ayako Kusui is known for her great attention to detail, a great volume of work and a commitment to excellence during her PhD studies. Her scaled-down tunnel testing provided the basis for an improved understanding of the mechanics of tunnel instability under dynamic loading.
- Dr Chris Drover comprehensively worked on de-stress blasting and analysis of micro-seismicity at the tunnel scale. His detailed PhD work at the CODELCO mine site proved the concepts and allowed technology transfer to the mines in Western Australia. He contributed significantly to Chapter 13.
- Dr Nadia Bustos provided ongoing innovative ideas on the integration of the geotechnical data, which allowed comprehensive studies of tunnel construction while taking into account all the geological, construction and resulting seismicity variables.
- Dr Ellen Morton contributed through her early work in mesh and shotcrete testing, which allowed the results and the boundary conditions to be put into perspective.

We are very grateful to several companies, organizations and individuals that supported the WASM's ground support technology research over the last 25 years, including the Development Mining at Great Depth research project. We would like to thank the following entities:

- Northern Star Resources, in particular Corin Arcaro, James Coxon and Vic Simpson, provided the research sponsorship and the tunnel testing sites at the Kanowna and Kundana mine sites. Without their support and confidence in our methods, we would not have been able to implement a change to a high energy dissipation ground support scheme, which is continuously being improved to date.
- CODELCO Chile, in particular Eduardo Rojas, Antonio Bonani and Octavio Araneda, provided the initial research sponsorship and test site at the El Teniente New Mining Level Mine. Their trust was critical to the development of the early high energy dissipation ground support design concepts and their early implementation.

- Geobrugg, in particular Andrea Roth, Roland Bucher and Erik Rorem, provided research sponsorship from the early concept stages until this day. Their support was critical to the development and improvement of the WASM Dynamic Test Facility.
- DSI Underground, in particular Derek Hird, supported and sponsored our research for decades. Their support allowed the development and comprehensive testing of the decoupled Posimix bolt, which is a fundamental part of the high energy dissipation schemes designed and implemented by WASM at several sites.
- Newcrest Mining, in particular our friends Eamonn Hancock and Dr Ayako Kusui, provided technical support and research sponsorship that resulted in the implementation of a high energy dissipation scheme at the Cadia Mine.
- BHP, in particular Rigoberto Rimmelin, provided research sponsorship that allowed some of the laboratory and theoretical developments presented herein to take place.
- Rio Tinto (Oyu Tolgoi), in particular Glenn Watt, Oleg Below and Ooi Johnson, provided research sponsorship that allowed dynamic testing of cable bolt reinforcement elements.
- MMG provided financial support during the early stages of our research project.
- INCHALAM Chile provided support in mesh strength and deformation research.
- MERIWA provided financial support for the design, construction, commissioning and comprehensive testing of the WASM Dynamic Test Facility in Kalgoorlie.

We are also grateful to the following individuals whose support has been critical to the conceptualization and completion of the research results presented herein:

- Dr Tao Li, then a member of WMC Resources and later of Newcrest Mining, provided encouragement and research sponsorship that allowed dynamic testing back in the early 2000s. His support of the WASM Rock Mechanics Chair is also gratefully acknowledged.
- Dr Dave Beck provided insight into detailed numerical modelling at both the global and detailed scales, which have provided input for full-scale tunnel experiments at the research sponsoring mine sites.
- Dr Rhett Hassell provided technical support and encouragement during the early stages of the implementation of high energy dissipation in Western Australia.
- Dr Italo Onederra, a friend, provided technical support with de-stress blasting during the early stages of the research in CODELCO and provided his support in the development of blasting optimization in Western Australia.

We are also grateful to several friends and colleagues who have worked with us at several mine sites over the last 25 years. Their support, practical approach and ideas have also helped us write this book. We would like to thank:

- Dr Cesar Pardo, Pedro Landeros and Washington Rodriguez of CODELCO;
- Roo Talebi and Simon Thomas of Northern Star Resources; and
- Dr Steve Webber, Dr Mostafa Sharifzade and Michael Burns of EKJV.

We also wish to thank all those who supported us throughout this very long undertaking, especially:

- The WASM rock mechanics technical staff, including Brett Scott, Lance Fraser, Pat Hogan, Jacopo Lopez and Myriam Sullivan;
- The master's students Erol Akdogan and Carlos Almanza for their thesis work in shotcrete and mesh testing, respectively;
- A number of undergraduate students for their time and efforts undertaking their research projects at the WASM Dynamic Test Facility, including Fred Wallefeld, Alex Cull, Martin Filar, Dave Pearce and Kyle De Souza;

- James Langdon and Glynn Cadby for their support with the instrumentation of the WASM Dynamic Test Facility;
- The financial and administrative support of Curtin University, the CRC Mining and Mining3; and
- Our families for their love, patience, tolerance and understanding during the long periods committed to writing this book.

1 Introduction

1.1 INTRODUCTION

Ground support technology has been used for many decades in civil and mining engineering projects involving both near surface and deep excavations. Over time, many strategies have been developed for internal reinforcement of the rock masses surrounding the excavations and also for support and retention of the newly exposed excavation faces. The effectiveness of a ground support strategy is important for two main reasons—safety to personnel and equipment and achieving the most economic extraction while maintaining safety at a reduced operating cost.

The types of support and reinforcement required at a particular location are dependent on several factors, including the purpose of the excavation, the prevailing rock mass strength, the geometry (shape and size) and orientation of the excavation, the induced stresses over the lifetime of the excavations, the blasting or construction practices used and the amount of water, alteration and the weathering processes involved.

Two stabilization techniques can be used to improve and maintain the load bearing capacity of a rock mass near the boundaries of an underground excavation (Windsor and Thompson, 1992a). Both techniques are capable of responding to a significant inward movement of the rock mass surrounding an excavation:

- **Rock Reinforcement**

Reinforcement is considered to be a system of components installed in boreholes drilled into a rock mass, such as cement- or resin-encapsulated threaded bar, friction stabilizers and cable bolts. The reinforcing elements are an integral part of a reinforced rock mass.

- **Rock Support**

Support is considered to be a system of components that are located within the exposed faces of excavations, such as plates, mesh, straps, shotcrete and steel arches. The supporting members provide surface restraint while enabling connection within a reinforcement pattern at an excavation boundary.

The reinforcing elements provide effective stabilization by enabling a rock mass to support itself (Hoek and Brown, 1980). This is achieved by preventing unravelling and enhancing the self-interlocking properties of a rock mass through keying, arching or composite beam reinforcing actions (Windsor and Thompson, 1992a). A reinforcement pattern strengthens the exposed rock mass surrounding an excavation by preventing the detachment of loose blocks and increasing the shear strength of the geological discontinuities intersected by the reinforcing elements. This results in a reinforced zone that minimizes the dilation of pre-existing geological discontinuities and helps redistribute stresses around the excavation.

Ground support comprises combinations of reinforcement and support systems (Figures 1.1 and 1.2). It is a normal practice to design the reinforcement to act with the support to form a ground support scheme (Windsor and Thompson, 1992a). That is, the support is restrained by a plate held in place by the reinforcement system. If this interaction at the collar of the reinforcement system fails, then the ground support scheme requires effective connection, usually by mesh, to other elements outside the instability. An important aspect of the ground support design is its overall response to the amount of rock mass deformation and the rate at which it occurs (Figure 1.3).

DOI: 10.1201/9781003357711-1

FIGURE 1.1   Example of support and reinforcement damaged by a violent failure with a shallow depth of instability.

FIGURE 1.2   Example of a high energy dissipation ground support scheme (a) before and (b) after an ML 2.4 seismic event.

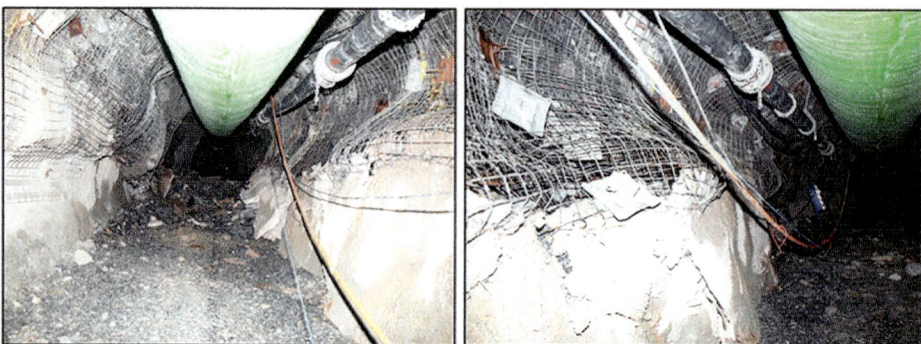

FIGURE 1.3   Example of a large deformation resulting in extensive excavation rehabilitation, significantly affecting productivity.

1.2 THE PROCESS OF SUPPORT AND REINFORCEMENT

A generalized approach to support and reinforcement is presented in the flow chart shown in Figure 1.4. Once an excavation has been created (or proposed), the first step is to consider the stability of the unsupported spans. The main general factors controlling unsupported behaviour are the excavation geometry (shape, size and orientation); the rock mass strength, including the effects of weathering and groundwater; the magnitude and orientation of the induced stresses; and the drilling and blasting practices. The potential failure geometries and their modes of failure, such as gravity fall, plane and wedge sliding, buckling, toppling, swelling and spalling or dynamic block ejection, must be determined at this stage.

An assessment of unsupported stability must also consider the likely shape and size of the potentially unstable blocks with respect to the exposed spans. In underground mining, the size of the equipment, such as the dimensions of the development mining jumbos and other production equipment, often controls the minimum size of the underground excavations. A critical relationship exists between the size of the potentially unstable blocks (i.e., potential falling/sliding) and the exposed

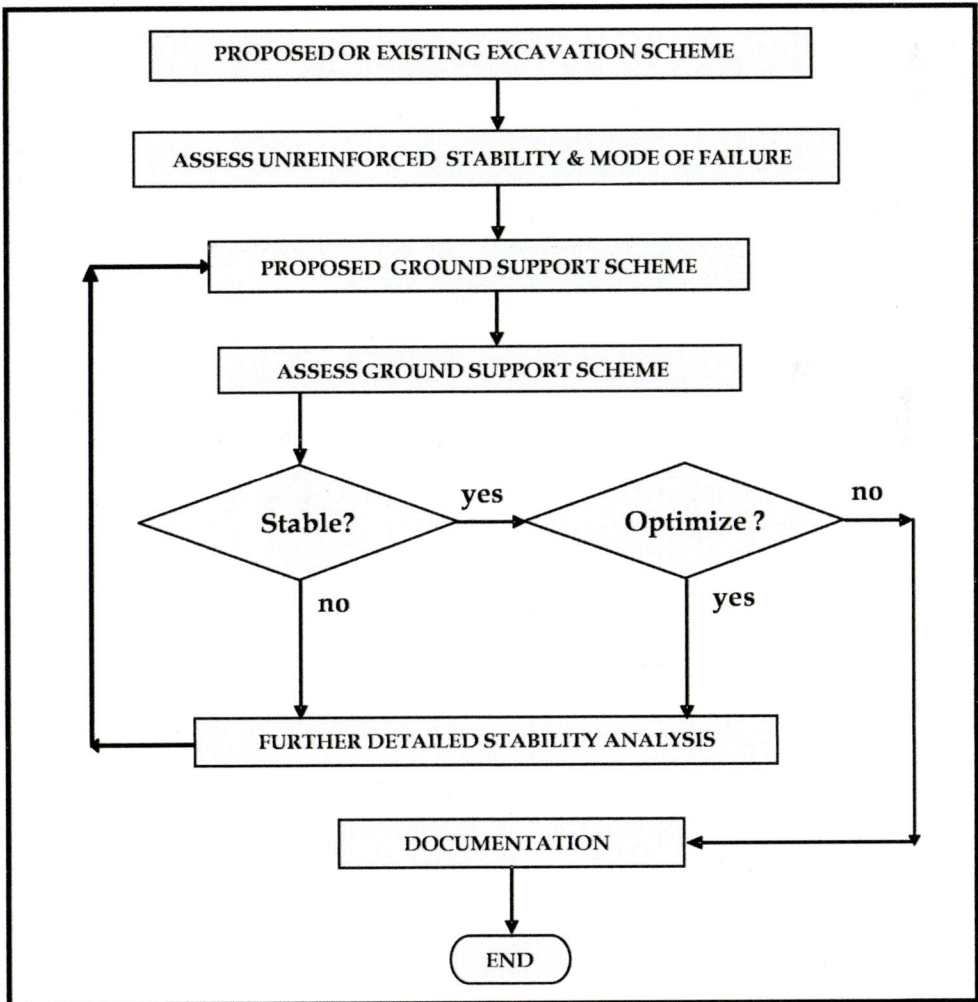

FIGURE 1.4 A generalized approach to rock support and reinforcement (modified from Windsor and Thompson, 1992a).

spans. The larger the span, the greater the probability of potential fully formed falling or sliding blocks with their mass controlled by the size, orientation and interaction of the geological structures. For static conditions, intact rock bridges are likely to limit the size of the potentially unstable blocks (Windsor, 1996). However, a rock mass can still fail violently under very high stress conditions, with the failure geometry completely controlled by the location and orientation of the geological discontinuities (Figure 1.5).

FIGURE 1.5 An example of violent failure controlled by geological structures having a limited area with a significant depth of instability.

When an excavation is created, the load originally supported by the excavated rock must be transferred to the surrounding rock mass. It causes an increase in the (tangential) rock stresses near the surface of the openings. The resulting damage depends upon the shape of the opening and the magnitude and orientation of the *in situ* stresses with respect to the rock mass strength. If a rock mass is weak, or if the excavation is at great depth, the load redistribution may increase the induced stress above a certain threshold leading to rock mass failure (Figure 1.6). Rock mass failure involves the progressive loosening of slabs and rock blocks, usually defined by geological discontinuities. Depending upon rock mass stiffness, the failure may be gradual, leading to a time-dependent, excessive deformation or could result in a sudden, violent ejection.

Even when a rock mass has failed, immediate collapse may not necessarily occur due to the self-interlocking effects of the broken rock pieces. Experimental work by Kusui (2015) has shown that moderately weak materials fail non-violently, with effectively zero ejection velocities. For moderately strong to very strong materials (Figure 1.7), the threshold for violent ejection can be given as the ratio of Uniaxial Compressive Strength (σ_c) to maximum induced tangential stress (σ_{max}), as shown below:

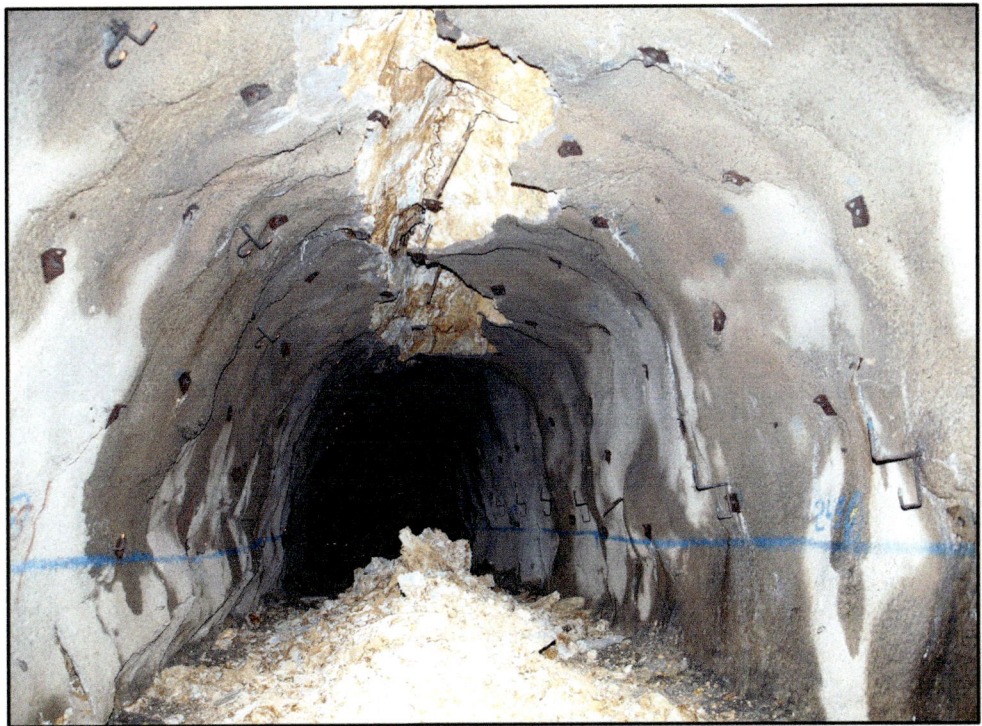

FIGURE 1.6 An example of stress-driven failure at shallow depth for an excavation located within a heavily altered, weak-strength rock mass.

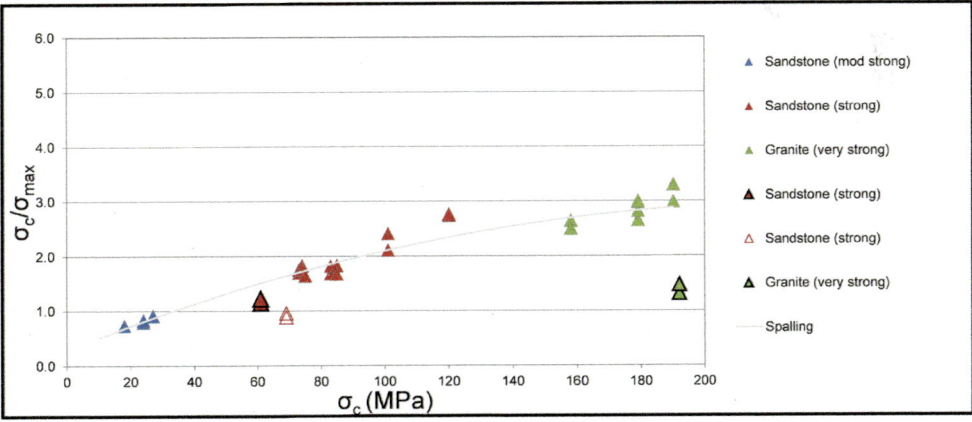

FIGURE 1.7 Unsupported tunnel spalling as a function of tangential stress and rock mass strength (Kusui, 2015).

$$\frac{\sigma_c}{\sigma_{max}} = -0.00005\sigma_c^2 + 0.0232\sigma_c + 0.2767 \qquad 1.1$$

Damage from development blasting can cause the opening of pre-existing discontinuities or cracking of intact rock bridges. This damage effectively decreases the rock mass strength and, in some cases, may lead to failures. Water permeating through the fracture pattern in a rock mass may also affect unreinforced stability, particularly if swelling of altered materials occurs near the excavation boundaries. The potential for corrosion of the reinforcement elements must be determined by analysing the water quality to determine if aggressive chemicals are present in a particular mining environment (Hassell, 2007).

If instability is predicted, the excavation can be redesigned (e.g., to have smaller spans, a differently shaped profile or a different excavation orientation axis). Alternatively, the excavation must be stabilized using ground support technology. In the latter, a ground support scheme with sufficient capacity to exceed the expected rock mass demand (while addressing the expected rock mass behaviour and failure mode) determined during the initial assessment must be selected. Any subsequent loading conditions from nearby or approaching mining fronts are critical. Factors such as long-term exposure to weathering must be defined in order to choose the best reinforcement and support scheme.

In the majority of mining tunnels, the basic method of reinforcement is based on a pattern that combines rock bolts and mesh. This scheme is compatible with the majority of current development mining strategies and represents the minimum level of ground support expected for excavation stabilization. Furthermore, a stabilization scheme cannot be selected without consideration of the drilling equipment and reinforcement installation technology available at a particular construction site. For example, for very deep tunnels, an optimal mechanized system must be able to install multiple rock bolts and mesh in a single pass in order to increase productivity and reduce the exposure of personnel during installation (Figure 1.8).

In situations where adverse geological structures are present or the excavation size is significantly increased, the potential for deep failures usually requires cable bolt reinforcement. Cable bolts provide deep anchorage and are used to stabilize large, potentially unstable areas where normal rock bolts do not possess a sufficiently deep anchorage. Cable bolts may also be needed in areas likely to undergo excessive loading from stress re-distributions or undergo significant damage from blast vibrations, such as draw point brows in the Sublevel Open Stoping and Sublevel Caving mining methods.

Following the selection of a reinforcement scheme, an assessment of stability is required. Empirical methods or experience under similar rock mass conditions can be used to provide an initial overall assessment of excavation stability with the chosen reinforcement scheme. The

FIGURE 1.8 An example of mechanized installation of multiple rock bolts and woven mesh support.

effectiveness of a particular reinforcement scheme can be assessed using empirical strategies, block theory, computational methods and geotechnical instrumentation. The first step is to determine the overall stability of an excavation, followed by a stability assessment of typical individual blocks according to data from a rock mass characterization program and the identified failure mode.

Empirical evidence from other mines with similar rock mass strength, failure mode and loading conditions, combined with empirical methods, can be used to establish the initial reinforcement design. Instrumentation of typical reinforcement arrangements coupled with observations of rock mass behaviour can also be used. Detailed stability of typical or worst-case scenario wedges and blocks can be carried out using three-dimensional computer programs, such as SAFEX (see Chapters 3 and 12). Data from detailed rock mass characterization programs can be used to determine the joint set characteristics needed to determine the likely shape and size of the largest rock block to be reinforced (Windsor, 1999).

The reinforcement scheme selected may be tested and optimized *in situ*. In cases where empirical evidence suggests stability, the interaction between the reinforcement scheme and the rock mass can be instrumented in order to achieve an optimal match with respect to demand, capacity and cost (Windsor and Thompson, 1993; Villaescusa and Schubert, 1999).

In cases where failure through reinforcement is experienced, information such as depth of failure, area, volume and mass of instability, along with the resulting out-of-balance pressure, must be determined (Figure 1.9). Other information, such as the likely failure mechanism and structural control of the failure geometry, as well as the observed reinforcement performance (Heal, 2010), may be used as an integral part of further detailed stability analyses.

Blasting practices, stress redistributions and the critically important operating practices during reinforcement installation must also be considered. Documentation of the ground support performance (Figure 1.10) is very important for future references and potential optimization of the design and practices for areas with similar conditions.

An actual failure must be inspected and documented by experienced underground rock mechanics personnel in order to gather the data needed to facilitate additional detailed stability analyses. Empirical evidence from failures may be used to improve the understanding of the interaction between the rock mass, the reinforcement schemes and the loading conditions from mining.

Detailed stability analysis may include calibrating a numerical model to simulate the extraction sequences and assess the role of stress redistribution in the rock mass (Wiles et al., 1994; Beck, 2008).

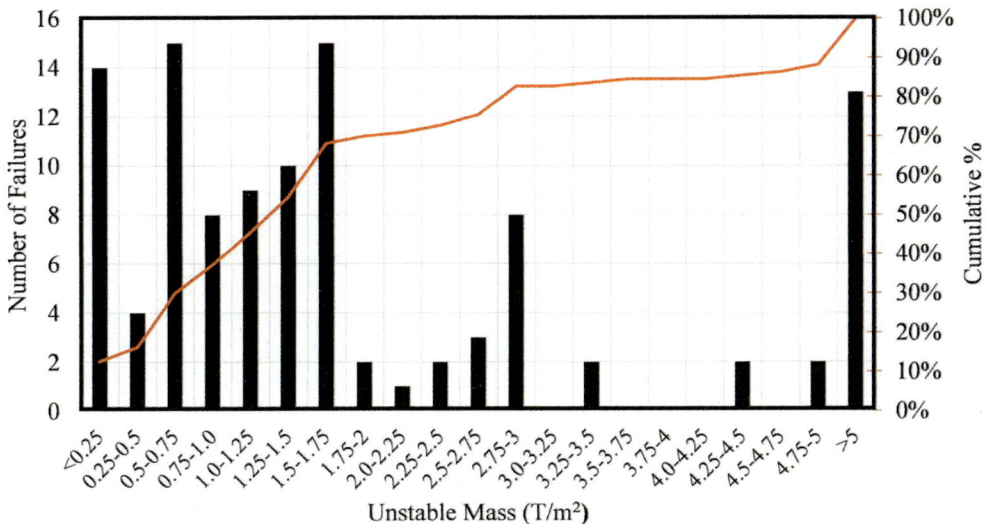

FIGURE 1.9 An example of unstable masses from back analysis of actual excavation failures.

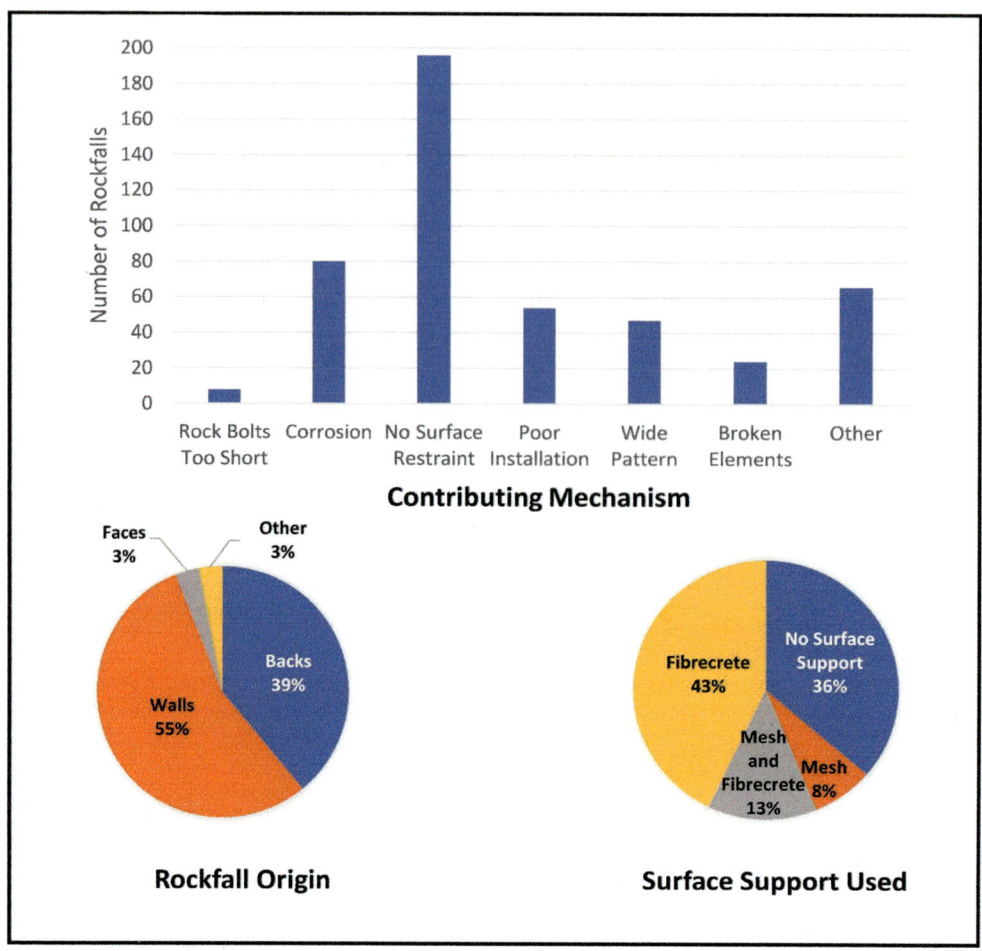

FIGURE 1.10   An example of documented ground support performance.

Information on the failure mechanism and the performance of the installed reinforcement is vital to determine the effectiveness of the support and reinforcement scheme. Evidence of corrosion or excessive damage to the bolts, cables, mesh or plates within the failure zone or the immediate area will provide an indication of the loading conditions contributing to the failure.

Information on the shape and size of the failure indicates whether appropriate reinforcement lengths were selected during the initial design. Any evidence of the failure trigger mechanism, including rock mass damage caused by local stress redistribution or production blasting, is critical if further failures of similar nature are to be prevented (Figure 1.11). In all cases, detailed documentation must be completed, and in some, a presentation of the results should be made to the excavation planning and construction personnel. The final documentation effectively closes the design process and represents an important step into the optimization of the design and practices. It also helps prevent any future failures.

The optimization of a reinforcement scheme should improve safety and increase productivity in most mining development situations. The optimization usually starts with an initial review of practices and quantification of the potential economic benefits likely to be achieved if practices can be improved. Field observations can be used to indicate areas of a mine that are being consistently over- or under-reinforced or where reinforcement and/or support selection is made without technical

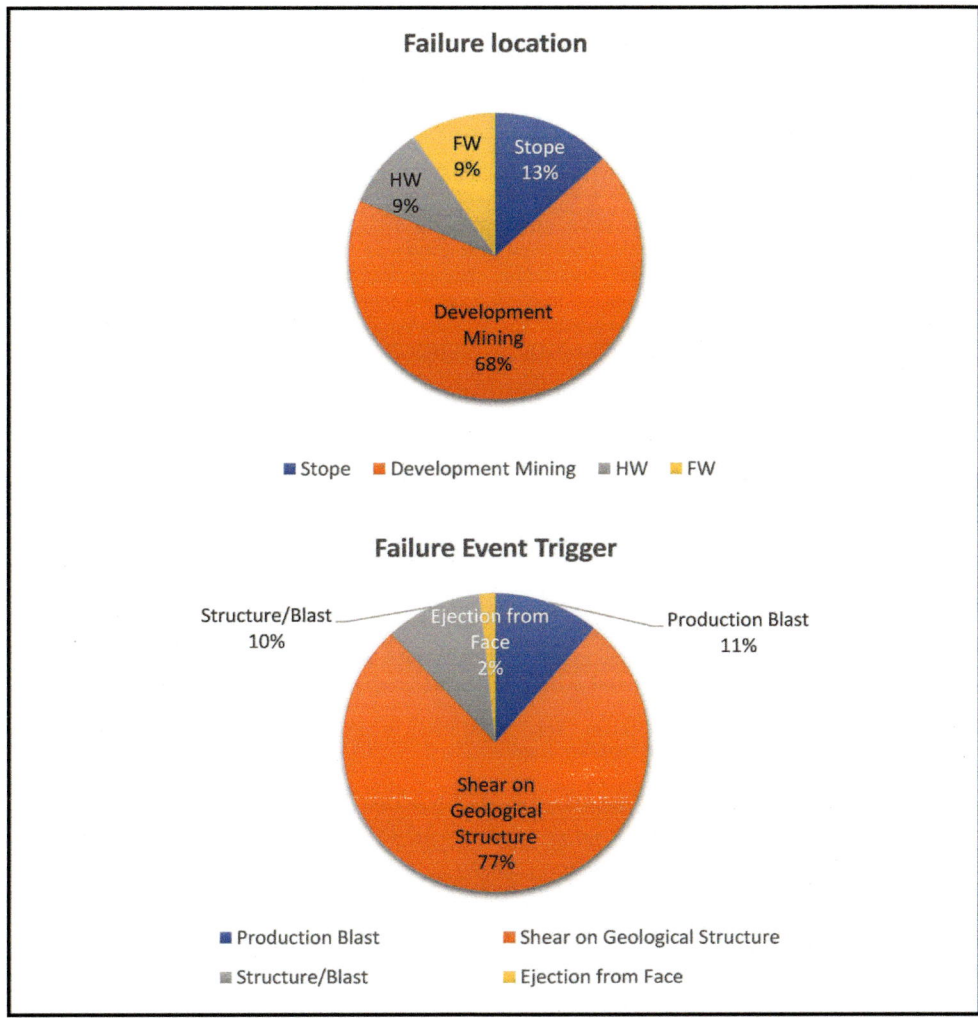

FIGURE 1.11 Example of historical data showing the location and failure mechanism of large seismic events, where HW = Hangingwall and FW = Footwall.

consideration. An example of the items that should be considered during the initial stages of the optimization process is shown in Table 1.1 (Villaescusa, 1999).

1.3 BRIEF HISTORY OF GROUND SUPPORT TECHNOLOGY

Over the years, several reviews of the development and availability of reinforcement systems have been published. The most comprehensive of these are a series of articles pertaining to ground anchors by Littlejohn and Bruce (1977), Littlejohn (1993) and Barley and Windsor (2000) and rock reinforcement applications in mining by Gardner (1971), Peng and Tang (1984), Windsor and Thompson (1993) and Brown (1999a, 2004).

The major developments in the early days were driven by the necessity to increase safety and productivity, firstly in tunnelling and subsequently in mining. Prime examples of the innovations in ground support were those associated with the Snowy Mountains Scheme in Australia in the 1950s and reported by Lang (1961). More recently, a mention of the pioneering work conducted in

TABLE 1.1

Items That Should Be Considered during the Initial Review of Ground Support Practices.

Support and Reinforcement Standards and Rationale

Excavation size (development and infrastructure, including intersections)
Excavation shape (rectangular, oval or semi-circular)
Mode of instability (time dependent, gravity driven, unravelling or sudden, violent failures)
Timing of ground support installation including shotcrete and cables
Geological structures (number, orientation, spacing, size and strength, seismically active)
Block shape and size (with respect to reinforcement pattern and spans)
Development mining practices (actual vs designed excavation profiles)
Reinforcement patterns for backs and walls (design vs implemented)

Equipment

Size and drilling limitations
Specialized scaling, shotcretingand cablebolting machines
Type of cement grout pumps (mono or piston based)
Cableboltplate installation devices

Methodology of Installation

Heavily mechanized or manual practices
Scaling methodology (mechanical, hydro-scaling, manual)
Drill, blast, muck cycles vs ground support installation including cables
Resin and cement grouting (resin cartridge, pumping)
Cablebolt plating (timing, greasing of cableboltbarrel & wedge fixtures) and tensioning

Long Term Effectiveness

Depth, area and volume of failure (T/m^2 of instability recorded for every failure)
Number of rehabilitation stages required
Corrosion
Cost of ground support per metredeveloped and tons mined

Management and Training

Induction modules for operators, supervisors and engineers
Minimum training foroperators (hazard recognition, implementation of design, minimum
    support standards)
Role of Underground Manager, Geologist, Supervisor
Role and independence of the Geotechnical Department

---

the Snowy Mountains has been documented by Lees (2009) and Mills (2009). These investigations involved improvements in the understanding of reinforcement mechanics and the inherent need to investigate the various aspects of rock mechanics reported by Brown (1999b, 2004) and involve both the rock and the reinforcement as a system. The previous widespread use of support systems, such as steel arches with timber lagging, allowed the rock to essentially fail. This type of ground support resulted in higher loadings than would have occurred if the rock failure was initially inhibited to some degree.

Current practice in mines and tunnels is to place a layer of shotcrete during the excavation and support cycle to prevent initial loosening. This layer of shotcrete is installed in conjunction with rock bolts and mesh and may be supplemented later by further support and reinforcement, including cable bolts (Villaescusa et al., 2016). Where it is feasible to generate high anchorage forces at depth from the excavation surface, high force capacity surface support, in combination with high force capacity reinforcement, are a recognized alternative to arches. It is also possible to form arches of locally thicker shotcrete with pre-assembled trusses of reinforcing bars. Note that in soils and soft rock, where indicated force demands are greater than those generated by reinforcement systems, a 'closed' ring may be necessary. A closed ring does not rely on the mechanical properties of the surrounding materials.

More specific accounts of the development of cable bolting for mining applications in Australia have been reported based on research by the Commonwealth Scientific and Industrial Research Organisation (CSIRO) and mining-related applied research at the major metalliferous mines in Broken Hill (e.g., Clifford 1974; Matthews et al., 1983), Cobar (e.g., Fuller 1983), Mount Isa (e.g., Bywater and Fuller 1983; Windsor et al., 1983; Greenelsh 1985; Hutchins et al., 1990; Villaescusa et al., 1992) and various other mines (e.g., Windsor 1992a, 2004). Other developments in various areas related to novel ground support applications and equipment were reported by Matthews et al. (1986), Thompson et al. (1987) and Thompson and Mathews (1989).

1.4   SCOPE AND CONTENTS OF THIS BOOK

In the next two to three decades, mining operations will be approaching depths and conditions in which the induced stress regimes will approach the strength of the rock masses surrounding the excavations. In such cases, failure may occur violently due to the release of energy stored within areas of high rock mass strength, where dynamic shear failure can occur due to sudden slippage on geological discontinuities intersecting the boundary of the excavations. The resulting failure depth, area and mass of instability are usually entirely controlled by the position, extent and orientation of the geological discontinuities mobilized by the shear failure.

In order to be prepared for such scenarios and to ensure safe and economical excavations in the future, the WA School of Mines (WASM) and a number of sponsoring companies have, over the last 25 years, conceptualized and undertaken a number of research projects in ground support technology. The projects ranged from static and dynamic laboratory testing of support and reinforcement elements to *in situ* assessment of ground support corrosivity. One of the main objectives was to understand the energy dissipation capabilities of reinforcement and support systems leading to the design of complete ground support schemes that are capable of maintaining integrity following the dynamic ejection of a mass of rock from an excavation boundary. Some of the research results have been documented in this book and the topics have been divided into the following thirteen chapters:

1. Introduction
2. Terminology
3. Probabilistic Determination of Rock Mass Demand
4. Ground Support Mechanics
5. Ground Support Design
6. Laboratory Testing
7. Energy Dissipation of Rock Bolts
8. Energy Dissipation of Cable Bolts
9. Energy Dissipation of Mesh Support
10. Energy Dissipation of Shotcrete Support
11. Dynamic Performance of Ground Support Schemes
12. Reinforced Block Analysis
13. Construction and Monitoring

This book was written to document 2.5 decades of research results obtained at the WASM, Kalgoorlie, Western Australia (Morton, 2009; Player, 2012; De Zoysa, 2015; Kusui, 2015; Drover, 2018; Bustos, 2022; De Zoysa, 2022). The research projects were sponsored by the Minerals and Energy Research Institute of Western Australia (MERIWA) (Villaescusa et al., 2005; Villaescusa et al., 2010; Villaescusa et al., 2015) and built into the research undertaken earlier by some of the authors at the Australian Commonwealth Scientific Industrial Research Organization (CSIRO). This book presents the current state of the art, including recent results from applied ground support research at the WASM, and hence, the book could be used for advanced postgraduate studies. Furthermore, mining and civil engineering practitioners and consulting engineers might also find this book useful.

2 Terminology

2.1 INTRODUCTION

A classification to describe the forms, functions, basic mechanics and behaviour of the different commercially-available rock support and reinforcement systems was initially developed by Thompson and Windsor (1992a). Their method classified the existing reinforcement systems by dividing them into three basic categories to explain the basic mechanisms of load transfer between the reinforcing elements and a rock mass. Additional research has identified that surface support systems may also be classified within a limited number of categories (Thompson et al., 2012). This classification is more complex given the greater number of variables associated with the loading and response of ground support systems than those associated with the response to loading of reinforcement systems.

As implied earlier, ground support technology involves the support and retention of newly exposed excavation faces and reinforcement of the materials surrounding the excavation. Accordingly, three basic terms are defined first:

Ground Support Scheme: Defined as a combination of support and reinforcement systems.
Surface Support System: Defined to be anything visually in contact with an excavation face. It is worth noting at this stage that some support systems are not installed to satisfy this requirement and only become an effective support system after movements of material adjacent to an excavation face lead to contact with the support.
Internal Reinforcement System: Defined to be anything that is embedded within the material surrounding an excavation. In the context of this book, reinforcement systems are associated with a borehole drilled into the rock mass.

In the subsequent text, the words 'surface' and 'internal' are no longer used as they become redundant following their formal definitions. Detailed descriptions are now given with regard to the basic components of support and reinforcement systems.

2.2 REINFORCEMENT SYSTEM RESPONSE

A description and comparison of devices within a particular category or between separate categories are facilitated by the methodology developed by Thompson and Windsor (1992). The categories are shown in Figure 2.1 and are described as Continuous Mechanically Coupled (CMC), Continuous Frictionally Coupled (CFC) and Discrete Mechanical and Frictional Coupled (DMFC). Some typical reinforcing devices are grouped according to this classification in Table 2.1. The response of the reinforcement system to rock loading involves several modes of load transfer between the various components, as shown schematically in Figure 2.2.

2.3 CONTINUOUS MECHANICALLY COUPLED

A continuous mechanically coupled (CMC) reinforcing element relies on a fixing agent, usually a cement or resin-based grout, which fills the annulus between the element and the borehole wall (Figure 2.3). The main function of the grout is to provide a mechanism for load transfer between the rock mass and the reinforcing element.

DOI: 10.1201/9781003357711-2

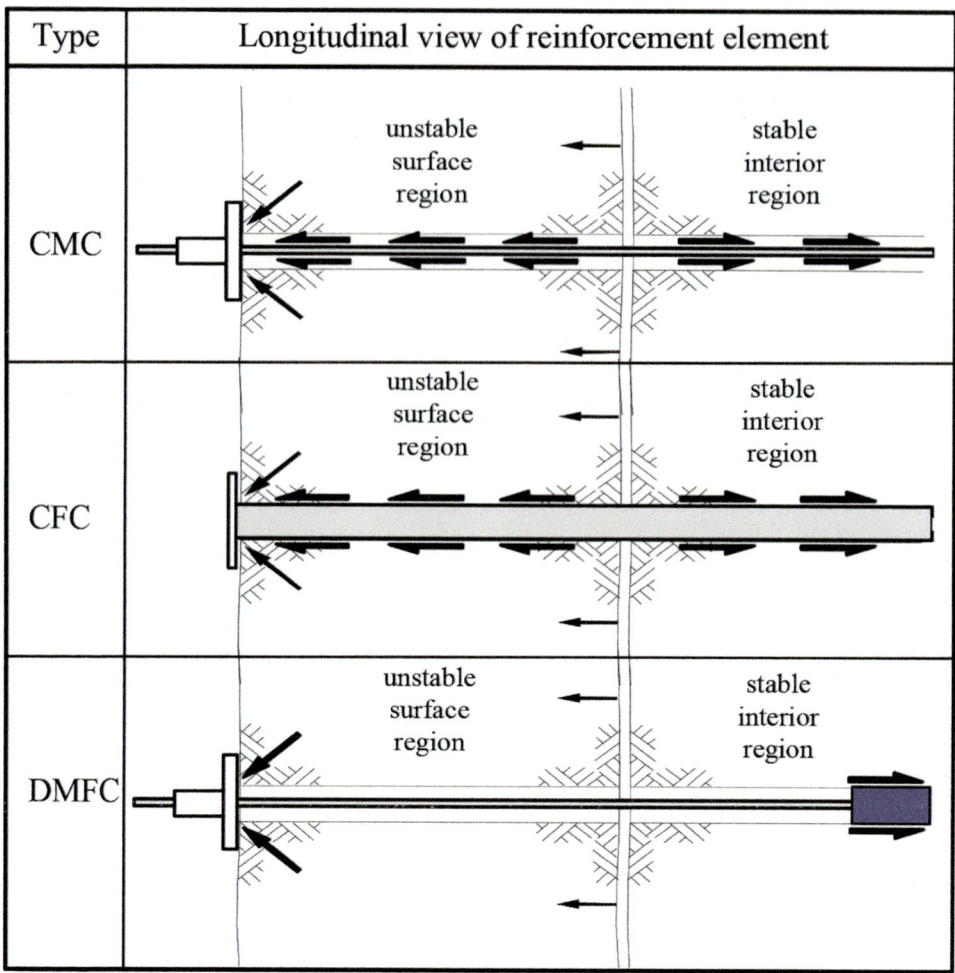

FIGURE 2.1   Classification of reinforcement action (after Thompson and Windsor, 1992).

TABLE 2.1

Classification of Typical Reinforcement Devices (Thompson and Windsor, 1992).

Type	Description
CMC	Full column cement-/resin-encapsulated bars (grouted CT bolt, deformed bar, threaded bar, fully coupled Posimix)
	Cement grouted cables (plain strand, modified geometry)
CFC	Friction stabilizers (split-set bolt, friction bolt, Swellex)
DMFC	Mechanical anchors (ungrouted CT and HGB bolts, expansion shell, slot and wedge)
	Single cement/resin cartridge anchors (paddle bolt, deformed bar)
	Decoupled Posimix (having a range of anchor and collar lengths)
	Decoupled cable bolts (having a range of anchor and collar lengths)

The reinforcing elements are usually manufactured with variable cross-sectional shapes in order to increase the element-to-grout interlock. A mechanical key is effectively created by the geometrical interference between the element and the grout along the entire reinforcement length. The element is defined as continuously coupled to the rock mass via an interlock with the grouting agent (Thompson and Windsor, 1992).

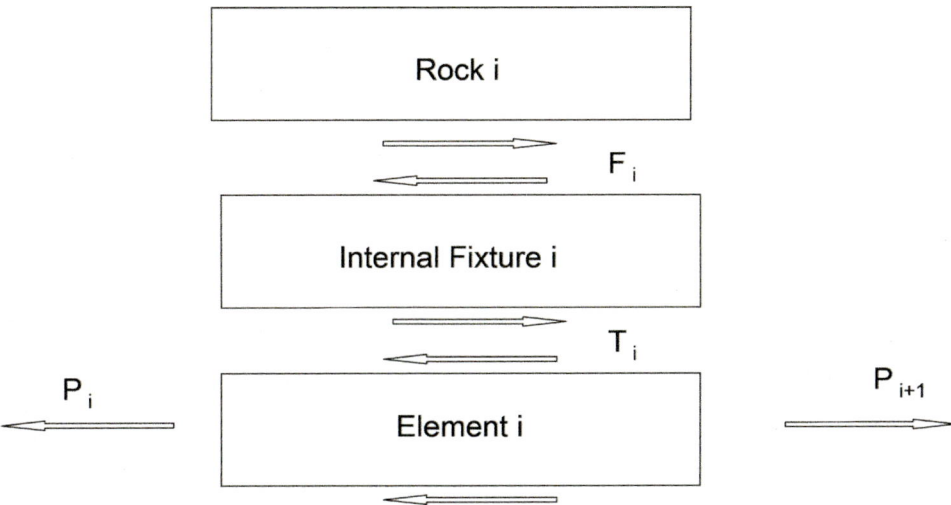

FIGURE 2.2 Schematic diagram showing the forces involved in load transfer between components in reinforcement systems.

FIGURE 2.3 Long section view of a resin-encapsulated (CMC) element.

The widely used term 'bond', associated mainly with early reinforced concrete investigations and used in the area of soil and rock reinforcement, is not deemed appropriate, as there is little evidence to show that true bonding (or adhesion) is relevant in most circumstances. The load transfer across interfaces is considered to involve mechanical interaction in the case of rough surfaces and friction following the development of a distinct, sliding interface due to the brittle failure of the relatively stiff materials used to form internal fixtures (Thompson et al., 2012).

2.4 CONTINUOUS FRICTIONALLY COUPLED

A continuous frictionally coupled (CFC) reinforcing element is installed in direct contact with the rock mass (Figure 2.4). The mechanism of load transfer is a function of the frictional forces developed between the reinforcing element and the borehole wall. The load transfer is limited by the radial pre-stress set-up during initial element installation. The bond strength is a function of the element shape diameter, the borehole diameter and any geometrical irregularities occurring at the borehole wall.

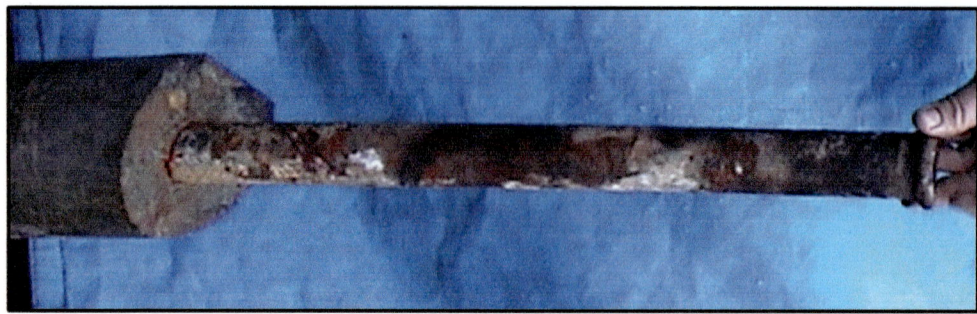

FIGURE 2.4   Example of a recovered friction stabilizer (CFC) element.

The radial stress can be related to a force along the length of the reinforcing element. It is achieved by deforming the cross-sectional area of the element to suit the borehole. It can be achieved by either contracting an oversized element section into an undersized borehole (friction stabilizer) or by expanding an undersized element section into an oversized borehole (Swellex bolt). This reinforcing action can be modified by cement grouting of the split-set bolts as described by Villaescusa and Wright (1997) or by adding a sleeve to the Swellex (Windsor and Thompson, 1992b).

2.5 DISCRETE MECHANICALLY OR FRICTIONALLY COUPLED

A discrete mechanically or frictionally coupled (DMFC) device transfers load at two discrete regions, namely the borehole collar region and the anchor point region, located at some depth into the borehole. The length of the element between the two discrete points is effectively decoupled from the rock mass. The load transfer is then limited to a relatively short length at the anchor end and at the collar. Load transfer at the anchor regions can be achieved by either mechanical (grouted anchor and collar region) or frictional means (some type of expansion shell at the toe anchor end).

The strength of an expansion shell may be limited by the strength of the rock at the bore-hole wall, and these devices are best suited to hard rock applications (Villaescusa and Wright, 1999). Grouted anchors may be used in soft and hard rock masses, where a high load transfer can be achieved over a short length, provided that gloving by the resin cartridge does not occur (Villaescusa et al., 2008).

2.6 THE LOAD TRANSFER CONCEPT

Reinforcement systems develop forces in response to material movements to limit deformations and discontinuity displacements. In order to be effective, a reinforcement system needs to connect to a stable zone beyond the volume of material undergoing deformation and displacements. The load transfer concept may also be used to define a qualitative relationship between length and force capacity for reinforcement systems. Thompson and Windsor (1992) showed that this could be applied to both underground and surface excavations where the force capacity required for reinforcement increases in relation to the volume of failure assumed.

The load transfer concept shown in Figure 2.5 can be used to understand the stabilizing action of all the reinforcing devices and their effect on excavation stability. The concept can be explained by three basic individual components (Windsor and Thompson, 1993):

1) Rock movement near the excavation boundary, which causes load transfer from the unstable rock, wedge or rock slab to the reinforcing element;
2) Transfer of load via the reinforcing element from the unstable portion to a stable interior region within the rock mass; and
3) Transfer of the reinforcing element load to the rock mass in the stable zone.

The separation of a rock block or a layer of rock being stabilized can occur during any one of the three separate components of load transfer. For example, it might occur because of insufficient steel capacity (rupture of the reinforcing element) or inadequate bond strength (slippage either within the unstable or stable region), as shown in Figure 2.6a and b, respectively.

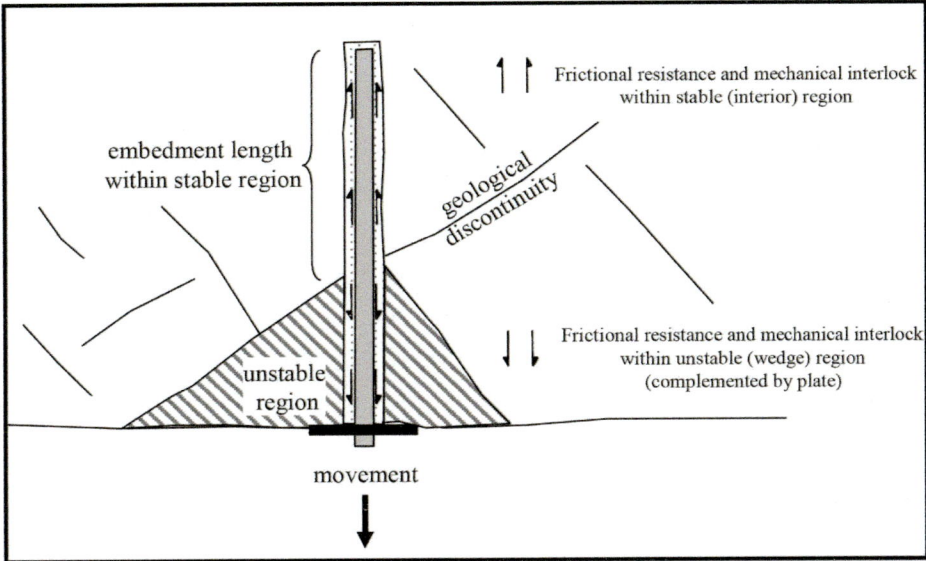

FIGURE 2.5   The concepts of load transfer and embedment length.

FIGURE 2.6   Instability due to (a) slippage within the unstable region and (b) slippage within the stable region.

2.7   THE EMBEDMENT LENGTH CONCEPT

Embedment length is the length of a reinforcing element on either side of an active geological discontinuity defining a potentially unstable wedge or rock block, as depicted in Figure 2.7. The critical embedment length is the minimum length of reinforcement required to mobilize the full reinforcing capacity of the system.

Short embedment lengths within an unstable region can be compensated for because a properly matched face plate and fixture provide sufficient surface restraint to mobilize the system capacity. Short embedment lengths within the stable region are more critical, especially when a reinforcement element is installed at an unfavourable angle with respect to the free surface.

2.8   MATERIALS BEHAVIOUR TERMINOLOGY

The following terminology can be used to describe materials and systems behaviour. Clearly, it is possible for systems to behave quite differently from their component parts.

2.8.1   Elastic

Elastic behaviour of materials or systems is observed when no permanent deformation results after a cycle of loading and unloading.

2.8.2   Plastic

Plastic behaviour of materials or systems is observed when permanent deformation results after a cycle of loading and unloading.

2.8.3   Brittle

Brittle is used to describe the behaviour of materials or systems wherein failure occurs with little or no preceding deformation.

FIGURE 2.7   Slippage within a stable region due to insufficient embedment length.

2.8.4 DUCTILE

Ductile is used to describe the behaviour of materials or systems wherein failure occurs after appreciable plastic deformation.

2.8.5 RESILIENCE

Resilience is the ability of a system to dissipate energy when it is deformed elastically, and it returns to its original condition when the driving force is removed. This behaviour is observed when the imposed transient forces are less than the yield capacity of the system (e.g., blasting, earthquake loadings).

2.8.6 TOUGHNESS

Toughness is the ability of a system to dissipate energy and deform 'plastically' (before 'rupture'). This behaviour is observed when the imposed forces exceed the yield capacity of the system and where the imposed forces reduce with displacements (e.g., in response to impact loading. This behaviour is also observed when the kinetic energy of loading reduces due to the reaction forces from a ground support scheme).

2.8.7 YIELD

Yield is associated with the level of loading beyond which permanent or plastic deformation occurs after the loading is removed from a material. The term cannot be used to describe the mechanisms of the behaviour of systems of materials wherein apparent yield is likely to be associated with slip, that is, relative displacement across an interface between two components.

2.8.8 STIFFNESS

Stiffness is a measure of the rate of change in response loading per unit of deformation. Material stiffness is a function of Young's modulus (E) and Poisson's ratio. The axial stiffness of a material is proportional to $A \times E$, where A is the cross-sectional area. In the case of bending, the rotational stiffness is proportional to $E \times I$, where I is the second moment of area.

2.9 PERFORMANCE INDICATORS OF CAPACITY

A number of parameters may be used to characterize the performance of different reinforcement and support systems. Figure 2.8 shows a generic force-displacement response with annotations of a number of system performance indicators. The response to loading for reinforcement may be axial or shear and that for support may include lateral shear and bending or in-plane combinations of shear with tension or compression.

The performance indicators of capacity may be grouped as follows:

- Force capacities

  F_{max} = Maximum force
  F_{res} = Residual force at maximum displacement

- Displacement capacities

  δ_p = Displacement at maximum force
  δ_{max} = Maximum displacement

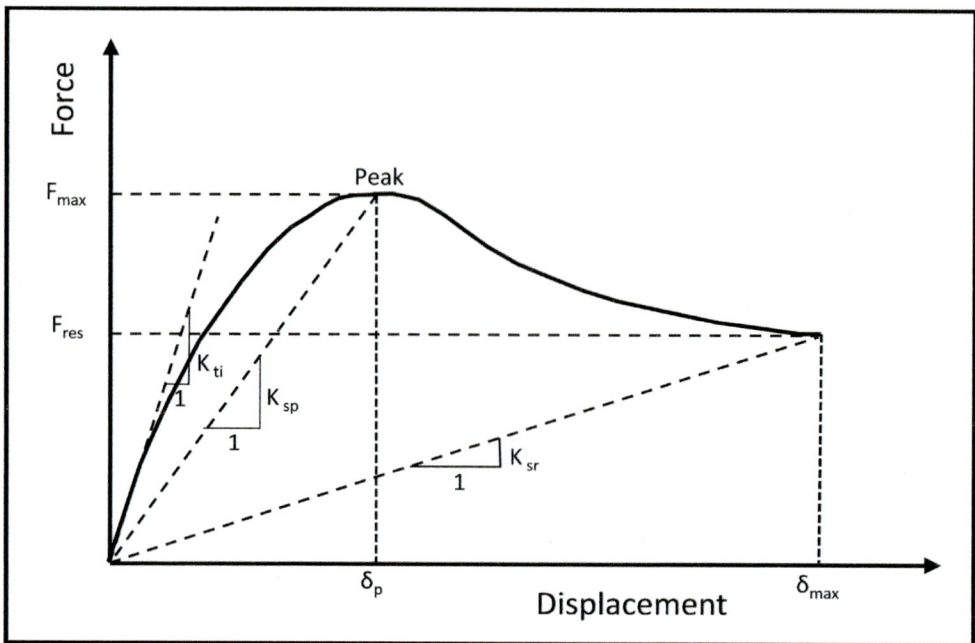

FIGURE 2.8 Force-displacement response for a generic reinforcement system subjected to axial loading (Thompson et al., 2012).

- Stiffnesses

    K_{ti} = Initial tangent stiffness
    K_{sp} = Secant stiffness at maximum force
    K_{sr} = Secant stiffness at maximum displacement

- Energy dissipation capacity

The energy dissipation capacity is equivalent to the area between the force-displacement curve and the displacement axis and is relevant to the performance of reinforcement subjected to dynamic loading.

    E_p = Energy dissipation to peak force
    E_r = Energy dissipation at maximum displacement

It is also necessary to consider the possible modes of loading when assessing a particular ground support component. One might note that the performance indicators δ_{max} and E_r may be related directly to the terms 'ductility' and 'toughness', respectively, defined earlier. Additionally, other parameters may need to be considered if a reinforcement system is loaded predominantly in shear. For example, it is known that a strand is more flexible than a solid bar when loaded in shear and can therefore sustain a higher shear displacement. The ability of a reinforcement system to sustain shear displacement is improved by decoupling the element from the grout as it allows for axial displacement of the element to be distributed over a longer length of the element across a discontinuity.

It must be noted that there is no need to use the term 'dynamic' when referring to ground support component systems, all of which have some deformability and energy dissipation capacity in response to vibration and impact loadings during the creation of excavations.

2.10   SURFACE SUPPORT SYSTEM RESPONSE

Support systems are inherently more difficult to classify than reinforcement systems. Firstly, support systems are available in various materials and configurations. Secondly, the geometry of excavation surfaces is three-dimensional compared with the essentially axial or co-axial nature of reinforcement systems. Finally, the loading from the materials surrounding the excavation triggers a variety of complex mechanisms of response in the support systems.

2.10.1   Type of Support Systems

The first step in classifying support systems is to take into account their surface extent, which can be considered to be one of the following:

- Point,
- Strip and
- Areal.

In most cases, the support systems rely on effective coupling with reinforcement systems. The major exception is in shafts and tunnels where a complete ring, comprising sprayed concrete or precast panels, is used to support the rock mass surrounding an excavation.

2.10.1.1   Point Support Systems

Most reinforcement systems contain a proprietary external fixture (e.g., nut and washer, barrel and wedge anchor) customized to a particular element. In the context of support, a point support system is considered to be a relatively small plate or a combination of plates that act locally at the collar of the reinforcement system (Figure 2.9). In this regard, the point support system acts independently of any other reinforcement systems in the surrounding pattern, implying that a ground support scheme cannot be formed.

FIGURE 2.9   An example of point surface support.

2.10.1.2 Strip Support Systems

Strip support systems, as the name implies, are linear systems that span between two or more reinforcement systems (Figure 2.10). Strip support systems may be flat or profiled steel sections, mesh-like or comprising single or multiple configurations of steel wire rope or strand. Steel sets and reinforced shotcrete arches may also be considered strip support systems.

2.10.1.3 Areal Support Systems

Areal support systems, as the name suggests, extend in two orthogonal directions to cover an area of an excavation face (Figure 2.11). Areal support systems include:

- A mesh that comprises apertures of various shapes and sizes formed between (usually steel) wires that may be woven or welded to form a continuous flat or profiled sheet or roll;
- Thin Sprayed Liner (TSL), one of many experimental proprietary products; and
- A thick layer of sprayed materials, such as mortar (gunite) and concrete (shotcrete).

2.10.2 LOADING MECHANISMS

The loading mechanisms for support systems result from the general convergence of the surrounding materials towards a newly created void. In more adverse conditions of stability, detachment

FIGURE 2.10 Buckling of 'strip' surface can be observed following a time-dependent, large deformation of an excavation wall.

FIGURE 2.11 Example of 'areal' support provided by a welded wire mesh.

of local structurally controlled or stress-induced failure can occur. The movements of materials mainly cause lateral loadings. However, for strip and areal support systems, loadings within the plane of support may also occur, leading to buckling. The responses of the support systems will depend on both the movements of the materials surrounding the excavation and the geometry of the support system. In the case of sprayed materials, the contact will have the same shape as the underlying surface, while the exposed surface will depend on other factors related to the spraying strategy. For prefabricated systems (e.g., mesh, straps, pre-cast panels and arches), the loading will be at discrete contact locations with the possibility, as discussed previously, of there being no contact at all until after significant excavation surface movements (Figure 2.12).

2.10.3 LOAD TRANSFER CONCEPTS

2.10.3.1 Point Support Systems

Point support systems may be associated with discrete reinforcement systems used in a massive rock where no areal support between reinforcement collars is deemed necessary but, more usually, to restrain strip and areal support systems. In the latter cases, it is necessary to consider the loadings from the support systems that may cause additional axial, shear and moment loadings at the collar. For static loading, tests have shown that mesh usually slips at the points of restraint; ultimately, after large displacements, one or more wires will engage with the shaft of the reinforcement element and may cause severe shear and bending loadings. In conditions of dynamic loading, the strength of typical welded wire mesh cannot secure an anchor point, as shown in Figure 2.13.

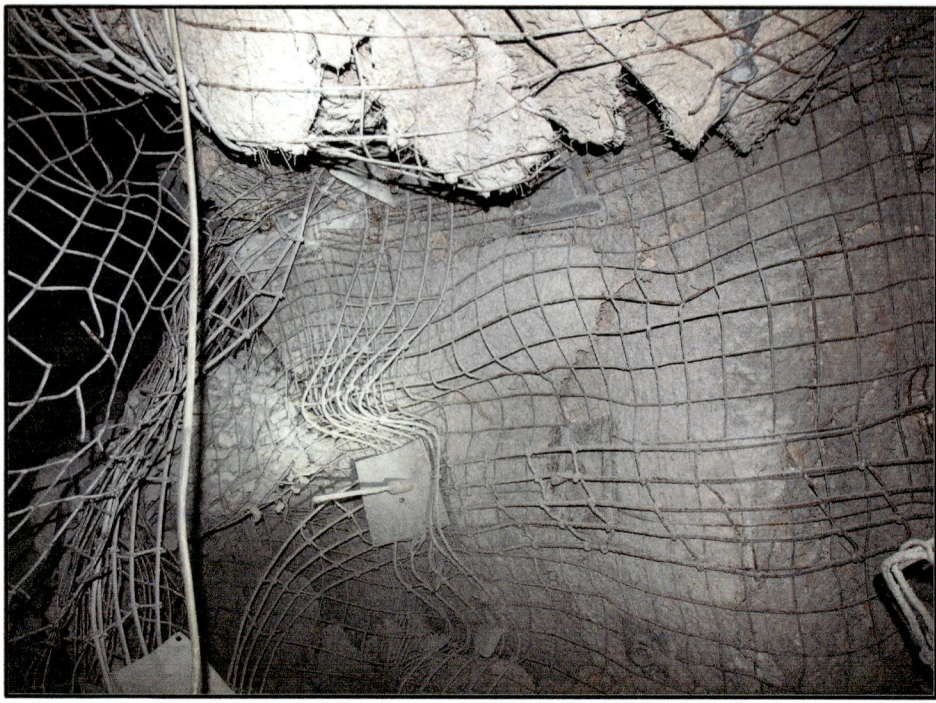

FIGURE 2.12 An example of mesh strap buckling following a large seismic event on a low-capacity ground support scheme.

FIGURE 2.13 Violent failure resulting in failure of welded wire mesh at the point of contact with the surface plate.

2.10.3.2 Strip and Areal Support Systems

For the purposes of discussion, strip and areal support systems are considered to behave similarly in regard to their association with excavation shapes. Accordingly, the surface support systems are classified as being 'thin' or 'thick', as shown by the schematics in Figure 2.14 and the images in Figure 2.15.

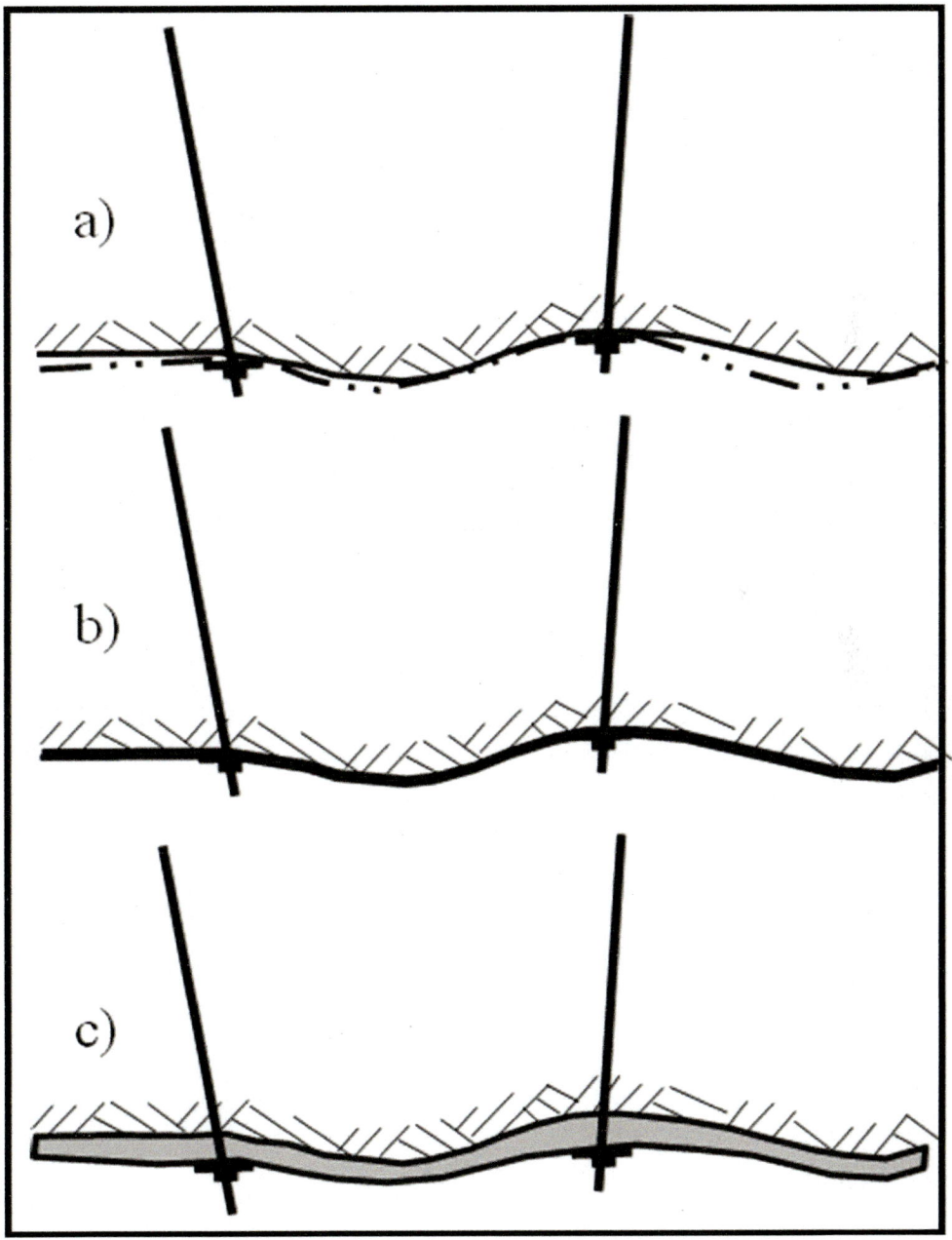

FIGURE 2.14 Schematics of support systems: (a) a thin non-contact mesh, (b) thin contacting sprayed layer and (c) thick contacting sprayed layer.

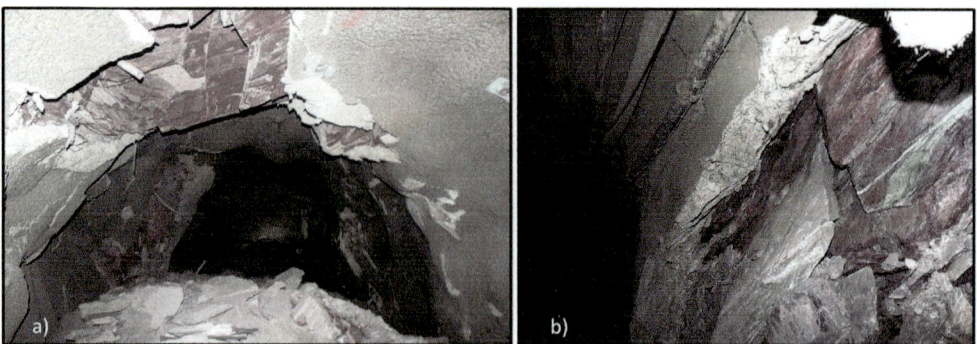

FIGURE 2.15   Examples of (a) thin and (b) thick sprayed shotcrete layer failure under slow, time-dependent loading conditions.

Strip support systems are generally used in conjunction with other surface support systems in order to provide additional restraint normal to the excavation surface between the reinforcement positions. Because the strip support is not continuous, the overall support performance is largely controlled by the areal support, which, in some cases, can be damaged before it is required to contribute to the stabilization (Figure 2.16).

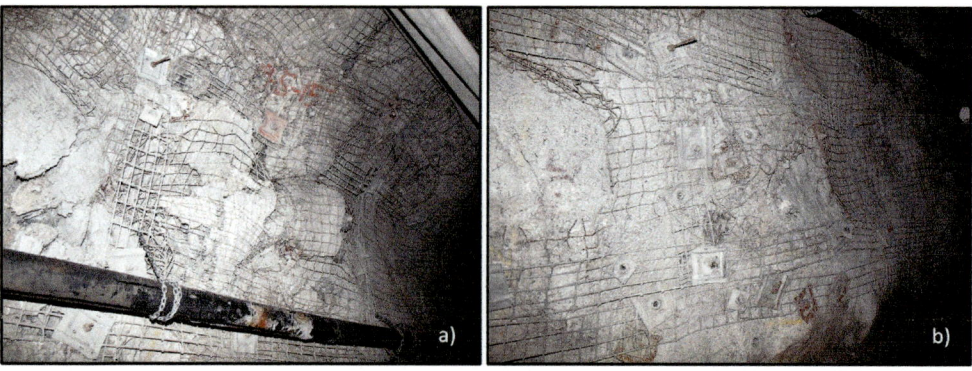

FIGURE 2.16   Examples of combined areal and strip support where (a) the majority of the deformation is constrained by the mesh and (b) mesh damage results in unsupported areas between the straps.

3 Probabilistic Simulation of Rock Mass Demand

3.1 INTRODUCTION

Development excavations at depths exceeding 800–1000 m are experiencing damage in a large number of tunnels and underground mines worldwide. The observations indicate that stress-driven, violent failures first occur as shallow, strain bursts. However, if the induced stress is increased, it can be followed by mobilization of geological discontinuities, potentially causing larger failures. Thus, the resulting depths of failure, area and volume of instability are largely controlled by the location, orientation and persistence of the geological structures close to the excavation surfaces. It shows that the failed block shapes and sizes are almost entirely defined by geological discontinuities (Figures 3.1 and 3.2).

The formulation of a rock mass structural model to assess the rock mass demand has been proposed by Windsor (1999). The design of reinforcement and/or support for deep excavation in structured rock requires a definition of the rock mass block assembly (Warburton, 1981; Goodman and Shi, 1985; Priest, 1985) and rock mass dynamic demand based on probabilistic block theory (Windsor, 1997). The computer program SAFEX described by Windsor and Thompson (1992c), and substantially developed and refined since then, can be used to calculate the probabilistic block assembly and to assess the instabilities associated with any excavation surface.

3.2 ROCK MASS CHARACTERIZATION

A great deal of effort has been devoted to the characterization of discontinuity networks and to modelling them quantitatively (Call et al., 1976; Hudson and Priest, 1979; Villaescusa, 1991; Windsor, 1995; Brzovic, 2010). Systematic collection of geotechnical data in conjunction with considerations of the induced stress and other geological factors is essential in designing deep, stable excavations. The structural data are initially utilized to develop an understanding of the various structural domains that can be used to predict the likely failure mechanisms and potential depths of failure during (and after) excavation.

The geotechnical data collected may be divided into two classes: major structures and minor geological features (Call et al., 1976). Major structures, such as faults, dykes, contacts and related features, usually have sizes of the same order of magnitude as that of the site to be characterized. Their position in space, physical properties and geometrical characteristics are usually established deterministically by the geological mapping process. Minor features represent, for practical purposes, a very large population in the region of an excavation. As a result, their geometrical characteristics and physical properties must be estimated by measurements of a representative sampled population. Several methods are available to determine the geological discontinuity set characteristics, including line sampling (Call, 1972; Priest, 1985), cell sampling techniques (Mathis, 1988), as well as combinations of both techniques (Villaescusa, 2014). Recent developments in automated data collection using remote sensing techniques have the potential to allow the systematic collection of geological data from unsupported excavation spans prior to the placement of shotcrete support (Andreas and Schubert, 2017).

DOI: 10.1201/9781003357711-3

FIGURE 3.1   Discontinuities defining the depth of failure and failed block assembly following a large seismic event.

FIGURE 3.2   Examples of back failure and floor heave controlled by geological discontinuity geometry and location.

3.2.1   GEOTECHNICAL MAPPING OF UNDERGROUND EXPOSURES

Geotechnical mapping of underground excavations can be used to determine the orientation, linear frequency, size and surface strength of the block forming geological discontinuities. The information is usually recorded on a tabular datasheet suitable for subsequent computer analysis. The dimensions of the observation window, regardless of the mapping technique used (Villaescusa, 2014), should be kept constant at each site and across sites since data from different locations are

Scanline Number	1			Location		5881		Date		31/05/2018	Mapped By:		LM& VB	
Bearing	130	deg		Plunge		0	deg	Face Dip		90	deg	Face Dip Direction	40	deg
Line Height	1.5	m		Height Above		1.5	m	Height Below		1.5	m			

| Location | | Type | Orientation | | | Joint Condition | | | | | | Persistence | | | | | Remarks |
Distance (m)	Number of End Points		Dip	Dip Direction	Rock Type	Roughness	Planarity	Infill Material	Thickness (mm)	Weathering		Termination Above T1	Above T2	Trace Length (m)	Termination Below T1	Below T2	
0.42	1	J	88	297	FW	S	R	UN	CA	1	F	IR	-	2.4	O	-	
1.00	1	J	86	276	FW	R	UN	CL	0	F	JH	-	2.3	O	-	
1.10	1	J	89	115	FW	S	UN	CA	2	F	JH	-	2.05	O	-	
1.14	1	J	85	284	FW	R	UN	CL	0	F	JH	-	2.15	O	-	
1.35	1	J	88	105	FW	R	P	CL	0	F	IR	-	2.2	O	-	
1.71	1	J	86	102	FW	S	UN	CL	0	F	JH	-	1.9	O	-	
2.00	1	J	86	275	FW	R	UN	CL	0	F	IR	-	2.7	O	-	
2.29	0	J	88	311	FW	R	UN	CL	0	F	O	-	3.15	O	-	
2.51	0	J	84	63	FW	S	UN	CA	1	F	O	-	3.2	O	-	
2.63	0	J	84	130	FW	S	P	CL	0	F	O	-	3.2	O	-	
2.87	1	J	90	121	FW	S	P	CL	0	F	JH	-	2.1	O	-	
3.15	1	J	87	305	FW	R	UN	CL	0	F	IR	-	2.6	O	-	

| Template | | Load | | Save | | Add Scanline | | Delete Scanline | | | OK | | Cancel |

FIGURE 3.3   Example template for geological discontinuity data collection.

usually grouped together. Basic information at each site should include the number, location, elevation, bearing and plunge of the reference line used to collect the frequency, the dip and dip direction of the rock exposure and the censoring levels of the convex sampling window (up and down from a reference observation line) (Villaescusa, 2014). The discontinuity set characteristics required to formulate a rock mass structural model are described below and tabulated in Figure 3.3.

1. **Distance** along a tape where the discontinuity intersects the sampling line. Discontinuity spacing is calculated from intercept distances, the mean set orientation and the orientation of the sampling line. Numerically, the individual apparent spacing values are defined by sorting the discontinuities by individual sets down the line and subtracting the distances between adjacent discontinuities of the same set.
2. **Endpoints** of the discontinuities intersecting the tape. When the discontinuities are observed through a convex window, three sets of observations are obtained: joints totally contained (two trace length endpoints observable), joints intersecting only one of the window boundaries (one trace length endpoint observable), and finally, joints transecting the window (no trace length endpoints observable). The mean trace length can be estimated using the following method proposed by Pahl (1981):

$$\text{Lmean} = \frac{h(2N1 + N0)}{2N2 + N0}$$
3.1

where N0 signifies the number of transecting joints, N1 is used to denote the number of discontinuities with one endpoint exposed, N2 indicates the number of intersecting discontinuities that are contained and h is the height of the observation window (or width for vertical lines).

3. **Type of structures**, naturally occurring features such as faults, shears, bedding, veins, joints, contacts, etc. For example, a joint is represented by 'J'.
4. **Orientation**, that is, dip and dip direction of features intersecting the tape.
5. **Rock type** recorded using a mnemonic system code compatible with the notation used by the local geologists. For example, 'BAS' for basalt.
6. **Roughness**, which is a qualitative measure of the small scale (2 cm or less) asperities on the discontinuity surface. Rough, smooth and slicken-sided categories are typically used.
7. **Planarity**, which is a qualitative indication of the geometrical nature of the discontinuities on a large scale. Planar, wavy and irregular categories are typically used.
8. **Infill material**, as in some discontinuities, such as faults, it may be gouge and slickenside, while in others, it may be a quantitatively defined mineralogical assemblage (Brzovic, 2010).

9. **Thickness**, the measured width between the two walls of a discontinuity.
10. **Alteration** of the discontinuity walls determined using scratching tests (Milne et al., 1991).
11. **Trace length**, which is measured as seen in the rock face. The trace length is the maximum measurable length of the resulting intersection between a discontinuity and a planar excavation in rock.
12. **Termination**, as observed at the top and bottom of a discontinuity in the dip direction, but only if the discontinuity is contained. Otherwise, the termination is artificially 'obscured' by the observation window or the excavation geometry. Discontinuities can terminate either against another feature or within an intact rock.
13. **Remarks**, which are used to describe other characteristics, such as alteration, observed hydrological conditions, etc.

The discontinuity data can be sorted into families to determine distributional nature and statistics for the joint set orientation, spacing, trace length and surface strength, following well-established procedures (Villaescusa, 1991; Windsor, 1995) and beyond the scope of this book.

3.3 ROCK MASS MODEL

The mutual intersection of discontinuities divides a structured rock mass into partially and fully formed blocks of rock. If an excavation cuts through this assembly of blocks, new sets of blocks are formed at the excavation surface. Some of these exposed or surface blocks will have a shape that will allow them to fall, slide or rotate into the excavation should the block driving forces exceed the block stabilizing forces. At great depth, where stress-driven, violent failures are experienced, failure surfaces are often associated with the ejection of blocks that could have been considered statically stable.

In order to understand how such a rock mass may best be stabilized, the assembly of blocks must be investigated. The ideal outcome would be to predict the exact shape, size and spatial position of each block that could form around an excavation. Depending upon the failure mechanism (static or dynamic), these block characteristics would then define the geometry to calculate the rock mass demand and provide the information needed to select and design trial reinforcement and/or support schemes.

The ability to properly define the block characteristics depends on the quality and quantity of the rock mass characterization data used to describe the rock mass surrounding a proposed excavation. Figure 3.4 shows an example of a rock mass structural model which for each set incorporates the orientation, spacing, trace length and shear strength parameters (i.e., cohesions and friction angles). Along with the excavation geometry and orientation, this represents the basic input for the calculation of rock mass demands—both static and dynamic.

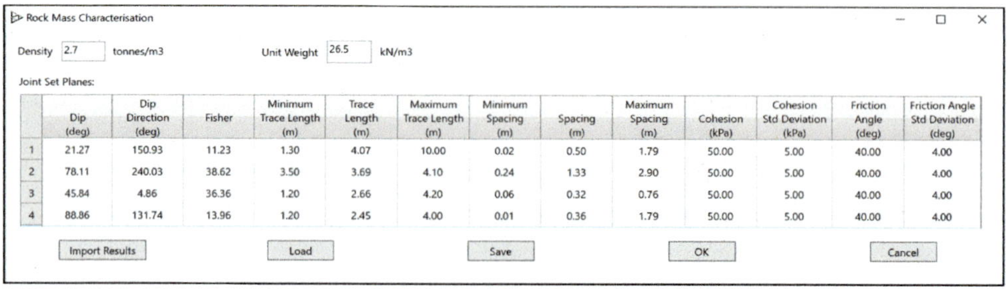
Rock Mass Characterisation

Density 2.7 tonnes/m3 Unit Weight 26.5 kN/m3

Joint Set Planes:

	Dip (deg)	Dip Direction (deg)	Fisher	Minimum Trace Length (m)	Trace Length (m)	Maximum Trace Length (m)	Minimum Spacing (m)	Spacing (m)	Maximum Spacing (m)	Cohesion (kPa)	Cohesion Std Deviation (kPa)	Friction Angle (deg)	Friction Angle Std Deviation (deg)
1	21.27	150.93	11.23	1.30	4.07	10.00	0.02	0.50	1.79	50.00	5.00	40.00	4.00
2	78.11	240.03	38.62	3.50	3.69	4.10	0.24	1.33	2.90	50.00	5.00	40.00	4.00
3	45.84	4.86	36.36	1.20	2.66	4.20	0.06	0.32	0.76	50.00	5.00	40.00	4.00
4	88.86	131.74	13.96	1.20	2.45	4.00	0.01	0.36	1.79	50.00	5.00	40.00	4.00

Import Results Load Save OK Cancel

FIGURE 3.4 An example of rock mass model formulation prior to rock mass demand calculations.

3.4 DETERMINISTIC ASSESSMENT OF ROCK MASS INSTABILITIES

Procedures based on 'key block' or 'keystone' theory have been used for over 35 years for the stability assessment of individual blocks near the surface and underground excavations (Warburton, 1981 and Goodman and Shi, 1985). The most popular assessment procedures are based on the identification and stability assessment of single blocks with simple shapes. For example, tetrahedral blocks formed adjacent to single excavation faces in underground excavations. These methods are often used in design to identify problematic blocks by assuming that the discontinuities are ubiquitous and form a block shape at some critical size, bounded by a roof or wall spans.

A procedure developed by Warburton (1993) involves defining a specific block assembly within a specific volume of rock surrounding an excavation. Warburton's algorithms have been implemented into the SAFEX software package (Thompson, 2002), together with the vector calculus procedures for defining the shapes, size and stability of all possible blocks that could occur at the excavation boundary (Windsor and Thompson, 2002c). With this software, it is possible to vary the orientation and persistence of discrete discontinuities together with the spacings for an arbitrary number of joint sets, as shown in Figure 3.4. An example of the deterministic block shapes formed and their potential failure modes are summarized in Figure 3.5. The block code indicates whether the block is (A)bove or (B)elow the discontinuity surface. The potential translation mode is one of free-falling (1), single plane sliding (2) or sliding on the intersection of two planes (3). A wide variation in the ratio of excavation face area to apex distance will influence the range of possible block sizes.

The trace-limited block volumes represent the largest blocks that can theoretically fully form, hence the significance of collecting unbiased data and implementing correction for trace length censoring (Pahl, 1981). A rating of importance for a particular block shape is based upon the scale-stability chart given in Figure 3.6 and the charts showing support pressure and reinforcement length demands versus block face area are shown in Figure 3.7 (Windsor, 1999). In both figures, the trace limited block size for each shape is shown by the dot at the end of the full line. Free-falling blocks are inherently unstable, and the candidate critical block shapes have the highest support pressure demand at the trace limited size. For example, in this case, block numbers 3, 8 and 9.

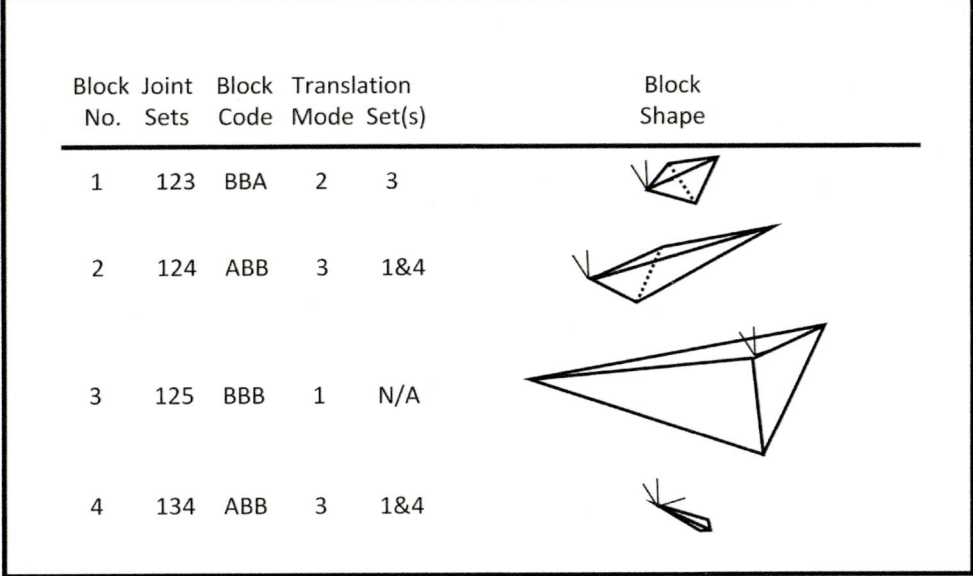

Block No.	Joint Sets	Block Code	Translation Mode	Set(s)	Block Shape
1	123	BBA	2	3	
2	124	ABB	3	1&4	
3	125	BBB	1	N/A	
4	134	ABB	3	1&4	

FIGURE 3.5 Example of basic tetrahedral block shapes and failure modes.

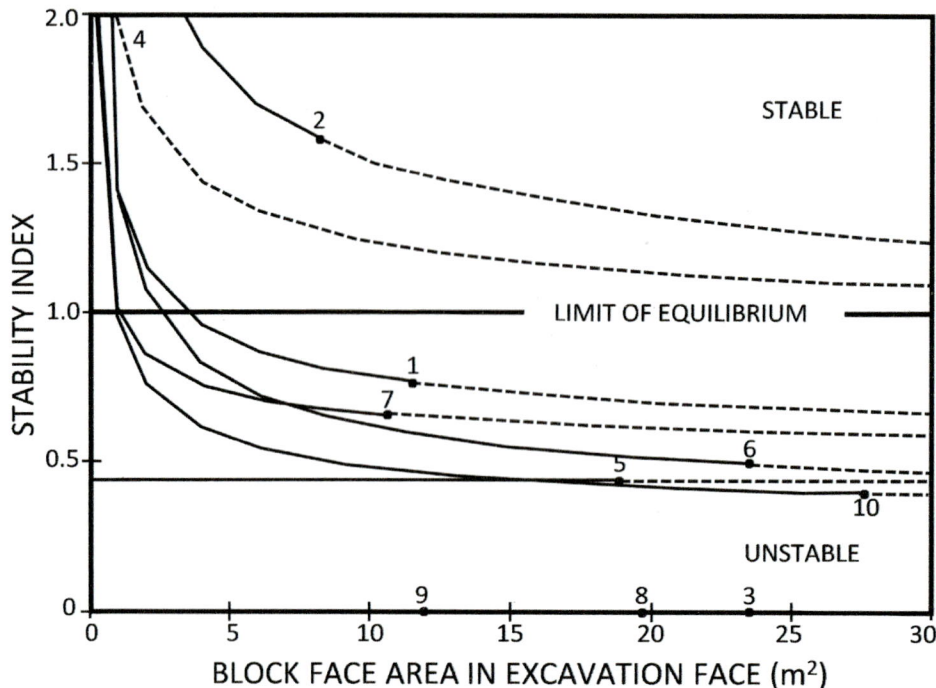

FIGURE 3.6   Scale-stability diagram showing stability index (i.e., the so-called Factor of Safety) versus block face area for each block shape (Windsor, 1999).

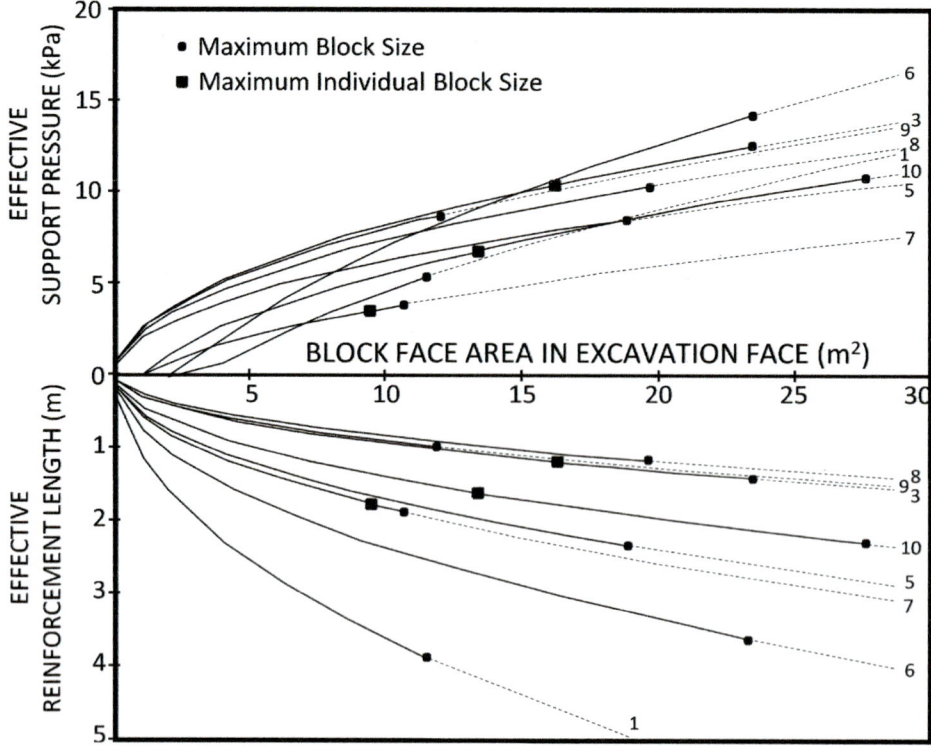

FIGURE 3.7   Support pressure and reinforcement length demands for the different block shapes at various sizes (Windsor, 1999).

3.5   PROBABILISTIC ASSESSMENT OF ROCK MASS INSTABILITIES

The methodology used here was first suggested by Windsor (1999) based on tetrahedral block real-izations resulting from using probabilistic analysis methods. Initially, a series of deterministic cal-culations are required to identify the block shapes and the range of sizes for each possible block shape that might possibly form. Here, the term 'relative probability' will be used—that is, the num-ber of tetrahedra of a particular size relative to the total number of all tetrahedra that might form. Relative probability is not a measure of the absolute probability that a tetrahedron of a particular size will form.

The most popular assessment procedures are based on the identification and stability assessment of single blocks with simple shapes. These methods are often used in design to identify problematic blocks by assuming that the discontinuities are ubiquitous and form a block shape at some critical size, bounded by a roof or wall span in an underground excavation.

The concept of ubiquitous joint sets assumes that combinations of discontinuities may occur everywhere and anywhere in relation to each other and with an excavation surface. This differs and is an advantage when compared with previous methods based solely on deterministic methods (e.g., Tinucci, 1992) because more potential instabilities are predicted to occur immediately adjacent to excavation surfaces. Windsor (1999) developed and reported the use of probability distribution functions for orientations and persistence to form a range of tetrahedral block shapes and sizes. The influence of different spacings within each set may be accounted for by using the method of Mauldon (1994). This work also incorporated the important concepts of a number of limiting block sizes previously defined by Windsor (1992b).

In the probabilistic method, the pertinent outcome from the analyses involving large numbers of simulations is a frequency distribution of sizes for each tetrahedral block shape formed from three discontinuities and an excavation face. This assessment requires that the orientation, persis-tence, shape, spacing and the strength and deformation characteristics of each discontinuity set be simulated in terms of probability distribution functions. It also requires that the position of each discontinuity relative to the other discontinuities and the excavation surface be simulated. A suf-ficient number of realizations must be conducted to properly account for the possible variations in discontinuity positions and characteristics. Importantly, the procedure assumes 'ubiquity' for dis-continuity positions and 'ubiety' for block vertex positions.

The outputs from the SAFEX program may be summarized as frequency distributions of block sizes for each block shape. The outputs are scaled identically to facilitate direct comparison and interpretation. The input requirements for the SAFEX simulation are as follows:

- The excavation orientation and dimensions
- The rock unit weight distribution and associated statistical parameters
- The number of discontinuity sets and, for each
  - The mean orientation and Fisher's constant
  - The trace length distribution and associated statistical parameters
  - The spacing distribution and associated statistical parameters
  - The joint friction distribution and associated statistical parameters
  - The joint cohesion distribution and associated statistical parameters

These variables may be simulated using the uniform, exponential, normal and log-normal distri-butions. A brief summary of the procedure is provided as a sequence of 14 steps (modified from Windsor, 1999) demonstrated with SAFEX output screens created using the data listed in Figure 3.4.

1. The orientation and geometry of the excavation surface are defined according to a global coordinate system.
2. A standard, deterministic block shape analysis is first conducted. This analysis uses the mean discontinuity orientations to identify the removable blocks (Figure 3.8).

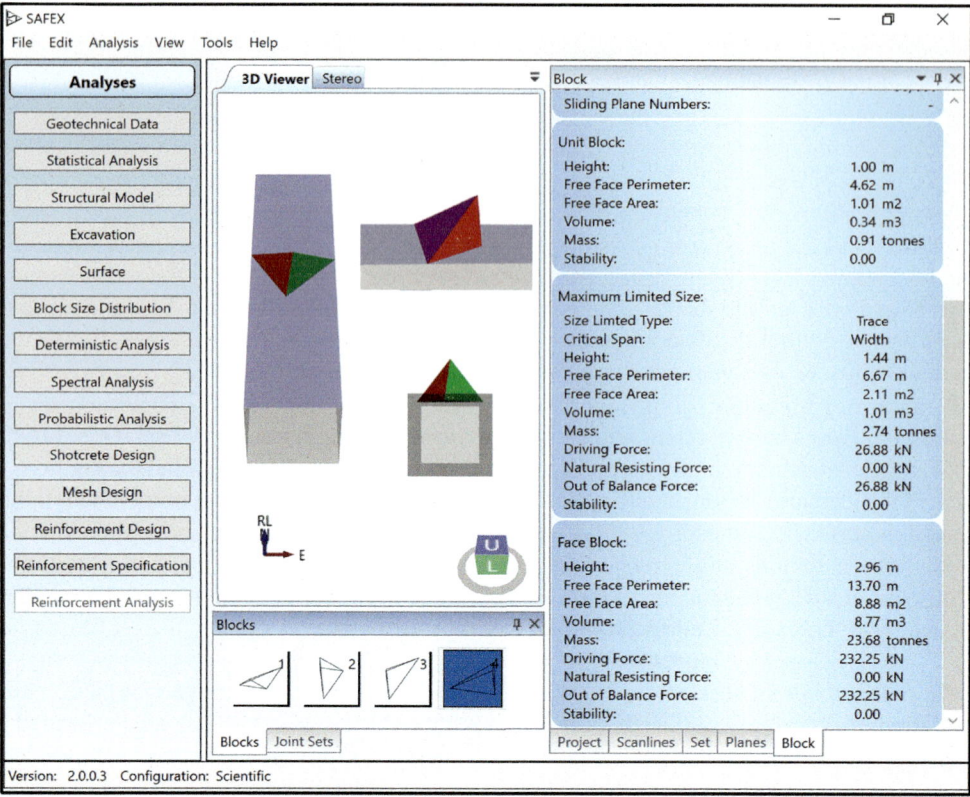

FIGURE 3.8 An example of identification of removable blocks with respect to the roof of an N-S oriented, 5-m wide excavation.

Steps 3–14 are conducted for each shape in the complete list of removable block list.

3. A 'possibilistic' block shape analysis is conducted (Windsor, 1999). This analysis uses the extreme dispersion boundary of the orientation distribution for each discontinuity set (Figure 3.9). The analysis is used to determine the shape extrema that bound all possible shapes associated with that block.
4. A 'possibilistic' block size analysis is conducted. This analysis uses the block shape extrema, the excavation dimensions and the trace length maxima to determine the extreme dimensions of the block. It results in a 'block existence zone' defined in the global coordinate system that bounds the shape and size for each combination of discontinuities involved in forming a removable block shape. There are two candidates for the block existence zone—the excavation limited block and the maximum trace limited block. The block existence zone is defined by the smaller of the two (or either if they are identical).
5. A point of nucleation is placed at random within the 'block existence zone'. This represents the intersection of at least three discontinuities of different orientations and coincides with the internal apex of a potential block.
6. A planar discontinuity from each relevant set is generated to intersect the point of nucleation. The orientation of each discontinuity is generated according to a given distribution function for the set. The shape and areal extent of the discontinuities are generated to comply with a given mathematical shape (Villaescusa and Brown, 1992) and with a dimension (e.g., diameter) selected according to a given distribution function for the persistence associated with each set. The coordinates of the centre of each discontinuity relative to the nucleation point are generated, according to a Poisson process (Figure 3.10), at a position

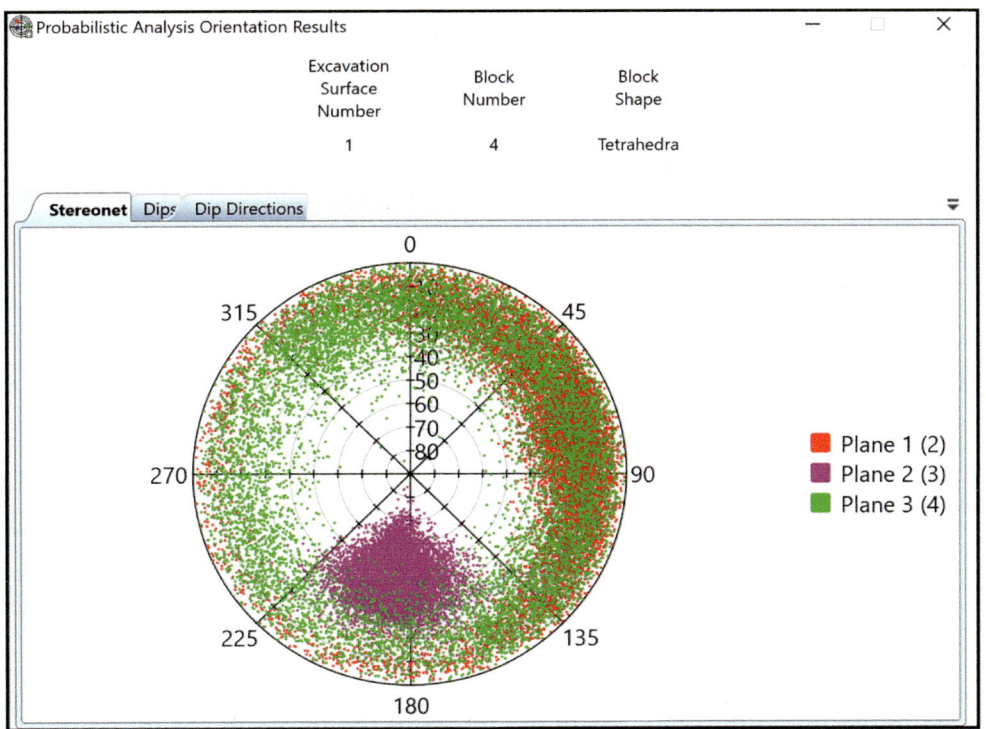

FIGURE 3.9 Dispersion of orientations for the three discontinuity families forming unstable Block Number 4 (in this case, families 2, 3 and 4).

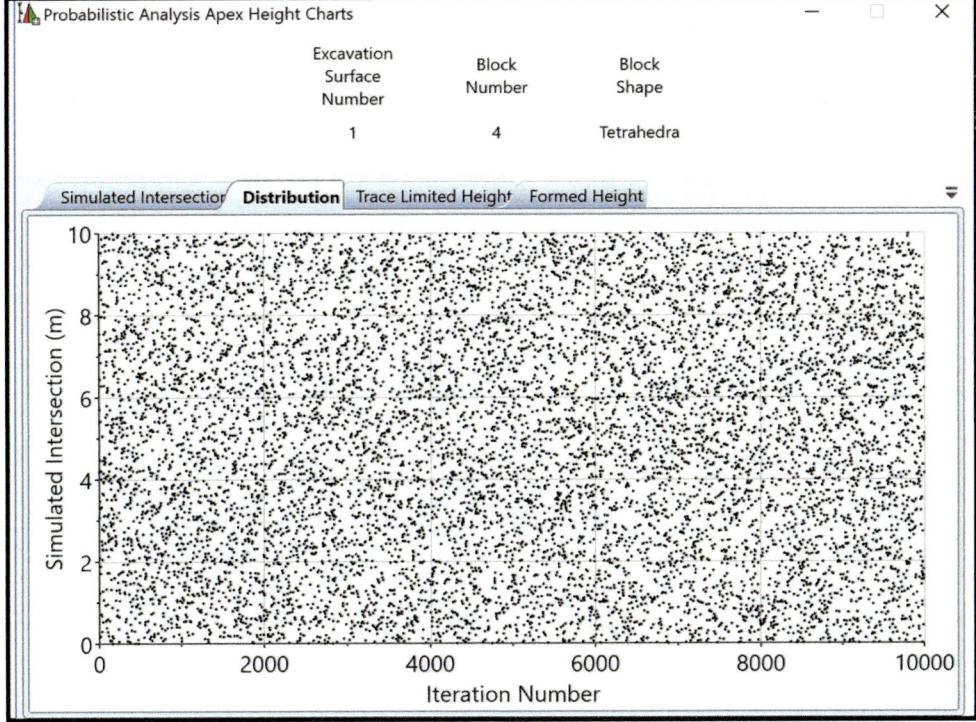

FIGURE 3.10 Distribution of nucleation points that simulate the intersection of the geological discontinuities.

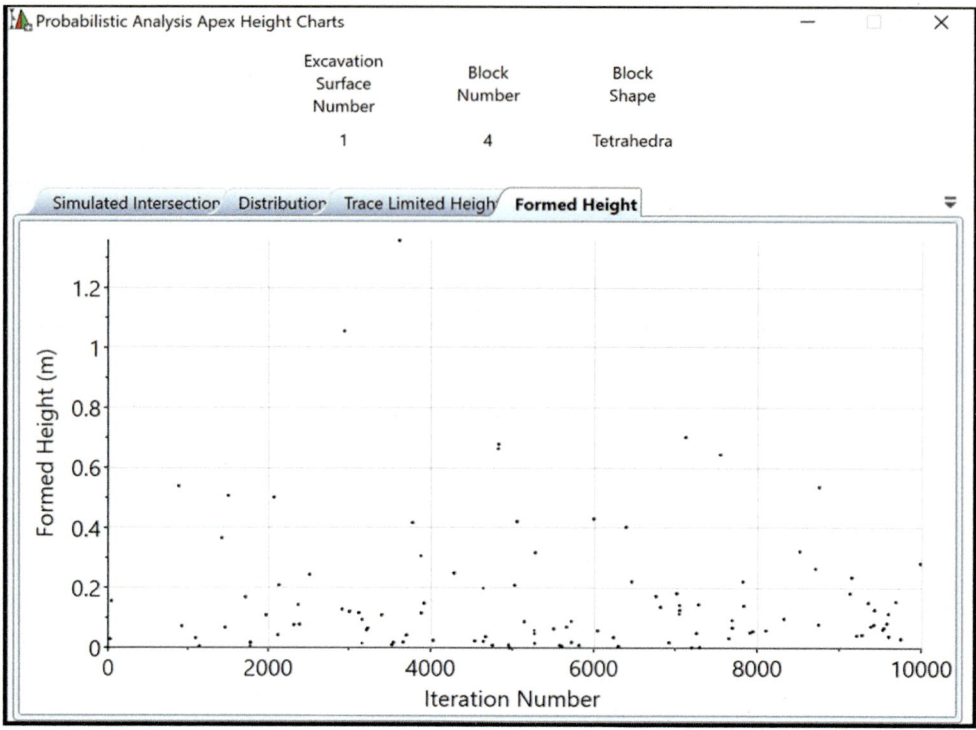

FIGURE 3.11   An example of the distribution of simulated apex heights.

within the plane of the discontinuity and constrained only by the shape and the areal extent of the discontinuity.

7. This step uses a number of complex algorithms based on vectors and vector algebra to identify a fully formed tetrahedral block that occurs when all three internal edges between two faces are predicted to be sufficiently long to intersect the excavation surface (Figure 3.11).

8. The static stability of the valid tetrahedral block is assessed using a static force equilibrium analysis. This analysis is conducted using the specific shape and dimensional characteristics of the block and data, simulated to obey the distribution functions associated with the strength characteristics of the discontinuities. An example showing the calculated modes of failure is shown in Figure 3.12, where it can be seen that all block shapes produce large proportions of free-falling blocks.

9. The details describing the shape, dimensional characteristics and, if applicable, the out-of-balance force, the mode and mechanism of instability are stored for later analysis.

10. Relative frequency histograms and cumulative frequency diagrams are prepared for each characteristic describing the block (e.g., apex height, face area, volume and static out-of-balance force).

11. Steps 5–10 are repeated until a sufficiently large number of realizations have been analysed. The simulation sequence is terminated on the basis of 'smoothness' and precision in the resulting relative frequency diagrams. After a certain number of simulations, these distribution curves become 'smooth' and do not change appreciably. In most cases, a total of 10,000 simulations are considered sufficient.

12. Steps 3–11 are repeated for each block shape appearing in the list of removable blocks.

13. The relative frequency histograms and the cumulative frequency curves are corrected for the relative probability of a nucleation point occurring in the block existence zone. This

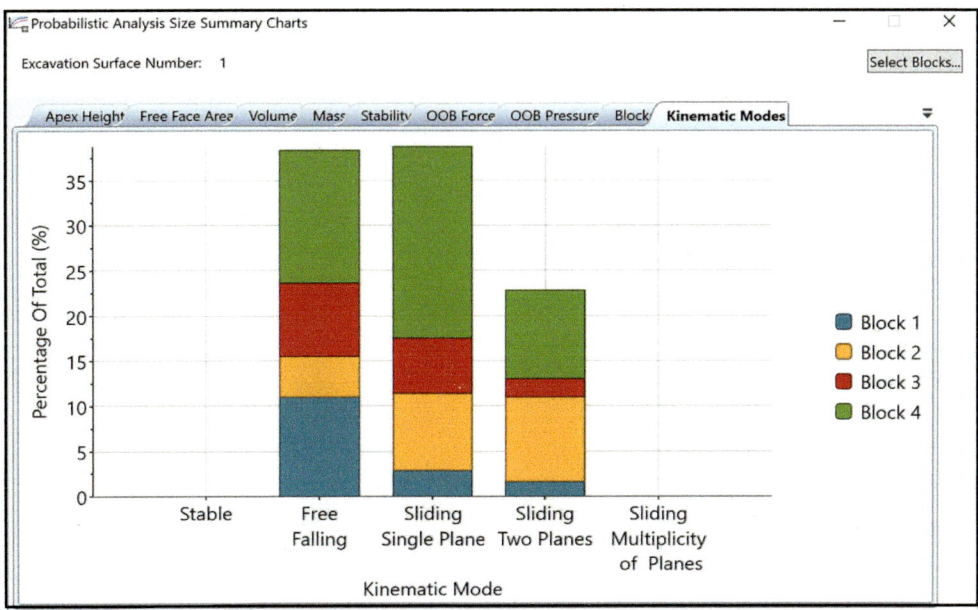

FIGURE 3.12   An example of kinematic modes of failure for all the wedge geometries.

correction takes into account the relative probability of a 'vector triple' occurring within the block existence zone (Windsor, 1999).

14. The corrected relative frequency histograms and the cumulative frequency diagrams for each of the removable blocks are redrawn on one diagram together with the total frequency distributions for the sum of all blocks. These diagrams are prepared for each block characteristic (e.g., apex height, face area, volume, mass and static out-of-balance force). This facilitates comparisons between the complete rock mass and the individual blocks.

Figure 3.13 shows the cumulative frequency of apex heights for all the removable blocks created from all four joint sets in Figure 3.4. The failure modes for all the blocks were identified earlier in Figure 3.12. The apex height provides an indication of the reinforcement length required to stabilize blocks, and hence the excavation surface.

Figures 3.14–3.16 show detailed geometrical results for the most unstable combination (Block 4) identified during the analyses. Three groups of data may now be readily identified for further analysis: the values at the 90% cumulative frequency, the values encompassing most of the data and the worst-case scenario (maximum value).

Table 3.1 summarises the calculated values for the unstable apex heights, failure masses and failure areas, such as Block Number 4. It should be noted that the probabilistic analysis was undertaken assuming a flat surface, which is slightly conservative. The output from the SAFEX program allows the probabilistic determination of the key parameter given in tons of failure Mass per m^2 of unstable area (T/m^2) to estimate the rock mass instability. In other words, this would be the unstable mass that the ground support scheme would need to stabilize if a dynamic event were to occur and mobilise the geological discontinuities.

3.5.1 Interpretation of Results

The following observations may be made with regard to a probabilistic size prediction of tetrahedral blocks adjacent to an excavation surface formed in a jointed, hard rock mass:

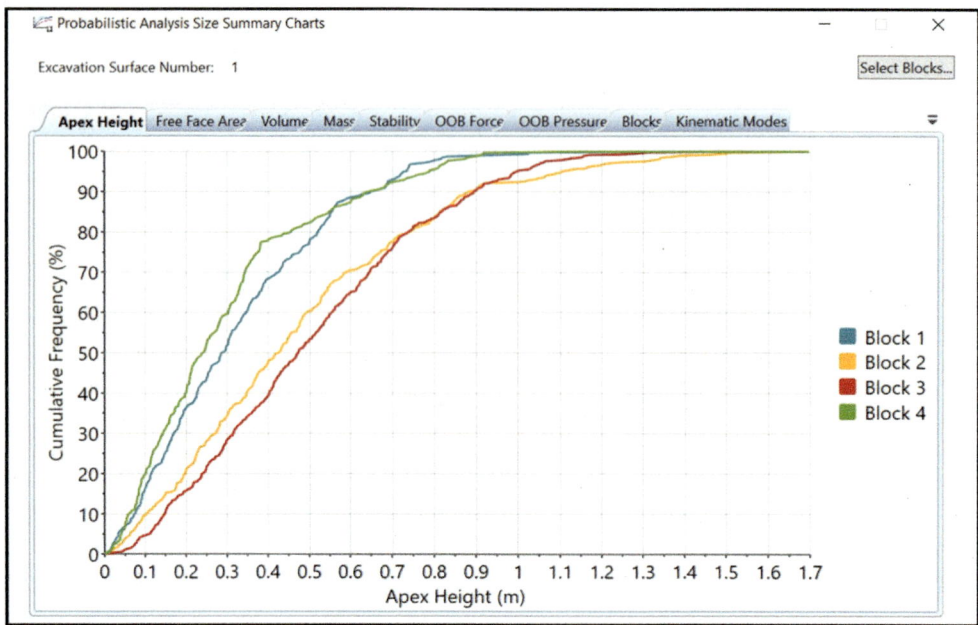

FIGURE 3.13 Example of the probabilistic frequency of apex heights for all removable blocks.

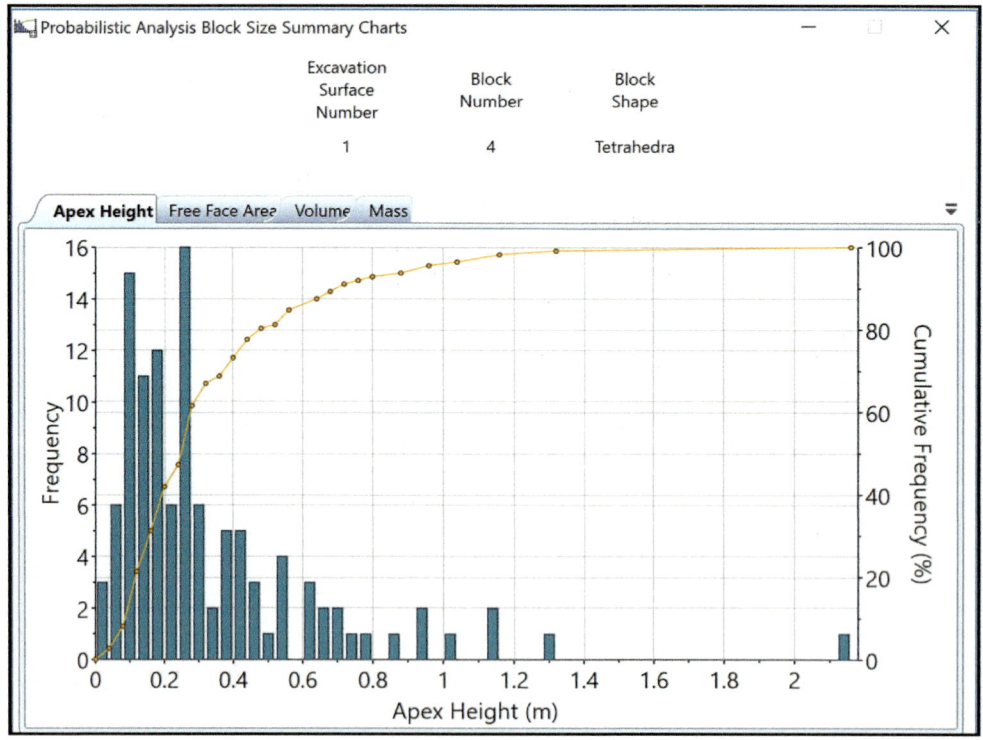

FIGURE 3.14 Example of the apex height distribution for all removable blocks—Block Number 4.

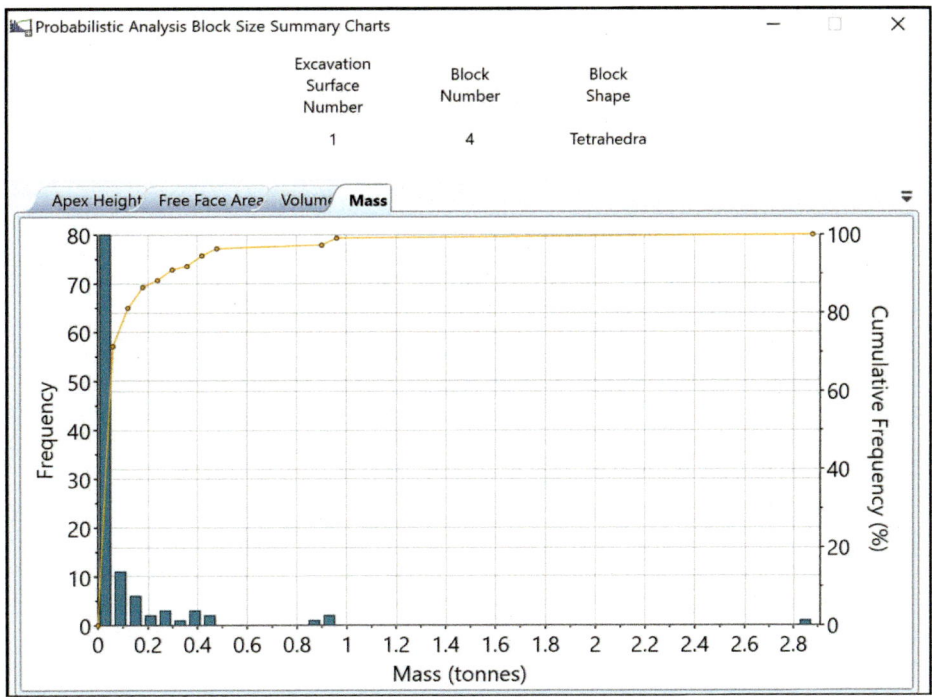

FIGURE 3.15 Example of the block mass distribution for all removable blocks—Block Number 4.

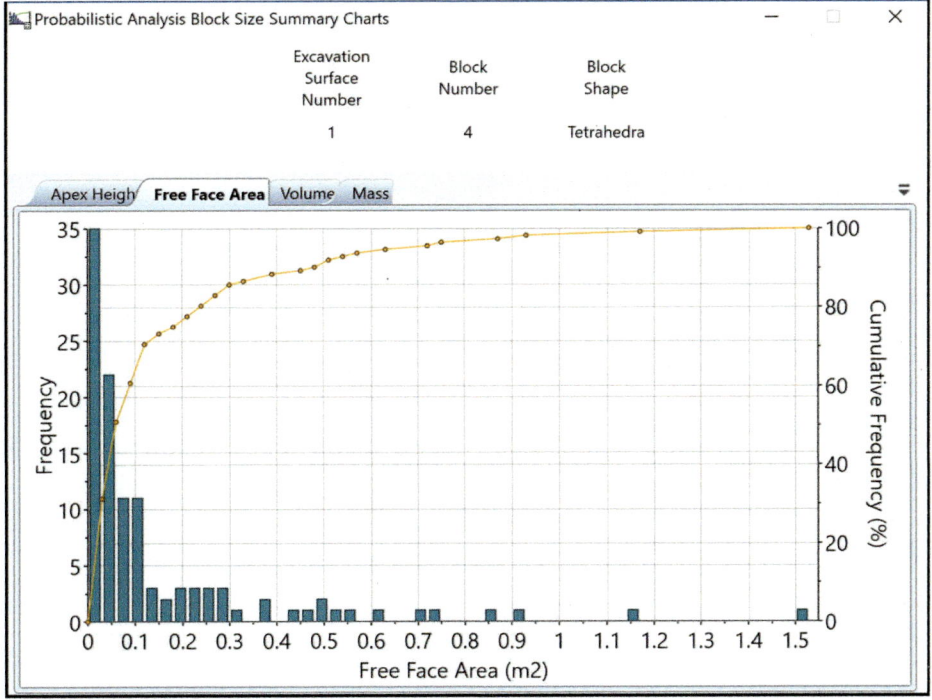

FIGURE 3.16 Example of the block area distribution for all removable blocks—Block Number 4.

TABLE 3.1

Probabilistic Determination of Block Instability.

Probabilistic Data	Apex Height (m)	Unstable Area (m²)	Unstable Mass (T)	Static Demand (T/m²)
90%	0.7	0.45	0.25	0.5
Most of the data	1.4	1.00	1.00	1.0
Worst case	2.1	1.50	1.80	2.0

- A large number of small blocks are formed.
- Relatively few large block sizes are formed.
- The largest block sizes are related to the block shape.
- The largest block sizes are limited by trace length.
- The range of sizes predicted probabilistically is larger than that predicted deterministically.
- A probabilistic analysis method predicts greater numbers of smaller and larger block sizes compared to a deterministic analysis.

An important point to consider is that since the joint set spacings are not considered within the probabilistic analysis, a trace limited block may comprise a number of smaller tetrahedra and other block shapes, as shown in Figure 3.17. Probabilistic analyses assume ubiquity of the joint set planes and ubiety of the block apex within the 'block existence zone' defined by the trace limited block formed by considering the extrema of the joint set orientations. Each block apex is simulated as a random point within the block existence zone without consideration of spacing.

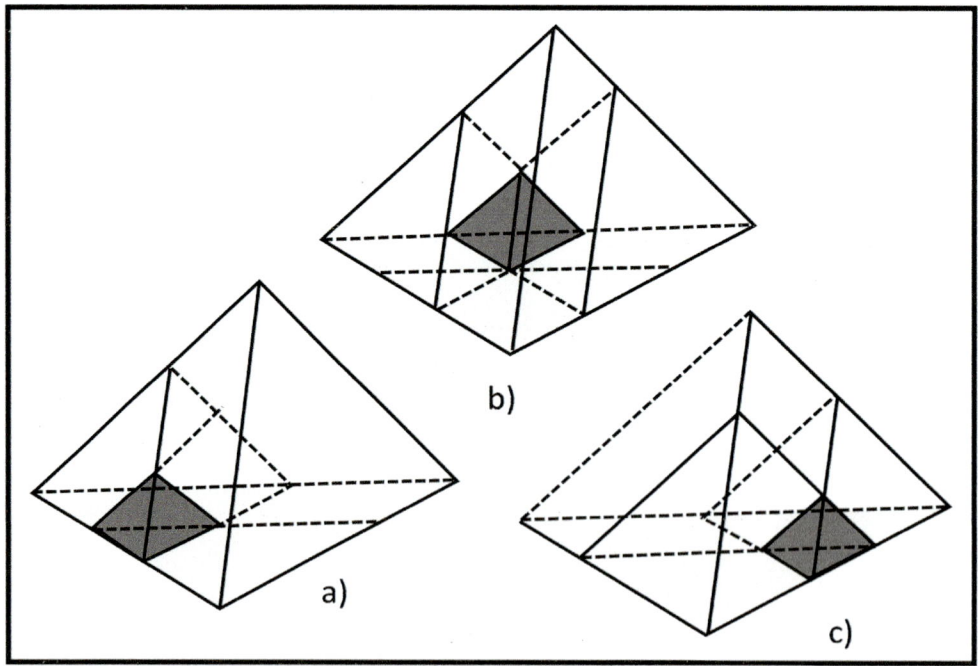

FIGURE 3.17   Example of smaller blocks of various shapes contained within the trace limited tetrahedral-shaped blocks (Windsor and Thompson, 1997).

Conventionally, deterministic analyses assume specific locations of the joint set planes within a simulated volume adjacent to the excavation face. Only tetrahedra adjacent to the excavation face are included in the block size distributions. It must be noted that if the deterministic analysis method could be modified to aggregate a number of smaller tetrahedra, pentahedra and hexahedra to form larger tetrahedral-shaped volumes, which would be more consistent with probabilistic predictions (Windsor and Thompson, 1997).

4 Ground Support Mechanics

4.1 INTRODUCTION

The performance of ground support as a scheme is essential in order to constrain failures occurring at the rock surfaces of deep or highly stressed excavations. As the incipient instabilities develop into significant failures, they are likely to first mobilize a number of reinforcement elements prior to loading the surface support layers. As the deformation increases, several reinforcement elements may fail either under slow, time-dependent deformation or violently, with the load being transferred to the surface support located immediately adjacent to any failed reinforcement elements. The surface support layers require sufficient toughness to displace and transfer the load to other reinforcement elements located in the periphery of the failure geometry (Figure 4.1). The surface support needs to be completely connected to all the reinforcement elements within and outside the failure for the load transfer to occur efficiently and contain the failed material.

It is critically important that the displacements allowed by the reinforcement and the surface support remain compatible and ideally below a range of maximum radial excavation convergence (~5%), such that following a violent failure, the excavation rehabilitation is minimal and the tunnel remains serviceable. If many reinforcement elements fail at low loads or allow very large displacements, then the surface support layers will likely be overloaded, leading to catastrophic failure (Figure 4.2). Excavation repair will then be required leading to direct and indirect costs, including loss of opportunity due to the excavation access not being available for its intended purposes.

4.2 REINFORCEMENT RESPONSE

4.2.1 LOAD TRANSFER WITHIN THE COLLAR REGION

For a particular failure event, the reinforcement elements are loaded by an unstable mass that can be probabilistically estimated, as described in Chapter 3. Regardless of the reinforcement action (CMC, CFC or DMFC), the mechanics at the collar loading are very similar. If an unstable depth with respect to the excavation surface is shallow, say, a depth of failure ranging from 0.25 to 0.5 m (Figure 4.3), then loading of the immediate bolt collar region will preferentially occur. In cases where failures are experienced, various results, including stripping of the bolt/nut, bending and pulling of the plate or bolt breakage, can be observed (Figure 4.4).

Collar instability for CFC systems is likely to test the capacity at the bolt-plate interface. This is due to the relatively low capacity per metre of embedment for such systems, resulting in a slip of the unstable region even for a shallow depth of instability. Such violent ejections can result in plate loading, which might fail at very low loads (Figure 4.5). Consequently, any static or dynamic pull testing program that preferentially loads a CFC element without mobilizing the bolt-plate interface does not account for the correct load transfer observed in the field (Figure 4.6). Significantly, shotcrete does not constitute a high energy dissipation layer, as it cannot guarantee the connection and eventual retention of the instability by other elements located outside the region of failure.

Due to their high load transfer capacity, fully encapsulated CMC systems are likely to mobilize the full element capacity even if loaded near the collar (Figure 4.7). Despite the element being fully encapsulated, the shallow, violent instability can also transfer the load to the collar surface fixtures leading to plate damage.

DOI: 10.1201/9781003357711-4

FIGURE 4.1 Examples of broken reinforcement elements that allow deformation into immediate mesh support with load transferred to other surrounding components.

FIGURE 4.2 Example of a tunnel with very low installed ground support capacity, leading to an excessive displacement of the lower walls.

Both experience and dynamic testing suggest that DMC systems (such as a decoupled Posimix bolt) require at least 0.3–0.4 m of effective collar length (plus a 0.2-m collar tail) such that load transfer near the boundary of an excavation can take place. Clearly, this region also needs to be fully encapsulated by either cement or resin. It would ensure that the performance at the collar is not only dependent on the inherent strength of the surface fixtures, which can be significantly less than the strength of the elements. For decoupled cable bolts, a load transfer region of at least 1 m is recommended at the collar region to ensure that the strength of the cable bolt surface fixtures is fully complemented.

FIGURE 4.3   Example of a shallow depth of a violent failure in which the support layer was unable to transfer the load outside the area of instability.

FIGURE 4.4   Example of broken, ejected bolts and damaged plate following a sudden violent failure.

FIGURE 4.5 Example of material slippage along CFC elements associated with shallow depth of failure (this led to violent plate failure).

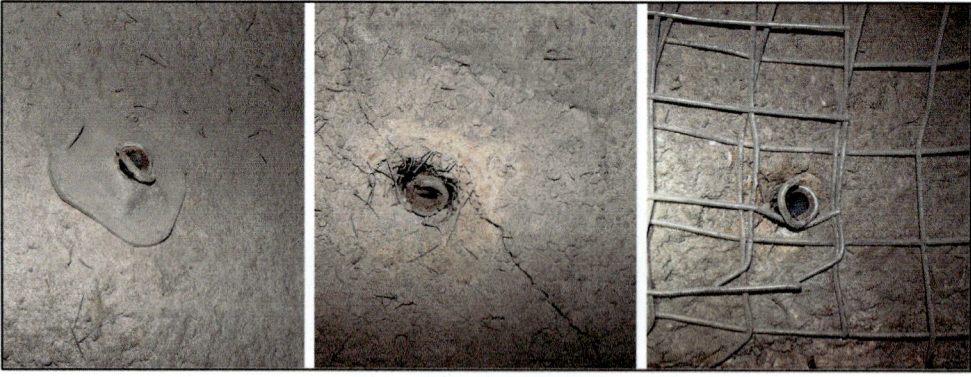

FIGURE 4.6 Examples of small surface displacement resulting in friction bolt plate failures.

4.2.2 Load Transfer along the Element and at the Toe Anchor Region

Once the collar region is able to transfer the load to the element, the performance along the element of these systems will vary according to the load transfer mechanics that apply (Windsor and Thompson, 1993). For CFC elements, slippage, as a function of embedment length, will likely occur (Player et al., 2009a). As the capacity is limited, the displacement could be very large, and, sometimes, the element may even pull out completely (Figure 4.8).

For instabilities with a depth of failure in excess of 1 m, the performance along the axis of CMC elements will depend upon the encapsulation media. Inspection following dynamic testing of fully

FIGURE 4.7 Examples of CMC reinforcement after a shallow, violent failure.

FIGURE 4.8 Examples of large displacement resulting from (a) dynamic loading of a 47-mm diameter friction bolt loaded at relatively low dynamic loads and (b) shallow depth of failure.

coupled cement-encapsulated elements shows that these are capable of rapidly developing a region of heavily fractured cement, which allows a rapid plastic stretch of the steel element (Figure 4.9). This release of the bolt at either side of an active failure discontinuity does not occur when resin is used as the encapsulation media (Figure 4.10). Consequently, the rupture of fully resin encapsulated bolts occurs at very low displacements, with minimal energy (<10 kJ) dissipation.

In order to increase the energy dissipation capacity, the elements can be configured with a decoupled region that allows deformation according to the inherent force-displacement properties of the material. The load transfer occurs at both the collar and anchor end regions, with the centrally located region allowing the majority of the deformation. A critical requirement is for the load to be transferred from the element anchor region to the rock mass, which, if designed correctly, must be located within a stable region beyond the instability. A typical toe anchor region where load transfer is required usually ranges from 0.6 to 1 m in length. For typical 2.4–3 m long reinforcement elements, decoupled regions ranging from 1 to 1.4 m can be achieved (Figure 4.11).

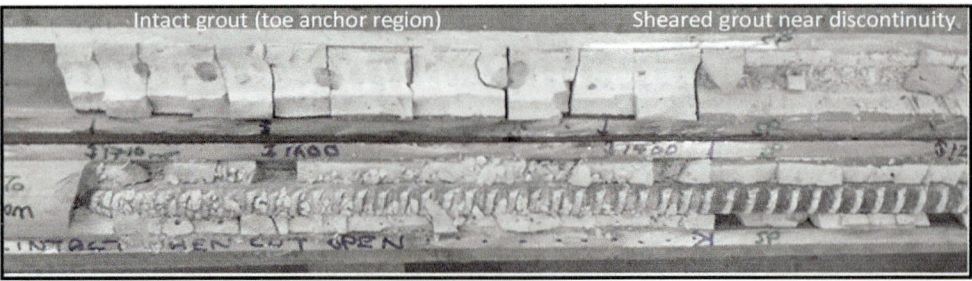

FIGURE 4.9 Inspection of cement-encapsulated threaded bar following dynamic loading showing a region of sheared grout within which the bar can elongate (Player, 2012).

FIGURE 4.10 Inspection showing a limited region of influence on either side of an active discontinuity for a resin-encapsulated threaded bar.

The collar region, in conjunction with the surface fixtures, needs to develop load transfer of a similar capacity to the toe-anchor region. It ensures that the element is able to dissipate large amounts of energy within the decoupled region (Figure 4.12—see Chapter 7 for a full description).

4.3 SURFACE SUPPORT RESPONSE

In practice, the amount of surface displacement will depend upon the depth of failure and the degree of geological structure mobilization, often resulting in a variable radial strain around the tunnel periphery. The resulting instability is heavily dependent upon the role of the geological structures. These can produce formed blocks that may simply displace into the opening or shear along an individual structure in a specific orientation (Figure 4.13).

If a violent failure occurs after a period of significant stress concentration and progressive discontinuity growth, then a number of slabs and pre-existing blocks may be agitated at the boundary of the excavation. It might result in a deep failure due to dynamic unravelling back to a larger geological discontinuity (Figure 4.14).

In most cases where geological structures are mobilized, a number of reinforcement elements may rupture locally, resulting in large load transfer to the surface support layers. If these layers are not in contact with the rock, a significant deformation will occur before their retention can be mobilized. In addition, if surface support cannot constrain large loads at the point of connection with the reinforcement, they will likely fail before they can transfer the load to the adjacent reinforcement elements (Figure 4.15).

The bulking and other related surface deformation can be reduced when a layer of shotcrete is immediately applied to the rock surface within the development cycle. In addition, the best

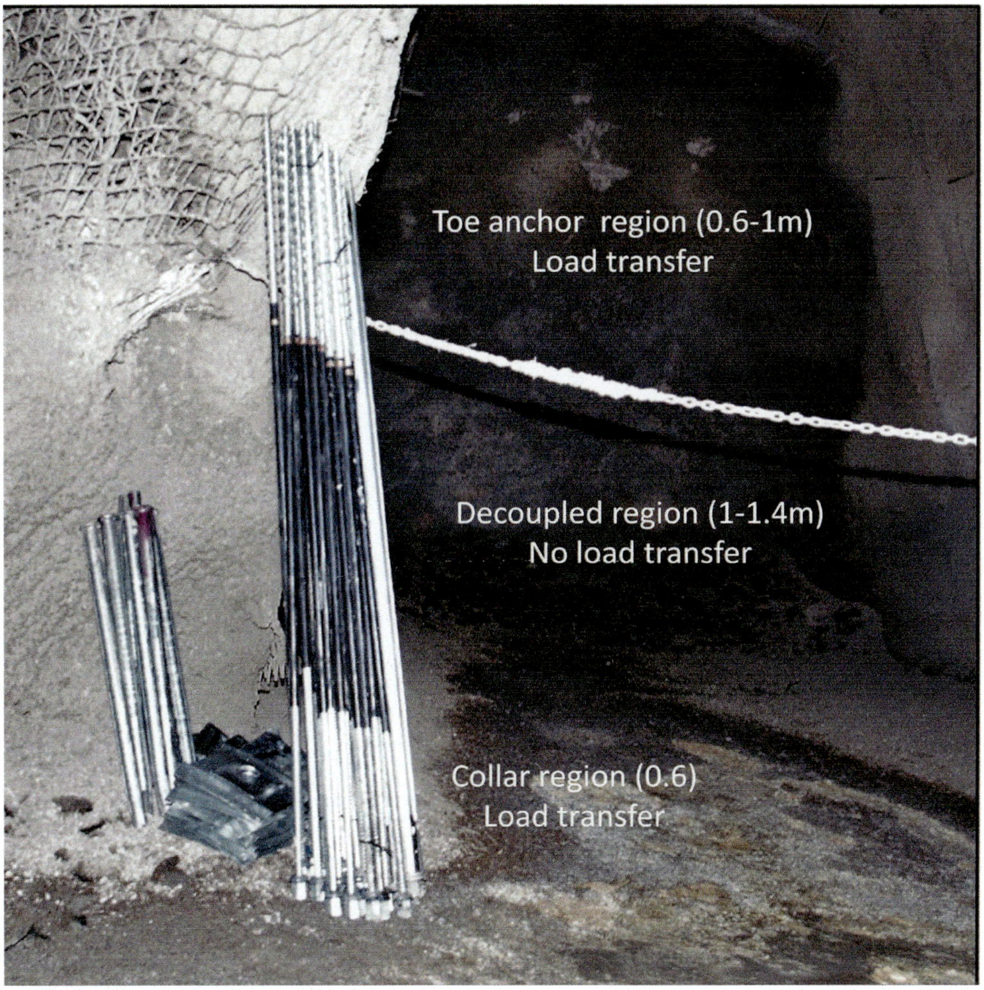

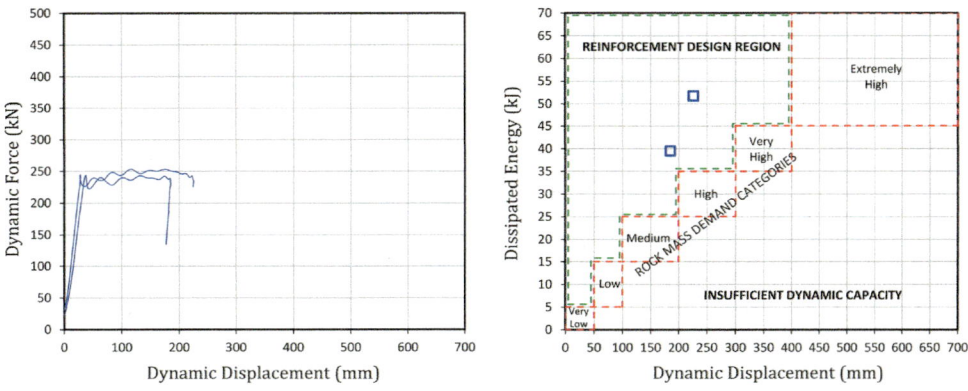

FIGURE 4.11 Example of typical decoupled Posimix bolts showing the load transfer regions as well as the centrally decoupled region.

FIGURE 4.12 Examples of dynamic force-displacement response for 3 m long decoupled Posimix elements.

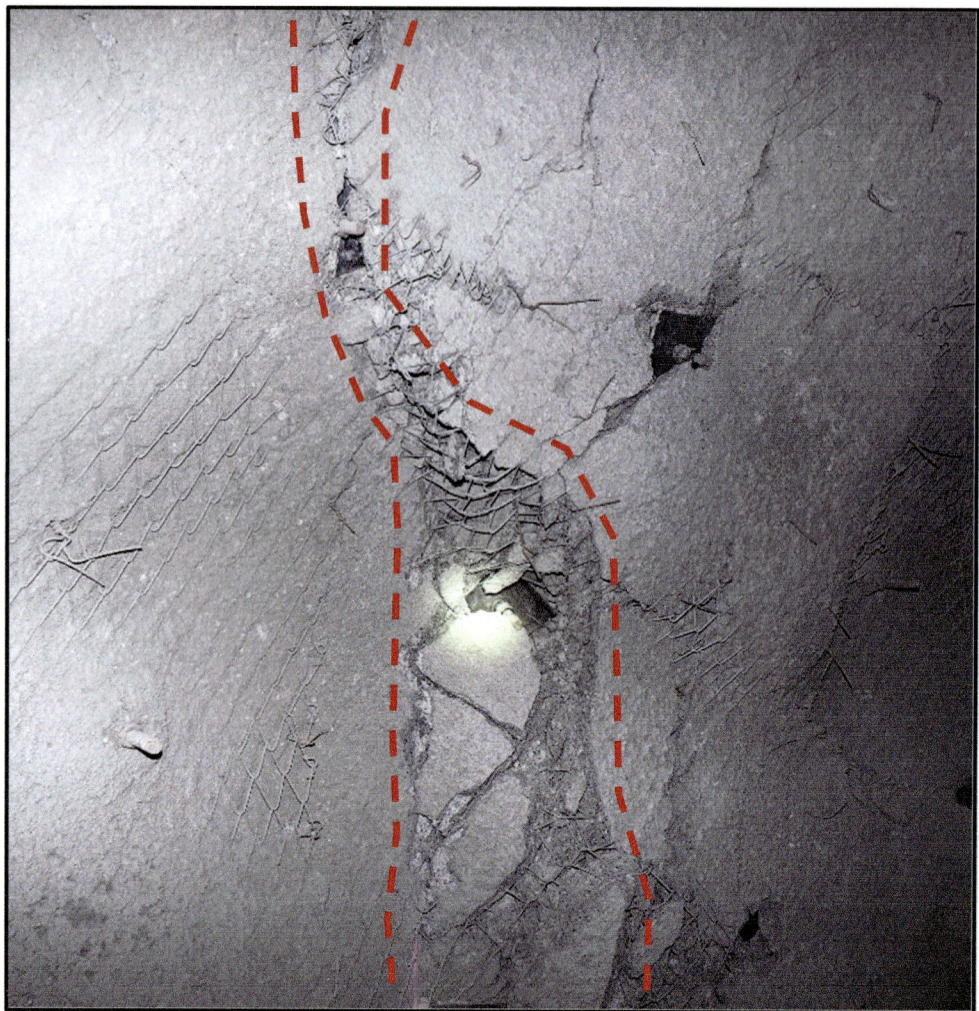

FIGURE 4.13 Example of dynamic shear rupture along an individual geological discontinuity.

performance is achieved when the shotcrete is reinforced with a layer of mesh to ensure that a minimal excavation deformation will occur before a reactive force from the mesh is mobilized. The objective is to ensure shotcrete to mesh contact while minimizing the sprayed thickness covering the mesh layer (Figure 4.16) so that any shotcrete ejection following dynamic impact is minimized (Figure 4.17).

4.4   FAILURE GEOMETRY

The depth of failure around an underground excavation will often depend upon the induced stress and the frequency and nature of any geological discontinuities. Damage can range from shallow spalling through intact rock to excavation profiles entirely controlled by block formation, resulting in overbreak defined by geological discontinuities. Although the size of the exposed spans may also play a role with respect to the actual out-of-balance pressure for a particular instability, the area, volume and mass of instability are largely controlled by the orientation, spacing and trace length of the geological discontinuities (Figure 4.18).

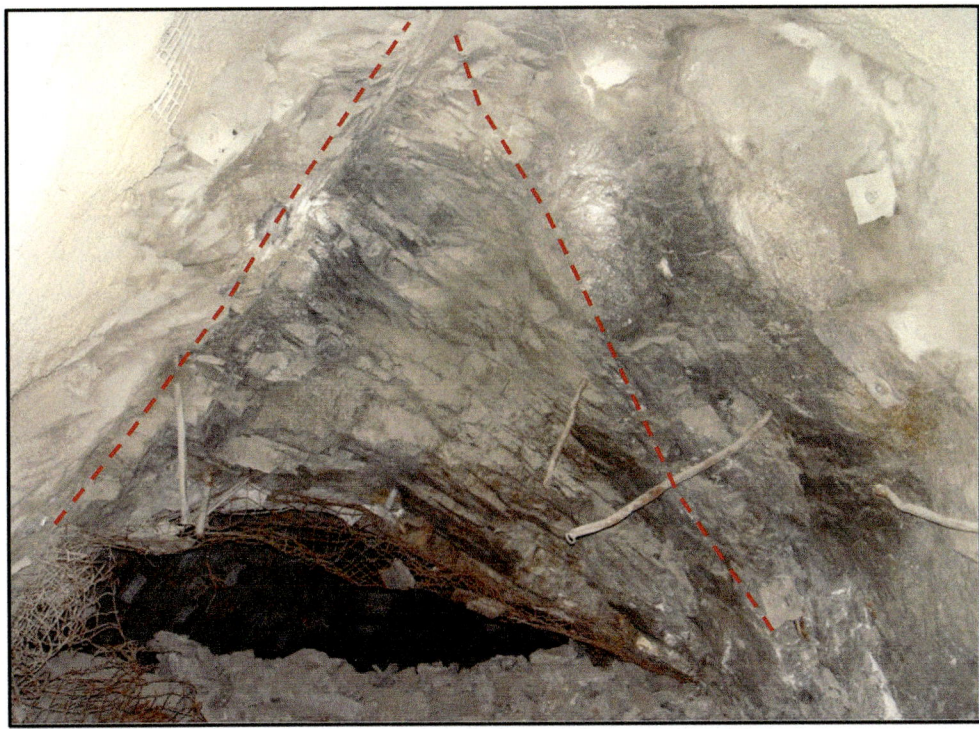

FIGURE 4.14 Slabbing due to sub-horizontal stress, which may unravel back to large geological structures.

FIGURE 4.15 Example of large weld mesh deformation leading to significant excavation damage that required substantial repair.

Figure 4.19 shows actual data from underground excavations where the observed failure mass confirms the importance of near-surface instability while suggesting an unravelling mechanism for the larger, deeper failures. In some cases, large failures can be a combination of an initial surface instability controlled by small blocks and slabs, followed by larger blocks released from deep within the failure (Figure 4.20). However, if a ground support scheme can contain the near-surface instability, then subsequent failure by unravelling is limited due to the unstable mass expanding and

FIGURE 4.16 Examples of mesh reinforced shotcrete with a minimal cover over the mesh to minimize ejection.

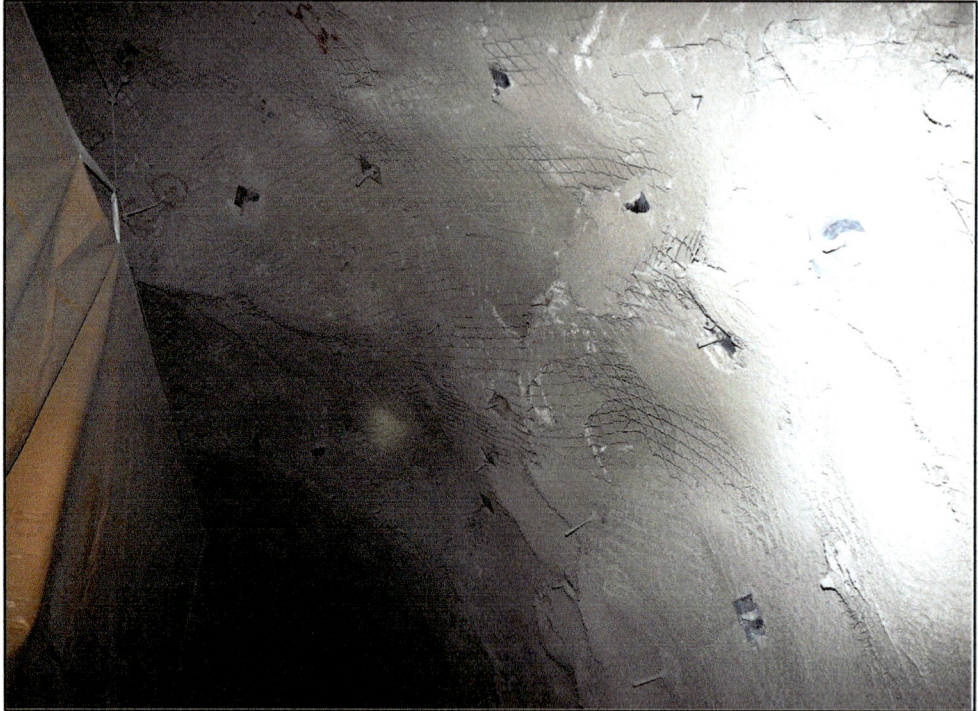

FIGURE 4.17 Example of minimal shotcrete ejection following a dynamic event. Mesh in contact with shotcrete ensures an immediate reaction to loading.

remaining in a place where interlocking is progressively mobilized to limit the resulting depth of failure (Warburton, 1993; Goodman and Shi, 1985).

4.5   SURFACE LAYER TOUGHNESS

In areas where geological discontinuities are mobilized by violent failure, they can result in very high localized ground support demand capable of causing loading reinforcement elements to rupture. As the reinforcement capacity is consumed and the area of instability increases, the surface

FIGURE 4.18 Example of violent block failure through fibrecrete support exposing the relief surface defined by geological structures.

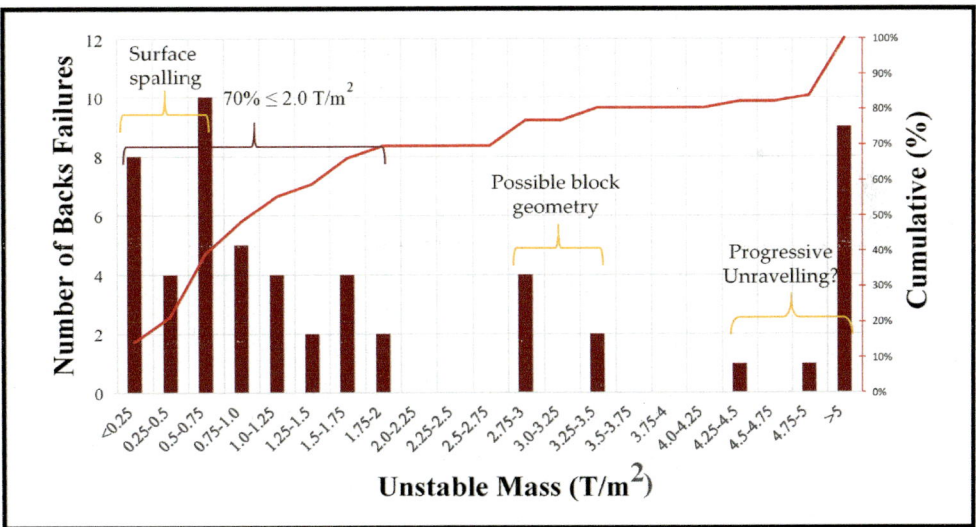

FIGURE 4.19 Example of observed mass of instability and related likely failure mechanisms.

support layers are progressively loaded and deformed, effectively transferring additional load to other reinforcement elements that still retain capacity.

The minimum surface support consists of a single layer of mesh installed on the surface of the excavation (Figure 4.21). Such an arrangement provides no active confinement to the rock mass while potentially allowing de-lamination due to stress-driven failures forming slabs immediately behind the exposed mesh. Furthermore, the flexible nature of the mesh in response to lateral loading results in a significant displacement before any instability can be contained

FIGURE 4.20 Example of a large potential wedge containing smaller blocks interlocked by reinforcing elements.

FIGURE 4.21 Example of an excavation stabilization scheme comprising rock bolts and mesh.

(see Chapter 9). Consequently, the depth of instability increases before the load can be transferred to the surrounding reinforcement elements. If the mesh energy dissipation capacity is exceeded, failure will occur through the mesh, with stable reinforcement elements nearby not adequately mobilized (Figure 4.22).

FIGURE 4.22   Example of a violent failure through weld mesh with a little load transfer outside and away from the region of instability.

A shotcrete layer in contact with a rock mass provides immediate local confinement such that progressive unravelling is minimized. The load capacity depends upon the thickness of the sprayed layer with crack propagation creating a shotcrete slab that can detach from both the surrounding shotcrete layer and the rock surface. The internal reinforcement at the boundary of those plates controls the load-displacement capacity of such a surface support layer. If fibre-reinforced, displacement compatibility at the crack or fracture surface is limited to half the length of the fibre. Observations show that in most cases, a fibre density of less than 1 fibre/cm^2 can be identified across failure boundaries. Dynamically, that typical fibre strength and density are insufficient to arrest crack propagation, with most fibres failing following dynamic loading (Figure 4.23).

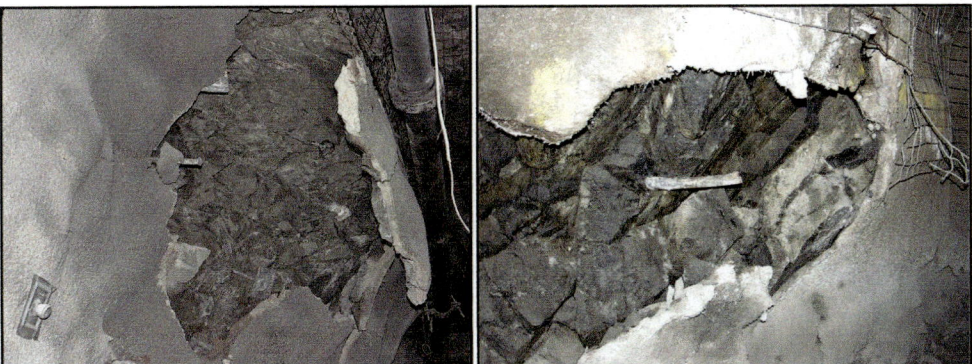

FIGURE 4.23   Shallow depth of instability through CFC reinforcement and a fibrecrete layer with this system unable to transfer the load to other reinforcement elements located around the instability.

In cases where welded wire mesh is used to reinforce the shotcrete layer, the layer performance depends largely upon the load-displacement capacity of the mesh (Figure 4.24). Mesh will fail according to mesh type and the energy dissipation capacity of the wire utilized. This is typically less than 2 kJ/m^2 for a 5.6-mm galvanized wire (Morton, 2009). Woven mesh (also known as chain link) mesh is articulated and can better accommodate any failure orientation and shear direction

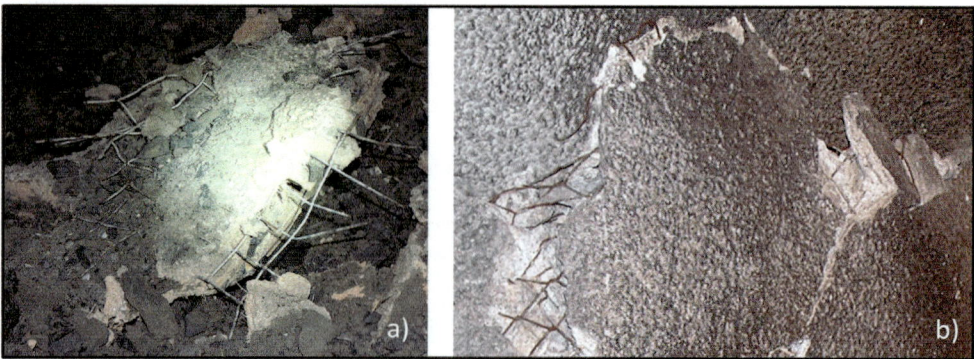

FIGURE 4.24 Load transfer at a shotcrete failure boundary: (a) weld mesh and (b) woven (chain link) mesh.

at the failed shotcrete plate boundary (Kusui, 2015). The capacity will depend on the type of mesh being utilized with values in excess of 20 kJ/m^2 calculated from back analysis of actual failures (Arcaro et al., 2021).

4.6 GROUND SUPPORT SCHEME

A ground support scheme must accommodate the expected rock mass behaviour and should be selected considering the available drilling equipment and also taking into account the excavation purposes. Importantly, the likely failure geometry and mechanism are critical, as well as the ongoing environmental effects, such as induced stress, weathering and, if applicable, equipment and blast damage. The requirements for ground support to perform as a scheme increases with elevated levels of rock mass demand. Historically, excavations located at shallow to moderate depths could be successfully stabilized with strategies ranging from scaling only to spot and pattern bolting with no surface support layers to connect the reinforcement devices. However, as a result of legislation (MOSHAB, 1999) and when the depth of mining increased beyond 500 m below the surface, ground support consisting of a combination of welded wire mesh and several types of rock bolts was introduced in Australia. In the early 1980s, wet-mix fibrecrete layers were introduced (as an alternative to welded wire mesh) and in conjunction with resin-encapsulated reinforcement to speed up excavation development rates (Figure 4.25).

FIGURE 4.25 Example of pattern bolting in conjunction with (a) weld mesh and (b) shotcrete.

Experience has shown that excavations that are stabilized using fibrecrete layers are only effective for small deformations. Consequently, any radial closure exceeding 50 mm results in cracking of the fibrecrete at the surface of the excavation. As the deformation increases, the width of the cracks increases until the fibres become ineffective in connecting the unstable fibrecrete slabs, which, in some cases, leads to violent failures even with a shallow depth of failure (Figure 4.26). It is impossible for a layer of fibrecrete to transfer the load to the reinforcement elements located outside the area of instability.

FIGURE 4.26 Examples of violently ejected fibrecrete slabs.

In cases where the deformation of fibrecrete surface support layer becomes incompatible with the underlying rock mass deformation, one common repair strategy is to install a layer of mesh in conjunction with additional reinforcement elements. For cases of moderate rock mass demand, the layer of mesh is installed on the shotcrete, with the two layers acting independently. In conditions of increased stress, this is ineffective if the resulting deformation becomes excessive, as the external mesh can be damaged and may buckle while requiring failure of the shotcrete to initiate the deformation (Figure 4.27).

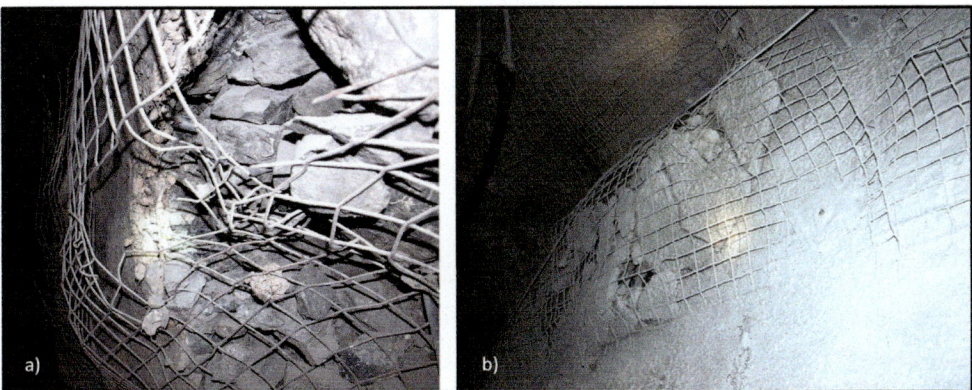

FIGURE 4.27 Independent mesh and shotcrete layers: (a) mesh separated and damaged and (b) buckling during excavation compression.

In cases where significant energy demand is expected, the shotcrete layer can be reinforced with a mesh layer, ensuring that both layers simultaneously respond to loading by being in close contact.

It is achieved by first placing a layer of shotcrete and then following with mesh and bolt installation within the development cycle. The position of the mesh within the shotcrete layer should be towards the external surface to ensure that ejection material is minimized. It often requires a second pass of shotcrete following the installation of the mesh. This ensures that the gap between the first layer of shotcrete and the mesh is infilled with shotcrete (Figure 4.28).

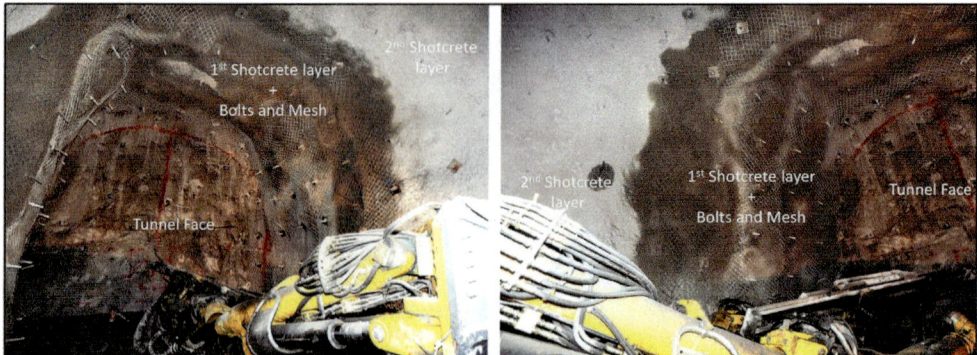

FIGURE 4.28 Example of a shotcrete layer reinforced with a high-capacity chain link mesh with a minimum cover over the mesh.

The choice of mesh to reinforce the shotcrete depends upon rock mass demand in a particular situation. The welded wire mesh reinforced shotcrete will have greater capacity and faster response than a separate layer of shotcrete and welded wire mesh. However, one disadvantage is that after a violent failure, the welded wire mesh can buckle and the external shotcrete ejects. In addition, the low values of strength and displacement capacity for the welded wire mesh (and related mesh straps) limit the energy dissipation of the combined surface support layer (Figure 4.29).

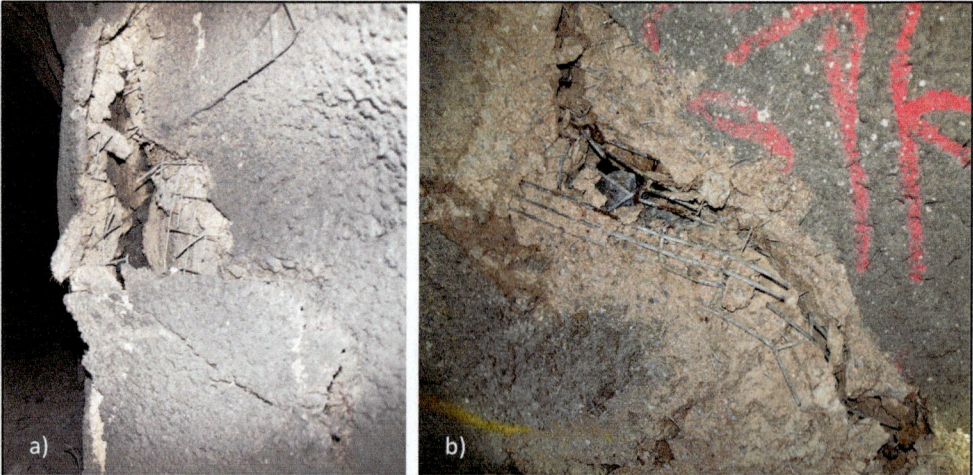

FIGURE 4.29 Examples of (a) broken mesh wires and (b) broken steel strap and welded wire mesh, resulting in ejected shotcrete following a dynamic event.

In situations where very large rock mass demand is anticipated, the mesh product can be changed from welded wire mesh to a high tensile chain link mesh. This mesh can articulate and adjust to rapid changes in excavation shear and/or compression (Kusui, 2015). Furthermore, the anchor points at the reinforcement elements can be secured and by mesh continuity, the load can be transferred to the reinforcement elements located outside the failure area (Figure 4.30). Again, the violent ejection of shotcrete can be minimized by placing the mesh very close to the outer shotcrete surface.

FIGURE 4.30 Example of performance for a high energy dissipation surface support layer comprising shotcrete reinforced by high tensile chain link mesh after an ML 2.4 seismic event.

Finally, in conditions of extremely high energy demand, two mesh layers may be required to provide additional energy dissipation capacity and surface support redundancy connected to deep reinforcement. The secondary layer of woven mesh surface support is installed externally to the primary surface support layer and is connected to deep reinforcement in the form of cable bolts reinforcement (Villaescusa et al., 2016). In case structurally controlled failures with large blocks overload the primary reinforcement and surface support layer, this external secondary mesh layer provides load transfer connectivity from the unstable to stable regions of the excavation (Figure 4.31). Another benefit of this external mesh layer is that it also minimizes the ejection of shotcrete slabs.

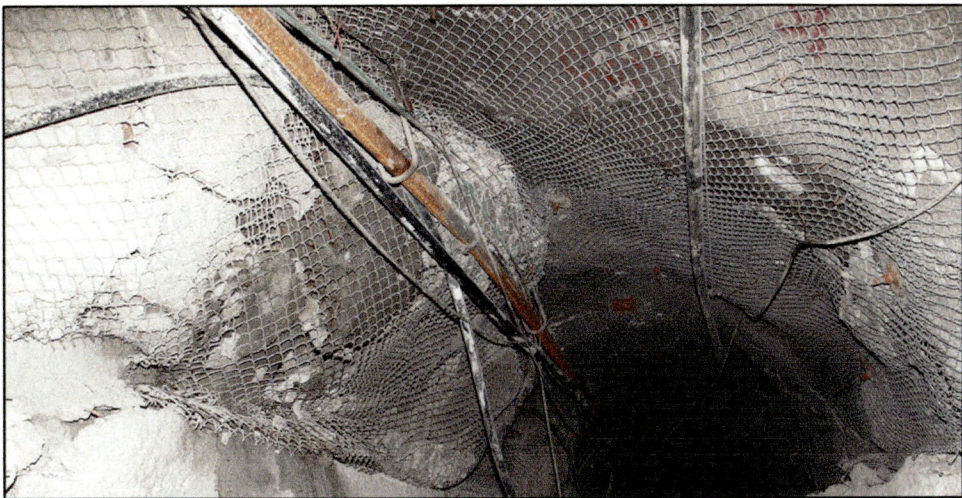

FIGURE 4.31 An external layer of high-capacity woven mesh surface support connected to secondary deep reinforcement provides load transfer from the unstable to stable regions of an excavation.

5 Ground Support Design

5.1 INTRODUCTION

Underground excavations reach depths below the ground surface, where high energy failure events frequently occur at the excavation boundary. In most cases, the depth of failure is less than the length of the rock reinforcement elements, which, for a typical modern mining or civil tunnelling excavation, commonly range from 2.0 to 4.0 m. In cases where the depth of failure is shallow (i.e., ≤ 0.5 m), violent failures often involve localized damage of a limited number of reinforcement elements as well as surface support, which, as discussed in Chapter 4, must have the capacity to transfer the load within a ground support scheme. In some cases, however, the violent failures mobilize geological structures, and the depth of failure can be more significant (i.e., ≥ 1.0 m), often resulting in a violent failure of a large number of reinforcement elements as well as significant load transfer to the surface support. This can occur when large-scale geological structures are sub-parallel to the tunnel walls or form kinematically unstable blocks that control the depth, area and volume of the failure (Figure 5.1). Excavation instabilities with a depth of failure ranging from 0.5 to 1.0 m may be transitional and may also involve a mixture of both reinforcement and support mobilization.

The load transfer mechanism for an excavation under high stress is very complex and depends on, among other things, the level of pre-existing rock mass damage and potential block size associated with the violent ejection. The larger the mobilized blocks, the more reinforcement action will be initially involved in dissipating energy. Conversely, small block size instability requires membrane support, such as that provided by combinations of shotcrete and mesh. Nevertheless, in all cases, the performance of the surface support is critical to achieving load transfer to the reinforcement elements located within and outside the area of instability.

5.2 ROCK MASS CHARACTERIZATION

The initial stage in ground support design is the rock mass characterization process. This is required to quantify the geometrical characteristics of the geological discontinuities (Chapter 3), the intact strength properties of the rock masses involved, the magnitude and orientation of the principal rock stresses and the induced rock mass strains arising from the extraction sequences being proposed. These factors must be investigated and characterized for each and every distinctive geotechnical domain that is identified in a particular mining precinct.

5.2.1 STRUCTURE

The structural characterization of a rock mass should be multi-scaled. That is, it must consider the geological discontinuities at the scale of the individual tunnel excavations (mainly joints) as well as the intermediate and large mine-scale, geological features (i.e., faults, shear zones and lithological contacts). The main purpose of characterizing the local scale discontinuities is to define a structural model for each geotechnical domain, which identifies the characteristics, such as orientation, frequency, persistence and joint condition of the features that define potentially unstable blocks (Chapter 3). The characterization of the intermediate and large mine-scale geological discontinuities is mainly focused on identifying their location and orientation with respect to the mine infrastructure as well as their potential to generate large magnitude seismic events. The study of the seismic response and related analysis of the large-scale geological discontinuities is beyond the

FIGURE 5.1 Failure of a ground support scheme following a violent, structurally controlled deep (≥1.0 m) instability.

scope of this book. Nevertheless, it would be expected that high energy dissipation ground support schemes would be installed in areas where the large-scale features are known to interact with the mining excavations.

5.2.2 STRENGTH

Another key stage in the ground support design process is the determination of high-resolution spatial models of laboratory-determined measurements of intact rock strength and deformational properties. Ideally, the rock strength data would be collected during orebody delineation (in advance of construction) at the locations of any planned development infrastructure via diamond drill core sampling (Villaescusa, 2014).

For each lithology, important properties include the range and spatial variability for parameters such as the uniaxial and triaxial compressive strength, tensile strength, Modulus of deformation and Poisson's ratio. Fracture toughness can also be collected within locations of core disking (Figure 5.2), as this test does not require large pieces of rock. Intense core disking within the exploration core indicates an unfavourable ratio of strength/induced stress during drilling. Additionally, experience at several mining locations indicates that the presence of stress-induced disking has been correlated to subsequent seismicity during mining development.

5.3 TUNNEL INSTABILITY

The ratio of intact rock Uniaxial Compressive Strength (UCS or simply σ_c) to the induced compressive stress tangential to the wall of an excavation (σ_{max}) has long been recognized as a critical factor controlling excavation stability (Barton et al., 1974; Mathews et al., 1980). As the ratio of

FIGURE 5.2 Example of massive rock drilled at great depth showing (a) occurrences of core disking and (b) a sharp change of lithology.

σ_c/σ_{max} reduces, excavation instability increases, as implied in Figure 5.3. The Stress Reduction Factor (SRF) on the vertical axis is part of the Q rock mass classification method (Barton et al., 1974) and given by Hutchinson and Diederichs (1996).

Data from many years of numerical modelling and observations of open stoping in hard rock at Mount Isa Mines (Villaescusa, 2014) have shown that as the ratio of σ_c/σ_{max} decreases below the value of 3, the instability increases markedly. In general, when excavating in rock with σ_c/σ_{max} ratios below 2, it can be expected that large deformations would be experienced in tunnels driven within a low-strength rock. In comparison, sudden, violent failures are more likely to occur in tunnels excavated in high-strength rock (Barla, 2014).

The relationship between the rock strength and the induced stress at the onset of failure has been investigated by several researchers using stress-strain data from UCS laboratory testing (Lajtai and Dzik, 1996; Martin, 1997; Diederichs, 2007). The research shows that failure initiates when the ratio of the uniaxial compressive strength obtained by laboratory testing (σ_c) to the stress magnitude causing the failure (σ_{max}) ranges between 2 and 3. Martin (1993) also conducted detailed investigations at an underground excavation at the AECL laboratory in Canada. Their work demonstrated that the crack propagation process initiated when the intact strength to the induced stress ratio reached approximately 2. A number of different laboratory and field test results (i.e., boreholes, raisebores and tunnels) suggest that the ratio of σ_c/σ_{max} is scale independent.

5.2.1 Spalling Failure

Spalling failure of an excavation under high compressive stress is characterized by tensile fracturing of rock in an orientation typically sub-parallel to both the major principal stress and the adjacent excavation surface, often with associated ejection of rock slabs. Spalling fractures may be sparse,

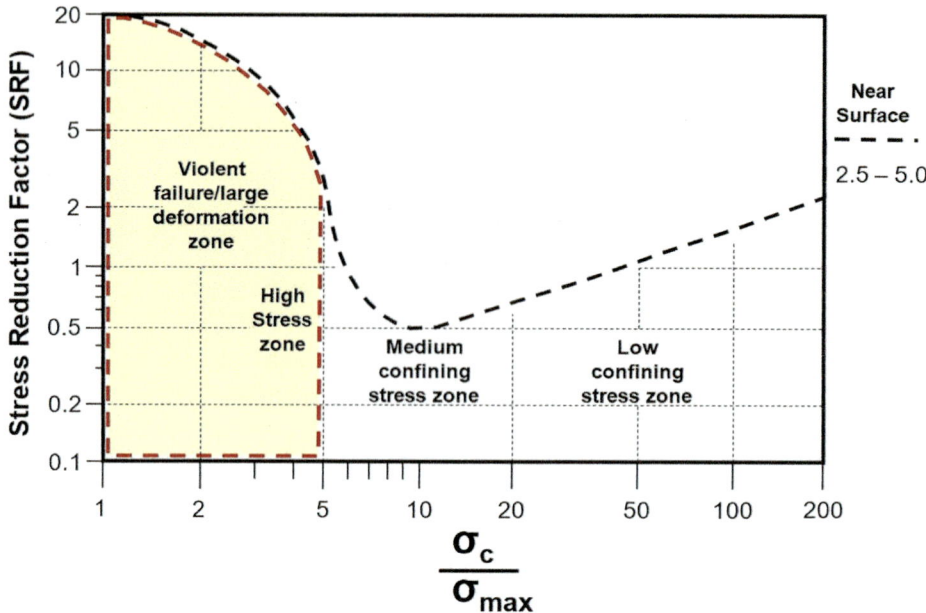

FIGURE 5.3 Excavation behaviour as a function of the ratio of compressive strength to the induced stress (after Barton et al., 1974; Hutchinson and Diederichs, 1996).

occurring through intact rock, or highly repetitive and closely spaced due to delamination of suitably oriented pre-existing discontinuity surfaces. Stress-driven damage in brittle rock often occurs initially as progressive violent spalling of the excavation walls and is localized within areas of maximum induced stress concentrations (Christiansson et al., 2012). Such failures typically result in an approximately v-shaped notch in the regions of violent ejection (Figure 5.4).

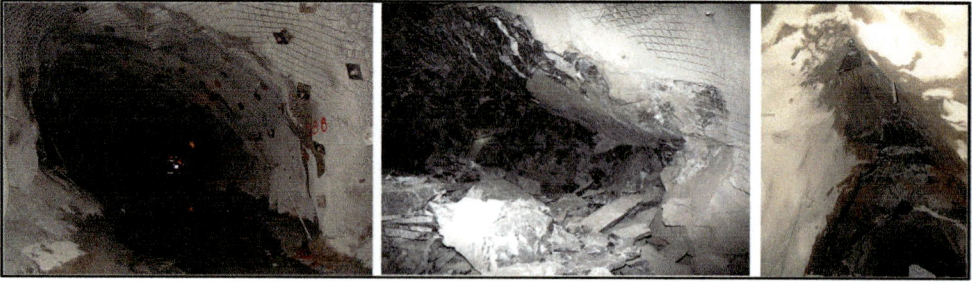

FIGURE 5.4 Examples of 'v-notch' formation due to brittle failure around underground tunnels.

Laboratory work by Kusui (2015) has calculated the ratio of compressive strength to maximum induced stress in the sidewalls of scaled-down model tunnels, which have been progressively loaded to failure. For the first stage of failure, where spalling was experienced, the ratio monitored was the value of compressive strength to maximum tangential stress (i.e., with effectively near zero confining stress normal to the excavation boundary). A total of 23 unsupported laboratory tests were performed for a range of intact rock strengths, the results of which are shown in Figure 5.5. The ratio of σ_c to σ_{max} was calculated using the Kirsch solution (Brady and Brown, 2004) and compared favourably with finite element modelling solutions using the program Abaqus (Kusui et al., 2016). This stress ratio value was calculated for both sides of the tunnel during wall spalling. Similar to previously published findings (Martin, 1993), violent ejection from the excavation walls occurred

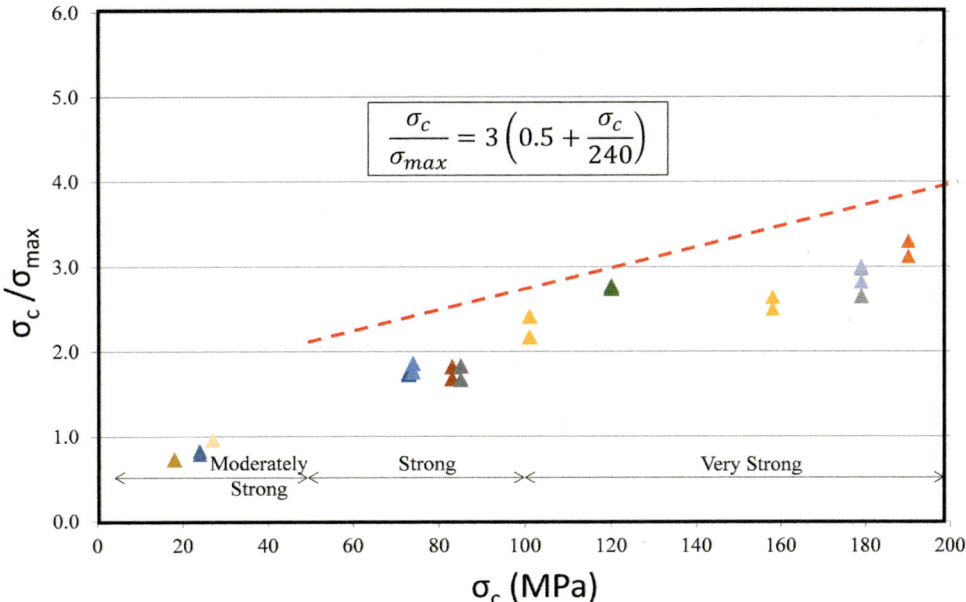

$$\frac{\sigma_c}{\sigma_{max}} = 3\left(0.5 + \frac{\sigma_c}{240}\right)$$

FIGURE 5.5 Unsupported tunnel spalling as a function of compressive strength and maximum tangential stress (Kusui, 2015).

prior to reaching σ_c. The strength/induced stress ratio at spalling and the uniaxial compressive strength show a strong correlation. The value of σ_c/σ_{max} at spalling ranges from 2 to 3.5 for strong to very strong rock. The red dotted line and related equation shown in Figure 5.5 represent the potential onset of failure, which could be interpreted as practical, safe limits prior to spalling failure. The samples of moderately strong rock ($\sigma_c < 50$ MPa) did not exhibit violent failure.

The laboratory results from Kusui (2015) correlate well with observations of 4–5 m diameter, full-scale unsupported tunnels excavated within a highly silicified rock mass. In that case, the rock had a UCS of 250–270 MPa, along with widely spaced, tightly healed geological discontinuities. The onset of stress-driven failure in these full-scale unsupported tunnels was experienced where the ratio of σ_c/σ_{max} was approximately 3.5. These mining tunnels were constructed using excellent drilling and blasting techniques and designed with semi-circular walls and a flat floor. Incipient failure of the tunnel back (roof) due to a sub-horizontally oriented major principal stress component can be just seen in Figure 5.6.

5.2.2 STRUCTURALLY CONTROLLED FAILURE

Testing of scaled-down tunnels by Kusui (2015) has also shown that under progressively increased loading conditions, an initial shallow spalling failure is followed by a deeper shear rupture in the tunnel walls. This is in accordance with laboratory testing of large-scale rock specimens with unfavourably oriented defects. As the induced stress increases, the failure transitions from stable to shallow spalling are ejected at high velocity to a catastrophic collapse of the block assembly defined by the discontinuities (Figure 5.7).

For the scaled-down tunnels tested by Kusui (2015), Figure 5.8 shows a typical load-displacement characteristic and acoustic emission count for an example of such progressive brittle failure of a tunnel under increased loading. After spalling, large shear cracks can propagate adjacent to the tunnel wall, with shear failure dominating the later stages of the failure mechanism. The results show the seismic response as loading gradually increased from zero until tunnel wall spalling and pillar

FIGURE 5.6   Full-scale unsupported semi-circular tunnels indicating the onset of brittle spalling at the centre of the excavation roof where the horizontal stress is highest.

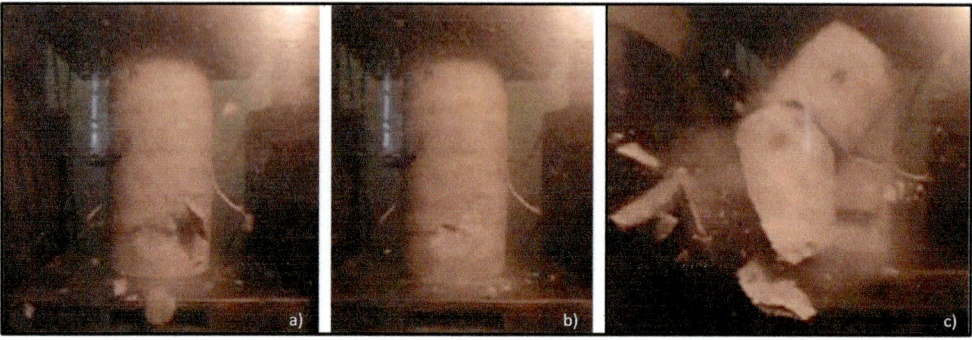

FIGURE 5.7   Example of (a) high velocity shallow spalling failure, (b) temporary stability under increased stress and (c) catastrophic failure along geological discontinuities.

crushing were sequentially experienced. The seismic activity initiates with the creation of vertical tension cracks in the floor and roof of the circular opening, as predicted by theory (e.g., Hoek and Brown, 1980). The rate of seismic activity clearly increased prior to the violent ejection of both walls. Significantly fewer acoustic emissions were monitored during this period between spalling and the onset of the shear failure mode of pillar crushing. Overall, a significant decrease in load-bearing capacity occurs following the final mode of shear failure.

5.2.3   DAMAGE AND DEFORMATION PRIOR TO VIOLENT FAILURE

Both the onset of tunnel damage and the progression to violent failure in massive rock were studied in detail by Kusui et al. (2016). As shown above, the complete excavation response has been determined as the induced stress near the excavation boundary was increased with respect to the intact rock strength. The critical levels of strength to induced stress related to the onset of visible instability, such as spalling on both walls and the start of pillar crushing, were calculated. The results show that the failure threshold values for both spalling and subsequent pillar crushing are both dependent upon the UCS. Higher values of σ_c/σ_{ave} at failure (where σ_{ave} is the average pillar stress) correlate to lower radial strain (Figure 5.9). This is in accordance with field data reported by Hoek (1999) and laboratory work on a critical strain by Li (2004). The complete data set is shown in Figure 5.10,

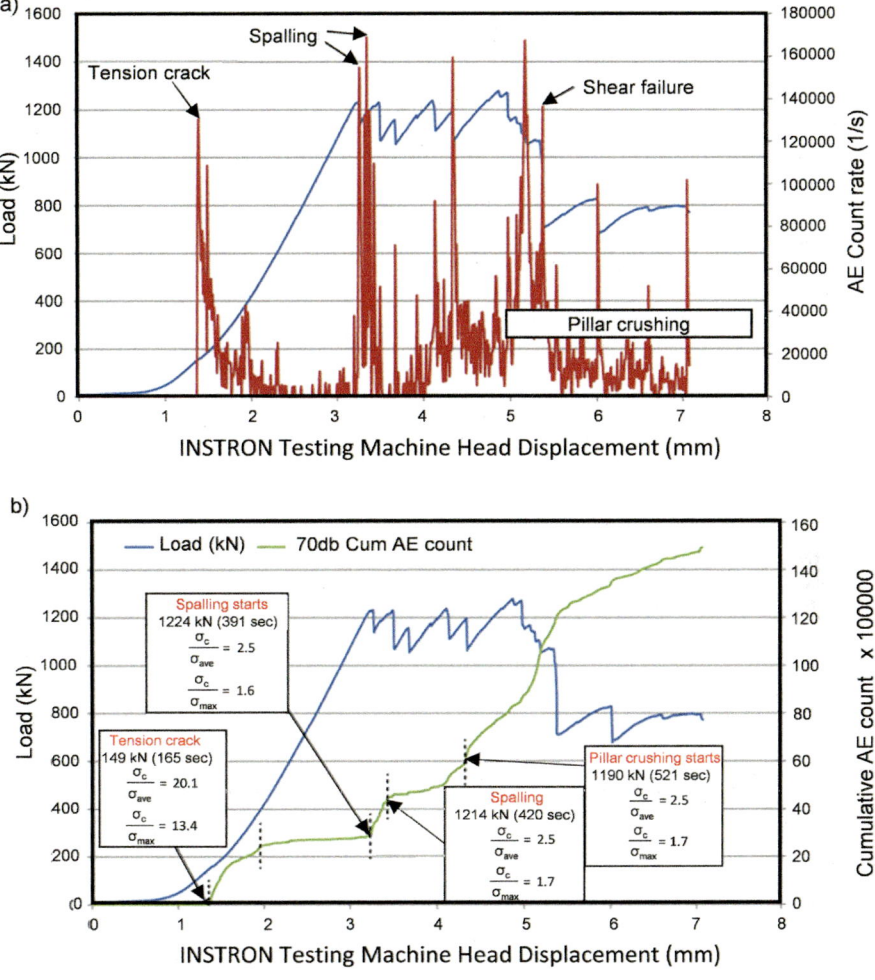

FIGURE 5.8 Example of (a) failure mode, rate of seismicity and (b) decreasing ratios of strength/induced stress for an unsupported tunnel in strong sandstone (Kusui, 2015).

wherein the progressive failure of the unsupported tunnels is shown as a function of the UCS of the materials.

5.3 EJECTION VELOCITY

For tunnels excavated in a very strong rock, violent ejection occurs simultaneously with crack propagation along the tunnel axis. Crack propagation is perpendicular to the orientation of the main principal stress. Once slabs detach from the tunnel surface, they are ejected violently with an associated rock fracture noise. Additional laboratory testing by Kusui (2015) established ejection velocities ranging from 3 to 10 m/s across a variety of rock types, which validates some of the back analyses of actual failures documented in underground mining (Ortlepp, 1993). A typical result for wall failure is shown in Figure 5.11, where an ejection velocity of 5.2 m/s can be determined from a background grid. Similar ejection velocities have been used at the WASM to conduct dynamic testing of rock reinforcement systems (Figure 5.12). The data indicate that an ejection velocity of 5 m/s is capable of rupturing most commercially available reinforcement systems.

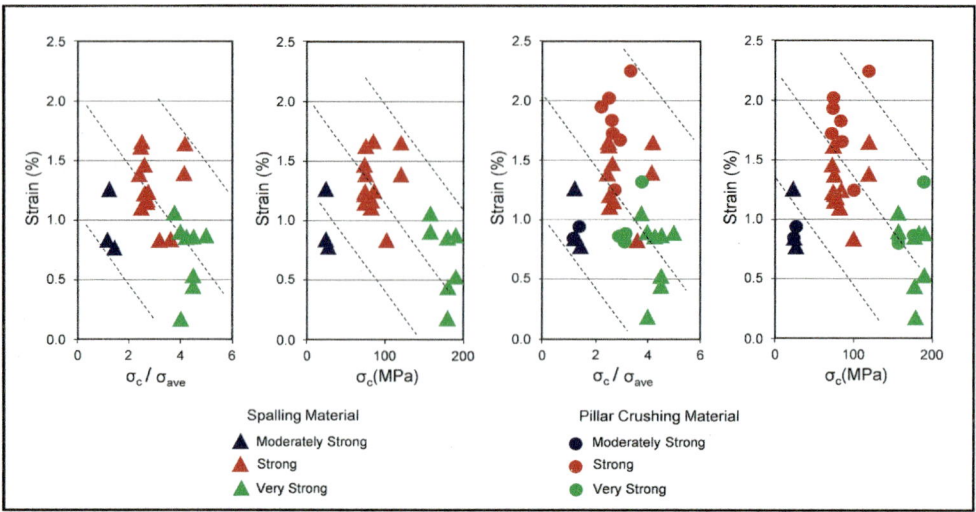

FIGURE 5.9   Radial strain at violent failure for a range of tunnel materials (Kusui et al., 2016).

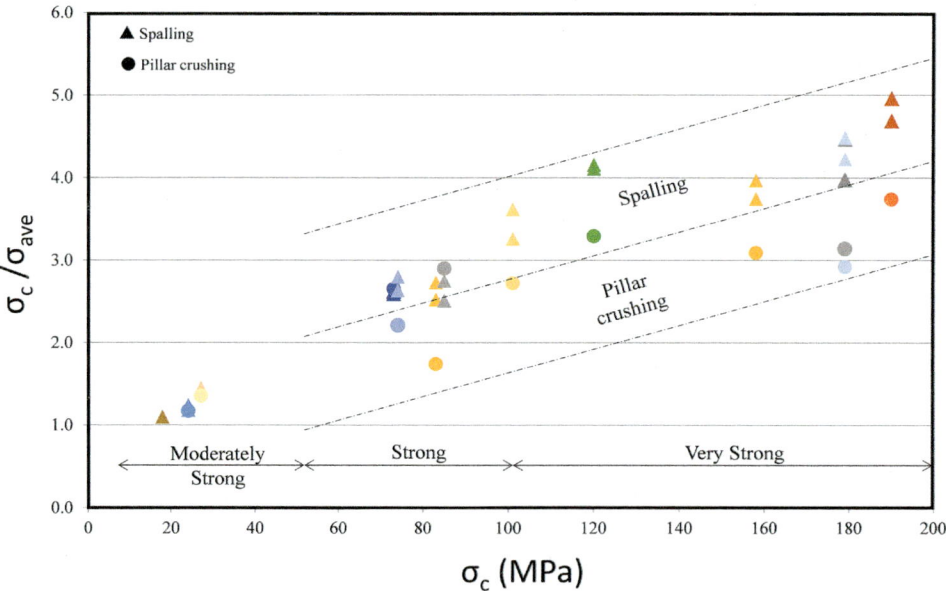

FIGURE 5.10   Progressive failure of unsupported tunnels as a function of material strength (Kusui, 2015), confirming the observations of Barla (2014).

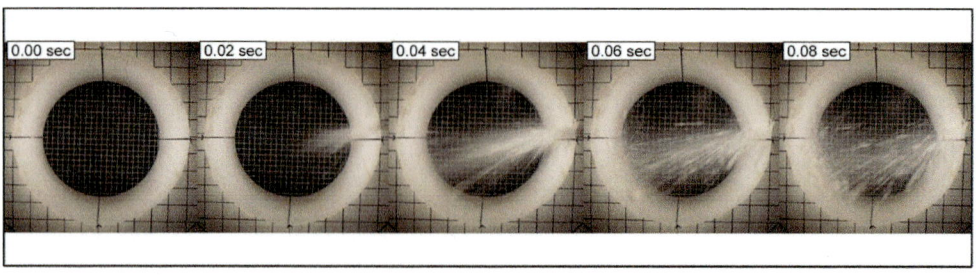

FIGURE 5.11   Determination of ejection velocity using a high-speed video camera.

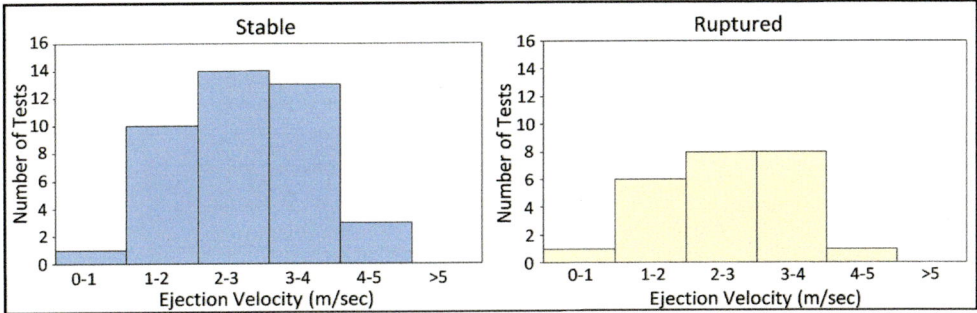

FIGURE 5.12 Range of ejection velocities required to mobilize the capacity of decoupled Posimix threaded bar at the WA School of Mines Dynamic Test Facility.

Theoretically, it can be shown that the initial velocity of ejection (V) is independent of scale and depends only on the failure stress (σ_F or earlier as σ_{max}), the rock Young's modulus (E) and the rock density (ϱ). The failure stress has been shown by many authors to be less than the UCS (between about 0.3 and 0.5) and the Young's modulus in the range of 200–500 times the UCS (Deere and Miller, 1966). Further, the initial velocity depends on the efficiency of the transfer of stored strain energy into kinetic energy of the failure volume, as indicated by previous research (Zuo et al., 2005). These concepts and assumptions may be used to formulate a relatively simple, empirically-based design chart of rock mass dynamic demand versus UCS for various failure depths as follows.

The stored strain energy per unit volume of rock (J) is given by:

$$J = \int_0^{\sigma_F} \frac{\sigma}{E}\, d\sigma \qquad\qquad 5.1$$

or

$$J = \frac{\sigma_F^2}{2\,E} \qquad\qquad 5.2$$

where

J is the stored strain energy (kJ/m³),
E is the Young's modulus (GPa),
σ_F is the failure stress, also known earlier as σ_{max} (MPa) and
σ_c is the UCS (MPa).

Figures 5.13 and 5.14 show the assumptions made for E and σ_F, respectively. The relationship between Young's modulus and UCS is based on a large database of values measured in unconfined compression tests performed according to International Society for Rock Mechanics (ISRM) Standards. The Young's modulus values were derived from the applied stress and the measured strain (measured by gauges bonded to the rock samples).

The equation for the relationship shown in Figure 5.13 is given by:

$$E = f_E \sigma_C \qquad\qquad 5.3$$

where

$$f_E = 1.05 / e^{\frac{\sigma c}{350}} \qquad\qquad 5.4$$

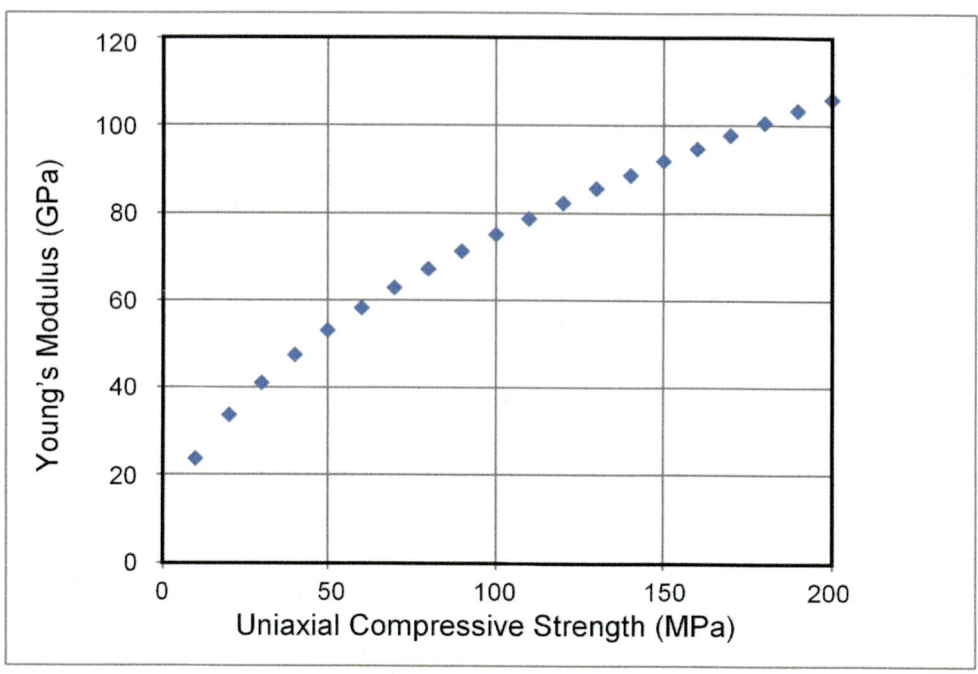

FIGURE 5.13   Assumed Young's modulus and compressive strength relationship for rocks.

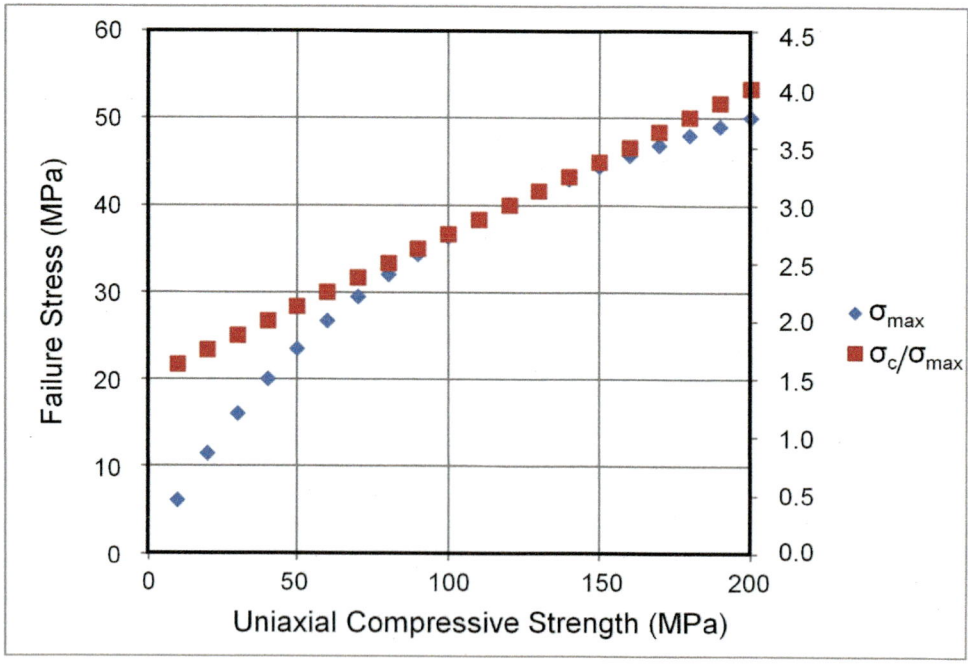

FIGURE 5.14   Assumed failure stress and compressive strength relationship for rocks.

Figure 5.14 shows the assumed relationships between σ_{max} and σ_c and the ratio σ_c/σ_{max} and σ_c. At this stage, it should be noted that the value of σ_{max} is assumed to be the average stress within the potential depth of failure and that the failure initiates at the excavation boundary and propagates to become an ejected volume of broken rock. Here, σ_c/σ_{max} is termed fs and defined as follows:

$$f_s = \frac{\sigma_c^{0.3}}{3} \qquad\qquad 5.5$$

For the cases of block ejection, it is assumed that a fraction (f_J) of the stored strain energy per depth of failure (t) is converted to kinetic energy. The stored kinetic energy per depth of failure becomes J_A (J/m^2), which is the stored energy per unit area given by:

$$J_A = J\,t \qquad\qquad 5.6$$

The corresponding mass per unit area is given by:

$$\delta t \qquad\qquad 5.7$$

where

$$\varrho \text{ is the density } (t/m^3) \qquad\qquad 5.7a$$

Thus, the kinetic energy of ejection is given by:

$$f_J J_A \qquad\qquad 5.8$$

Thus,

$$\frac{\delta t\, V^2}{2} = f_J\, J\, t = f_J\, \frac{\sigma_F^2}{2E}\, t \qquad\qquad 5.9$$

Hence, the ejection velocity is given by:

$$V = \frac{\sigma_C}{f_S} \sqrt{\frac{f_J}{\rho E}} \qquad\qquad 5.10$$

Considering Equations 5.3 and 5.4:

$$V = \frac{1}{f_S} \sqrt{\frac{f_J\,\sigma_c}{\rho\, f_E}} \qquad\qquad 5.11$$

The dynamic factors ('f') used here are termed:

f_s = The failure stress factor,
f_E = The deformation factor and
f_J = The energy release factor.

This can be used to conclude that ejection velocity is independent of scale and, with some simplifying assumptions, depends only on UCS. Figure 5.15 shows example plots of failure stress as well as Young's modulus as a function of the unconfined compressive strength. Similarly, Figure 5.16 shows example charts with the energy efficiency release factor (f_J) versus depth of failure as well as the resulting ejection velocity as a function of the UCS.

In addition to the theoretical calculations shown previously, Kusui (2015) calculated average ejection velocities for unsupported tunnels with 200 mm diameter as a function of the UCS. Ejection velocities ranging from 2 to 10 m/s were determined, which accords with back analyses of

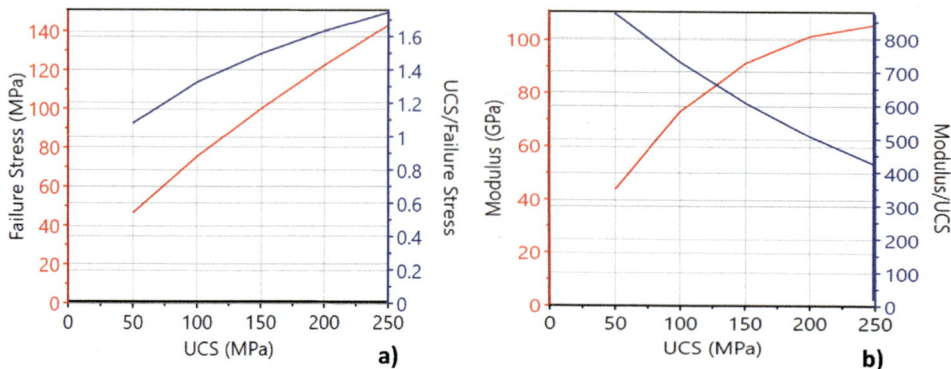

FIGURE 5.15 Example of SAFEX software calculation showing (a) failure stress and (b) modulus as a function of the UCS.

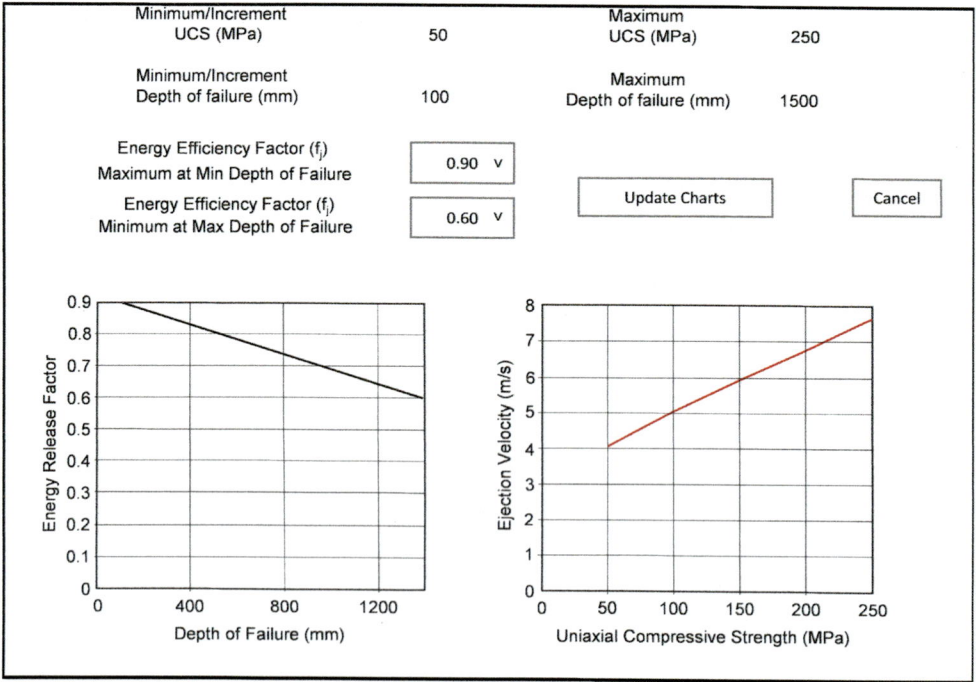

FIGURE 5.16 Example of SAFEX software calculation showing (a) energy efficiency release factor (f_J) and (b) ejection velocity as a function of the UCS.

actual failures in underground mining (Ortlepp and Stacey, 1994, Drover and Villaescusa, 2015b). Generally, the higher the UCS, the higher the measured ejection velocity. Such a positive correlation between the intact rock strength and the velocity of ejection is consistent with the tunnel observations of Broch and Sørheim (1984). Kusui (2015) also tested scaled-down tunnels that were stabilized using several ground support schemes (Kusui and Villaescusa, 2016). Figure 5.17 compares the average ejection velocity results for unsupported and supported tunnels. The results show that the ejection velocity is reduced when a surface support system is installed. The results also show that the reinforcement element installation pattern also influences the ejection velocity, with staggered patterns proving to be more effective than square patterns for the same reinforcement scheme and spacing. The ejection velocity results for staggered patterns (i.e., 2.6 and 4.4 m/s) are in the low range of the measured values. Additionally, when a mesh layer was installed, the ejection velocity was lower than when shotcrete was exposed as the final layer. In summary, it was found that:

- Ejection velocity is independent of scale and depends on UCS alone.
- Ejection velocity from similitude models accords with back analyses of full-scale excavations (Figure 5.18).

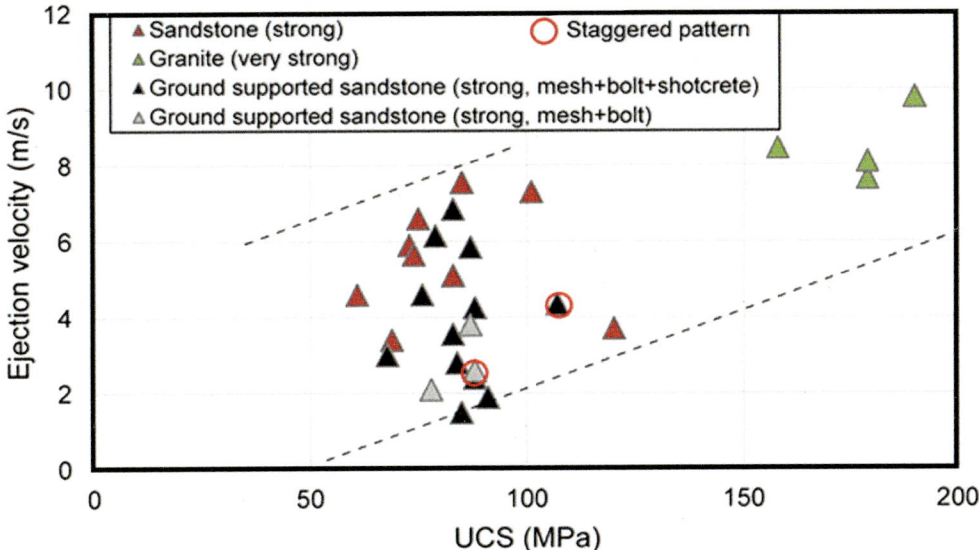

FIGURE 5.17 Ejection velocity as a function of UCS for unsupported and stabilized tunnels (Kusui, 2015).

5.4 GROUND SUPPORT DEMAND

Ground support design is often based on previous experience from similar geotechnical environments and work practices. Challenging this approach is the fact that the rock mass conditions usually change with the progress of a mine, and the ground support performance may become unacceptable over time (Brown, 2004). That is, when the induced stresses increase due to greater depth of mining or increased global extraction, the installed reinforcement and support capacities may not be able to address the rock mass demand. Geological models of rock strength variability, *in situ* stress measurement data and three-dimensional non-linear numerical stress modelling may be used to investigate potential changes in excavation loading conditions ahead of the development horizons. Routine collection and analysis of such data are necessary for continuous ground support design verification throughout a mine life.

FIGURE 5.18 Example of violent spalling through a shotcrete layer exposing the rock surface (Velocities ranging from 2.7 to 3.6 m/s were back-calculated).

The generic procedure for ground support design consists of several distinct steps (Thompson et al., 2012):

1. Identify a mechanism of failure.
2. Estimate the areal support demand.
3. Estimate the reinforcement length, force and displacement demand.
4. Estimate the energy demand of the complete scheme.
5. Select appropriate reinforcement and support systems.
6. Propose an arrangement of reinforcement and support systems and evaluate.
7. Specify the complete ground support scheme.

This procedure may need to be applied to several different observed mechanisms and depths of failure, which, in turn, will control the ground support demand. In general, the energy demand for a particular failure is controlled by the amount of mass that becomes unstable and the velocity of its ejection. The unstable mass can be expressed in t/m², reflecting the maximum mass in tons of unstable rock that can be ejected per unit area of the excavation surface where failure occurs. For the purposes of design, demand may be quantified in terms of the kinetic energy of the ejected rock. This design approach considers that strain energy released by the rock mass during a violent stress-driven failure is partially converted into the kinetic energy of the ejected blocks. These blocks load the ground support scheme dynamically, causing a force-displacement response. The mass of unstable rock that is violently ejected and the initial velocity of its ejection are the critical input variables in the kinetic energy calculation.

The mass of instability may be quantified via probabilistic analysis of the local structural geological data, as shown in Chapter 3. For structurally controlled mechanisms of failure, it may be reasonably assumed that the unstable mass is controlled by the largest possible tetrahedral wedge that can be formed by the discontinuities present. For pure spalling mechanisms, observational experience indicates that the mass of instability is limited by a depth of failure of approximately less than 0.5 metres or so in moderately strong rock.

The initial velocity of rock ejection in the field may be estimated most conveniently from the UCS of the host rock. Kusui (2015) recently demonstrated the dependence of ejection velocity on intact rock UCS via a series of scaled-down laboratory tests of circular excavations in hard rock. These experiments revealed an approximately linear relationship between UCS and ejection velocity, from which a first-degree polynomial trend line equation was derived. Considering this relationship, plotting the kinetic energy equation solutions for a range of rock types and instability scenarios yields the chart shown in Figure 5.19. This figure shows the estimated kinetic energy demand on a ground support scheme imposed by a range of unstable masses as a function of the ejection velocity. This demand chart has been calibrated with a back analysis of actual violent failures occurring in a range of material strengths. It may also be used as a design tool to estimate energy demand on a ground support scheme, considering the potential modes of failure and site-specific mechanical characteristics of the rock mass.

Additionally, the dynamic energy demand can also be calculated using the ejection velocity determined using Equation 5.11 and the same range of unstable masses. The results have been calculated using the SAFEX software and are shown in Figure 5.20. The range of predicted energy demand is very similar between the two methodologies.

5.4.1 Shallow Depth of Failure (Spalling)

For most rock masses experiencing seismicity and related dynamic ejections, the depth of failure is shallow, with instability often limited to a depth range of 0.25–0.5 m or so (Figure 5.21). This spalling mechanism of failure is likely to preferentially load the surface support, including shotcrete and mesh layers, as well as rock bolt and cable bolt plates. The mass of instability associated with spalling style failure ranges mostly from 0.25 to 0.75 t/m² for most hard rocks (Figure 5.22). This type of failure would most likely test the surface retention capacity of the chosen ground support scheme. In terms of energy demand, for a range of hard rock conditions, spalling failure mechanisms could be expected to generate a maximum of 15–20 kJ/m² demand on the ground support scheme (Figure 5.23). Energy demand above this threshold is more likely to be structurally controlled.

5.4.2 Structurally Controlled Depth of Failure

In cases where violently mobilized geological structures approach the span width of an excavation, the depth of failure may be very large, sometimes exceeding 2 m. Such instability can occur due to dilation and shear along major structures, which, in some cases, intersect the excavation boundaries where reinforcement and support have been installed. The presence of geological structures continuous on the scale of the tunnel walls increases the depth of failure. The observed damage frequently comprises shear failure along structures, resulting in a sudden and violent ejection of large blocks. Because the large face and release surface area are a consequence of a large depth of failure, it is likely that the ejection will damage many reinforcement elements, followed by the destruction of the surface support layers (Figure 5.24). Structurally controlled violent failure is expected to result in the mass of instability exceeding 1 T/m² with energy dissipation often well in excess of 25 kJ/m². In some difficult conditions, the energy demand could locally exceed 60–80 kJ/m² (Drover and Villaescusa, 2015a).

Geotechnical observations and monitoring are required to identify if large-scale instability may develop within the unsupported spans that are exposed immediately after development blasting and prior to the installation of ground support layers, such as in-cycle shotcrete (see Chapter 10).

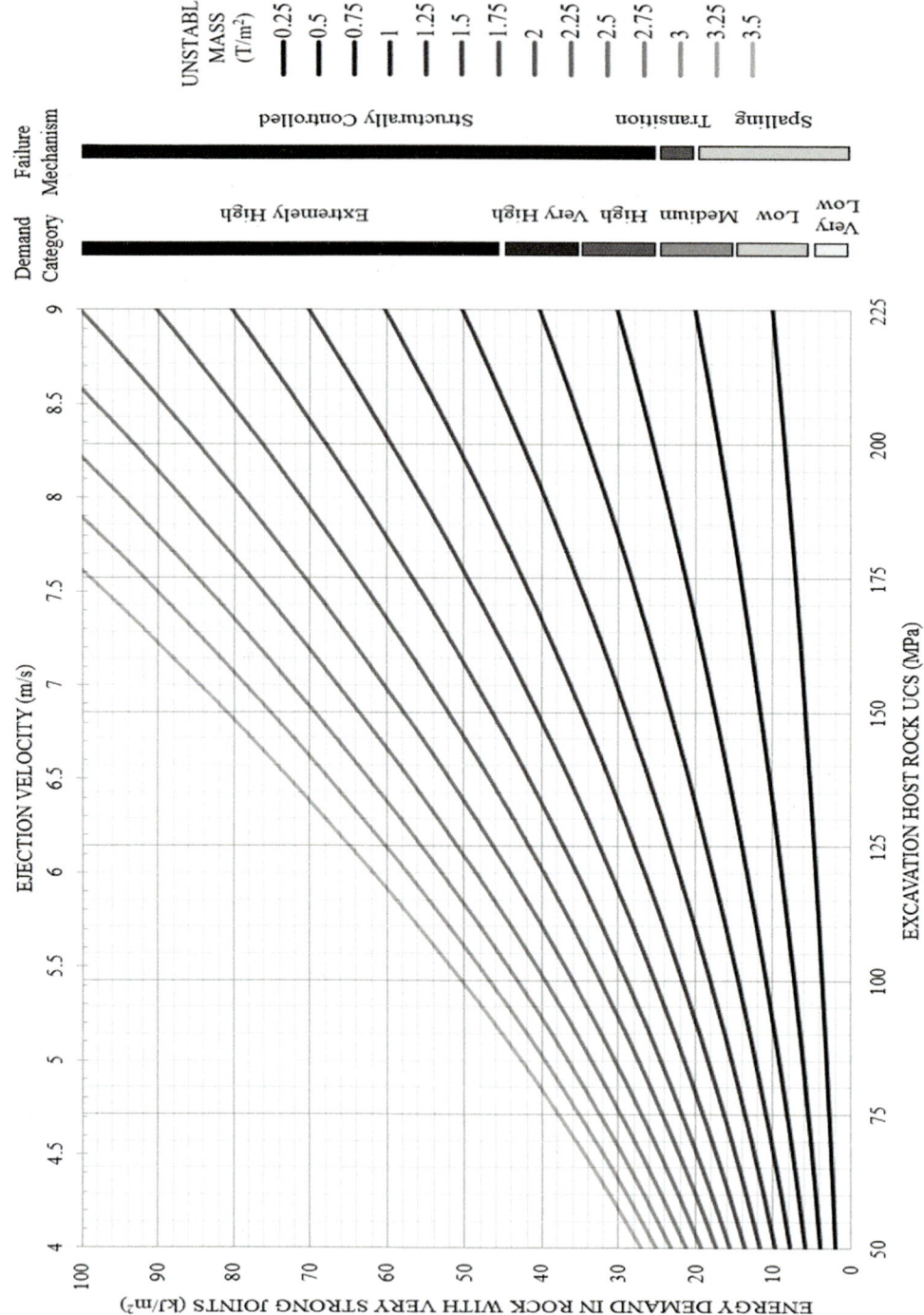

FIGURE 5.19 Range of energy demand on ground support and related mechanisms for stress-driven failures in hard rock.

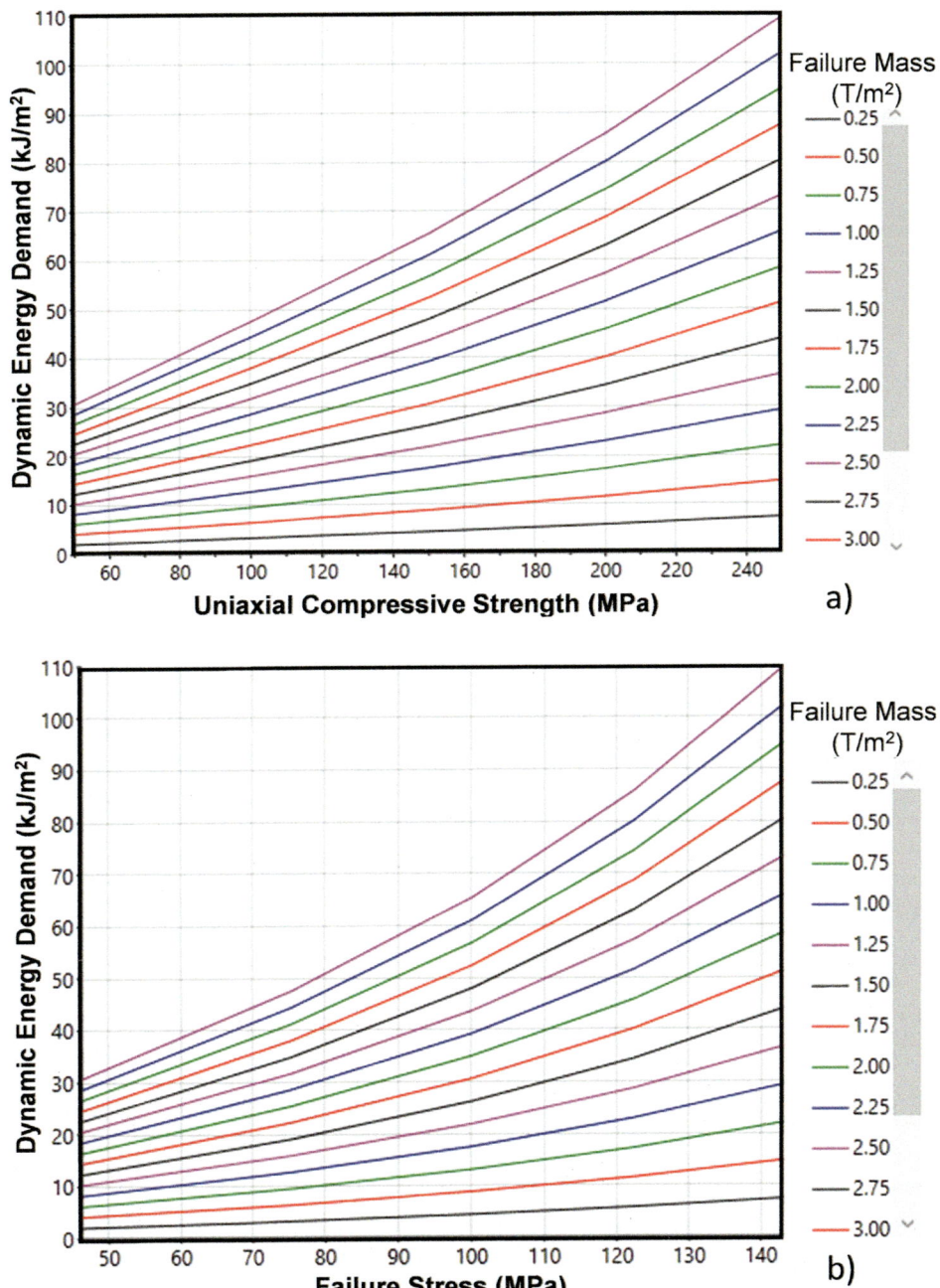

FIGURE 5.20 Dynamic energy demand as a function of (a) UCS and (b) failure stress for a range of material strengths and various depths of failure.

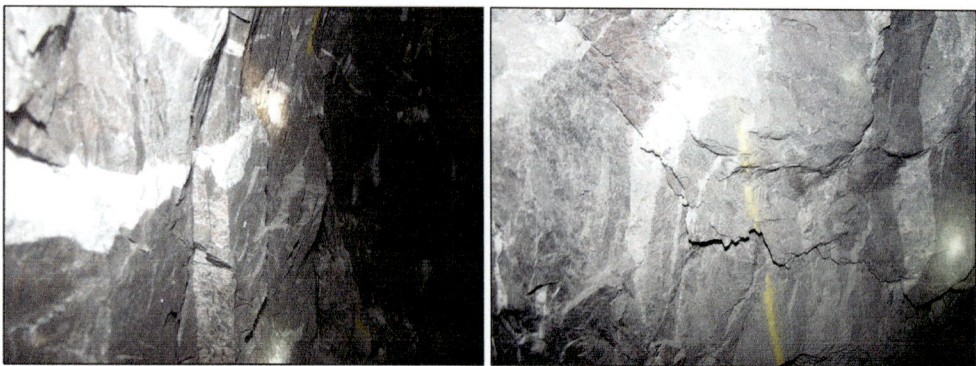

FIGURE 5.21   Example of shallow spalling failure geometry in a highly stressed tunnel.

FIGURE 5.22   Example of violent but shallow depth of failure preferentially loading the surface support.

Event Number	Date	Location	TM	Thickness [m]	Length [m]	Width [m]	Area [m^2]	Vol [m^3]	Ton	[Ton/m^2]	Estimated Vel [m/s]	Energy [KJ]
			-								-	
99	13/09/2016	Face	2117.5	0.2	2	1	2	0.4	1.08	0.54	6	9.72
100	15/09/2016	Face	2119.5	0.1	0.5	1	0.5	0.05	0.135	0.27	6	4.86
102	23/09/2016	Face	2131.4	0.1	0.5	1	0.5	0.05	0.135	0.27	6	4.86
103	24/09/2016	Face	2133.4	0.1	0.4	0.6	0.24	0.024	0.0648	0.27	6	4.86
104	24/09/2016	Face	2133.4	0.2	3.5	2.5	8.75	1.75	4.725	0.54	6	9.72
107	5/10/2016	Face	2151.4	0.2	1	2.5	2.5	0.5	1.35	0.54	6	9.72
109	8/10/2016	Face	2155.4	0.5	4	3	12	6	16.2	1.35	6	24.3
115	27/10/2016	Face	2173.7	0.2	1	1.5	1.5	0.3	0.81	0.54	6	9.72
118	3/11/2016	Face	2183.2	0.3	5.5	4	22	6.6	17.82	0.81	6	14.58
125	14/11/2016	Face	2199	0.2	4	2.5	10	2	5.4	0.54	6	9.72
127	17/11/2016	Face	2205	0.3	2.3	1.7	3.91	1.173	3.1671	0.81	6	14.58
			Averages	0.22	2.25	1.94	5.81	1.71	4.63	0.59	6.00	10.60

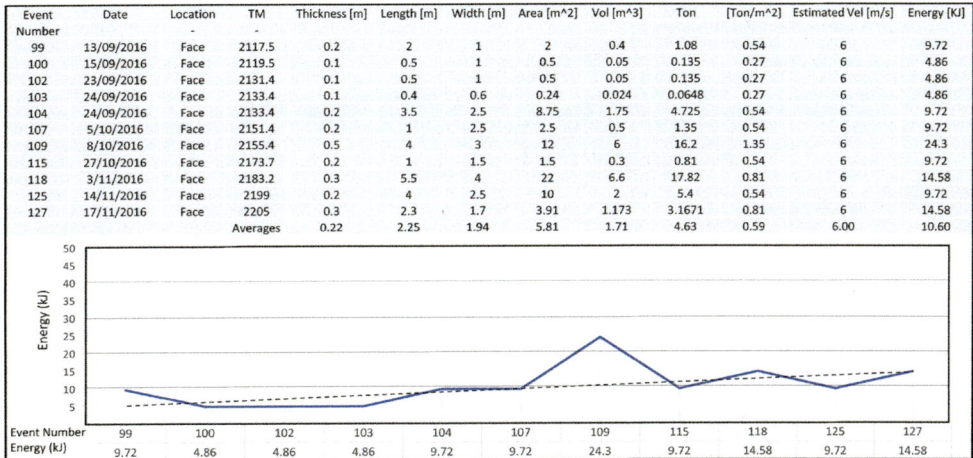

Event Number	99	100	102	103	104	107	109	115	118	125	127
Energy (kJ)	9.72	4.86	4.86	4.86	9.72	9.72	24.3	9.72	14.58	9.72	14.58

FIGURE 5.23 Example of back-calculated energy demand following face bursting during tunnel construction.

FIGURE 5.24 Example of violent, structurally controlled failure that has broken reinforcement elements and ruptured the surface support layer.

Information from structural geotechnical mapping can be used as an input for probabilistic determination of large-scale instability, considering all possible block geometries likely to become unstable (Figure 5.25). This information can be used to design the length of high energy dissipation reinforcement elements, such as cement grouted and plated plain strand cable bolts. These may also be installed in conjunction with layers of high-strength woven steel mesh (Kusui and Villaescusa, 2016).

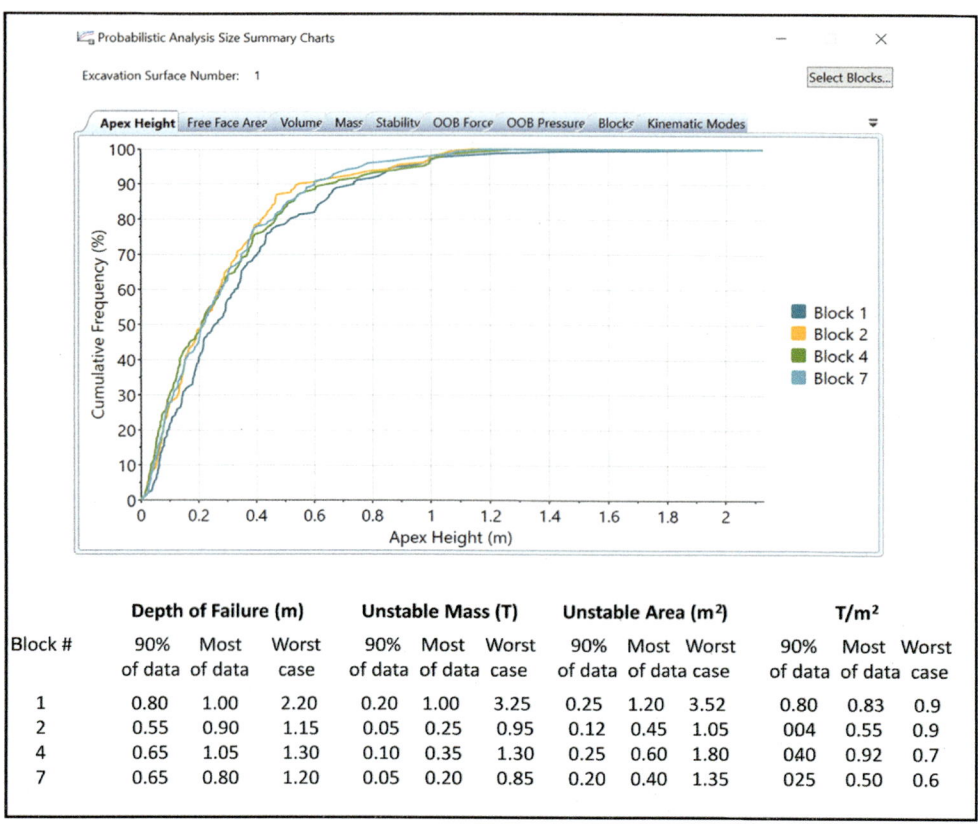

FIGURE 5.25 Example of probabilistic prediction of depth, mass and area for a group of potentially unstable tetrahedral block shapes.

Observations from underground failures can also be used to estimate the energy demand imposed upon the ground support schemes. Precise survey data from actual instabilities can be used to indicate if the depth, area and volume are increasing as the excavations become deeper (Figure 5.26). If the calculated mass of instability (T/m²) and related potential ejected energy increases beyond the installed capacity of the ground support, catastrophic failures might occur. Figure 5.27 shows data from actual failures, showing instabilities greater than 2 m for a 5-m wide tunnel with a horseshoe profile.

5.5 GROUND SUPPORT CAPACITY

The installed energy dissipation capacity of a ground support scheme should exceed the rock mass demand imposed upon the excavation by at least 10 kJ/m². Depending upon the depth of failure, this rock mass demand may be applied directly from the rock mass or through support layers that are retained by the reinforcement elements. Villaescusa et al. (2014) have defined rock mass demand in terms of ranges of allowable displacement and energy (Table 5.1) and combined it with the WASM dynamic reinforcement capacity database (Player, 2012; Villaescusa et al., 2016). An example design chart using these ranges and the capacity data for a particular rock reinforcement type is shown in Figure 5.28 (also see Chapter 7). For each rock mass demand category, the corresponding ranges of displacement and energy were used to define a region, shown as a dashed box. These ranges are labelled very low, low, medium, high, very high and extremely high energy demand. For each region, the acceptable ground support scheme components should achieve displacement

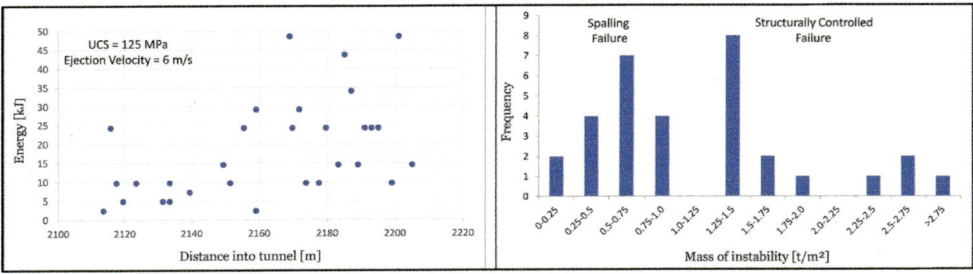

FIGURE 5.26 Example of violent tunnel instability data as an access tunnel is deepened.

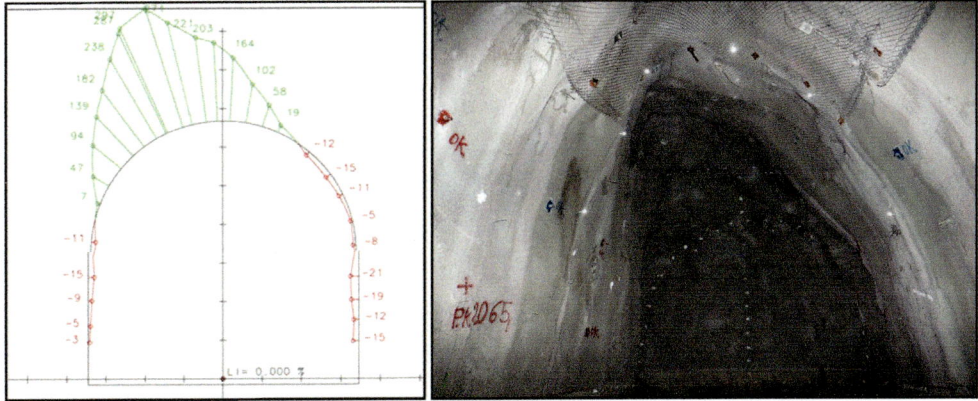

FIGURE 5.27 Example of stress-driven, violent failure with a depth of failure exceeding 2 m.

TABLE 5.1

Typical Rock Mass Demand for Ground Support Design.

Demand category	Reaction pressure (KPa)	Surface displacement (mm)	Energy (KJ/m^2)
Very Low	<100	<50	<5
Low	100-150	50-100	5-15
Medium	150-200	100-200	15-25
High	200-400	200-300	25-35
Very High	400-500	300-400	35-45
Extremely High	>500	>400	>45

compatibility while providing higher (>10 kJ) energy dissipation capacity than the required demand. As a comparison, Jager (1992) suggested that the minimum capacity for a scheme should be 25 kJ/m^2. However, he also suggested that 50 kJ/m^2 was a better specification in cases where the yielding elements are expected to withstand at least 25 kJ/m^2. The recommendation from the authors here is that in all cases, the chosen reinforcement must plot within the green design region, as shown in Figure 5.28.

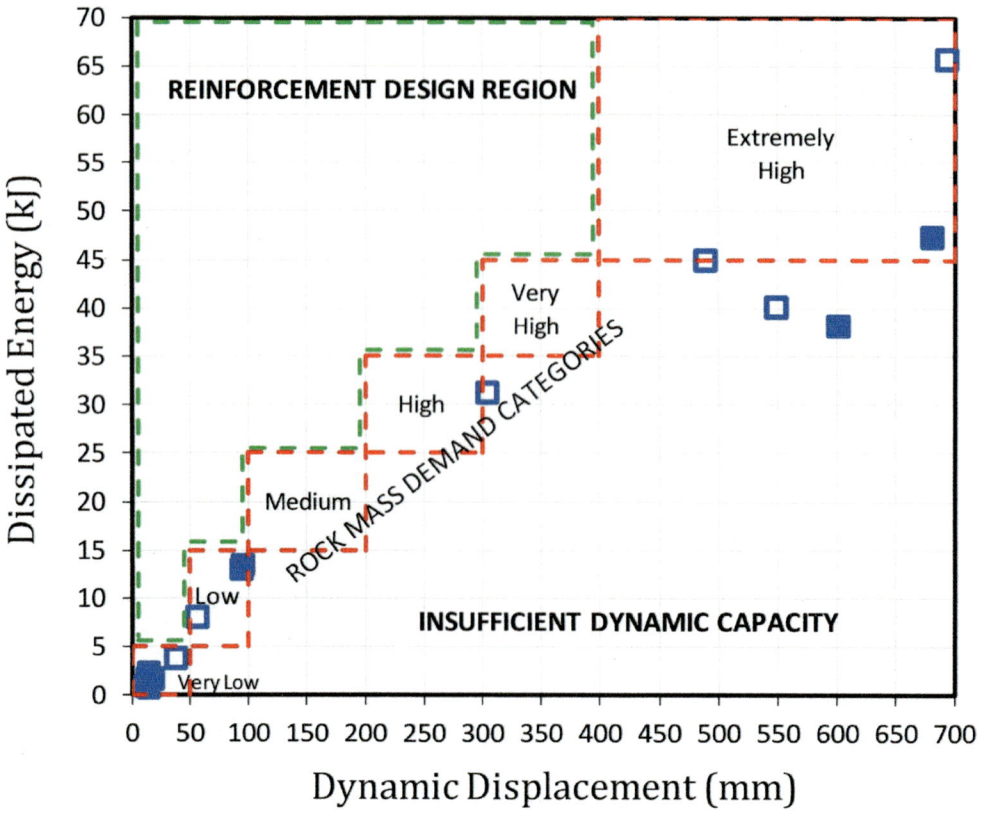

FIGURE 5.28 Dynamic energy dissipation versus displacement at failure for CFC reinforcement elements tested at WA School of Mines Dynamic Test Facility and plotted (blue symbols) on the design chart.

5.5.1   GROUND SUPPORT DESIGN FOR EXTREMELY HIGH DEMAND CONDITIONS

In future years, as the depth of mining and tunnelling operations generally increases worldwide, energy demand conditions exceeding 45 kJ/m² (extremely high demand) can be expected to be encountered more frequently. During the development stage of a project, leading tunnels can enter conditions of extreme energy demand. Failure at this stage of the project can lead to significant scheduling difficulties, costing millions of dollars in lost revenue or rehabilitation expenditure. Similarly, large failures occurring during the production phase of a mine or service phase of a civil excavation can also result in financial losses. Already a small number of mining operations have experienced failure mechanisms, which generated in excess of 60–80 kJ/m² in sudden and violent energy demands on the ground support schemes (Drover and Villaescusa, 2015a). This level of demand may be localized to areas where the rupture of a significant geological structure coincides with an excavation. In such cases, severe ground support scheme damage can occur, and effective load transfer outside the failure surfaces is required. As such, in order to ensure continuity of tunnelling operations at great depth, it is necessary to formulate a ground support scheme arrangement that can adequately manage this level of extreme energy demand.

Laboratory experiments at WASM and field observations of extremely high energy demand failure events at several mines in Chile and Australia indicate that a superior ground support scheme arrangement includes a multi-layered, integrated scheme of very high energy dissipation capacity components. The first stage of a ground support scheme for extremely high energy dissipation includes a primary shotcrete layer of approximately 75-mm thickness, internally reinforced with high tensile woven steel mesh. This type of mesh is preferred because its energy dissipation capacity significantly exceeds that of the common-variety mild steel welded wire mesh (see Chapter 9). The ability of high tensile woven mesh to articulate and constrain post-fracture of the shotcrete and its superior stiffness and displacement performance under load also support its selection for extreme demand conditions (Figure 5.29). Fibres are not required to be included in shotcrete that is internally reinforced with woven mesh in this context due to the relatively negligible strength performance benefit that fibres provide under extreme loading, both pre- and post-fracture (Drover and Villaescusa 2015b).

FIGURE 5.29 Example of a dynamic event causing severe damage to the layer of shotcrete with the load transferred by the mesh to the reinforcement.

A 75-mm thick, internally mesh reinforced shotcrete layer of this construction can be expected to dissipate approximately 20–25 kJ/m^2 of energy demand if using, say, a 4.6-mm diameter high tensile wire, 80-mm aperture woven mesh product, together with 25–35 MPa (28-day) unconfined compressive strength shotcrete. This surface support system can be installed in conjunction with a primary reinforcement pattern, utilizing 20–25-mm diameter, resin-encapsulated (Posimix bolts with an internal decoupled region exceeding 1.4 m), 550 MPa tensile strength steel-threaded rebar in a DMC arrangement. WASM testing data (Chapter 7) indicate that these elements can dissipate up to 30–40 kJ dynamically. A square pattern at 1 m × 1 m spacing of the reinforcement elements is standard, but this spacing may be optimized depending on the demand estimated from probabilistic structural analysis of potentially unstable blocks. The benefit of a staggered pattern is its ability to arrest and contain fracture propagation within bolt spacing (as evident earlier in Figure 5.17). This prevents fractures from extending across multiple bolt spacings, with associated increased displacements, as would be observed for a square pattern where larger drops of load-bearing capacity are experienced (Figure 5.30). This first pass of surface support and reinforcement constitutes the primary ground support scheme sequence.

In conditions of extremely high energy demand, it may be required to install an overlapping secondary ground support scheme (Figure 5.31). This secondary layer is installed externally onto the primary surface support layer and provides additional energy dissipation capacity and mechanical redundancy. The secondary layer is often connected to deep reinforcement in the form of plain strand cable bolts approximately 5–7 m in length. To be more effective, the cable bolts are positioned in rows, spaced evenly between the primary rock bolt pattern (Figure 5.32).

If structurally controlled failures involving large blocks overload the primary reinforcement and surface support layer, a secondary mesh layer provides load transfer connectivity from the unstable to stable regions of the excavation via the external mesh. Deep reinforcement helps anchor the unstable material to deeper stable rock. An external layer of high tensile woven mesh with 4.6 mm

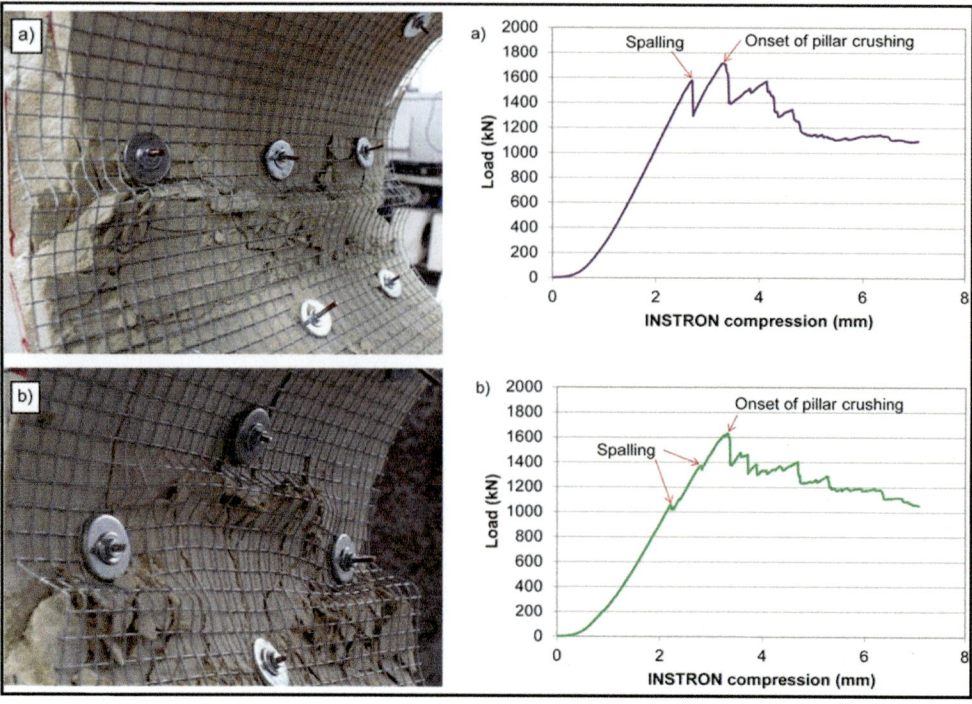

FIGURE 5.30   Square (a) versus staggered (b) reinforcement pattern performance during failure (Kusui, 2015).

FIGURE 5.31   Example of secondary, external mesh layer near a bench stoping brow.

FIGURE 5.32 Example of rock bolt and cable bolt arrangement within a high energy dissipation ground support scheme.

wire diameter and 80 mm aperture can be expected to dissipate up to 20–25 kJ/m^2 dynamically, whereas single-strand plain cables (15.2–17.8 mm diameter) may each dissipate about 30–40 kJ under dynamic loading. A high energy dissipation (HED) ground support scheme arrangement is illustrated in Figure 5.33. For deep failures, which initially mobilize reinforcement, this scheme can be expected to dissipate more than 60 kJ/m^2 under extremely high demand conditions.

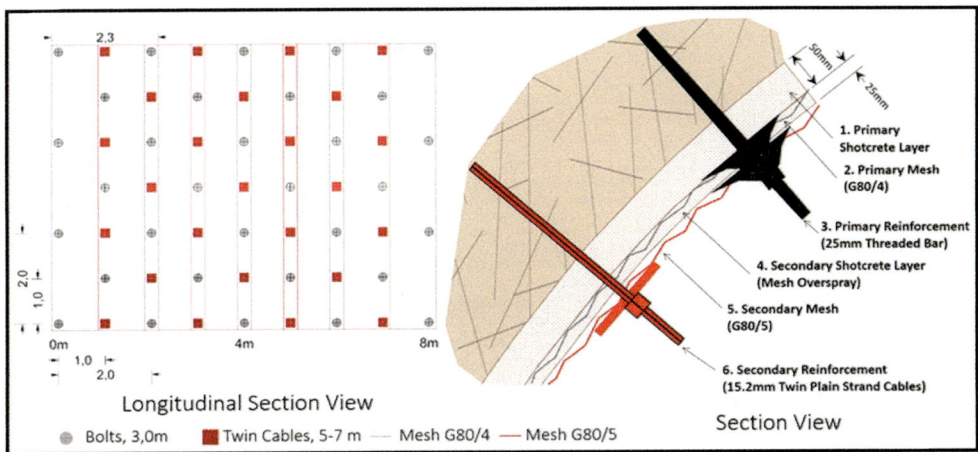

FIGURE 5.33 Example of a multi-layered, integrated ground support scheme for extremely high energy dissipation capacity (Villaescusa et al., 2016).

In civil engineering practice, statically designed, multi-layered rigid surface support arrangements anchored deep in the rock mass using ground anchors are called diaphragms (Barley and Windsor, 2000). Thus, the design shown in Figure 5.33 could properly be termed a 'dynamic diaphragm'. It is not necessarily rigid but designed to achieve force equilibrium and displacement compatibility at the excavation surface under dynamic loading.

6 Laboratory Testing

6.1 INTRODUCTION

The need for ground support testing arises mostly due to uncertainties associated with one or more of the hardware components and the interactions between them. Often, the rock mass is not recognized by hardware designers, suppliers and their clients as the most critical component influencing the overall performance of a ground support scheme (Figure 6.1). The results from a number of case studies will be presented in this chapter to highlight some of the issues that have been identified with various reinforcement components and systems. In some instances, a detailed examination of the behaviour mechanisms and associated theoretical calculations are used to illustrate and interpret the testing results.

The case studies to be presented have been divided into component testing and system testing. The component testing case studies demonstrate the test configurations used to define the mechanical properties of basic elements, internal fixtures and external fixtures (see Chapter 2). The systems testing case studies involve both laboratory simulations and *in situ* performance evaluations.

6.2 TESTING DOCUMENTATION

The basic load transfer principles outlined in Chapter 2 can be used to design appropriate tests to measure the required component properties and their interactions within a reinforcement system. The results of testing programs are often poorly documented. It is probably true to say that over-reporting of the details of testing is preferable to the omission of crucial information required for interpretation and use at a later time. The documentation required for laboratory and field test programs is essentially the same.

The documentation should include a complete description of each of the four generic components given in Figure 6.2, together with their physical and mechanical properties and their configuration for the test. The equipment and methods used for sample preparation and installation prior to testing should all be detailed. An important aspect of documentation is the date of sample preparation. The equipment used for loading and instrumentation for recording data should be documented in detail. Again, the date of testing is a crucial piece of information, especially for reinforcement systems that involve materials that have properties that change with time. Finally, the results should be presented in an unambiguous way, especially if the information is to be used in analysis software or design where the system configuration is different from the test configurations (Thompson et al., 2013).

As an example, Table 6.1 provides a comprehensive checklist for the information to be documented for component and system testing in the laboratory and *in situ*.

6.3 COMPONENT TESTING

The details of component testing can be found in various codes (e.g., ASTM, 2010; Standards Australia, 2007a, 2007b; British Standards, 2007). It should be expected that reinforcement component manufacturers and suppliers have conducted tests prior to making components that are commercially available. However, the reality in some cases is that components with apparently similar dimensions may have vastly different mechanical properties. In extreme cases, for example, for threaded devices, there have been instances where the physical dimensions have varied to make it impossible to assemble threaded components or threads have failed at forces much less than the

FIGURE 6.1   Unravelling of large blocks initially supported by a thin shotcrete layer and friction bolts.

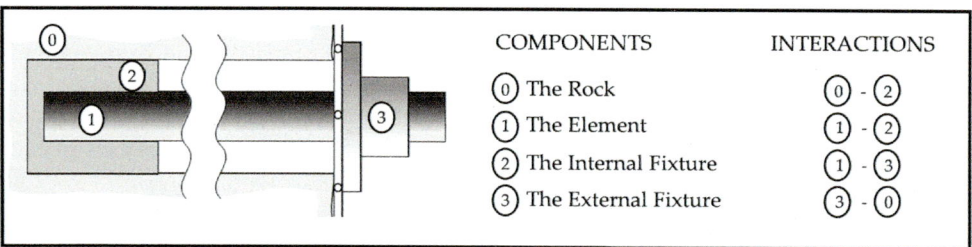

FIGURE 6.2   Components and interactions within a generic rock reinforcement system.

rated force capacity of the element. Consequently, dimensional checks and mechanical tests should be undertaken when any changes are made in the metallurgical composition or manufacturing process of any components of a reinforcement system. This may include the change in the supplier or type of resin or cement used for producing a grout for encapsulation. For example, a change in the cement particle size distribution will completely change the physical properties of the grout in the fluid state at a particular water/cement ratio (WCR) and the strength of the cured and hardened material.

In addition to testing following any changes that may result in variation of physical and mechanical properties, quality assurance testing of components should be undertaken at regular

TABLE 6.1

Checklists for Laboratory or *In Situ* Component and System Testing.

Basic Information

System Components
    Element
    External Fixture
    Internal Fixture
Component Materials
    Physical Properties
    Mechanical Properties
Component Testing
    Element
    External Fixture
    Plate

Laboratory Testing	***In Situ* Testing**
Borehole Formation	Borehole Formation
Pipe material and dimensions	Rock type and properties
Borehole diameter and bit size	Borehole diameter and bit size
Specimen and Test Configuration	Specimen and Test Configuration
Lengths in double embedment test	Anchor length and free length
Date of Sample Preparation	Date of Installation
Date of Testing	Date of Testing
Equipment Used	Equipment Used
Testing machine	Hydraulic cylinder and pump
Monitoring Devices	Monitoring Devices
Load and displacement	Load and displacement
Data Recording	Data Recording
Manual or data logger	Manual or data logger
Test Procedure	Test Procedure
Loading or Displacement Rate	Loading Rate

Testing Results

Raw Data
Data Processing
Presentation of Results
    Force-displacement response
Comments on Test Results (including any unexpected or unusual observations)

intervals. In the case of steel pre-stressing strand used to form cable bolts, tests are conducted on each coil of material and may result in rejection at this stage. ASTM F432–10 specifies that 'The manufacturer shall select and test a minimum of two bolts, threaded bars, threaded slotted bars, bearing and header plates, frictional anchorage devices and washers from each discontinuous turn or each 24 h of continuous production.' The frequency of testing installed rock reinforcement systems is much less prescriptive in mining than in civil engineering practice (Barley and Windsor, 2000). Some regulatory authorities may suggest testing 5% of bolts, but it is unlikely that this number of bolts is routinely tested. The frequency of testing for ground anchors used in civil construction is governed by various codes of practice (e.g., British Standards, 2009; Standards Australia, 1973).

6.4 ELEMENT TESTING

Simple axial testing of the element is the most common test performed for reinforcement systems. However, rarely are force-strain responses provided by manufacturers and distributors despite the importance associated with element extension during testing.

6.4.1 PRESTRESSING STRAND

Prestressing strand used for cable bolts is the same as the strand used in prestressed concrete construction and is therefore subject to stringent code specifications for minimum ultimate force and deformation. Despite these codes, sometimes strand is suspected of non-compliance and tests are requested by either suppliers or clients. One problem often observed in testing a strand is the rupture of one wire prior to the minimum specifications being reached. This is usually caused by either gripping the strand in the tapered jaws (often with saw tooth grip surfaces) of a universal testing machine or using barrel and wedge anchors that cause excessive stress concentrations at the tips of the wedges, as shown in Figure 6.3.

This problem was solved by having a custom-designed barrel with two supplementary tapered wedges, as shown in Figure 6.4. The anchor is designed so that the wedges do not get jammed between the barrel and strand. Figure 6.5 shows the difference between non-complying results and a complying test with ultimate strain >3.5%. An additional benefit is that the set of wedges intermediate between the barrel and standard three-part wedges has an outer taper that allows for easy release of the anchor following a test; otherwise, the strand may need to be cut to be removed from the testing machine.

6.4.2 SOLID THREADED BAR

Many threaded bars used for rock reinforcement have high elongation prior to rupture, making ordinary foil strain gauges unsuitable for direct strain measurement. It is not recommended to use

FIGURE 6.3 Excessive wedge penetration leading to premature strand failure.

FIGURE 6.4 Special barrel and wedge for strand testing.

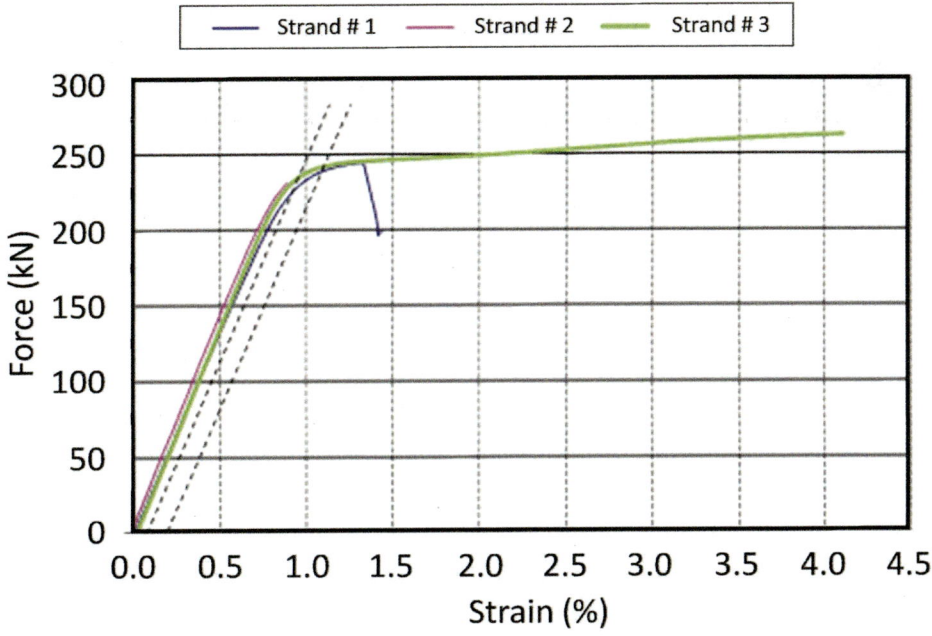

FIGURE 6.5 Force-strain curves for 15.2-mm diameter strand resulting from non-compliant (Strand #1 and Strand #2) and compliant (Strand #3) testing.

the machine jaw displacements to derive strains, as, usually, the strain will be over-estimated due to spurious displacements being recorded. Fortunately, all types of steel (including multi-wire steel strands) have a Young's modulus value that lies in the range of 200 ± 10 GPa, and a force-strain response may sometimes be corrected to within an acceptable tolerance. However, in order to properly measure displacement, transducers with a 'gauge length' of at least 500 mm need to be mounted onto the actual specimen, as shown in Figure 6.6.

FIGURE 6.6 Electronic displacement transducers with a gauge length of 500 mm mounted on a deformed threaded bar.

Figure 6.7 shows the results for three tests on high elongation steel threaded bar. Note that consistent results were obtained, and the initial stiffness can be derived using the known cross-sectional area in conjunction with an expected Young's modulus of ~200 GPa. The 'plateau' in force as the elongation increase from <1% to >3% strain corresponds with plastic elongation of the bar between the transducer mounts. This plateau will not exist for work-hardened steels that will also have a much smaller total elongation at rupture.

It is noteworthy that bonded strain gauges used for field instrumentation studies will be limited to the measurement of <3% strains. For the measurement of >3% strains, special wire strain gauges or mechanical devices are required to measure the relative displacement between anchor points located apart by a known gauge length (Hyett et al., 1992). However, as detailed by Thompson and Windsor (1993), the average strain over the gauge may under-estimate the peak strain for a continuously coupled reinforcement system, especially if the gauge does not happen to span across a dilating discontinuity in the rock mass.

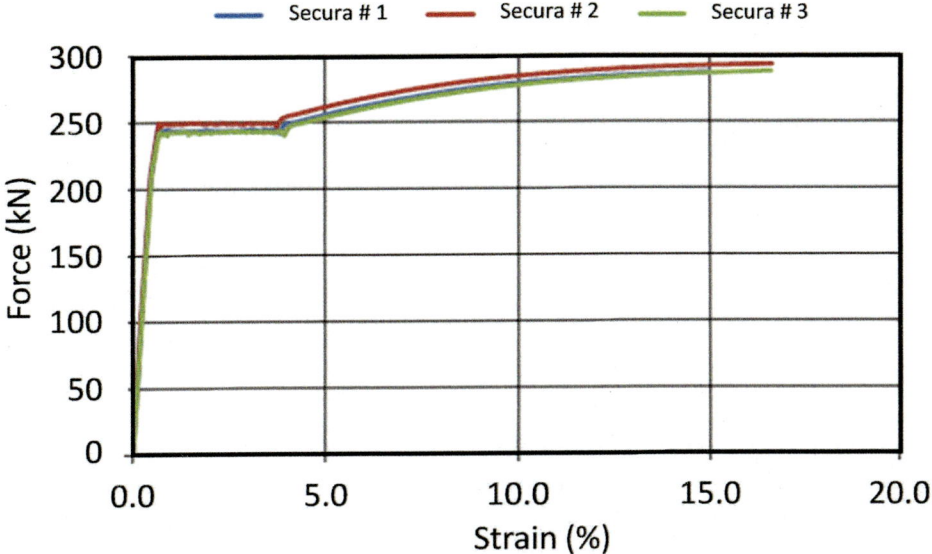

FIGURE 6.7   Example of force-strain responses for a high elongation threaded bar.

6.5   INTERNAL FIXTURE TESTING

Testing an internal fixture usually involves other components of the reinforcement system, such as the element and the borehole wall.

6.5.1   SLOT AND WEDGE ANCHORS

Slot and wedge anchors were one of the first types of anchors to be developed. They were found to be acceptable in very hard rock and were introduced during the 1950s in the Snowy Mountains Scheme in Australia. The Kiruna bolt is a later example of a bolt developed in the northern region of Sweden. This bolt is designed to be inserted in a borehole that has been pre-filled with cement grout with the objective of being a 'one-pass' system with the slot and wedge anchor providing immediate reinforcement action prior to curing of the cement grout (Potvin et al., 1999). The anchor sections are shown in Figure 6.8.

The authors have examined the feasibility of using either of these bolts with some of the conventional mechanized drilling and installation equipment available to the mining industry. One part of the study looked solely at the geometry of slot and wedge anchors and boreholes. A complementary program of testing was used to assess the strength of the slot and wedge anchors.

The first change to both bolts was a decision to increase the recommended borehole diameter from 32 to 36 mm. The consequences of this change became quickly apparent simply by examining the geometry of the anchors and the wedge movement necessary to cause the element to expand radially and interact with the borehole. For 36-mm diameter boreholes, it was predicted that the slot lengths needed to be >90 mm (for the Kiruna bolt) and >105 mm (for an alternative bolt) compared with the available slot lengths of 103 and 150 mm, respectively. However, the formed boreholes were actually larger than recommended, and so, in soft rock, the wedge displacement also needed to be greater than the available slot lengths.

Pull tests were performed on both bolt types. The maximum forces recorded for the Kiruna bolt varied between <5 kN in soft rock to ~12 kN in hard rock. The maximum forces for the alternative bolt were 35 kN in soft rock and up to ~70 kN in hard rock. In both cases, these initial anchorage values were deemed unacceptably low, and neither bolt could be safely implemented.

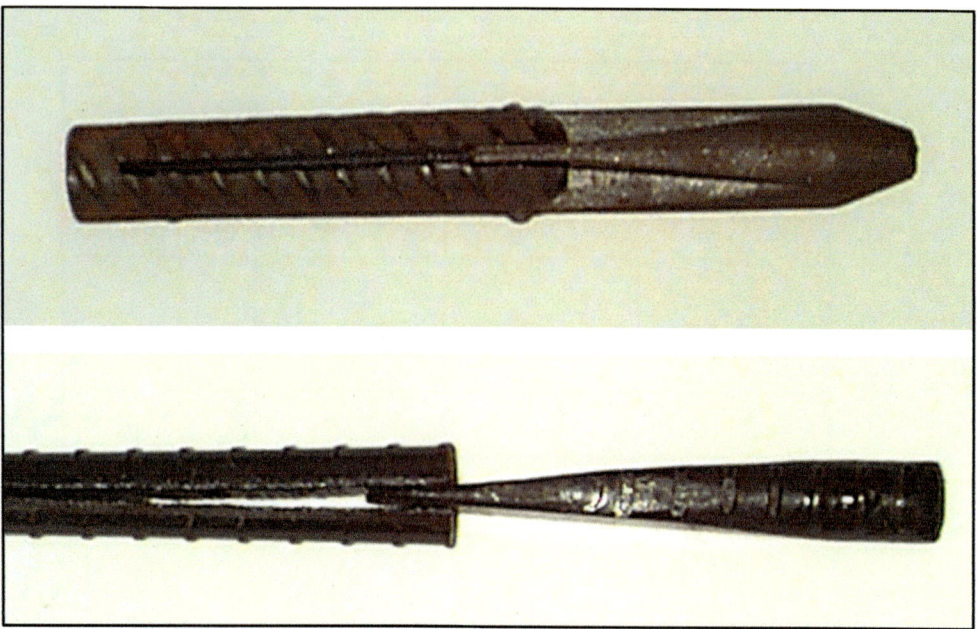

FIGURE 6.8   Example of Kiruna and alternative bolt slot and wedge anchors.

6.5.2 EXPANSION SHELL ANCHORS

Various configurations of expansion shell anchors were developed at the Snowy Mountain Scheme Australia research laboratories in the 1950s and subsequently patented. The expansion shell anchors were found to provide superior installation and performance characteristics than the slot and wedge anchors. These anchors rely on a tapered cone attached to the element being pulled through 'leaves' attached to bail. The movement of the cone towards the collar caused by pushing, pulling or tension in the element results in radial expansion of the leaves such that they grip the borehole wall. Similar to slot and wedge bolts, expansion shell anchors are best suited to high-strength rocks. One point of contention is the method used to ensure the anchor properly engages with the borehole wall. The original method was stated by Lang (1961) as follows:

> The anchorage is expanded by applying torque to the bolt. This may be done as a 'pre-set' operation before the bolt is tensioned or by expanding the anchorage and setting the bolt at the same time.

The understanding of the authors is that this method of installation requires the bail to be pushed to the end of the borehole by the element, as shown in Figure 6.9, so that element rotation causes the cone to displace and expand the leaves. This method of the initial set means that the bail is only required to sustain forces necessary to resist the initial expansion to contact the borehole wall and create the 'pre-set' mentioned in Lang (1961). Further rotation is then used to tension the element between the internal and external fixtures.

Subsequently, Hoek et al. (1995) stated that:

> Once the assembly is in place, a sharp pull on the end of the bolt will seat the anchor.

This method of initial pre-set may be suitable for handheld installation but cannot be achieved by jumbo drilling or rock bolt installation machines. Problems of mechanized installation will be exacerbated by the further suggestion by Hoek et al. (1995) that:

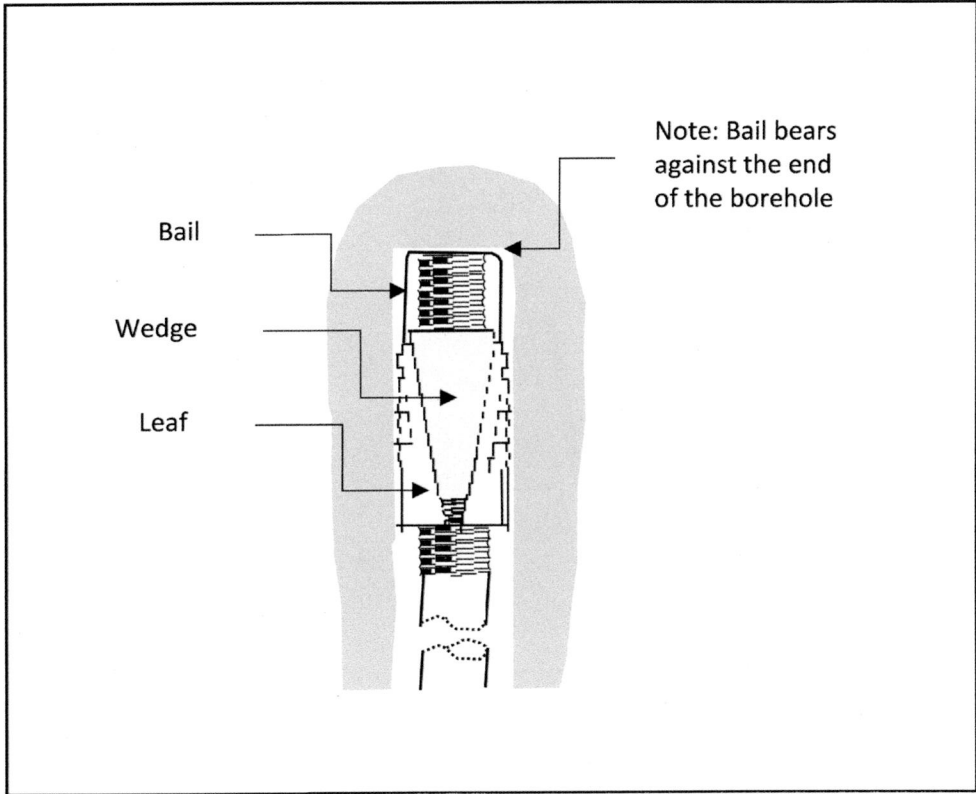

FIGURE 6.9   Schematic of expansion shell anchor and element configuration.

The length of the hole should be at least 100 mm longer than the bolt; otherwise, the bail will be dislodged by being forced against the end of the hole.

The authors suspect that element rotation with the bail located without a reaction from the end of the borehole may be responsible for instances of untensionable bolts due to breakage of the bail and misalignment of the leaves; or possibly one or both leaves sliding past the cone.

Assuming that the anchor is set properly within the borehole, other considerations as to the effectiveness of the anchor need to be taken into account. Equations involving the anchor geometry and equilibrium of the forces shown in Figure 6.10 can be used to show that:

$$\phi_b \leq \phi_r - \alpha \qquad\qquad 6.1$$

where α is the taper angle of the cone/leaf interface, ϕ_b is the friction angle at the cone/leaf interface and ϕ_r is the friction angle at the leaf/borehole rock interface.

Equation 6.1 indicates that the friction angle for the interface between the cone and the inner leaf surface must be less than the friction angle for the interface between the outer shell and the rock by an amount greater than the taper angle of the cone and inner leaf interface. In many cases, this condition is unlikely to be achieved because anchors are manufactured from cast components with an inherent roughness. The friction angle associated with this less than smooth cast surface is then often increased if a corrosion protection coating (e.g., zinc) is applied.

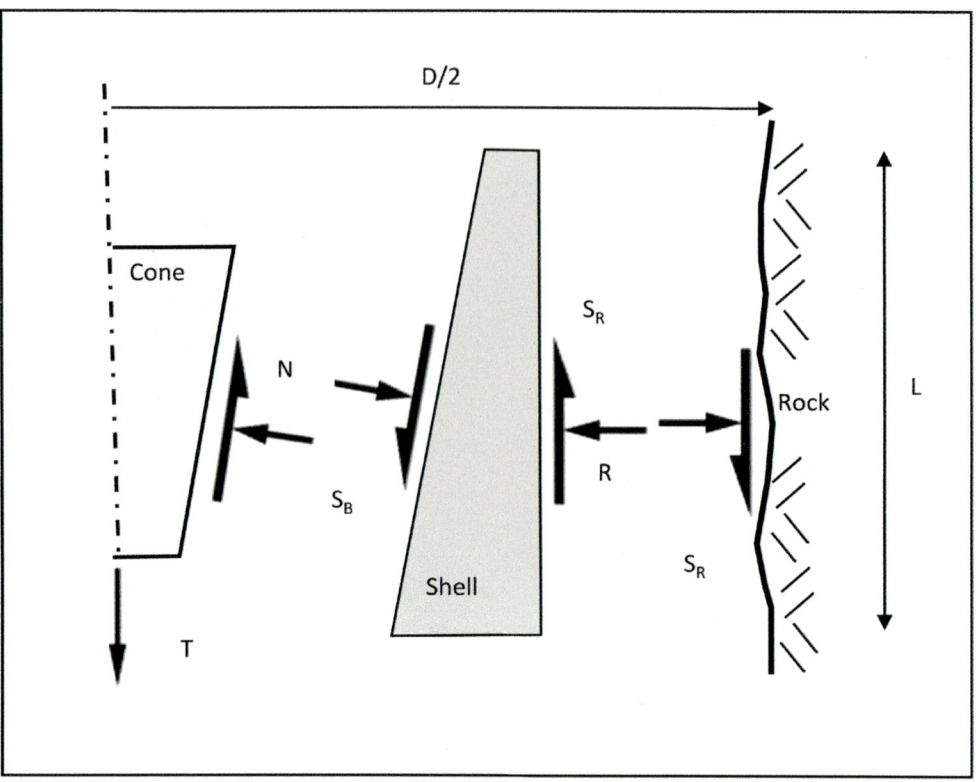

FIGURE 6.10 Expansion shell anchor geometry showing the internal and external forces.

The radial (R) and longitudinal (S_R) forces at the interface between the shell and the rock shown in Figure 6.10 can be converted to approximate equivalent normal (σ_r) and longitudinal (τ) stresses using the following equations:

$$\sigma_r = \frac{T}{\pi DL \tan(\alpha + \phi_b)} \qquad\qquad 6.2$$

$$\tau = \frac{T}{\pi DL} \qquad\qquad 6.3$$

where T is the tension in the bolt, D is the nominal diameter of the borehole and L is the length of the shell in contact with the rock.

The radial stress predicted by Equation 6.2 assumes that the force is equally distributed around the circumference of the borehole over the total length of the leaves. In reality, the stress will be higher than this estimate due to the non-uniform distribution of the radial stresses. Also, in hard rock, the teeth in the leaves will initially be in contact with the rock, and the contact stresses will be much higher, potentially resulting in local failure at the borehole surface.

In situ tests reported by Villaescusa and Wright (1999) for CT bolt anchors confirmed that anchor load transfer depends on the rock surrounding the borehole, with pull-out tests results ranging from 50–75 kN in schist to 150–190 kN in stronger rocks. These tests indicated gross slip (several tens of millimetres) of the anchors at approximately constant applied load. Following encapsulation with cement grout as required for proper installation and corrosion protection, the bolts were able to sustain loadings in excess of 225 kN (equivalent to ~200 kN/m of encapsulation).

Figure 6.11 shows an expansion shell with an 'end-spear' being used by CODELCO Chile to mechanically insert a threaded bar into a hole full of cement grout. The setting of the expansion shell allows for an immediate response by the rock bolt before the eventual curing of the cement grout.

FIGURE 6.11 Example of a high-capacity threaded bar to be installed into a hole filled with a cement grout prior to the element insertion.

6.5.3 CEMENT GROUTS

Cement-based grouts will be described in Chapter 8 in the context of cable bolt reinforcement. The main factor influencing strength is the WCR and curing conditions, as described by Thompson and Windsor (1998). There has been a trend to use low-heat (LH) cements to prolong workability and pumpability. However, the strength gain of LH cement grouts has been found to be slower that Ordinary Portland Cement grouts and may result in increased curing times before post-tensioning.

6.5.4 RESIN GROUTS

In cases where immediate reinforcement is required, resin encapsulation can be implemented. In order to install rock bolts that are coupled with resin along their entire lengths, it is necessary to insert multiple resin cartridges with a sufficient volume of resin to fill the annulus between a rock bolt and a borehole. Several problems, including gloving by the resin cartridge plastic and lack of resin at the collar, have been observed in practice and proven by bolt overcoring and will not be discussed here (Hassell, 2007; Villaescusa et al., 2008). Quality assurance based upon pull testing of essentially CMC sections is complicated by the full element capacity being mobilized by short embedment lengths of good quality resin mix anywhere along the element axis, making pull testing at hole collars essentially meaningless. In the future, resin pumping prior to bolt insertion is likely to significantly improve productivity while achieving full encapsulation and optimized load transfer along the entire element.

6.6 EXTERNAL FIXTURE TESTING

External fixtures have an important role to play in restraining surface support, such as plates, straps, mesh and shotcrete. It is desirable that plates have a capacity that is at least equal to that of the element, especially for reinforcement systems that do not progressively transfer load between the element and the rock (i.e., in the DMFC category). The following case studies show that the external fixtures, or the component plates with which they are associated, fail at less than the element capacity.

6.6.1   Split Tube Rings

In underground mining operations, plates are often observed to be missing at the collars of split tube bolts. Figure 6.12 shows a split tube ring prior to loading. The failure shown in Figure 6.13 occurred at ~75 kN compared with the split tube axial strength of 170 kN. The eccentric loading produced a lower failure load than the maximum load of 150 kN attained when the top of the split tube was loaded directly by a flat platen. The lower value of capacity is deemed more appropriate for *in situ* performance, where the ring is often damaged during installation, especially when driven oblique to the rock surface (Figure 6.14).

6.6.2   Barrel and Wedge Strand Anchors

Barrel and wedge strand anchors have been the subject of many investigations (Thompson, 1992, Thompson and Windsor, 1995, Thompson, 2004). In these investigations, theoretical studies were complemented by testing. Two aspects of barrel and wedge anchors have been examined:

     i. Installation
    ii. In-service performance

With regards to installation, it was shown that the tension produced in the strand between the external and internal fixtures was a complex function of:

- Taper angle of the wedge,
- Friction between the outer wedge surface and the inner surface of the barrel,
- Friction/mechanical interference between the strand and serrations on the wedge,

FIGURE 6.12   Set up of split tube ring and combined large domed plate prior to tension testing.

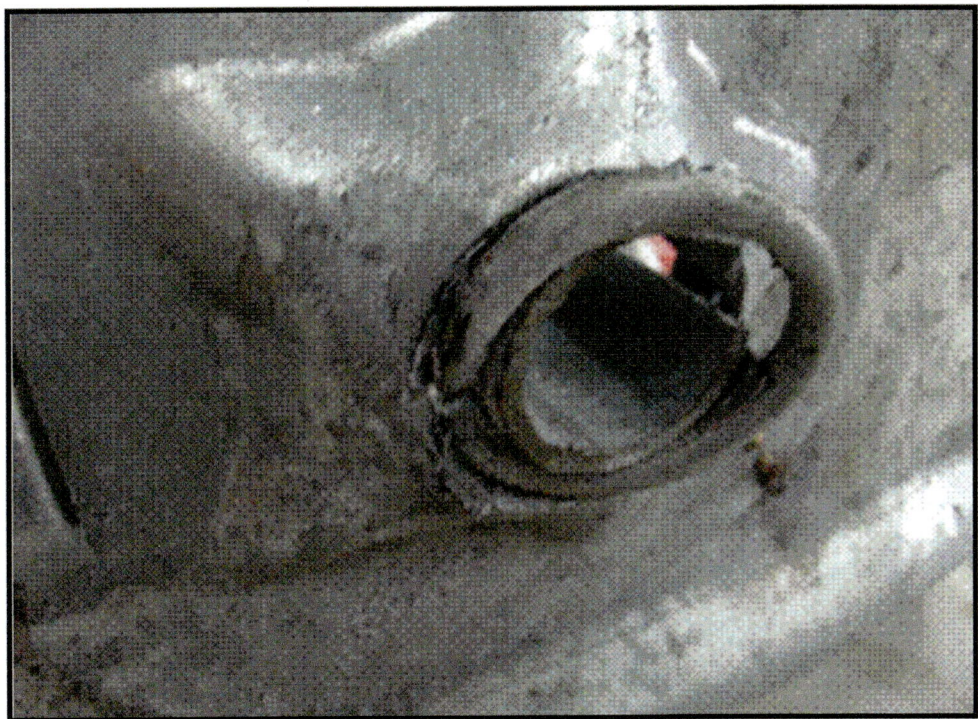

FIGURE 6.13   Failure to split the tube ring.

FIGURE 6.14   Example of damage to the split tube ring during bolt installation.

- External installation force and
- Force applied to the wedge.

The tension (T_i) produced during the application of a hydraulic piston force (P_i) was derived as:

$$T_i = K_i P_i \qquad\qquad 6.4$$

$$K_i = 1 - \frac{P_W / P_i \left(1 - \tan\alpha \, \tan\phi_B\right) \tan\phi_C}{\tan\phi_B \left(1 - \tan\alpha \, \tan\phi_C\right) + \tan\alpha + \tan\phi_C} \qquad\qquad 6.5$$

where

α = wedge taper angle
ϕ_B = friction angle between barrel and wedge
ϕ_C = friction between cable and wedge
P_W = force applied directly to the wedge

In particular, theoretically and by measurement, it was shown that the maximum tension achieved is only about 40% of the applied peak force when the installation force is applied directly to the wedges. For instance, the tension in the strand and between the plate and the rock surface was about 36 kN after applying an installation force of 100 kN.

In some cases, the in-service performance of barrel and wedge strand anchors has been poor. This is attributed to the inability of the wedge to slide relative to the barrel as a result of atmospheric corrosion (Thompson, 2004; Hassell et al., 2006). Figure 6.15 shows results from an experimental set-up implemented by Hassell (2007) where greased versus non-greased barrel and wedge anchors were placed into a chamber to accelerate corrosion. Inspections following pull-out testing showed that for the greased anchors, the internal surface of the barrel and wedge remained relatively corrosion-free, with the wedge teeth in good condition. On the other hand, the non-greased anchors showed significant corrosion on the internal surface of the barrel, with evidence of teeth shearing due to cable slippage. The laboratory and field experience show that when the anchors slip relative to the strand (Figure 6.16), they provide little resistance to rock movements and can fail at forces less than about 20% of the strand force capacity (Figure 6.17). In some observations on post-seismic event loadings, the anchors were actively missing resulting in complete loss of restraint (Figure 6.18). However, when the anchors are greased and properly installed, they can mobilize the full strength of the strand, even under conditions of violent and repetitive dynamic impact (Figure 6.19).

6.6.3 PLATES

Previous discussions indicate that in many circumstances, a plate is a critical necessity, and its mechanical performance needs to be properly defined. Figure 6.20 shows a flat plate setup for loading by an external fixture (in this case, a strand fitted with a barrel and wedge anchor). The underlying larger plate is configured with a central 100-mm diameter hole as required by ASTM F432–10. The results from three tests are shown in Figure 6.21. Note that the maximum force capacity of ~220 kN is less than the minimum strand capacity of 250 kN as shown in Figure 6.5.

6.7 SYSTEMS TESTING

First, it is important to state here that it is virtually impossible to measure the overall *in situ* response of an installed reinforcement system. The main reason for this is that it is impossible to cause and measure the dilation of an internal discontinuity. In a conventional *in situ* axial pull test, the measured

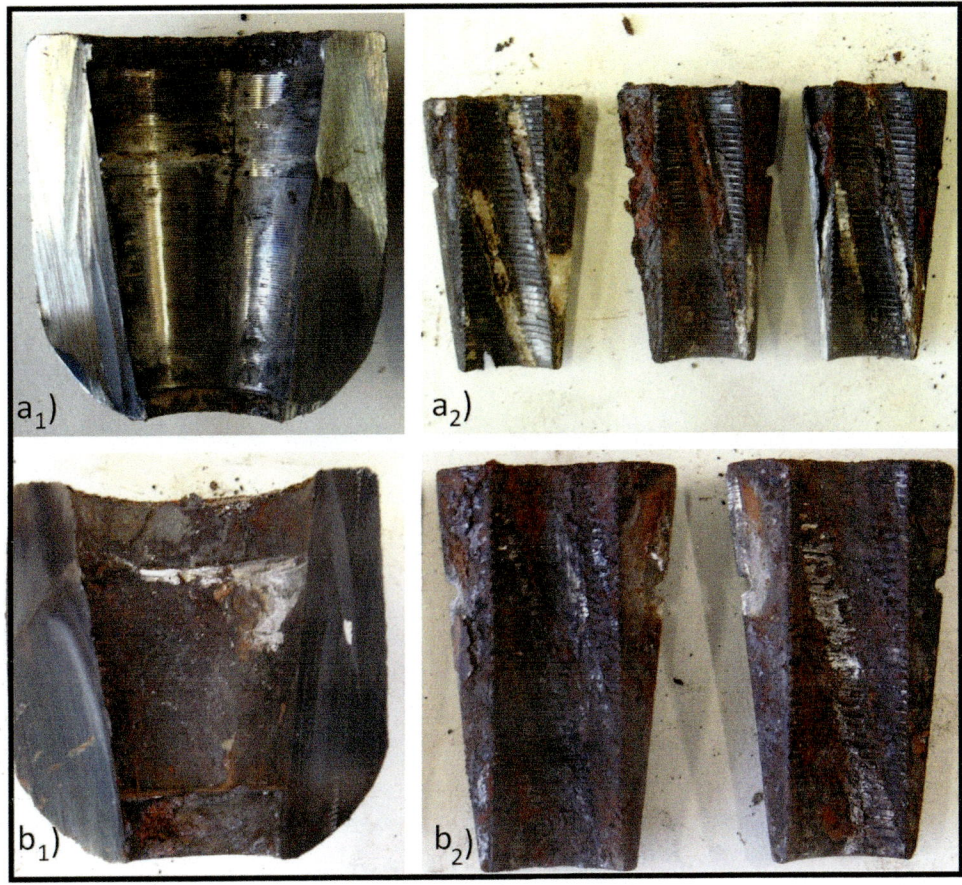

FIGURE 6.15 Some experimental outcomes from corrosion chamber tests showing a greased anchor with (a_1) clean internal barrel and (a_2) wedge teeth in good condition. Also shown is a non-greased anchor with (b_1) corroded internal barrel and (b_2) sheared teeth due to cable slippage.

response is a combination of element displacement relative to the rock in the toe anchor region and the extension of the element between the anchor region and the point of application of the loading.

On the other hand, in the laboratory, it is possible to configure a test, as shown in Figure 6.22, in which the overall system response can be measured. It implies that the test must evaluate the response of both the collar and toe regions while allowing for tension to be applied to the surface restraint. It generates the correct load transfer to the reinforcement system and allows for failure to occur at whichever is the weakest location, that is, at the surface hardware connection to the reinforcement element, at the collar region, in the reinforcement element at the simulated discontinuity or slip from the toe anchor region. This type of test is a simplification of a much larger test that used concrete blocks rather than steel tubes (e.g., Stillborg, 1994). It was first used in the late 1970s and early 1980s in Australia (Fuller and Cox, 1975, Hutchings et al., 1990), and has since been used routinely (Villaescusa and Wright, 1999, Thompson et al., 2004). The test has become known as the double embedment or split-pipe test and has subsequently been adopted by other organizations in several countries, such as the USA (Goris, 1990; Part 1 and 2), Canada (Bawden et al., 1992), Finland (Satola and Aromaa, 2004) and UK (BS 7861–2:2009). Although the test configuration was ideally suited to CMC systems, the WASM has developed procedures for manufacturing simulated boreholes in which friction rock stabilizers, point anchored systems and resin-encapsulated decoupled rock bolts have been installed and tested.

FIGURE 6.16 Example of barrel and wedge anchors installed without grease. This has resulted in the anchor sipping along the strand.

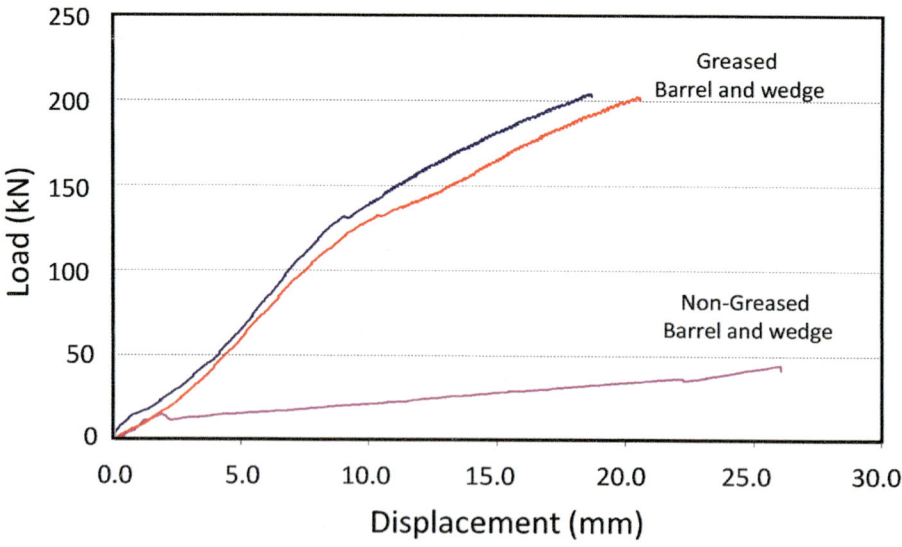

FIGURE 6.17 Example of pull-through resistance for greased and non-greased barrel and wedge strand anchors (Hassell et al., 2006).

FIGURE 6.18   Example of dynamic loading resulting in widespread failure of cable bolt external fixtures.

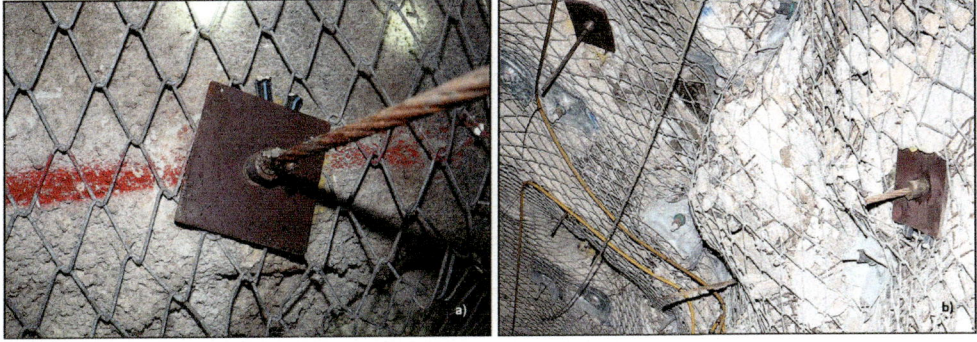

FIGURE 6.19   Examples of barrel and wedges installed using grease (a) just after installation and (b) after loading following an ML 2.4 seismic event.

FIGURE 6.20   Flat cable bolt plate before and after loading by a barrel.

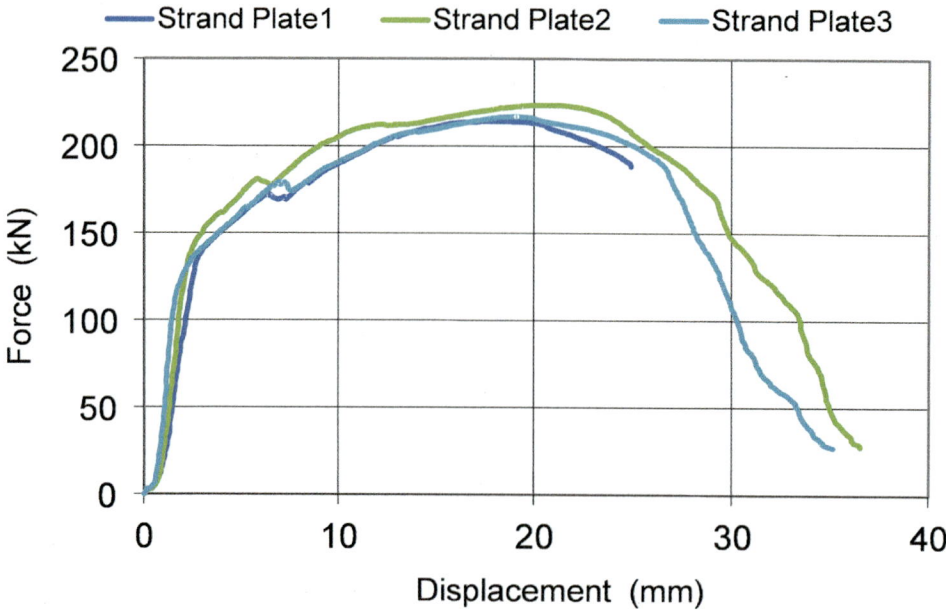

FIGURE 6.21   Results from some tests on flat plates that are used with cable bolts.

6.7.1 Cable Bolt Pull Tests

In the early years of pre-stressed concrete reinforced testing, results were presented as a peak bond stress. It was more appropriate for the circumstances than it is for rock reinforcement systems due to the larger relative displacements between the components. One phenomenon that needs to be appreciated is that average 'bond' stress (force per unit interface area) is a poor indicator of the actual behaviour of a reinforcement system. Average 'shear' stress resistance reduces with increasing embedment length, and therefore, it is not appropriate to extrapolate results from short embedment lengths to longer lengths. The reasons for the reduction in average shear stress might be attributed to the relationship between shear resistance and relative displacement at the interface between the element and internal fixture. There will be a transition from high shear resistance at small displacements to lower shear resistance at larger displacements due to the extension of the element within the encapsulated length and, eventually, the gross slip of the entire element.

In order to account for the reduction in average shear stress, a test program needs to include one embedment length sufficient to result in element rupture. Three tests with shorter embedment lengths wherein a failure occurs by the gross slip of the element are also required in order to determine the critical embedment length (i.e., the minimum embedment length that will result in the rupture of the element in a tension test).

It is important that test results are presented in a consistent and meaningful format. This requires that the raw force and displacement measurements made during a test are processed to give a consistent meaning to the response given in charts. Firstly, in a single-ended test (i.e., with one embedded length and a free length), an estimate should be made of the extension over the free length. It implies that the strain at a particular force is multiplied by the free length and subtracted from the measured displacement at the loading point. Secondly, for double-ended, equal embedment length tests, it is necessary to divide the total measured separation by a factor of two. By presenting the results in these forms, the responses represent the same mechanism of response for the reinforcement system and enable valid performance comparisons to be made.

It has been found, both experimentally and theoretically, that the variation of force-displacement responses with different embedment lengths may be characterized as shown in Figure 6.23. The

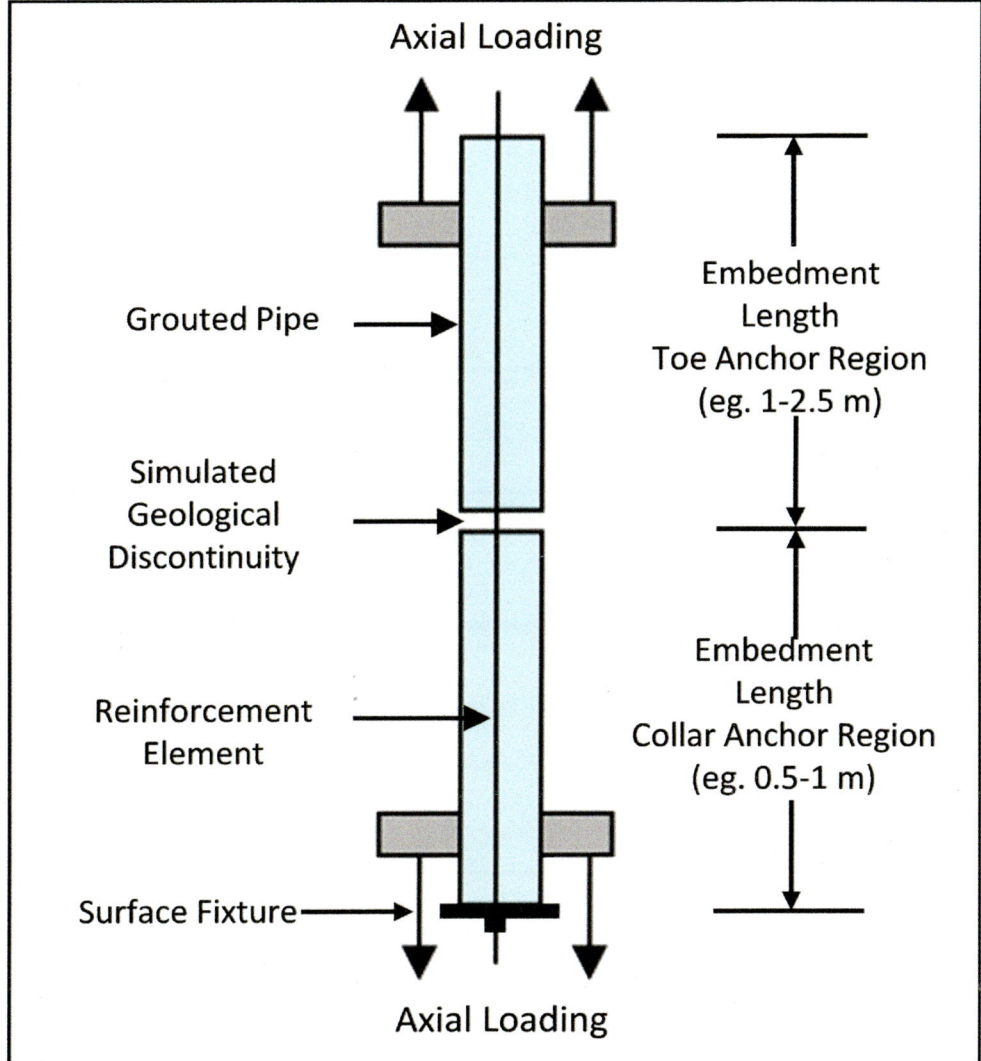

FIGURE 6.22 Schematic of double embedment or split-pipe test configuration used for reinforcement system testing.

characteristic responses include that the initial stiffnesses of the responses are the same for all embedment lengths. In this chart, all displacements are consistent with those described previously.

The results from a testing program may also be plotted as peak force versus embedment length, as shown in Figure 6.24. The critical embedment length may then be determined immediately by extrapolating the curve to intersect the element strength line.

6.7.2 RESIN-ENCAPSULATED REINFORCEMENT

The typical resin-encapsulated bolts currently being used in underground hard rock mines have been modified from the bolts used in the coal mining industry. The modifications have been necessary due to the need to drill larger borehole diameters with the type of mining equipment used for installation. The bolt modifications are mainly in the form of paddles or the use of a spiral steel

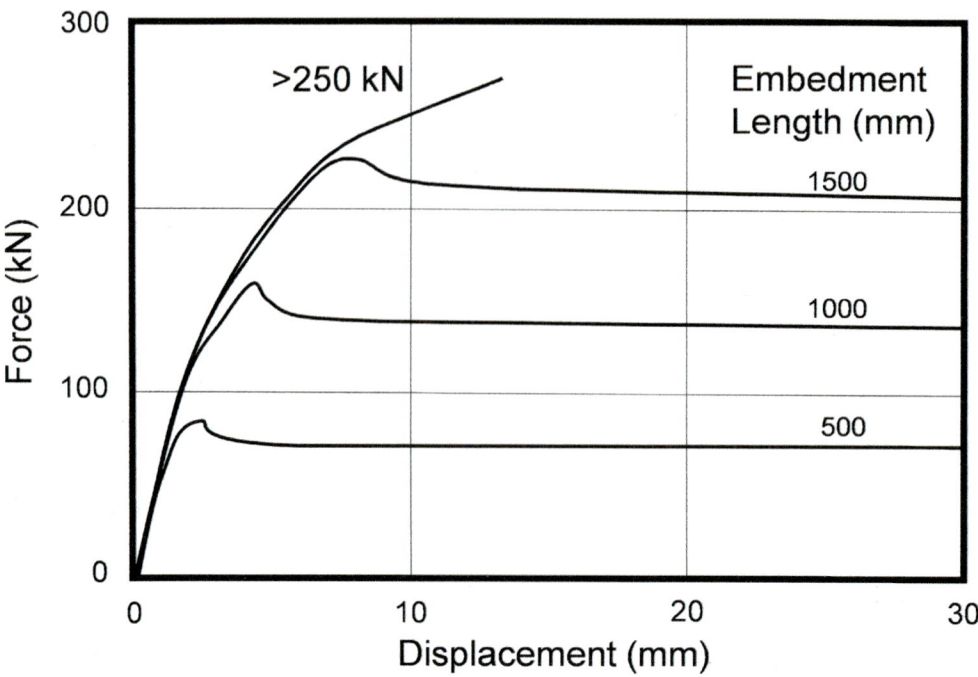

FIGURE 6.23 Example of CMC reinforcement system test responses for different embedment lengths.

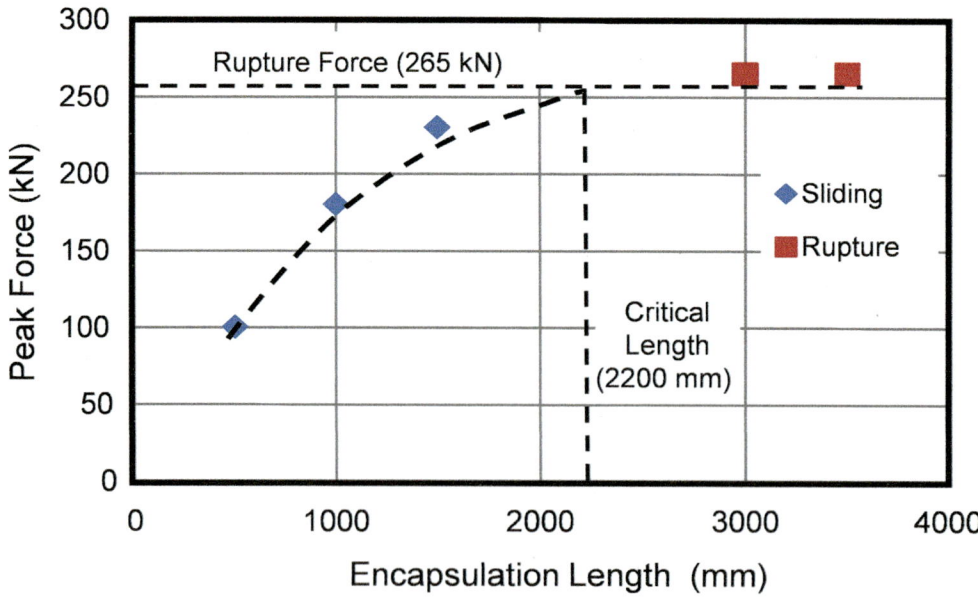

FIGURE 6.24 Example of the extrapolation used to predict the critical embedment length for cement-encapsulated, single-strand cable bolts.

wire welded to the end sections of the bolts. The modification for a 24-mm diameter Posimix bolt is shown in Figure 6.25.

FIGURE 6.25   Example of a Posimix bolt showing the spiral arrangement for resin cartridge mixing.

Despite the modifications to enhance mixing, problems associated mainly with larger diameter boreholes still occur. These problems may be poor mixing, often near the toe of the borehole due to over-drilling the length, gloving due to the bolt penetrating the cartridge or over spinning resulting in damage to the partially cured resin. Many of these problems are not visible and cannot be identified from *in situ* pull-out tests. Consequently, a purpose-built drill rig (Figure 6.26) capable of overcoring reinforcement elements within a production mining environment was developed in order to undertake a systematic investigation of the entire length of fully encapsulated rock bolts *in situ*. The investigations and results reported, in detail, by Hassel (2007) involved both visual inspection by splitting the recovered rock cores containing the bolts and laboratory (tension and push/shear) tests on samples, as shown in Figure 6.27.

FIGURE 6.26   The WASM bolt overcoring machine in operation.

FIGURE 6.27   Example of an overcored sample following push testing at the WASM laboratory.

The overcoring data showed that with resin cartridges, the best resin mixing and bolt encapsulation occurs within the middle region of the bolt for the majority of the currently used bolt and borehole size combinations, reinforcement systems and installation practices common in the Australian hard rock mining industry (Villaescusa et al., 2008). In all cases of poor load transfer, poor resin mixing was identified as the main cause. In addition, the majority of the overcored bolts had no resin at the collar region, indicating that the system capacity near the excavation surface must be controlled by the plate-bolt interaction. Resin near the collar minimizes a violent ejection of broken bolts following large dynamic events, as demonstrated in Figure 6.28. Ongoing research into resin pumping has the potential to achieve full bolt encapsulation and continuous load transfer along the entire reinforcement length.

6.8   WASM DYNAMIC TEST FACILITY

In the past 20 years, the WASM has undertaken research to quantify the dynamic energy dissipation capabilities of full reinforcement systems, including a large number of commercially available rock bolt devices. WASM secured mining industry and Western Australian government sponsorship to develop a dynamic loading simulation test facility using a unique methodology (Player et al., 2004; Thompson et al., 2004; Villaescusa et al., 2005; Villaescusa et al., 2010; Villaescusa et al., 2015).

FIGURE 6.28 Example of partial resin-encapsulated collar regions retaining broken bolts following a seismic event.

The WASM Dynamic Test Facility uses thick-walled steel tubes to simulate the rock and borehole and the reinforcement system by having (Figure 6.29):

- A toe anchor region comprising a beam and the upper tube length,
- A collar region comprising a loading mass and lower tube length and
- An element that spans between the upper anchor and the lower collar zones.

A typical test involves raising a beam with the installed reinforcement system to be tested to heights of up to 5 m. A loading mass of up to 3000 kg is attached to the lower end of the specimen. The complete assembly is then dropped. Loading of the reinforcement system is affected by the relative displacement between the loading mass and drop beam. On initial contact, the beam is decelerated by buffers, and the momentum of the loading mass causes a relative velocity to develop between the mass and the beam, imposing a dynamic axial force on the reinforcement element system. The ability of the reinforcement system, as a whole, to successfully decelerate the loading mass depends on its response characteristics. The response characteristics typically reflect how much it deforms when subject to load as well as both the force and displacement capacities. Under the influence of this force, the reinforcement system will either yield, slide or break while dissipating the kinetic energy. With the measured response characteristics, it is possible to calculate the energy dissipation (Thompson et al., 2004). This provides the criterion by which different reinforcement systems may be compared directly.

The loading mass in Figure 6.29 simulates the detachment process that occurs *in situ* when a single block or fragmented mass of rock is ejected from the surrounding rock mass towards the excavation, loading the reinforcement system. This loading is not instantaneous. It takes a finite time that will, for a remote event, be related to the seismic wave velocity and amplitude, the acceleration pulse of the dominant frequency and fracture velocity within the rock mass. For a local event, the loading is a 'pulse' resulting in a mass (M) of rock moving at a particular velocity (v). Consequently,

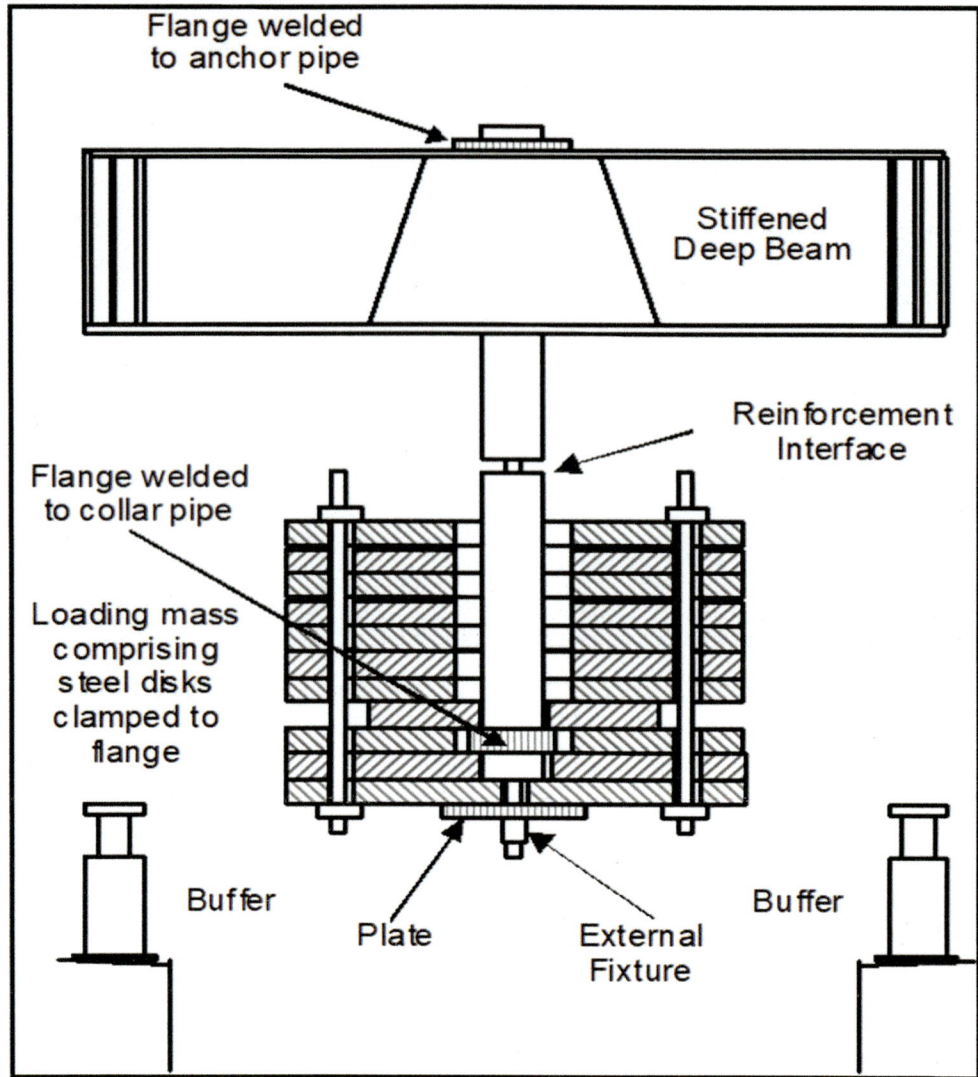

FIGURE 6.29   WASM dynamic test facility components.

the failure volume will have a change in momentum that is related to a force (F_t) acting over a short period of time (t), as given by the following impulse equation:

$$\Delta Mv = \int F_t dt \qquad\qquad 6.6$$

In this simulation of ground support loading, it is clear that prior to any event, the rock mass and ground support are stationary. After an event, the rock mass accelerates to a certain velocity in a short but finite time. Hence, in a test facility, it is not appropriate to apply the load instantaneously, but rather it should be applied rapidly, and the rate of load application should be measurable and similar to that considered to cause damage to underground mine ground support. Due to the different reinforcement load transfer mechanics (CMC, CFC and DMFC), it is not always correct to apply the load simply to the external fixture at the collar of a system.

It must be noted that all the WASM dynamic tests are carried out under pure axially loading conditions. Hence, they are not representative of any shear loading that may occur underground.

FIGURE 6.30 An example of a reinforcement pattern largely radial to an excavation boundary, ensuring a large component of axial loading dominates during dynamic failure.

Shear loading (both dynamic an static) of a reinforcement system *in situ* is particularly difficult to define. The performance will be significantly affected by the quantity of dilation on a particular shearing structure, how intact the rock activating the shear loading remains and the rate of debonding/fracturing between the reinforcement element and its encapsulating medium. An important influence on performance is whether the shear displacement is concentrated at one location or at multiple locations along the element axis. Concentrated shear will result in failure at lower shear displacement. With axial loading, the assumption is that sufficient elements are radially installed in a reinforcement pattern (Figure 6.30), such that a large component of axial loading always dominates following a dynamic event.

6.8.1 Force Transfer and Displacement

The selection of instruments and their locations at the WASM Dynamic Test Facility is based directly on load transfer and displacement of the different test configurations. The equilibrium of each component in a test configuration forms the basis for the engineering analysis. The results are expressed in terms of a force-displacement response and an energy dissipation. Figure 6.31 shows the load transfer mechanics in play and the locations and symbols of force transfer in a reinforcement system dynamic test. It is the free-body diagram that forms the basis of the equilibrium analysis initially presented by Thompson et al. (2004).

6.8.2 Simulated Boreholes and Sample Installation

An important component of a reinforcement system is the rock surrounding a borehole. In particular, the interface between the internal fixture and the borehole wall influences the rate of load transfer. Player (2012) found that load transfer for reinforcement systems encapsulated within cement grout is

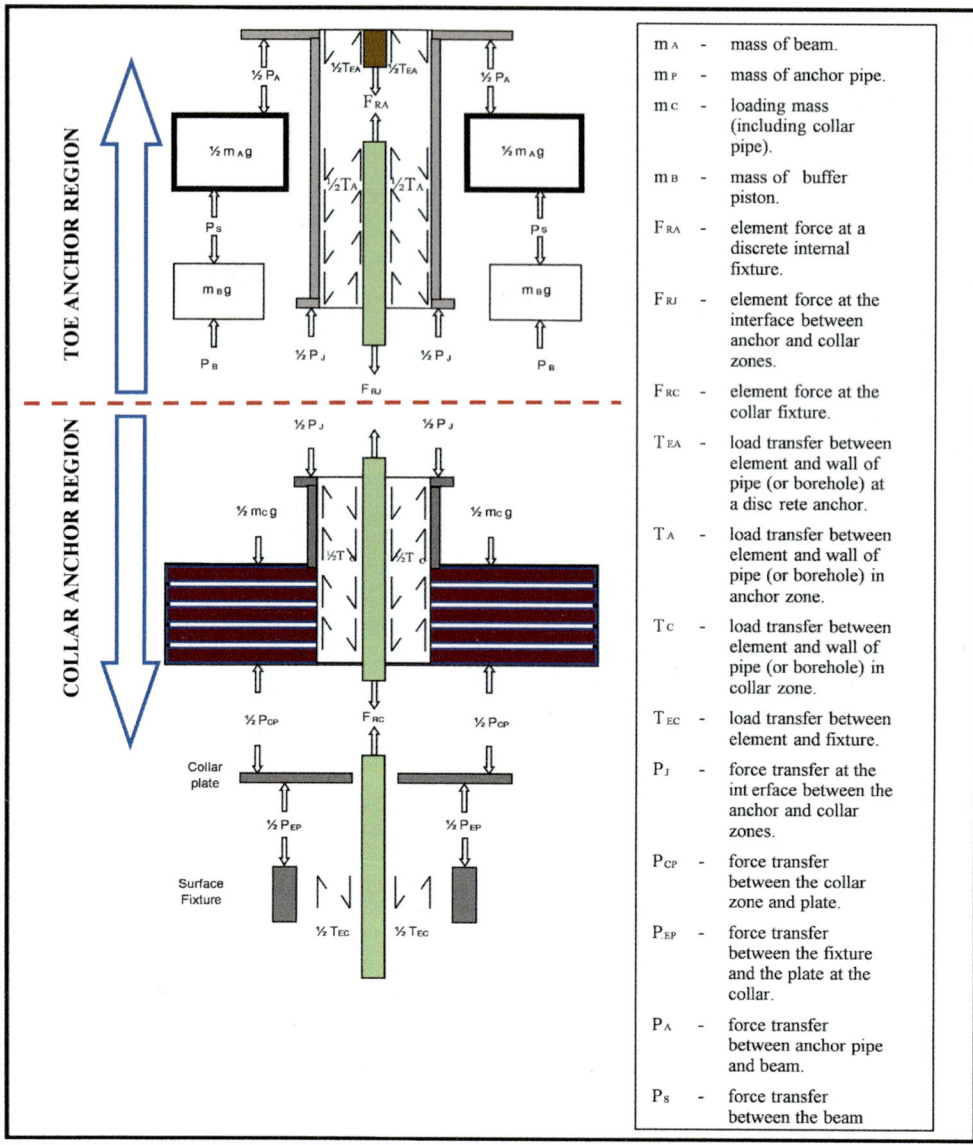

m_A	-	mass of beam.
m_P	-	mass of anchor pipe.
m_C	-	loading mass (including collar pipe).
m_B	-	mass of buffer piston.
F_{RA}	-	element force at a discrete internal fixture.
F_{RJ}	-	element force at the interface between anchor and collar zones.
F_{RC}	-	element force at the collar fixture.
T_{EA}	-	load transfer between element and wall of pipe (or borehole) at a discrete anchor.
T_A	-	load transfer between element and wall of pipe (or borehole) in anchor zone.
T_C	-	load transfer between element and wall of pipe (or borehole) in collar zone.
T_{EC}	-	load transfer between element and fixture.
P_J	-	force transfer at the interface between the anchor and collar zones.
P_{CP}	-	force transfer between the collar zone and plate.
P_{EP}	-	force transfer between the fixture and the plate at the collar.
P_A	-	force transfer between anchor pipe and beam.
P_S	-	force transfer between the beam

FIGURE 6.31   Free-body diagram showing load transfer mechanisms within a reinforcement system.

controlled mainly by the interface between the grout and the element. For reinforcement systems such as those within the CFC category, load transfer occurs directly between the element and the borehole wall. These systems have not been tested in the laboratory to the same extent as encapsulated systems. Therefore, WASM has developed a methodology to create a material that simulates the *in situ* rock. The borehole simulation and installation of the reinforcement are undertaken at the WASM Rock Mechanics facilities in Kalgoorlie and the several sponsoring mining companies in Western Australia. The cementitious material that simulates the rock around the borehole is prepared using a GP2000A mixer (Villaescusa, 2014). A non-shrink iron-reinforced construction grout (Masterflow 4600) is used at the recommended WCR. In addition, basalt aggregate (20% passing 4-mm chip) is also used in the mix, as it can be easily mixed and pumped with the GP2000A machine.

Figure 6.32 shows details of borehole drilling in simulated rock using a Tamrock Axera jumbo at a mine site in Western Australia. An appropriate drill set-up was conceptualized and implemented using a specially designed drilling platform to minimize the drift of the long (up to 3 m) holes. These are drilled within heavily confined concrete (> 90 MPa strength at the time of drilling). Typical drilling times ranged from 5 to 6.5 minutes. After drilling each hole, the bit is measured to ensure no wear is experienced that would reduce the size of subsequent holes. Figure 6.33 shows typical results for the simulated boreholes prior to bolt installation. In the majority of cases, the boreholes showed little to no deviation within the nominal 150-mm internal diameter, 7 mm thick wall steel pipes. Figure 6.34 shows the typical borehole diameters measured after drilling. These variations with borehole depth are similar to borehole diameter measurements measured in boreholes drilled in various rock types at operating mines. A typical CFC bolt installation is achieved using a jumbo, as shown in Figure 6.35.

6.8.3   DYNAMIC TEST RESULTS

A number of indices can be used to characterize the performance of reinforcement systems. These indices allow comparison between different reinforcement systems. In particular, it has been found

FIGURE 6.32   Rigid set-up arrangement to ensure accurate, co-axial drilling of simulated boreholes using a mining jumbo.

FIGURE 6.33   Front end and rear end views of the 3 m long boreholes drilled in high-strength grout and heavily confined by thick pipes.

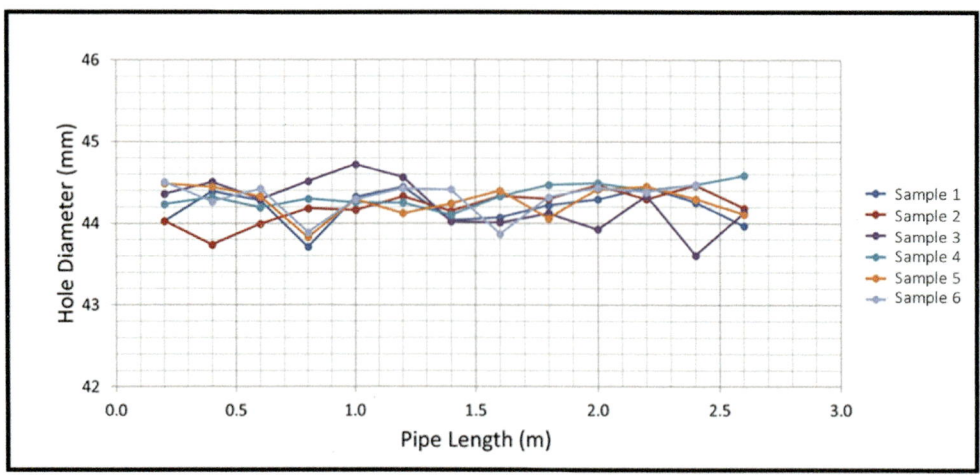

FIGURE 6.34 Six examples of the resulting borehole diameter with depth within the simulated boreholes.

FIGURE 6.35 Example of bolt installation using a mining jumbo.

that it is insufficient to only report dissipated energy. Other parameters, such as total displacement, peak velocity and acceleration, are required to assess performance. Testing shows significant differences in performance when applying multiple loadings on a single reinforcement system to the failure point compared with a single loading that fails the reinforcement system. Most importantly, the energy dissipated from multiple smaller loadings cannot be simply summed to produce the ultimate capacity achieved by a single loading event. This is due to progressive interface damage that changes the condition of the load transfer mechanism and the system configuration. That is, this change in bonding at the interface between a bar and its encapsulant and sliding friction are velocity- or strain rate-dependent. Therefore, the results from the first loading are preferentially used in characterizing the response of a reinforcement system.

The results from the WASM Dynamic Test Facility are given in the form of a Standard Test Report that has two major sections:

- A Test Summary section that is used for general distribution and discussion on the testing arrangements
- A Detailed Analysis and Interpretation section that contains details on how the analysis was performed, errors encountered and the shapes of the filtered and processed waveforms recorded during testing. A lengthy discussion on waveform analysis is well beyond the scope of this book

The Test Summary includes the reinforcement system's specification and configuration, the test specification, the overall system performance (physical measurements and response to the load), photographs and analyses of the failure or yielding mechanism. Finally, the Test Summary reports include:

- Energy dissipation and balance for all system components,
- Load-displacement curve at the simulated discontinuity,
- Acceleration-time curve for the loading mass and
- Velocity-time curve for the loading mass.

An example of the test information recorded for a reinforcement system specification is shown in Figure 6.36. This specification documents:

WASM Dynamic Test Results – Reinforcement Bolt No 319

Company	Kanowna Belle Mine	Tested by	Ushan de Soyza
		Sample	3.0 m Posimix Bolt
Test Date	2019-08-02	Bolt Type	DMFC

Reinforcement System Specifications							
Total Length	Toe Anchor Region	Collar Anchor Region	Simulated Depth of Failure	Length Decoupled Region	Length Exposed Tail	Bar Diameter (Nominal)	Steel Grade Specification
3000 mm	600 mm	280 mm	1000 mm	1800 mm	320 mm	22 mm	-

Simulated Borehole	3m long, Jumbo drilled, rough simulated borehole 150 XHV
Simulated Rock	Master Flow 4600 grout, drilled used a 38mm bit
Surface Hardware	150 x 150 x 6 mm and nut size – 36 (45mm long)
Reinforcement Encapsulation	Resin Cartridges
Collar Tension to Plate	Installed by a Mining Jumbo at the KB UG Workshop

FIGURE 6.36 Example of a reinforcement system test specification.

- The test operator, data and the type of reinforcement system under test;
- Classification of the system, whether CFC, CMC or DMFC;
- The encapsulated or coupled distance of the collar and toe anchor regions;
- The total length of the reinforcement system;
- The presence, positioning and length of any decoupling component;
- Steel grade specification;
- A description of the simulated borehole;
- A description of the surface hardware; and
- The applied collar tension or the torque applied to a nut.

The test specification section documents the following details (also tabulated in Figure 6.37):

- The conditions of the borehole that may influence the test results (i.e., the internal diameter, variation in diameter and equivalent radial stiffness);
- The masses of components that contribute to energy calculations, the loading mass and the borehole collar mass (i.e., reinforcement system plus the simulated borehole);
- The length of the collar pipe used to simulate the discontinuity;
- The impact velocity of the beam onto the buffers;
- Nominal energy input; and
- The presence of a load transfer ring (which defines whether the loading mass is integrated with the outside of the borehole).

Test Specifications			
Equivalent Radial Stiffness	Loading Mass	Collar Borehole Mass	Toe Borehole Mass
38 GPa	2994 kg	84 kg	149 kg
Impact Velocity	Input Energy	Load Transfer Ring	Drop Number
6.1 m/s	55.9 kJ	Yes (Integrated)	First Drop

FIGURE 6.37   Example of a dynamic test specification.

Figure 6.38 shows a typical test performance summary for a reinforcement system. The energy summary shows the total input energy and the energy dissipated by the loading mass plus collar pipe mass, buffers, beam and reinforcement system.

The energy summary is used to determine the difference between the input energy and the energy dissipated by the buffers and reinforcement system. The peak energy dissipated by the reinforcement system is reported in a summary table. The dynamic force-displacement response is calculated at the simulated discontinuity. This can be quite different from the anchor and collar load cell response due to load transfer within the reinforcement system. The calculated peak force and displacement from this curve are also reported. The loading mass acceleration-time chart displays the average of the filtered responses of the loading mass as the reinforcement system responds to bring the mass to rest from the initial impact velocity. The peak deceleration is also reported.

The energy dissipation capacity of a reinforcement system can be estimated if one of the components of the system fails. If the system does not fail, then the test indicates that the energy dissipation capacity is greater than that calculated.

6.8.4   ANALYSIS OF DYNAMIC TEST RESULTS

The results from the WASM Dynamic Test Facility have been interpreted, in detail, by Player (2012), De Zoyza (2015) and Villaescusa et al. (2015). The resulting energy dissipation from laboratory

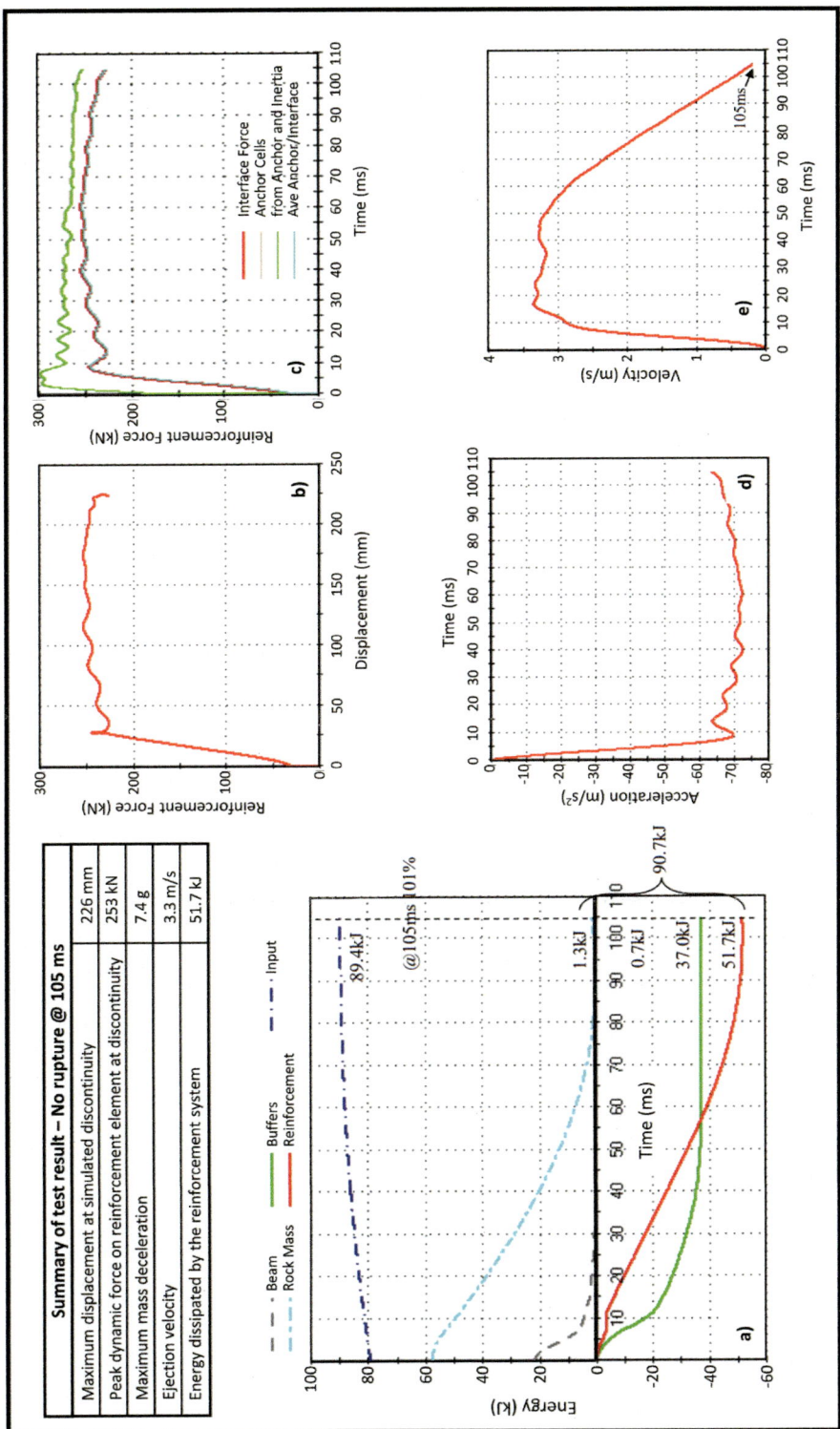

FIGURE 6.38 Example of reinforcement system performance response: (a) energy summary, (b) reinforcement force with displacement, (c) reinforcement force with time, (d) rock mass acceleration with time and (e) reinforcement relative velocity with time.

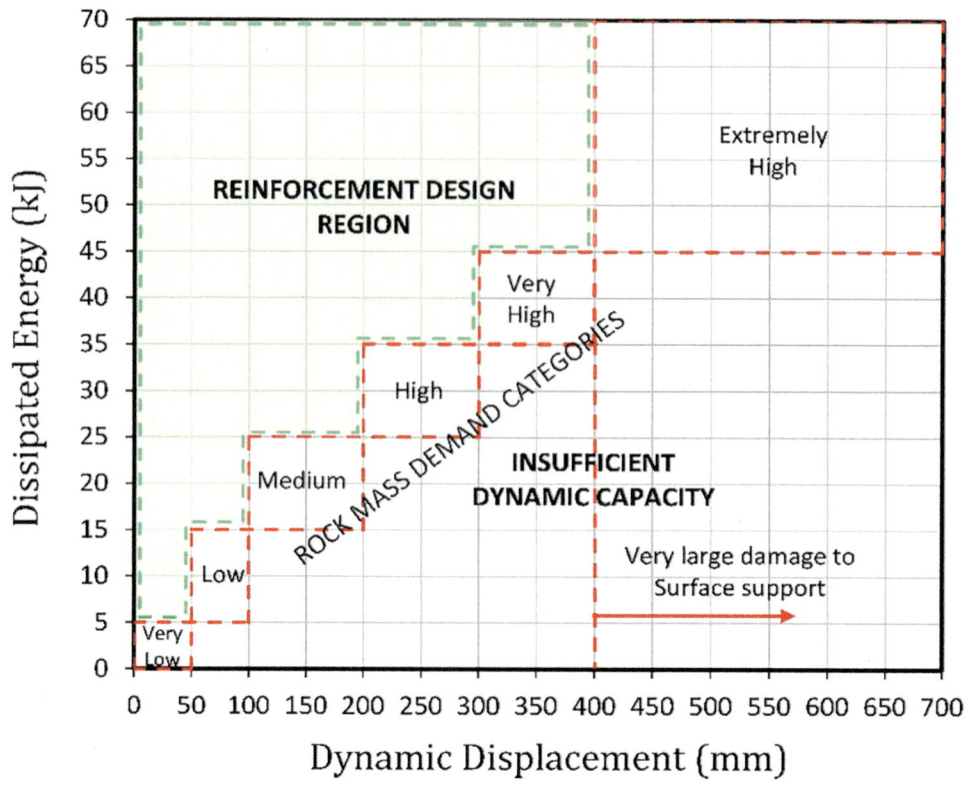

FIGURE 6.39   Regions of dynamic energy dissipation versus displacement at failure.

FIGURE 6.40   Example configuration used for dynamic testing of different types of mesh.

testing is compared with the potential rock mass demand likely to be imposed upon an excavation. Villaescusa (2014) defined rock mass demand in terms of ranges of allowable displacement and energy (shown previously in Table 5.1). For each rock mass demand category, the corresponding ranges of displacement and energy can be used to define a region, shown as a box labelled as very low, low, medium, high, very high and extremely high energy rock mass demand. For each region, the acceptable reinforcement elements should maintain displacement compatibility while providing higher energy dissipation capacity than the required demand. That is, for each demand region, the recommended reinforcement should plot within the design region shown in Figure 6.39. This plot can be used to compare the different reinforcement systems and ground support scheme responses illustrated in Chapters 7, 8 and 11.

6.8.5   DYNAMIC TESTING OF MESH AND SHOTCRETE LAYERS

In order to be able to dynamically test and compare the load-displacement capacity of surface support systems, such as mesh (welded wire sheet mesh and square- or diamond-shaped chain-link mesh), shotcrete panels and membranes, require samples of each to be tested under the exact same boundary conditions. For mesh and shotcrete panel testing, a frame to restrain the support layers is bolted to the drop beam, as shown in Figure 6.40. Prior to testing, the layers are held in place using bolts, shackles and eye bolts (Morton, 2009). These test boundary conditions attempt to simulate the continuation of the material beyond the limited sample boundary.

The basic principles for dynamic testing of reinforcement systems also apply to the dynamic testing of support systems. The drop beam is arrested by the buffers, and the momentum of the loading mass is restrained by the mesh or shotcrete elements. Some example results will be shown in Chapters 9, 10 and 11.

7 Energy Dissipation of Rock Bolts

7.1 INTRODUCTION

An understanding of the dynamic energy dissipation capabilities of rock reinforcement systems is an essential component of the design of complete ground support schemes to maintain rock mass integrity following dynamic events. The results presented here have been determined using the WASM Dynamic Test Facility, as described in Chapter 6. The main objective of research sponsored by the MERIWA and the mining industry was to establish databases of measured dynamic responses for different types of reinforcement and support systems (Villaescusa et al., 2005; Villaescusa et al., 2010; Player, 2012).

The data discussed include testing results for most commercially available rock bolt systems, including their internal fixtures, external fixtures and face restraints. The elements were contained within a steel pipe (double embedment) and the force-displacement response at the interface between the two embedment lengths was established (see Chapter 6). This follows the established practice of assessing force-displacement curves and load transfer in quasi-static performance testing of ground support elements.

7.2 MOMENTUM TRANSFER

The momentum transfer concept was used to determine the energy balance during rock reinforcement testing (Thompson et al., 2004). It required the design of instrumentation and selection of monitoring points to provide multiple independent measurement methods of key parameters with redundancy (Player, 2012). Knowledge of the energy consumed by elements as well as the test facility impact surface allowed a calculation of the maximum input energy and the relative energy split between the simulated ejected block and the impact surface. All the laboratory test results presented here were performed using axial loading to measure the force-displacement response from which an energy dissipation was derived for each system analysed. Details of the WASM Dynamic Test Facility procedures, data analysis, engineering calculations and simulation to determine the energy balance during testing have been detailed by Villaescusa et al. (2005) and Villaescusa et al. (2010) and are not repeated here.

7.3 REINFORCEMENT SYSTEM LOAD TRANSFER

The results presented here are divided based on the modes of load transfer between the element and the rock mass as described by Windsor and Thompson (1992a). This classification was discussed in Chapter 2 and resulted in three basic classes of reinforcement systems, namely:

1. Continuously Mechanically Coupled (CMC) Systems
2. Continuously Frictionally Coupled (CFC) Systems
3. Discretely Mechanically or Frictionally Coupled (DMFC) Systems

It can be easily demonstrated that all commercial reinforcement systems fit within one of these three classes, including those systems that have yet to be developed. The load transfer and distribution of force for reinforcement systems within each of the three classes differ greatly in their responses and abilities to sustain dynamic loading. Figure 7.1 shows the conceptually expected force distributions within each class. These conceptual force distributions can be used as the basis for analysis and to identify where to instrument reinforcement systems in both the field and in the laboratory (Thompson and Windsor, 1993). These load distributions accord with the strain distributions measured by special strain profiling instruments in cement grouted boreholes (Windsor, 1995).

From previous experience based on testing and theoretical considerations, it is known that CMC systems produce a 'stiff' response, CFC systems are low strength and produce a 'soft' response and may displace excessively under moderate loads. DMFC systems are less stiff than CMC systems because, in the former, the element can extend over a decoupled element length. Decoupling means

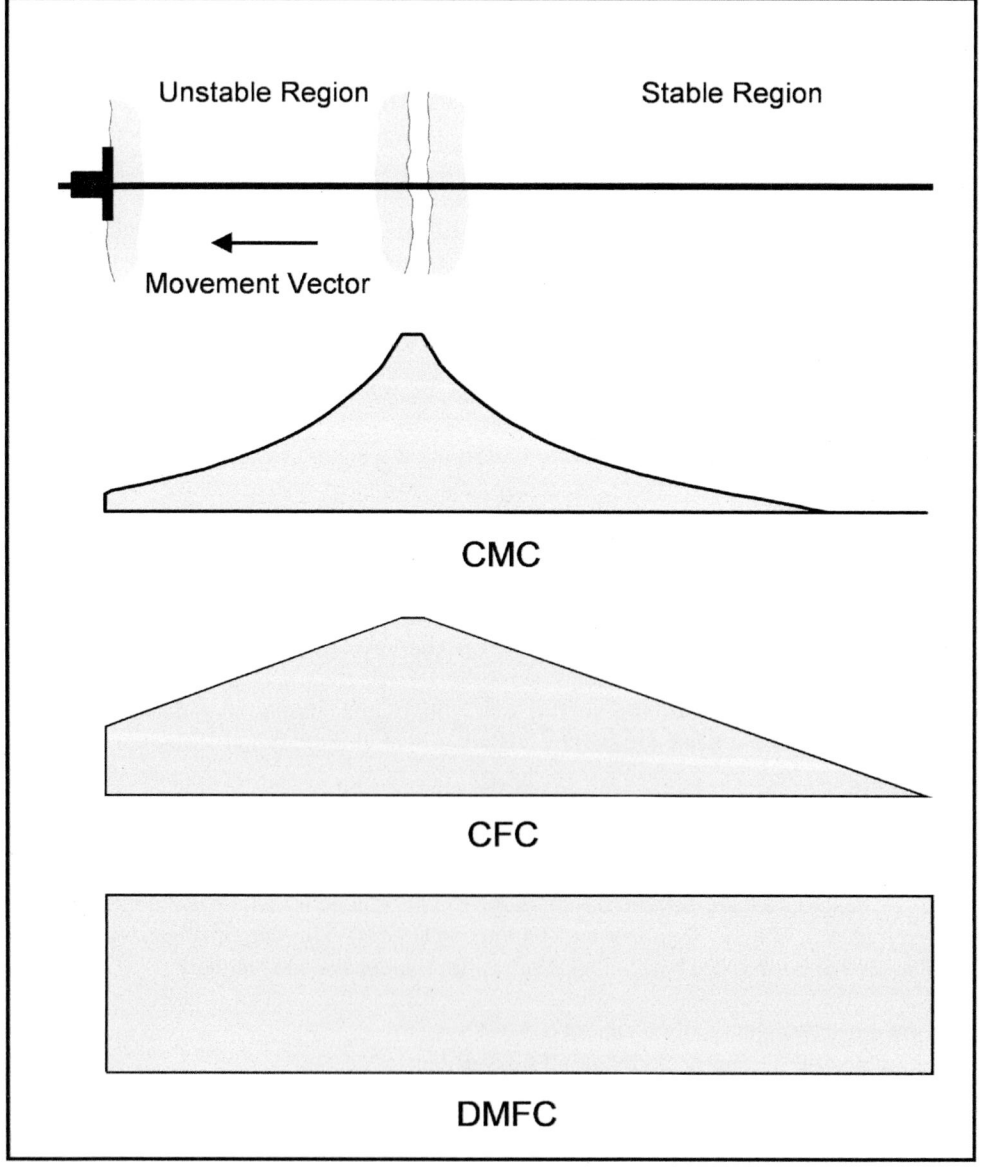

FIGURE 7.1 Schematic showing the different load force distributions within each of the three classes of reinforcement systems.

that the reinforcement system can be engineered for deformation. It is also within the DMFC class that most newer reinforcement systems have been proposed for use where high energy dissipation is required.

All these systems rely on the displacement of the element relative to the internal fixture within the toe and collar embedment regions. Accordingly, the results from the testing discussed here will be presented within one of these three classes.

7.4 CONTINUOUSLY FRICTIONALLY COUPLED

CFC elements rely on the load transfer resulting from friction between the reinforcement element and a borehole wall. The actual strength per metre of embedment length of a CFC element is determined by any radial pre-stress set up during installation. Proper testing requires the development of rough simulated boreholes of similar dimensions and texture to the holes drilled underground as well as similar installation by bolting equipment as described in Chapter 6. Static pull tests are also required prior to dynamic testing to compare capacities with conventional underground pull test results.

The results presented here include friction rock stabilizers, commonly known as Split Sets, where an oversized c-section is driven into an undersized drill hole. Results for inflatable, closed steel tubes that operate on the same principle as the Atlas Copco Swellex (Li, 1997) are also included. In addition, results for the relatively recent development of a solid bar installed within a friction stabilizer (hybrid friction bolts) are also presented.

7.4.1 FRICTION ROCK STABILIZER

This type of rock bolt comprises a hollow rolled tube with a slot along its entire length, which is driven into a drilled hole of a smaller diameter (Figure 7.2). It relies on friction between the tube and the rock to provide the reinforcement action. Although easy to install, the bolts have a very low initial bond strength per metre of embedment. A static capacity of approximately 4–5 tons per metre of embedment has been established for 46–47-mm diameter elements (Villaescusa, 2014). This may be insufficient to guarantee effective dynamic reinforcement of blocks and slabs potentially formed within the immediate backs of excavations. These non-encapsulated bolts are prone to corrosion (Hassell and Villaescusa, 2005) and failure at the bolt-plate interface.

Given that bolt performance is sensitive to the equipment and operator skills employed during installation, the simulated borehole specimens were taken to mine sites to have the friction rock stabilizers installed as close to the usual Western Australia mining practice as possible

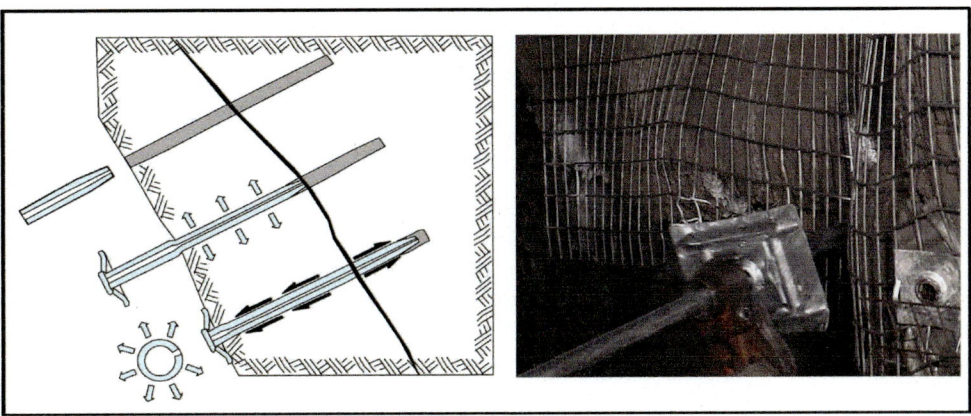

FIGURE 7.2 Schematic of the installation process for friction rock stabilizers.

(Player et al., 2009a). This was achieved by drilling an oversized hole into a tunnel wall to enable the placement of the simulated borehole assembly. A jumbo was then used to drive the bolt into the simulated rough borehole, as shown in Figure 7.3. The simulated borehole and its installed bolt were recovered and returned to the laboratory for static and dynamic testing. A more recent procedure to simulate the borehole and secure the bolt installation was described previously in Chapter 6.

FIGURE 7.3   Installation of a friction stabilizer bolt into a simulated rough borehole at a mine site.

7.4.1.1 Static Testing

A number of the friction stabilizer bolts had pull rings fitted to allow static pull testing. Static pull tests were undertaken prior to dynamic testing in order to compare the loads with conventional underground pull tests. The data set comprises bolts from different manufacturers, including galvanized and non-galvanized variants. The elements nominally had 47-mm diameter and were driven into nominal 45-mm diameter holes. The target strength during quality assurance testing for 2.4 m long (effectively 2.28 m when allowance is made for the taper at the end of the bolts plus a pull ring at the collar) friction stabilizer bolts when installed *in situ* is usually 80–120 kN. Figure 7.4 shows the results for friction stabilizers installed within the simulated boreholes and the comparison with field test data. All tests were terminated once a collar load of 120 kN was reached or when sliding was initiated. A static pull test result on a split tube bolt at the collar would be expected to show an increase in load until sliding starts. Sliding occurred at a reasonably constant load, and the values were consistent with *in situ* pull tests. This important check classified them as suitable for dynamic testing at the WASM Dynamic Test Facility.

7.4.1.2 Dynamic Testing

Tests were undertaken on 19 friction rock stabilizer samples installed in simulated rough boreholes for a total of 36 dynamic impacts. Observations show increased variability from the initial loadings compared to the second or tertiary loadings. This could be caused by a change in the condition of the bolt, changes in the roughness of the borehole, or inconsistent response of friction stabilizers to dynamic loading. This is not unusual given that results from static pull testing (e.g., Villaescusa and Wright, 1997) commonly indicate a large scatter, as shown in Figure 7.5.

The dynamic results shown here are purposefully restricted to the initial loadings, with the results summarized such that the chosen vertical and horizontal axes scales enable comparisons with other reinforcement systems. The displacement data is restricted to 700 mm for clarity. A solid

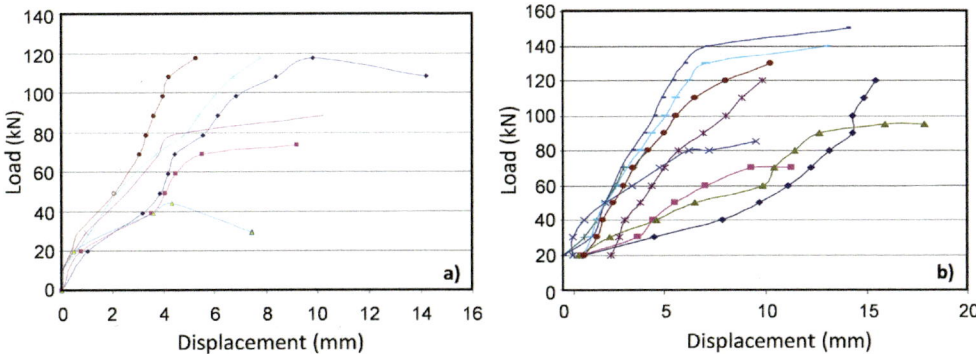

FIGURE 7.4 Comparison of static collar load versus displacement for 47-mm diameter friction stabilizers installed (a) within simulated boreholes installed using a jumbo and (b) from common *in situ* pull tests.

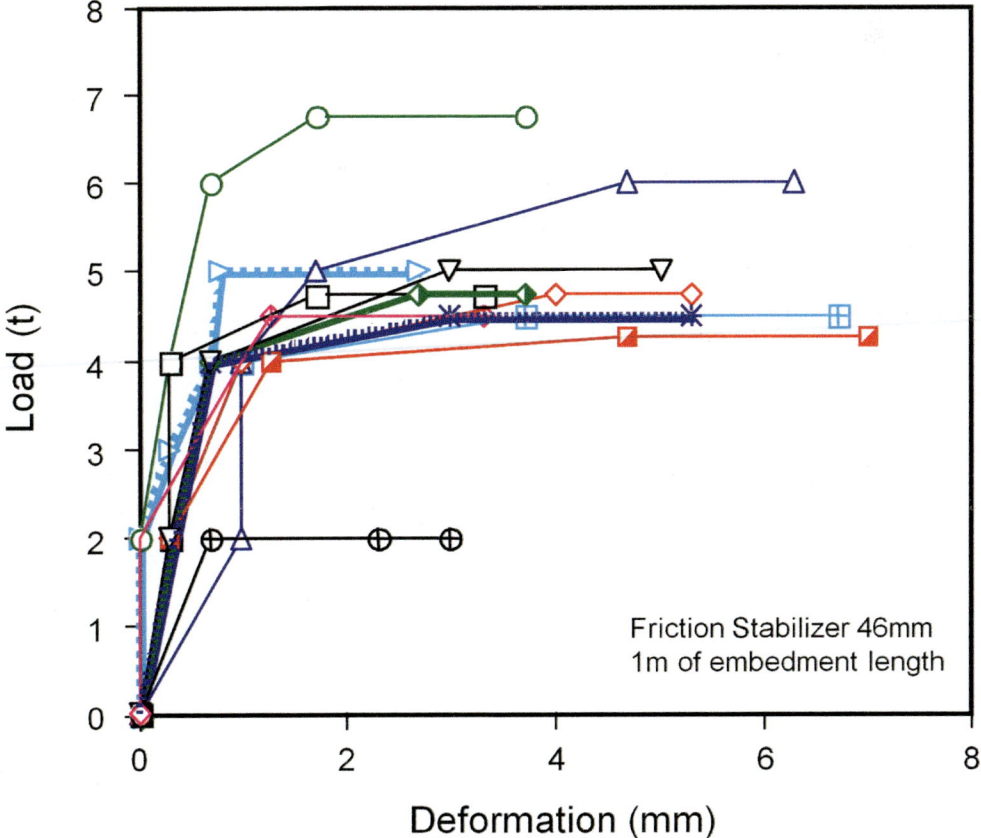

FIGURE 7.5 Results from static pull testing of friction stabilizer bolts (Villaescusa and Wright, 1997).

data point within the legend indicates that the loading mass, and hence the reinforcement system, failed (i.e., the loading mass impacted the bottom of the test pit).

Figure 7.6 shows the dynamic force and the related dissipated energy versus dynamic displacement for several friction stabilizers. The results highlight the variability in the response of these reinforcement systems. The data show a higher resistive response to lower input energies and a lower resistive response to high input energy. However, in all cases, the data plot within the insufficient

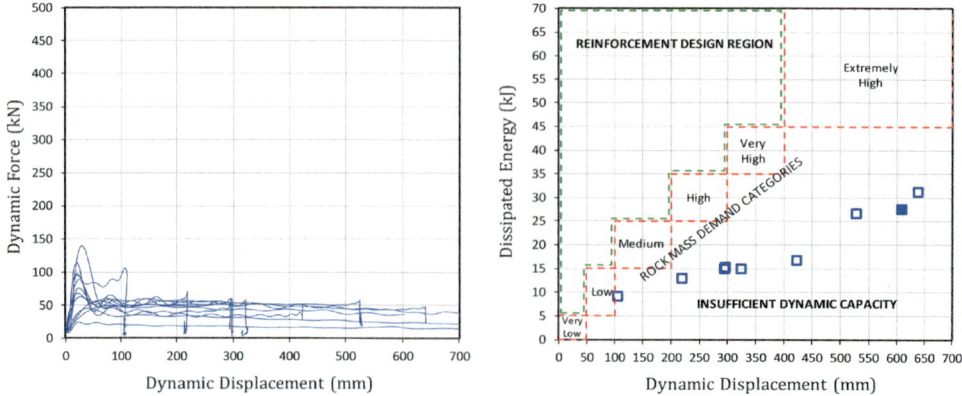

FIGURE 7.6   Dynamic force and related energy dissipation capacity for 47-mm diameter friction stabilizers (unstable collar region of 0.45 m, stable toe anchor region of 1.85 m).

dynamic capacity region. Note that a shallow depth of failure, with an average depth of instability of 0.45 m, was tested. When considering the 0.1 m tapered end of the elements, this represents a 1.85 m embedment length within the toe, anchoring region.

The 47-mm diameter friction stabilizers indicate poor dynamic performance with excessive displacements associated with sliding. In some cases, the bolt detached completely from the toe anchoring region, with little evidence of the loading mass being arrested. Figure 7.7 shows results in which the input energy significantly exceeded the capacity of the bolt.

FIGURE 7.7   Examples of complete slip of a friction rock stabilizer from within the toe anchoring region for (a) laboratory testing result and (b) rock fall following a seismic event.

Due to the known low capacity of the bolts, the input energy and the collar anchor length were reduced to simulate shallow failures, such as those occurring in spalling. Despite this, failure by a slip of the element from within the long toe anchor region was observed. Figure 7.8 shows an example wherein the loading mass was arrested after nearly 500 mm of pull out from the upper embedment. This amount of displacement will be unacceptable in some mining situations.

7.4.2 EXPANDED TUBE BOLTS

Expandable bolts also develop load transfer resulting from friction between the reinforcement element and the borehole wall. The reinforcement capacity is dependent upon the radial stress between

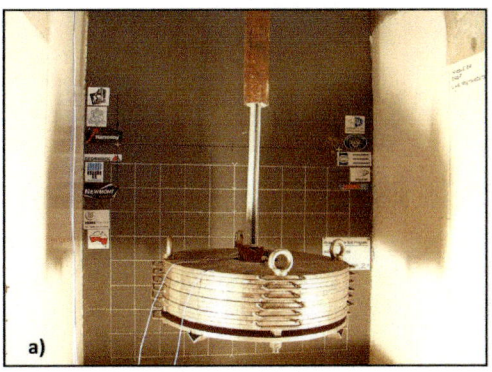
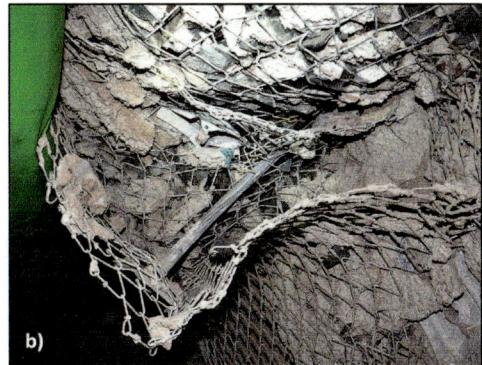

FIGURE 7.8 Example of excessive slip of friction rock stabilizer from within anchoring regions for (a) laboratory testing and (b) following a seismic event.

a fully inflated element and a borehole wall following expansion. Most theoretical equations for pull-out resistance assume a full circular-shaped contact for inflated elements. However, this may not be achieved in practice due to the inability to fully expand the 'tongue' of the bolt, and thus, only partial contact is achieved with the borehole wall along the inflated portion of the element (Figure 7.9).

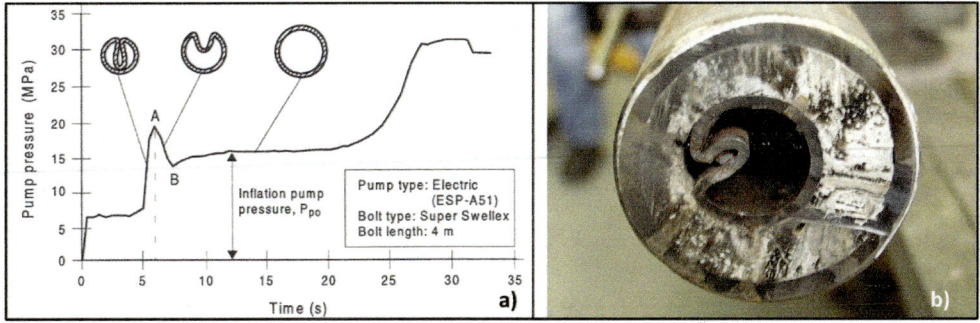

FIGURE 7.9 (a) Pump pressure versus time when a 4-m long Super Swellex rock bolt is expanded in free air (Li, 1997) and (b) example of a partially inflated Swellex bolt within a simulated borehole.

The inflatable elements are simply installed by pushing them into a rough simulated borehole and then attaching an inflation nozzle. The pump inflates to a set pressure of 280–300 bar and holds for up to 15 seconds. The results reported here were from testing two types of tubular elements with a 34.5–37.5 mm diameter prior to inflation. The bolts were fitted with a 38.3–40.6 mm diameter and a 67.5–70 mm long ferrule with a 47.2–49.8 mm lip on which load transfer from the plate can develop (Figure 7.10).

The bolts were installed through 5.38 mm thick, 150-mm square plates with a 17-mm high central dome. An elliptical hole (47.5 mm major and 38.5 mm minor axes) enabled easy installation through the plate. Drilling of the simulated boreholes in very high strength grout prior to bolt installation was done as described in Chapter 6. Typical drilling times ranged from 5 to 6.5 minutes. A 51.8-mm bit was used for drilling the simulated holes for inflatable bolt installation. After drilling the simulated boreholes, the drill bit was measured with little, if any, wear detected.

7.4.2.1 Static Testing

Installed inflatable elements within simulated boreholes can be cut transversely into 1-m long pieces, which allows examination of the toe region during static pull testing (Player et al., 2009a).

FIGURE 7.10   Detailed view and measurements of a typical expandable bolt.

The bolts were tested using a standard 1 m embedment length in order to compare with existing data from other continuously frictionally coupled elements. The collar of the bolt is pulled with readings taken of collar load, ram extension and toe displacement. The 200 mm of collar-free length allows elastic stretch, followed by sliding of the bolt. The main objective is to evaluate the effectiveness of the simulated borehole and the static load transfer development once sliding initiates, as shown in Figure 7.11.

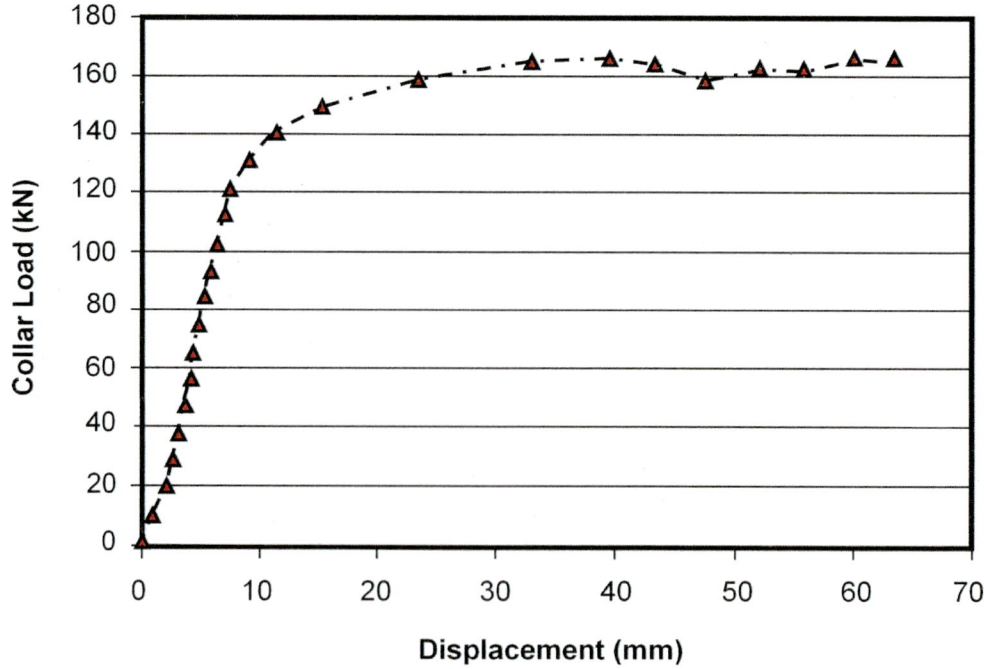

FIGURE 7.11   An example pull test result for an inflatable bolt with a one-metre length of embedment.

7.4.2.2 Dynamic Testing

The inflatable bolt results presented here were conducted over several test programs. Figure 7.12 shows several results for an average depth of failure (viz. unstable collar region) set to 0.5 m and a toe (viz. stable anchor region) length set to 1.9 m. The results show that the resistive force slowly decreases with increasing displacement, and the dissipated energy does not strictly fall within the reinforcement design region that has been proposed here. The individual results show less variability and higher capacity than a similar loading arrangement for the friction stabilizer data shown in Figure 7.6. It is worth noting the advantage of the standard axes scales in both Figures 7.6 and 7.12 and in subsequent reporting.

Figure 7.13 shows a pair of results from the same testing program for inflatable bolts dynamically loaded but in this case with a depth of failure of 1 m, which simulates blocky rock mass failure. The results show that the bolts failed catastrophically with a steady decrease of resistive force resulting in the loading mass impacting the bottom of the test pit.

This simulated blocky mass failure (i.e., 1 m unstable collar region but now with an anchor region of 3 m) was repeated using testing devices from another country and from two different manufacturers. The results are shown in Figure 7.14, where higher initial resistive forces were determined.

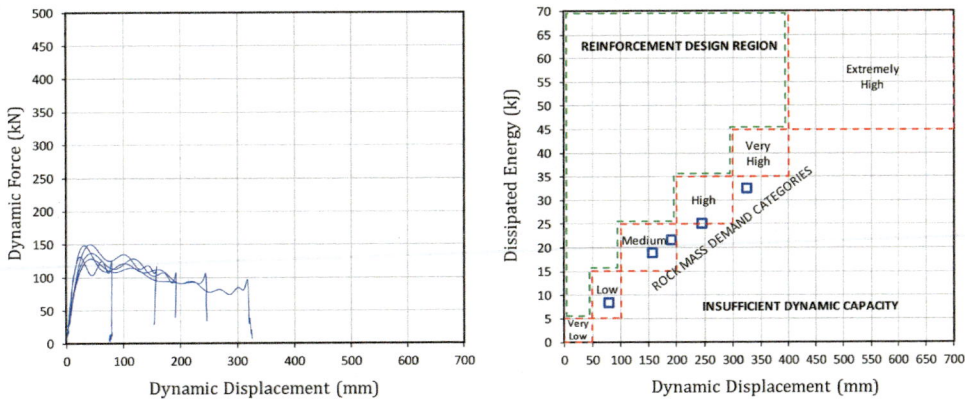

FIGURE 7.12 Dynamic force and related energy dissipation capacity for five 50-mm diameter inflatable bolts (unstable collar region of 0.5 m, stable toe anchor region of 1.9 m).

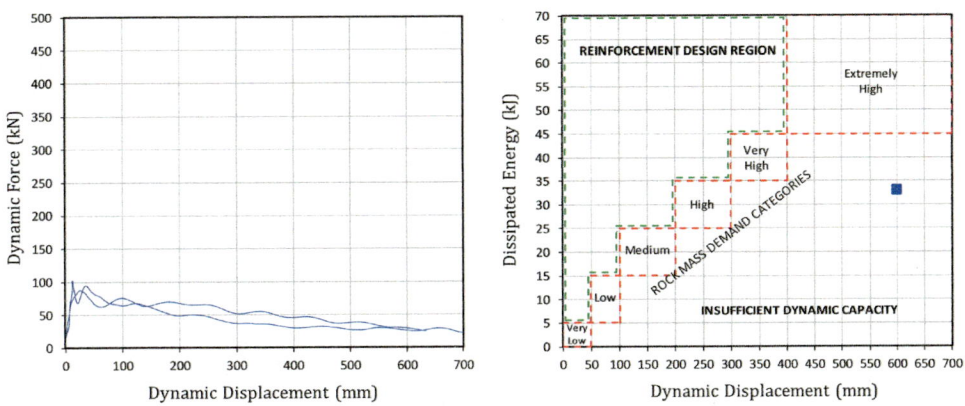

FIGURE 7.13 Dynamic force and related energy dissipation capacity for a pair of 50-mm diameter inflatable bolts (unstable collar region of 1 m, stable toe anchor region of 1.4 m).

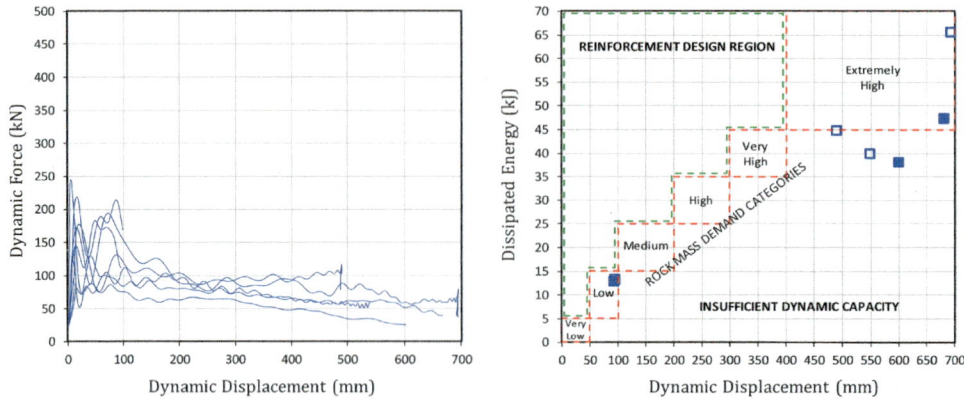

FIGURE 7.14 Dynamic force and related energy dissipation capacity for seven 50-mm diameter inflatable bolts (unstable collar region of 1 m, stable toe anchor region of 3 m).

However, the sliding friction was very similar to previously tested inflatable elements and, again, the resistive forces reduced with increased displacement.

An important requirement of dynamic testing is the assessment of the complete reinforcement system response. For CFC load transfer, shallow depths of instability are likely to significantly load the element-plate interface, as suggested by Figure 7.1. For inflatable bolts, which may 'lock-in' within the stable toe anchor region, the loading of the plate may be even more significant.

Consequently, laboratory testing to simulate short, unstable collar lengths is required to quantify this particular mechanism. The results shown here are for 2 m long inflatable elements in which the engaged unstable length was set at approximately 0.5 m. Inspection after testing showed that the 1.4 m effective toe anchoring length remained in place following the ejection of the unstable collar region. Figure 7.15 shows an example of plate inversion at an energy dissipation of 7.9 kJ. The test was considered stable but with plate collapse, which would leave the system very susceptible to any additional dynamic loading.

FIGURE 7.15 WASM Dynamic Test Facility test set-up: (a) beam and loading weight prior to testing; (b) displacement across simulated discontinuity; (c) bolt and plate in full contact; (d) stable, but damaged (inverted) plate.

For similar shallow depth of failure mechanisms, but in cases where the bolt is not installed exactly perpendicular to a tunnel wall, the plate can bear on and load the ferrule lip at an angle. Figure 7.16 shows an example of violent failure at a low energy dissipation of 1.7 kJ.

FIGURE 7.16   Results following dynamic testing showing (a) violent failure and ejected mass and (b) bolt shearing within an unstable collar and severe damage to the ferrule lip.

Laboratory testing has been used to confirm that inflatable bolts with toe anchoring regions exceeding 1.5 m and tested with shallow depths of failure show movement and instability at the plate-bolt interface where plate inversion and rupture can occur. Figure 7.17 shows a field observation in which a violent failure occurred by loading an inflatable bolt plate by a short unstable depth of failure following a seismic event. A replication of this failure in the laboratory was duplicated with a simulated discontinuity at 250 mm, giving an effective collar length of 180 mm and with a 1.7 m engaged toe anchor length. Following the testing, the anchor length remained in place, but the collar pulled off, similar to that observed in the field. The collar instability involved a sequential failure by flattening, inverting, and then punching through the plate. Analysis of the data indicates a 0.8 kJ dissipated energy capacity at the plate, which is significantly lower than the element capacity.

FIGURE 7.17   Example of inflatable bolt plate failure associated with a shallow and violent depth of failure: (a) field observation, (b) the element tested in the laboratory element and (c) the resulting inverted plate.

Figure 7.18 shows the combined data for the inflatable bolts tested at the WASM Dynamic Test Facility. The results pertain to the testing of products from different manufacturers using a wide range of input energy. All show that the dissipated energy is highly correlated to the dynamic displacement at the simulated discontinuity. However, the inflatable bolts do not plot into the rock reinforcement design region proposed here, where, ideally, the capacity is required to exceed the rock mass demand region by at least 10 kJ.

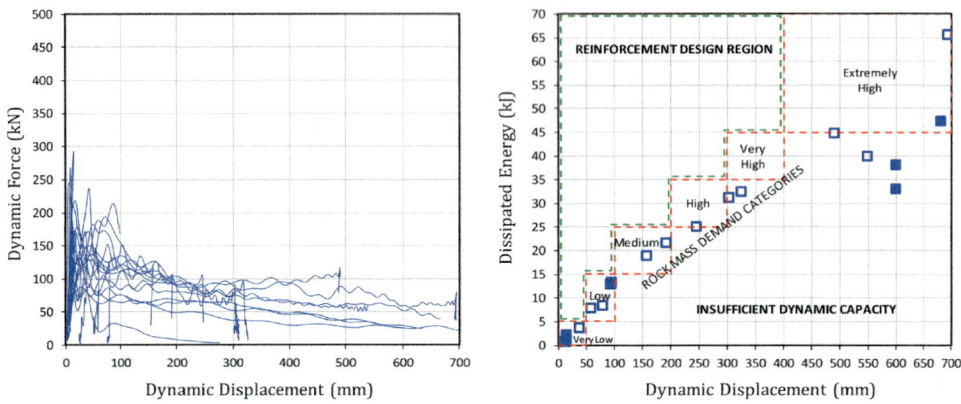

FIGURE 7.18   Dynamic force and related energy dissipation capacity for a suite of 50-mm diameter inflatable bolts (unstable collar region of 0.25–1.8 m, stable toe anchor region of 1.4–3.0 m).

7.4.3   Hybrid Point Anchored Bar and Split Tube Bolts

A hybrid reinforcement system comprises a point-anchored element that is installed inside a split tube arrangement (e.g., in this case, a friction stabilizer). The point anchor is an expansion shell that can be set either inside the toe region (Figure 7.19) or can be set within the rock borehole immediately in front of the end of the split tube. The plate is installed as part of the point anchored system, and the nut is used to provide torque and set the expansion shell. Like a friction stabilizer, hybrid reinforcement systems can be installed mechanically using a conventional mining jumbo. A number of commercially available devices exist and have been published in the literature (e.g., Carlton et al., 2013; Darlington et al., 2019).

7.4.3.1   Borehole Simulation and Installation

The borehole simulation and jumbo installation employed to produce the samples and the test results presented here were undertaken as described in Chapter 6. The simulation of a high-strength grout followed the established procedures used with other CFC elements, such as inflatable bolts, that achieve very high load transfer per metre of embedment. The concrete for the simulated boreholes is prepared using a GP2000A mixer using a non-shrink Fe-reinforced construction grout (Masterflow 4600 from BASF) at the recommended WCR of 0.12. In addition, a basalt 20% passing 4-mm chip aggregate is used such that a very high compressive strength is achieved. When constructing the boreholes, at least two samples are taken from each batch and prepared for unconfined compression testing at 7–21 days. The results are given in Table 7.1. One sample is cured at an elevated temperature and high humidity curing chamber simulating underground conditions, and the second sample is simply sealed in a bucket. The simulated boreholes are also sealed within plastic bags to allow complete curing of the cement grout.

Figure 7.20 show the details of borehole drilling and reinforcement element installation using a jumbo at a mine in Western Australia. A set-up has been conceptualized and implemented using a

FIGURE 7.19 Long section view of the MDX hybrid reinforcement system.

TABLE 7.1

UCS Results for the High-Strength Concrete After a 12-Day Curing Period.

Batch Number	Curing Chamber sample (MPa)	Curing Chamber sample (MPa)
16	97	81
17	108	89
18	103	89
19	107	78
20	92	88
21	108	72
Average	103	83

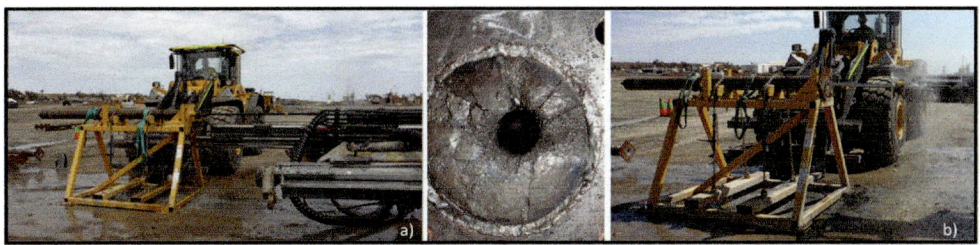

FIGURE 7.20 Details of (a) borehole drilling and (b) hybrid bolt installation using a mining jumbo.

specially designed drilling platform to minimize the drift of 2.4–3.0 m long holes, which are drilled within the heavily confined (38 GPa radial confinement and UCS > 90 MPa strength at the time of drilling) concrete. A 44.5-mm diameter bit is used, and the typical drilling times range from 5 to 6.5 minutes. After drilling each hole, the drill bit is measured to ensure no wear is experienced during the drilling of any of the simulated boreholes. In most cases, the boreholes show little deviation within the nominal 150-mm internal diameter, 7 mm thick steel pipes.

Figure 7.21 shows a typical borehole diameter profile measured after drilling. These variations with borehole depth are similar to borehole diameter profiles taken from boreholes drilled in various rock types at several operating mines in which the hybrid reinforcement systems have been implemented.

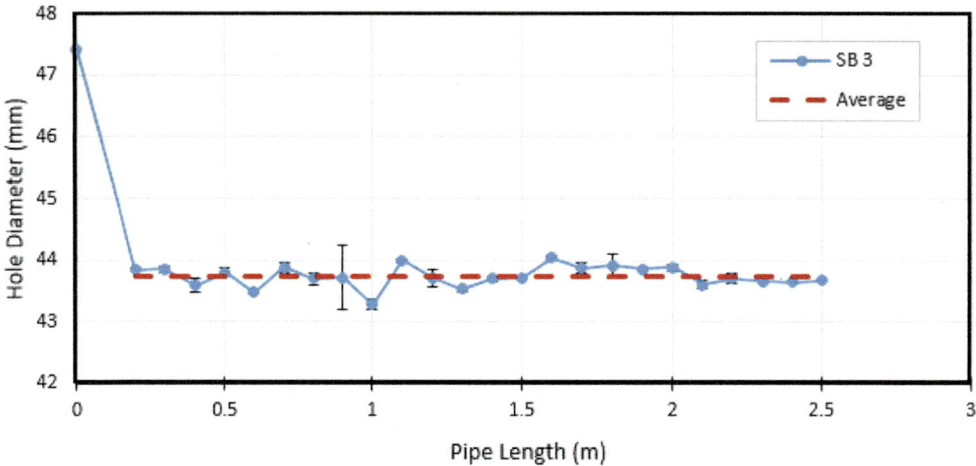

FIGURE 7.21   Typical results following simulated borehole drilling.

One of the limitations of this reinforcement system in mining is the amount of torque that can be delivered by a mining jumbo, which typically ranges from 200 to 400 Nm. In order to set up an anchor, the expansion shell needs to expand either inside the split tube or against the borehole wall partially in front of the bolt. This is very difficult to achieve in hard rock with UCS ranging from 100 to 250 MPa. Conceptually, the initial energy imposed on a point-anchored reinforcement system must be dissipated by plate loading followed by anchor engagement at the rock frictional device interface. If the anchor holds, the element must elongate between the points of contact, i.e., the plate and the anchor point, which, in theory, must be stronger than the bar selected (ASTM F-432–95). Consequently, any energy dissipation results that are reported for the complete system based solely on pulling the element-end anchor component (i.e., Carlton et al., 2013; Darlington et al., 2019) cannot be correct. A point-anchored reinforcement system requires the plate to load first before the load can be subsequently transferred to the anchor region via the steel element (Thompson and Windsor, 1993). Therefore, it is not recommended that the energy response of a component (bar-end anchor) be assigned to the entire system, especially when the energy balance of the testing arrangement is not reported (e.g., Carlton et al., 2013; Darlington et al., 2019). Mechanically, the plate is a key component of a point-anchored reinforcement system, and its behaviour must be included in every pull testing experiment.

7.4.3.2 Dynamic Testing

Figure 7.22 shows the details of the Kinloc system, which is one of the several commercially available hybrid reinforcement products. Two types of systems have been tested, namely ringed and ringless (i.e., with and without a ring collar fitted to the split tubes) and a driving ring. The Kinloc

FIGURE 7.22 Details of the Kinloc hybrid reinforcement system.

locates the anchor point immediately in front of the toe end of the friction stabilizer to allow full engagement with the rock mass, independently of the friction stabilizer.

The results show that both systems pulled out at impact loadings greater than 26 kJ. For lower input energy, the ringless reinforcement systems survived after experiencing large deformation at the simulated discontinuity (Figure 7.23). Examples of system performance for the first impact loading are tabulated in Table 7.2. All test results had an acceptable energy balance, indicating a low error in the determined response. The processed curves show a consistent initial loading profile, as the point anchor locks in until slippage occurs. The responses from the Kinloc reinforcement system are best understood by comparing the anchor cell record shown in Figure 7.24. The degree of similarity is low, demonstrating the importance of proper engagement of the point anchor for initial load transfer followed by sliding of the friction bolt once the anchor has been mobilized.

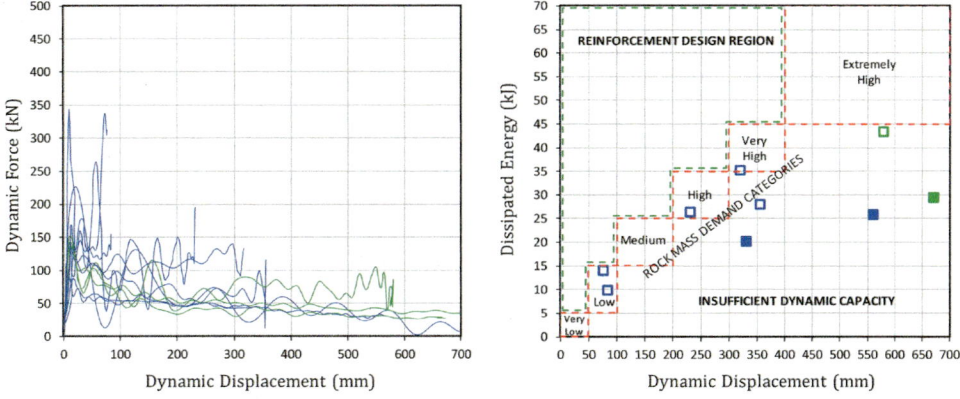

FIGURE 7.23 Dynamic force and related energy dissipation capacity for 47-mm diameter hybrid.

TABLE 7.2

Tabulated Results for a Kinloc Dynamic Testing Program.

Kinloc Type	Ringless	Ringless	Ringless	Ringed	Ringed	Ringed
Input Energy (kJ)	25.4	20.4	22.8	29	24.2	19.2
Loading Mass (kg)	1510	1510	1510	1510	1510	1510
Impact Velocity (m/s)	5.8	5.2	5.5	6.2	5.7	5.0
Displacement at Discontinuity (mm)	837	212	356	726	580	672
Peak Dynamic Force (kN)	227	198	168	137	152	142
Ejection Velocity (m/s)	4.3	2.4	1.3	4.7	3.6	3.6
Mass Deceleration (g)	13.5	11.7	9.8	7.8	8.7	8.1
Dissipated Energy (kJ)	41	14.5	26.8	37.9	43.3	29.5
Pulled-out (P) Stable (S)	P	S	S	P	P	P
Test Time (s)	270	120.7	159.8	174	253	180
Energy Balance (%)	104	85	93	108	119	111

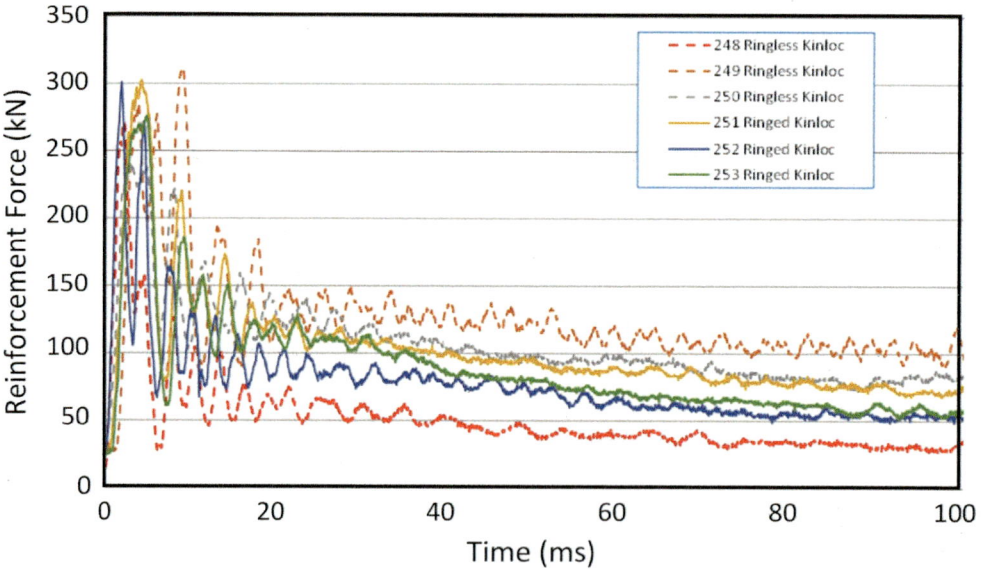

FIGURE 7.24 A comparison of Kinloc reinforcement system anchor cell records with force curves synchronized to impact time.

Following initial test findings at WASM, the Kinloc bolt was modified by the bolt manufacturer, with the subsequent dynamic test results shown in Figure 7.25. The results show a large improvement with the energy dissipation in some cases exceeding 35 kJ of energy dissipation and displacement limited to less than 200 mm. Figure 7.26 shows that the simulated borehole and the jumbo installation set-up can engage both the anchor and plate, allowing the test to utilize the full capacity of the bar to rupture.

The improvement, due to proper anchor engagement, for the new modified Kinloc design can clearly be seen in Figure 7.27, where the individual tests from all testing campaigns are compared.

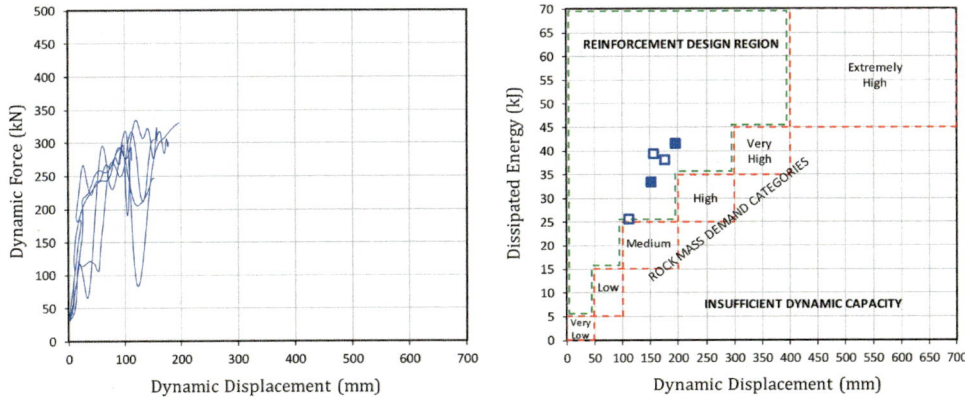

FIGURE 7.25 Dynamic force and related energy dissipation capacity for 47-mm diameter hybrid Kinloc bolts (unstable collar region of 1.0 m, stable toe anchor region of 1.4 m).

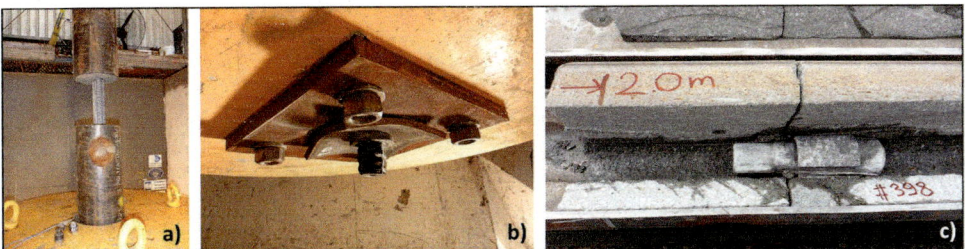

FIGURE 7.26 Example of Kinloc bolt displacement at (a) the simulated discontinuity, (b) plate deformation and (c) anchor engagement with no slippage.

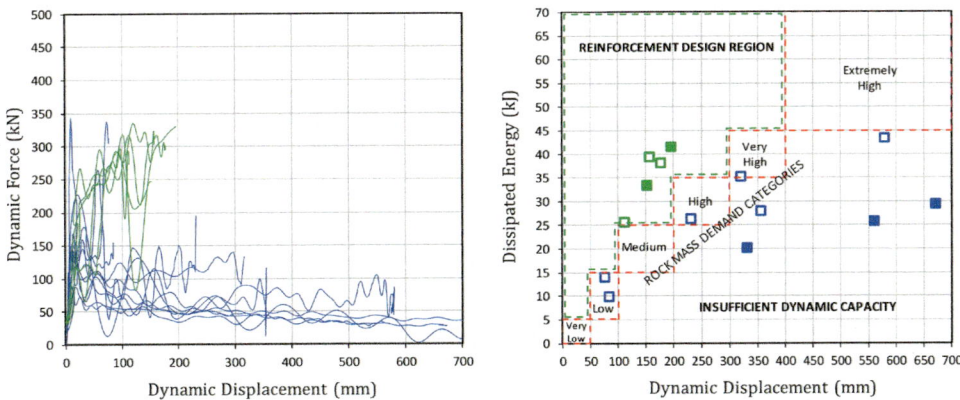

FIGURE 7.27 Dynamic force and related energy dissipation capacity for 47-mm diameter hybrid Kinloc bolts (new modified design in green, unstable collar region of 0.5–1.0 m, stable toe anchor region of 1.4–1.9 m).

Figure 7.28 shows another commercially available hybrid reinforcement product (called the Hybrid Bolt) that was installed at a mine site and dynamically tested following WASM procedures described earlier. Figure 7.29 shows the performance for testing undertaken with short unstable collar zones simulating near-surface instabilities (0.3–0.5 m). Although the results do not plot within the preferred reinforcement design region, the performance is good enough to suggest that

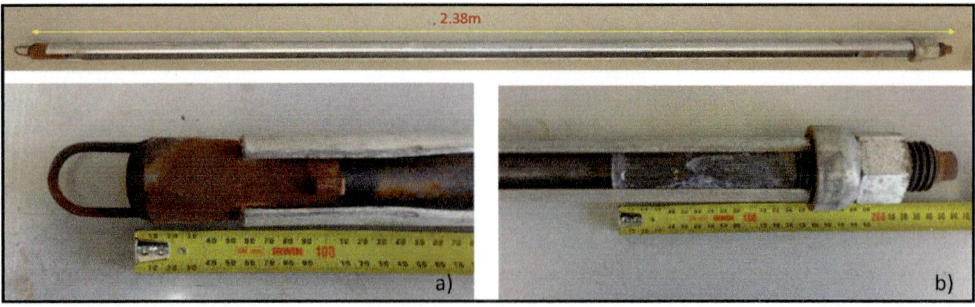

FIGURE 7.28 Long section views of a Hybrid Bolt reinforcement system: (a) expansion shell within toe end region and (b) collar details.

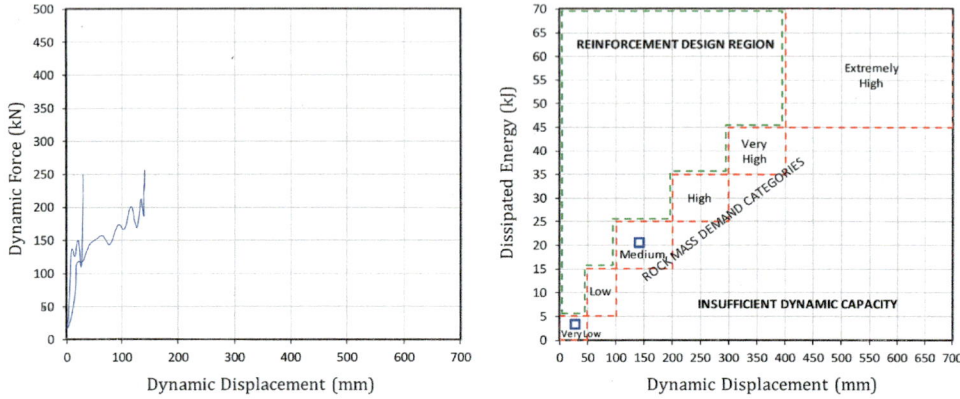

FIGURE 7.29 Dynamic force and related energy dissipation capacity for two 47-mm diameter Hybrid Bolt reinforcement systems (unstable collar region of 0.5 m, stable toe anchor region of 1.9 m).

if the bolts are used in conjunction with weld mesh, they could stabilize near surface instabilities (i.e., very low to low demands).

Figure 7.30 show the test results for collar instabilities with a 1-m depth of failure. Here, the reinforcement system was selectively loaded with a range of loads in order to understand this failure mechanism. The results show that very large displacements were experienced. As with earlier graphs, a solid symbol represents catastrophic failure following dynamic testing.

Other commercially available hybrid bolts include the MD and MDX reinforcement systems (Carlton et al., 2013; Darlington et el., 2019). WASM was commissioned to undertake testing of these reinforcement systems in research sponsored by a mining company. Borehole simulation and installation were undertaken following the procedures described earlier. The results are shown in Figure 7.31. Again, the responses do not plot in the reinforcement design region being proposed here.

Finally, Figure 7.32 combines the responses obtained for several commercially available 'hybrid-type' bolts tested under consistent, repeatable conditions at the WASM Dynamic Test Facility. The results showed that the initial energy imposed on the bolt is dissipated by plate loading and anchor slippage at the rock-frictional device interface. That is, the initial contact (and hence, pull-out resistance) is mainly limited to the anchor length region. If the shear strength of this mechanical interface is exceeded, the bolt appears to work only marginally better than a conventional continuously frictionally coupled device. Figure 7.33 shows *in situ* performance where large deformations are facilitated by design issues with some of these hybrid-type reinforcement systems.

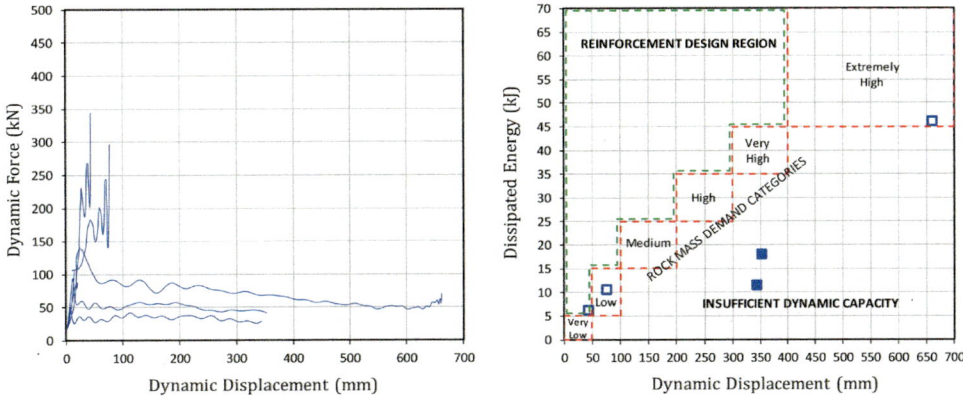

FIGURE 7.30 Dynamic force and related energy dissipation capacity for five 47-mm diameter hybrid reinforcement systems (unstable collar region of 1.0 m, stable toe anchor region of 1.4 m).

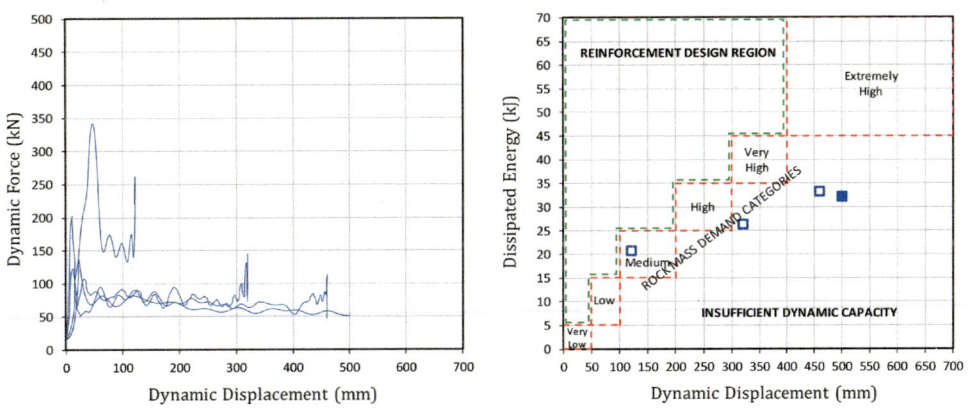

FIGURE 7.31 Dynamic force and related energy dissipation capacity for four 47-mm diameter hybrid MD and MDX bolts (unstable collar region of 1.0 m, stable toe anchor region of 1.4 m).

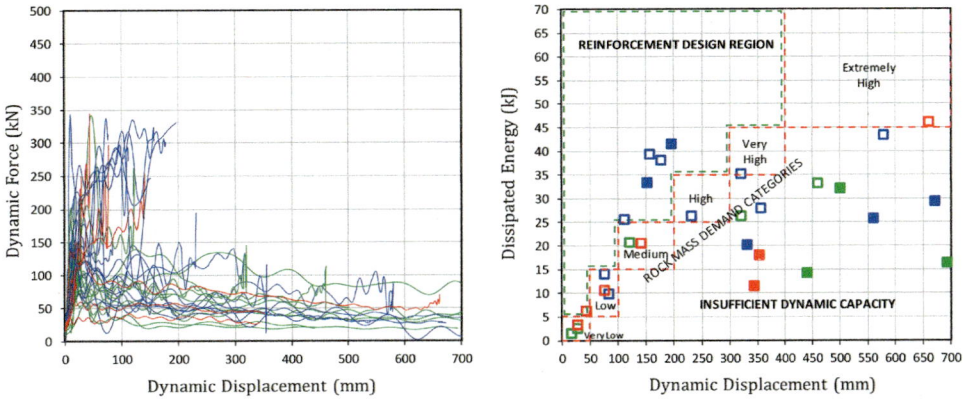

FIGURE 7.32 Dynamic force and related energy dissipation capacity for several commercially available 47-mm diameter hybrid bolts (unstable collar region of 0.5–1.0 m, stable toe anchor region of 1.4–1.9 m, where Kinloc is blue, MD & MDX is green and Hybrid is red).

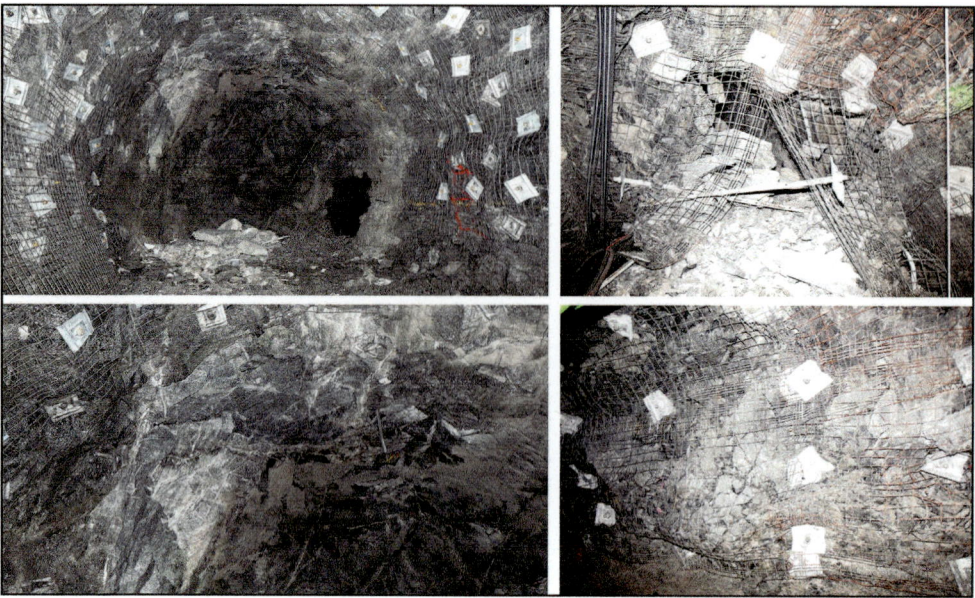

FIGURE 7.33 Examples of closely spaced hybrid bolt patterns allowing excessive deformation and eventual failure following a seismic event.

7.5 CONTINUOUSLY MECHANICALLY COUPLED

These reinforcement systems rely on a grout encapsulant that fills the annulus between an element and a borehole wall. The strength of the system is a function of the nominal capacity of the element, the grout strength and the active embedment length. The encapsulants considered here comprise both cement and resin-based grouts.

7.5.1 CEMENT-ENCAPSULATED THREADED BAR

A typical cement-encapsulated threaded bar comprises a 2–3 m long, 20–25-mm diameter corrugated bar that is grouted along its entire length. The bolts are usually manufactured with a variable cross-sectional shape to provide effective geometrical interference between the grout and the bolt surface. The geometrical interference creates a mechanical interlock that extends over the entire length of the element (Figure 7.34). The cement grouts typically use a WCR varying from 0.35 to 0.40 depending upon any additives and the pump type being used for installation.

Threaded bars have a long history of use in civil engineering projects, particularly for long ground anchors where the ability to join short straight lengths axially is achieved using thread couplers. Threaded bars are also widely used in mining applications for the control of static and dynamic loads and are often referred to as rebar or Gewi bar (a proprietary product developed by Dywidag Systems many years ago). The results presented here are for threaded bar (20–25 mm diameter) products common in the Australian and Chilean civil and mining industries (Tables 7.3 and 7.4).

It must be noted that Chilean regulation fixes the minimum property values while allowing some percentage of error in population samples. For example, the weight error allowance is 6% for one individual bar and 3% for a sample analysed.

Threaded bars, when fully encapsulated and loaded by displacement across a single discontinuity, will have load transferred as a CMC system, as summarized in Figure 7.1. The effective load transfer rate and hence system performance will be influenced by the shear strengths, strain rate capabilities

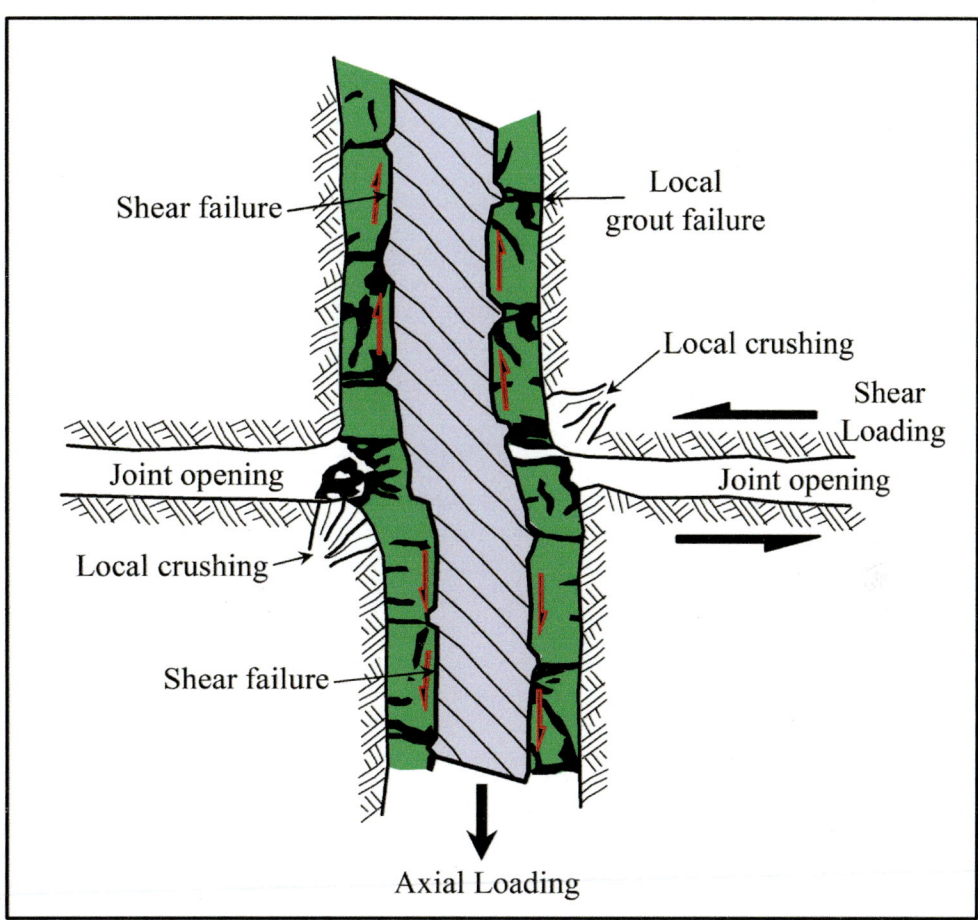

FIGURE 7.34 Cross-sectional views of a threaded bar. Failure occurs by slippage at the bolt-grout interface.

TABLE 7.3

Physical and Mechanical Properties of 20-mm Australian Threaded Bar.

Property	Minimum	Average
Core Diameter (mm)	19.3	19.5
Cross-Sectional Area (mm^2)	293	299
Yield Strength (MPa)	500	550
Yield Force (kN)	147	165
Ultimate Tensile Strength (MPa)	600	640
Tensile Force (kN)	175	191
Elongation (%)	12	21

and the interface conditions between the steel, grout and rock components. Dynamic failure occurs by rapid shearing propagating at the steel-cement grout interface (Player, 2012). This releases the steel from the grout allowing it to elongate, dissipating the energy over a longer reinforcement length.

Dynamic testing by Player (2012) has shown that energy dissipation by a fully encapsulated threaded bar requires plastic deformation of the steel bar at a simulated discontinuity. The threads

TABLE 7.4

Physical and Mechanical Properties of 22- and 25-mm Chilean Threaded Bar.

Property (Minimum)	Threaded Bar Type		
Core Diameter (mm)	22	22	25
Steel Type	A-440	A-630	A-630
Cross-Sectional Area (mm^2)	380	380	491
Yield Strength (MPa)	280	420	420
Yield Force (kN)	106	160	206
Ultimate Tensile Strength (MPa)	440	630	630
Tensile Force (kN)	167	239	309
Elongation (%)	16	8	8

of the bar set up a geometric interlock, enabling the shaft to break under critical loading conditions. The critical loading conditions are related to the rate at which the energy is consumed in the plastic deformation of the steel bar compared with the fracture growth between the steel bar and grout interface (Figure 7.35). At a sub-critical loading velocity, the plastic deformation along the shaft of the bolt and the fracturing of the grout allow a free length to develop away from the simulated discontinuity towards the toe and collar of the bolt. The fracture process fills the partial threads on the bolt with a pulverized grout. This effectively reduces the embedment length towards the toe and collar while increasing the central length of the bolt over which deformation can occur.

7.5.1.1　Australian Threaded Bar

The Australian threaded bar used in the fully encapsulated tests was hot dipped, galvanized and encapsulated in a WCR of 0.45 grout with 0.2% Methocel to control bleed and segregation in the grout. The threaded bars were pushed through a grout-filled thick wall pipe and centred within the grout column. The pipes had a 49.5 mm internal diameter and 60 mm external diameter, resulting in an equivalent radial rock stiffness modulus of 49 GPa (Hyett et al., 1992). The tested threaded bars were 2.4 m in length, and all samples were configured to have a 1-m collar unstable section. Due to variations in installation, the exposed collar length varied between 0.1 and 0.2 m, giving stable toe anchor lengths of 1.2–1.3 m. Surface hardware consisted of a 150-mm square 8 mm thick domed plate, a washer and a 32-mm long nut (T20 × 10.0 pitch, LH thread). The nut was tensioned with a torque wrench during the sample set-up in the test facility. A load cell was used between the nut and a dome ball washer when space allowed. Figure 7.36 shows the results for the fully coupled, cement-encapsulated Australian galvanized bars. A solid symbol indicates a catastrophic failure of the loading mass into the test pit. The results show that a fully coupled cement-encapsulated galvanized rebar is inadequate for high-energy dissipation tasks.

In general, the results show forces greater than the average static yield force of 165 kN. Malvar and Crawford (1998) have shown that for strain rates approximating one strain per second, a dynamic strength factor of approximately 1.3 in yield and ultimate strength capacities can be realized. That is, for a 20-mm diameter threaded bar with a nominal 550 MPa yield stress, an increase of the average yield load from 165 to 213 kN can be achieved in agreement with the dynamic yield load assessed at the WASM Dynamic Test Facility.

7.5.1.2　Chilean Threaded Bar

An important difference with a Chilean threaded bar is that a significantly lower yield stress steel grade is used compared with the threaded bar utilized within Australia. Although the Chilean product has been designated with a 22 mm diameter, it actually has the same bar area as the Australian 20-mm bar. Results from static tests undertaken on some samples are shown in Figure 7.37. An

FIGURE 7.35 Example of fractured cement grout allowing large deformation following dynamic loading of a fully encapsulated threaded bar.

important aspect of using lower yield stress steel is that when subjected to dynamic strain, it has a significantly plastic strain capacity, as shown in Figure 7.38. The results are consistent, and the majority of them plot within the design region preferred here.

Figure 7.39 shows the entire database of cement-encapsulated, fully coupled Chilean steel bars, including Grade A-440 and Grade A-630 steel types (see Table 7.4). This plot also includes results for 25-mm diameter bars, which show energy dissipation capacity exceeding 35 kJ for approximately 200 mm of displacement. Most of the results plot within the design region preferred here.

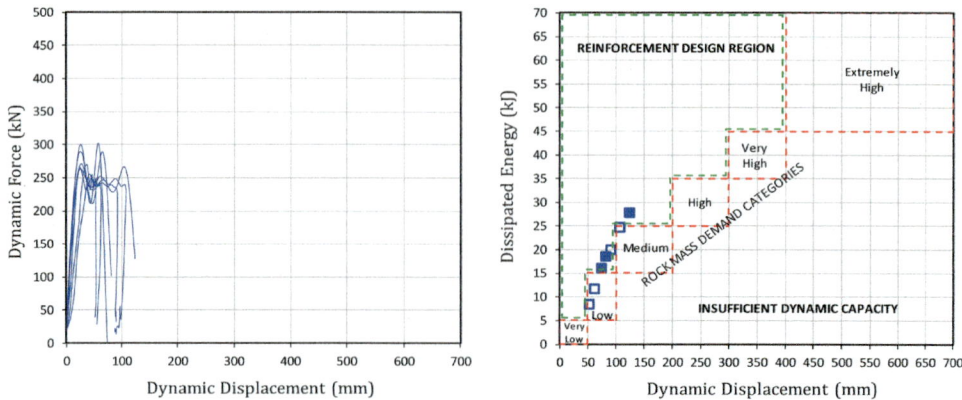

FIGURE 7.36 Dynamic force and related energy dissipation capacity for Australian 20-mm diameter, fully coupled, cement-encapsulated galvanized threaded bolts (unstable collar region of 1.0 m, stable toe anchor region of 1.2–1.3 m).

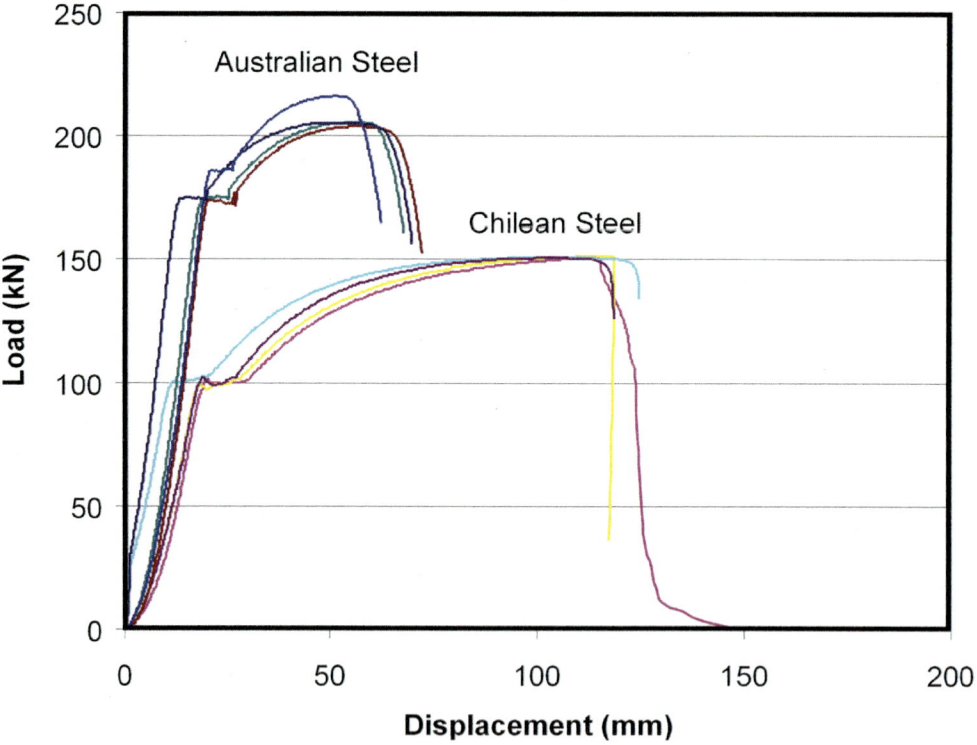

FIGURE 7.37 Comparison of static responses of Australian and Chilean 20-mm diameter threaded bars.

7.5.2 RESIN-ENCAPSULATED THREADED BAR

In recent years, fully encapsulated resin anchored bars have been implemented as an alternative to friction stabilizers for long-term reinforcement (Varden, 2005). The typical bolts being used in underground hard rock mines have been modified from the bolts used in the coal mining industry. The modifications have been necessary due to the need to drill larger hole diameters with the type of equipment used in the hard rock metalliferous mines. The modifications are mainly in the

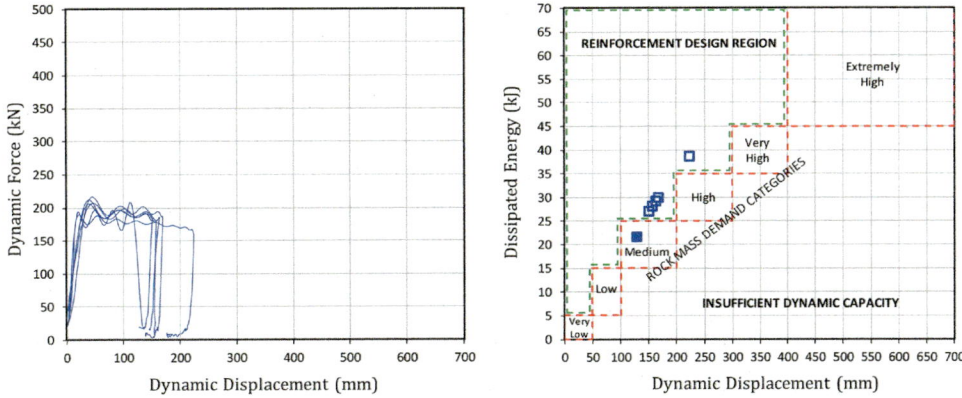

FIGURE 7.38   Dynamic force and related energy dissipation capacity for Chilean 20-mm diameter, fully coupled, cement-encapsulated threaded bolts (unstable collar region of 1.0 m, stable toe anchor region of 1.2–1.3 m).

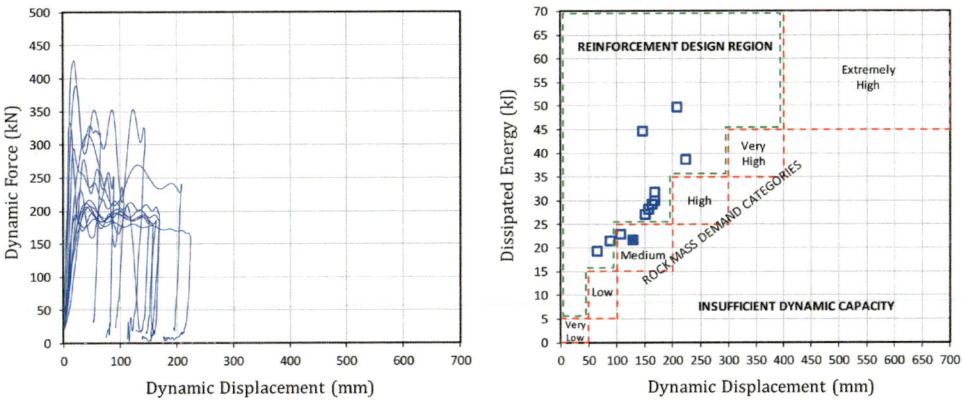

FIGURE 7.39   Dynamic force and related energy dissipation capacity for Chilean 20- and 25-mm diameter, fully coupled, cement-encapsulated threaded bolts (unstable collar region of 1.0 m, stable toe anchor region of 1.2–1.3 m).

form of paddles or the use of a mixing spiral welded onto the end section of the bolts as shown in Figure 7.40a and b, respectively. Figure 7.40a shows the anchor section for a 24-mm Posimix bolt with the spiral arrangement. The Posimix wire is 3 mm in diameter and has a length of 500 mm. The paddle width for the Secura bolt shown in Figure 7.40b is 29.2 mm, and they have been sheared into the end of the bolt for the purpose of mixing resin. Figure 7.41 shows the results for fully encapsulated Secura bolts. The results show that although the bolts mix the resin consistently, they fail at very high load and with very little displacement, resulting in very low energy dissipation capacity.

The laboratory results agree with field observations where *in situ* pull testing shows that for resin-encapsulated systems, high transfer loads can be achieved over short embedment lengths. However, systems that depend on cartridge mixing may suffer from either under-spinning or over-spinning. Under spinning results in poor mixing and low resin grout strength, often at the critical toe anchor region or at the collar of the hole. In some cases, the resin will never set. Overspinning during installation can result in the shearing of the partially cured resin. This results in the reduced bonded area and lower load transfer. In addition, gloving of the rock bolts by the plastic packaging may occur, completely eliminating load transfer along sections of the bolt axis (Mould et al., 2004;

FIGURE 7.40   Posimix and Secura bolts showing (a) spiral and (b) paddle arrangement for resin mixing.

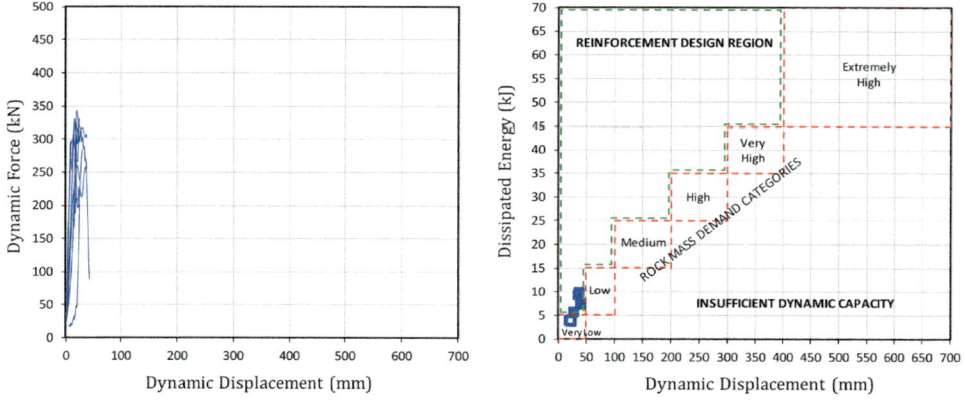

FIGURE 7.41   Dynamic force and related energy dissipation capacity for 20-mm diameter, fully coupled, resin-encapsulated Secura bolts (unstable collar region of 1.0 m, stable toe anchor region of 1.2–1.3 m).

Villaescusa et al., 2008). Figure 7.42 shows the results for fully resin encapsulated J-Tech bolt, where large variations in capacity can be attributed to inconsistent resin mixing (Figure 7.43).

The performance and ultimate capacity of a reinforcement scheme can be affected by sub-standard installation practices. However, in CMC schemes, faulty installations are difficult to detect, given that the only visible parts of an installed element are the plate, nut and a short length of the bolt indicating the orientation of installation with respect to an excavation wall. Thus, for a fully resin encapsulated threaded rebar, it is very difficult to determine the position and actual bonded length (bolt encapsulation) along the axis of the bolt. In addition, because the full steel capacity may be mobilized with very short embedment lengths of good quality resin, pull testing of exposed collar lengths within a fully encapsulated element is almost meaningless. Pull testing only provides an indication of resin effectiveness at the collar or at the first (unknown) location along a rock bolt axis where the resin is working effectively. It only provides a definite indication of poor installation in cases where the entire length of resin-encapsulated reinforcement fails well below its designed capacity.

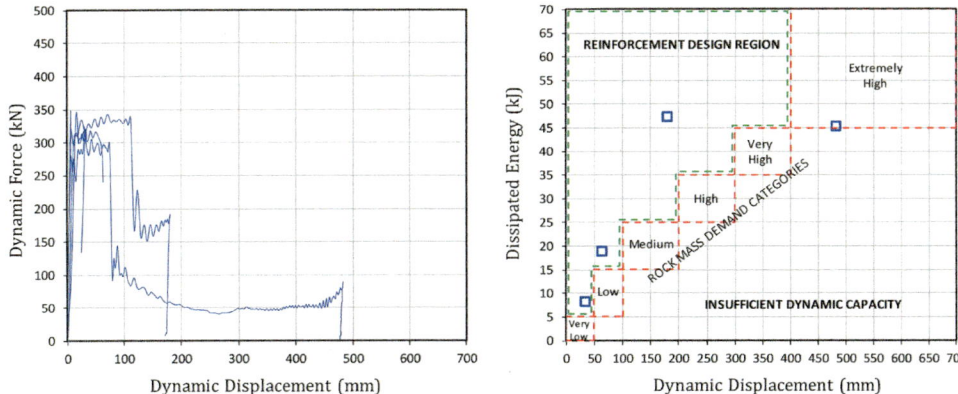

FIGURE 7.42 Dynamic force and related energy dissipation capacity for 25-mm diameter, fully coupled, resin-encapsulated J-tech bolts (unstable collar region of 1.0 m, stable toe anchor region of 1.2–1.3 m).

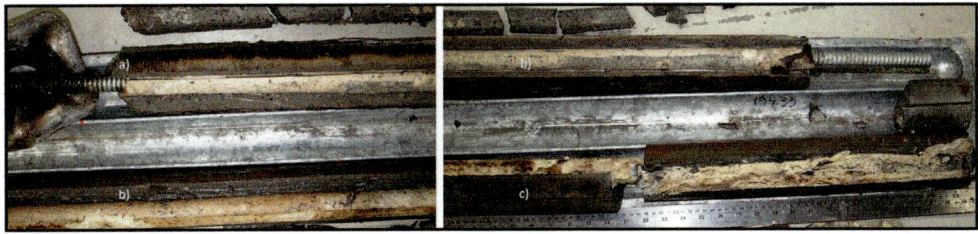

FIGURE 7.43 Dissection of J-Tech bolts showing full encapsulation with well-mixed resin at (a) collar and (b) middle region but (c) poor mixing at toe anchor region.

Examination of the entire length of a fully encapsulated rock bolt *in situ* can be achieved by overcoring of the reinforcement element (Hassell and Villaescusa, 2005; Villaescusa et al., 2008). Rock bolt overcoring not only allows the recovery of the element but also provides a clear view of the surrounding rock mass and a better understanding of the rock bolt system/rock mass interaction. It also provides a range of information, including location and frequency of geological discontinuities, overall rock mass conditions, bolt encapsulation, load transfer along the bolt axis and corrosion effects.

7.6   DISCRETELY MECHANICALLY OR FRICTIONALLY COUPLED

A DMFC device transfers load at two discrete regions, namely near the borehole collar and at an anchor region, which is located at some depth into the borehole. The length of the element between the two discrete load transfer points (collar and toe regions, respectively) is decoupled from the rock mass. Compared with a fully coupled device, the load transfer is limited to relatively short regions, and the collar becomes critical for the overall performance of the system. Historically, the load transfer at the toe anchor region was achieved by either mechanical (e.g., short length of grout annulus) or frictional means (e.g., expansion shell) and at the collar by a plate and associated fixtures compatible with the element. The original DMFC systems exposed the plate and fixture to direct rock loading. Recently, the load transfer capacities at both the toe and collar regions have been improved by implementing reinforcement systems that incorporate full encapsulation from toe to collar but have a centrally decoupled region. Cement or resin grouts may be used to encapsulate the elements with similar installation equipment and methods used for fully coupled elements. This significantly limits the system's deterioration by corrosion. An added benefit of decoupling is that it allows for greater shear movements across the axis of a borehole compared with fully encapsulated systems.

7.6.1 CEMENT-ENCAPSULATED DECOUPLED THREADED BAR

Figure 7.44 shows the non-galvanized, partially decoupled threaded bars investigated here (Player et al., 2009b). The elements had a nominal length of 3.0 m, with a PVC tube clamped on the central 1.6 m, leaving a 1 m toe anchor region and a 0.4 m collar region. Load transfer can only occur at the toe and collar regions, and the 1.6-m decoupled region can stretch, hence dissipating more energy under dynamic load when compared with a fully encapsulated threaded bar that is geometrically keyed into the encapsulant. For cement encapsulation, a toe anchor region of 1 m is the minimum recommended, along with a collar region of 0.4 m, slightly short of the expected load transfer to reach the plate. This short region of load transfer at the collar allows a detailed analysis of the plate-nut-bar interaction under dynamic loading. Variations in installation and manufacturing resulted in bolt tails with different lengths. These, together with other test configurations, are listed in Table 7.5.

The elements were installed using a grout with a WCR of 0.40 and tested between 22 and 56 days after set. The unconfined compressive strength determined from tests on cylindrical samples was >40 MPa in all cases. The thick-walled steel pipe used for borehole simulation had the same dimensions as the fully encapsulated threaded bar configuration presented in Section 7.5.1.1. The simulated discontinuity (equivalent to a simulated depth of failure) was located at a standard 1 m from the collar. Two surface hardware configurations were investigated. The first used a separate 32 mm long nut (T20 × 10.0 pitch, LH thread) and a domed ball washer. The second comprised an integrated nut and dome ball washer with a continuous thread 45 mm in length. Both types of tests used a 150-mm square, 8 mm thick domed plate (Figure 7.45).

FIGURE 7.44   Details of 3.0-m long decoupled threaded bar indicating (a) 1.0 m toe anchor region, (b) 1.6 m decoupled region and (c) 0.4 m collar region.

TABLE 7.5

Cement-Encapsulated Decoupled Threaded Bar Test Configurations.

Nut Type	Bolt Tail (mm)	Collar Region (mm)	Decoupled Length (mm)	Toe Anchor Region (mm)
Nut & washer	213	187	1600	1000
Nut & washer	172	240	1600	1000
Integrated	141	249	1605	995
Integrated	162	238	1615	945

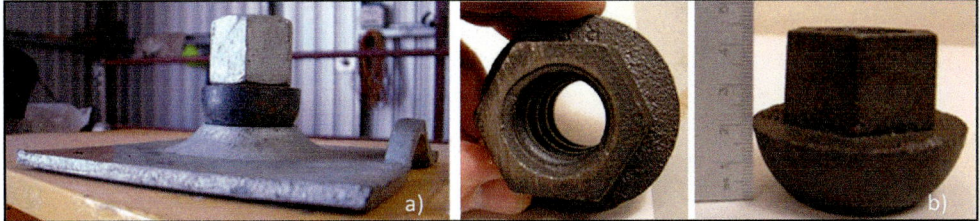

FIGURE 7.45 Example of surface hardware: (a) nut and domed washer and (b) integrated nut.

Figure 7.46 presents the results for the decoupled threaded bars in cement grout. The results indicate plate loading followed by plastic deformation of the steel in the decoupled region. Loading of the plate was excessive due to the short length of the collar where load transfer could occur, and failure occurred at the plate-nut interface. This shows that the critical functionality of a decoupled threaded bar depends upon a correct selection of the surface fixture and the correct length of collar grouting encapsulation. If the plate is loaded, failure can occur at lower loads compared with the bar strength. The failure mode can be by plate punching or by partial shearing along the shaft of the bolt and shearing of the nut threads, as shown in Figure 7.47.

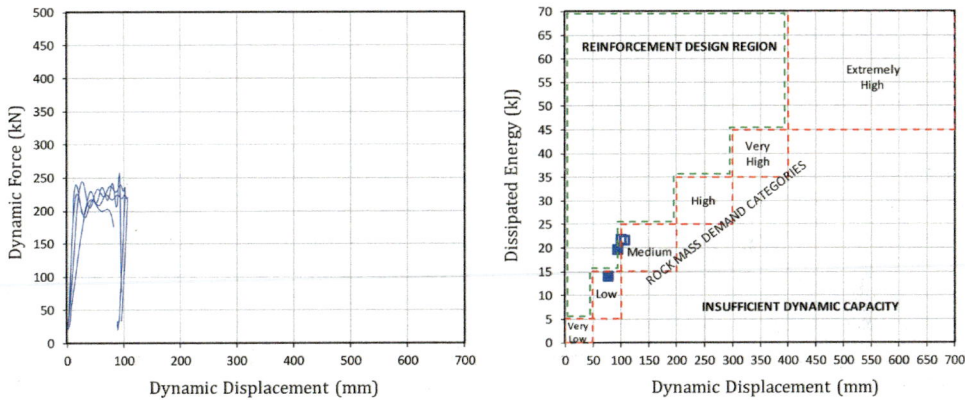

FIGURE 7.46 Dynamic force and related energy dissipation capacity for 20-mm diameter, centrally decoupled, cement-encapsulated threaded bar (1.0 m deep instability with coupled collar region of 0.23 m, decoupled region of 1.6 m and coupled toe anchor region of 1.0 m).

7.6.2 RESIN-ENCAPSULATED DECOUPLED POSIMIX

The 20-mm diameter decoupled Posimix is shown in Figure 7.48. The bolt can be installed using standard mining jumbos in holes ranging from 35 to 38 mm diameter in conjunction with resin cartridge systems. The reinforcement system incorporates a spiral resin mixing device that partially centralizes the bolt while acting as an Archimedes screw pump forcing the resin cartridges towards the back of the hole. The spiral profile shreds to minimize the accumulation of the resin cartridge plastic as the bolt rotates and is pushed towards the end of the hole. The bolt can be made of either galvanized or black steel with a 25-mm diameter bar also available.

Typical decoupled lengths range from 1.0 to 1.6 m depending on the total bolt length, which ranges from 2.4 to 3.0 m. Typically, the toe anchor region is set to 0.8 m, while the collar anchor regions are set to 0.6 m to allow for a nominal 0.2 m bolt tail that enables proper installation of the external fixture. Ideally, the minimum collar region required to achieve a very high load transfer is

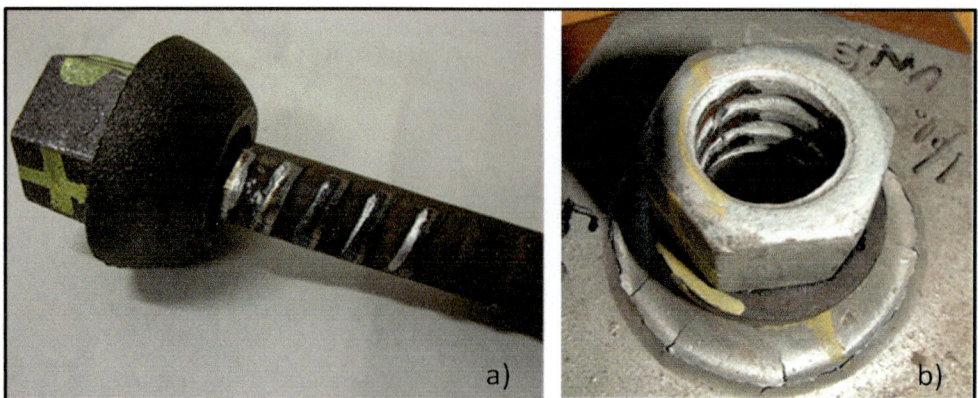

FIGURE 7.47 Example of plate loading showing (a) shearing of the threaded bar and (b) shearing of the nut.

FIGURE 7.48 Details of jumbo installation for a decoupled Posimix typically having (a) toe anchor region with mixing spiral of 0.8 m, (b) a decoupled length of 1.0 m and (c) collar region length of 0.6 m.

set to 0.35–0.4 m. This ensures that the capacity at the excavation surface does not depend solely on the plate-nut interface. Furthermore, underground observations have shown that if resin encapsulation can be assured at the collar, the violent ejection of broken bolts is minimized.

Figure 7.49 presents dynamic test results for a very long decoupling length aimed at defining the energy dissipation capacity of a 5 mm thick plate. The catastrophic results are shown with a solid

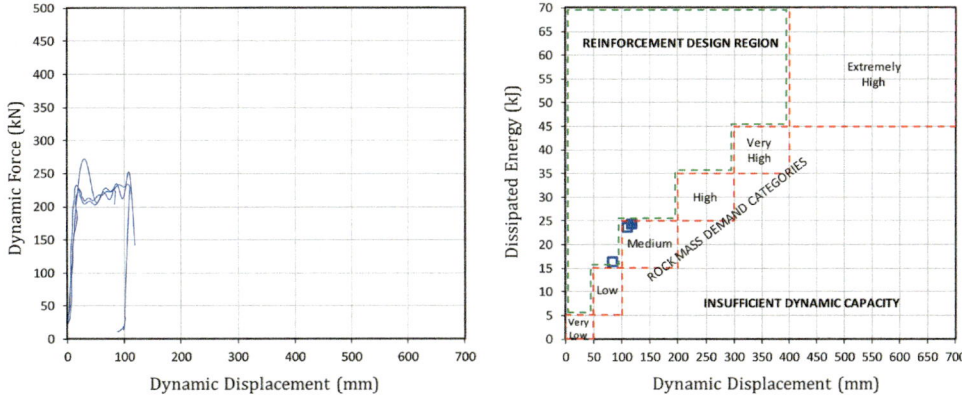

FIGURE 7.49 Dynamic force and related energy dissipation capacity for 20-mm diameter Posimix bolt (1.0 m deep instability, bolt decoupled to the collar, collar tail of 0.136 m, decoupled region of 1.7 m and coupled toe anchor region of 0.5 m).

symbol, and the data suggest a surface fixture (bolt-plate-nut) capacity of approximately 24 kJ. The failure mode was by stripping of bolt threads, as shown in Figure 7.50. The internal threads of the 46 mm long nut were also partially stripped, with a high-speed video showing nut rotation during the violent failure.

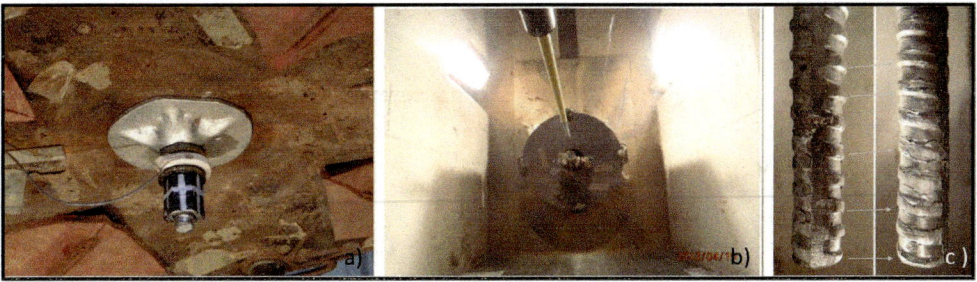

FIGURE 7.50 Details of bolt-plate interaction (a) prior to testing, (b) failure at the bolt collar and (c) stripping of the bolt threads.

Figure 7.51 presents the results for three 2.4 m long decoupled Posimix installed in the field using a jumbo and resin cartridge system. Dissection of the simulated boreholes has shown that a bolt can be installed with a very high degree of encapsulation from toe to collar (Figure 7.52). Tensile failure occurred within the 1 m decoupled region with no damage to the surface fixtures and with the reinforcement system achieving its full capacity, exceeding 25 kJ.

An increased decoupling region results in higher energy dissipation capabilities, as shown in Figure 7.53, where results for a 1.4 m long decoupling encapsulated in cement grout are presented. This was undertaken to ensure that the collar region was encapsulated and to investigate the effect of an increased decoupling region. The results showed that the reinforcing system is capable of exceeding 40 kJ with 180 mm of displacement at the simulated discontinuity. Not all the tests were undertaken with similar input energy, and the results that plot around 30 kJ indicate those systems still have a large reserve capacity. The results show a significant improvement compared with a fully coupled Posimix (where no decoupling region was implemented), which failed violently at approximately 7 kJ and with less than 50 mm of displacement (Figure 7.54).

Figure 7.55 shows additional results for 20-mm diameter resin-encapsulated Posimix bolts installed in simulated boreholes using a mining jumbo. The results are for bolts fully decoupled to

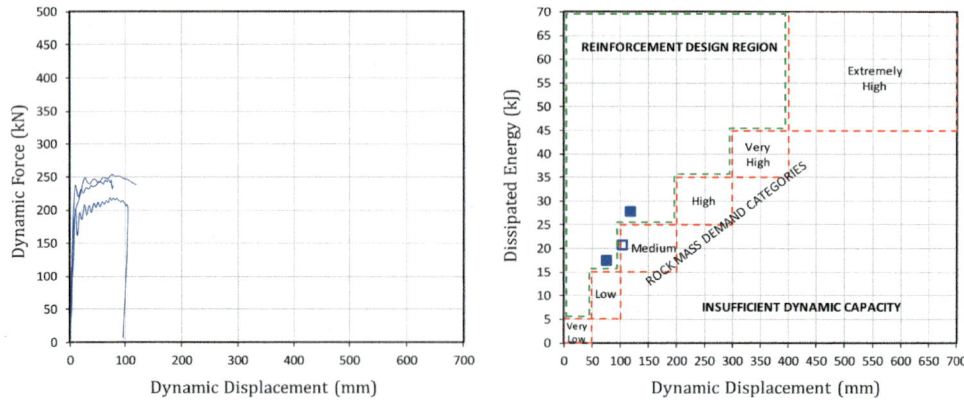

FIGURE 7.51 Dynamic force and related energy dissipation capacity for 20-mm diameter Posimix bolt encapsulated in resin grout (1.0 m deep instability, collar tail of 0.215 m, coupled collar region 0.385 m, decoupled region of 1.0 m and coupled toe anchor region of 0.8 m).

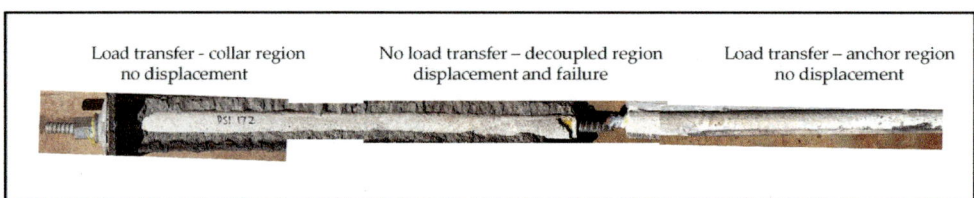

FIGURE 7.52 Long section view of a 2.4 m long decoupled Posimix bolt, fully encapsulated in resin with the installation using a mining jumbo and resin cartridges.

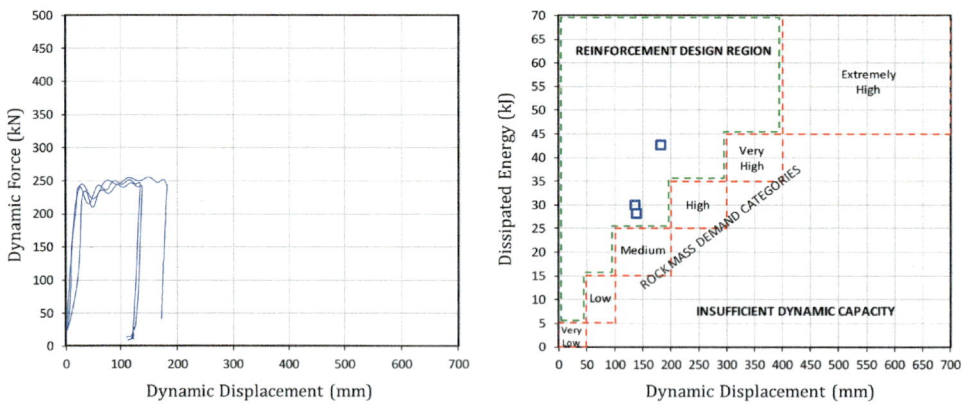

FIGURE 7.53 Dynamic force and related energy dissipation capacity for 20-mm diameter Posimix bolt encapsulated in cement grout (1.0 m deep instability, collar tail of 0.146 m, coupled collar region 0.254 m, decoupled region of 1.4 m and coupled toe anchor region of 0.6 m).

the collar in order to test the strength of a 6 mm thick plate. These plates have performed very well in the field, where bolt rupture is achieved without plate punching failure (Figure 7.56). The results show that a 6 mm thick plate is capable of matching the 20-mm diameter bar. The plot indicates that during the dynamic impact, the load increases gradually until the bar is broken at 27.5 kJ; however, the plate is not punched through.

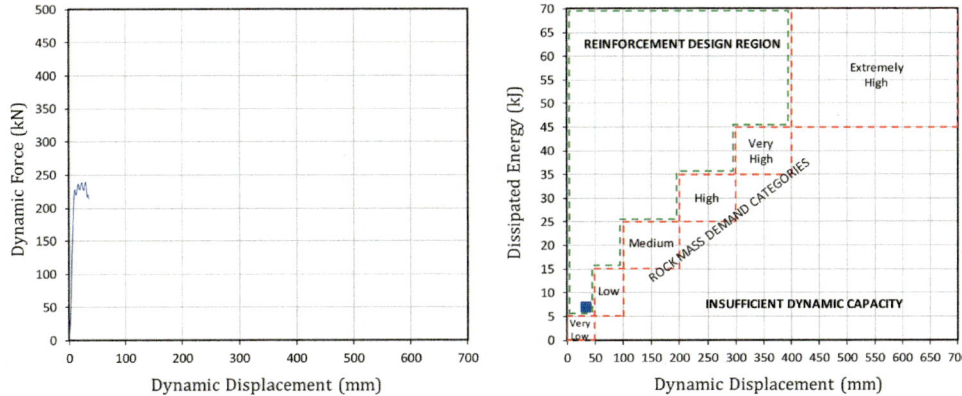

FIGURE 7.54 Dynamic force and related energy dissipation capacity for 20-mm diameter Posimix bolt encapsulated in resin grout (1.0 m deep instability, fully coupled).

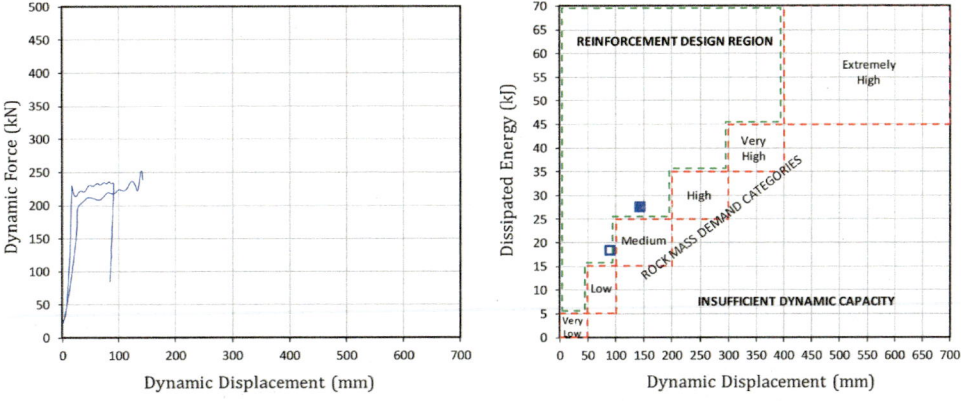

FIGURE 7.55 Dynamic force and related energy dissipation capacity for 20-mm diameter Posimix bolt encapsulated in resin (1.0 m deep instability, collar tail of 0.230 m, decoupled (to the collar) region of 1.57 m with 6 mm thick plate and coupled toe anchor region of 0.600 m).

FIGURE 7.56 Examples of loaded 20-mm diameter Posimix bolts installed with 6 mm thick plates following a large seismic event: (a) loaded but stable, (b) broken element ejected and (c) with plate inversion but without fixtures punching through.

Figure 7.57 presents the results from testing two jumbo-installed, 2.4 m long decoupled Posimix in conjunction with a resin cartridge system. The reinforcement system was able to dissipate 30 kJ at a displacement of 150 mm. Dissection of the bolts following testing indicates that in both cases, the resin was not well mixed within the collar region (Figure 7.58). The observations also indicated that the bolts yielded during the test and that the 6 mm thick plates were able to safely dissipate the load.

Figure 7.59 shows the results for 3 m long elements in which the decoupling region has been set to 1.8 m. The system was installed using a jumbo and resin cartridges. The system was able

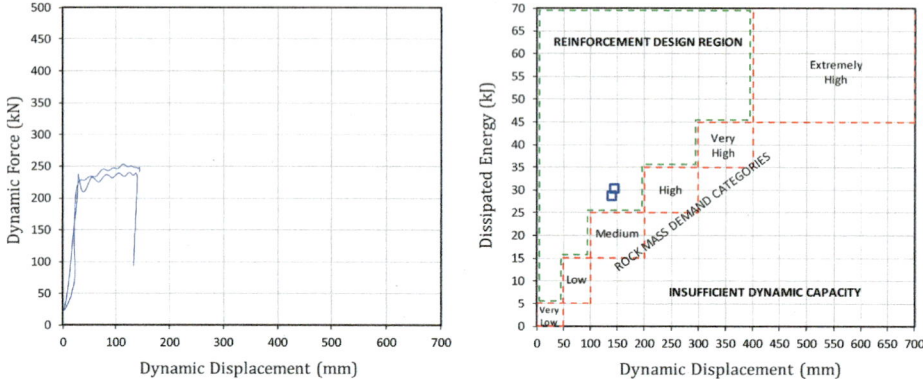

FIGURE 7.57 Dynamic force and related energy dissipation capacity for 20-mm diameter Posimix bolt encapsulated in resin (1.0 m deep instability, collar tail of 0.228 m, coupled collar region of 0.362 m, decoupled region of 1.2 m and coupled toe anchor region of 0.6 m).

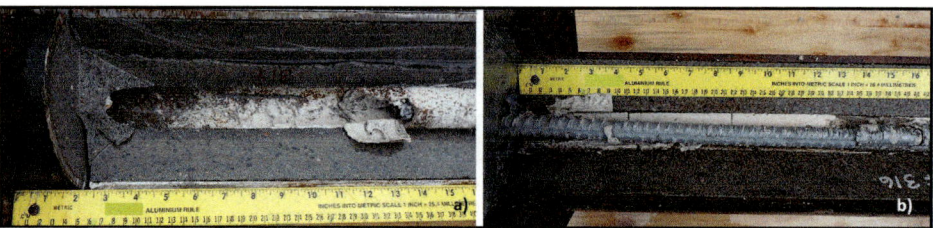

FIGURE 7.58 Dissection of reinforcement system following testing showing loading to the plate due to (a) no resin near the collar and (b) poorly mixed resin.

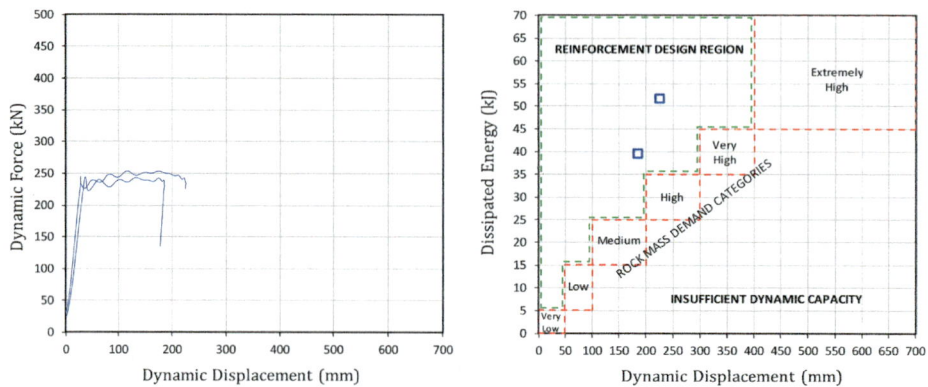

FIGURE 7.59 Dynamic force and related energy dissipation capacity for 20-mm diameter Posimix bolt encapsulated in resin (1.0 m deep instability, collar tail of 0.260 m, coupled collar region of 0.330 m, decoupled region of 1.8 m and coupled toe anchor region of 0.6 m).

to dissipate over 50 kJ of energy without plate damage. Inspection of the bolts following testing showed that the collar region was fully encapsulated by the resin in both cases (Figure 7.60).

Figure 7.55 shows the results of two Posimix reinforcement systems that were purposely decoupled from the start of the toe anchor region extending to the collar. This test was conducted to test the plate-bolt configuration, with the results indicating that failure occurred by reinforcement element rupture at 27 kJ. Figure 7.61 shows a cross-sectional view of the collar and anchor region, showing complete resin encapsulation of the toe anchor region and decoupling extending to the collar. This agrees with many other observations from bolt test dissections, which indicate the Posimix spiral to be very effective in mixing resin cartridges.

Figure 7.62 combines all test data from all 20-mm diameter, decoupled Posimix bolts installed within 38 mm holes using a jumbo and resin cartridges. As expected, the energy dissipation and

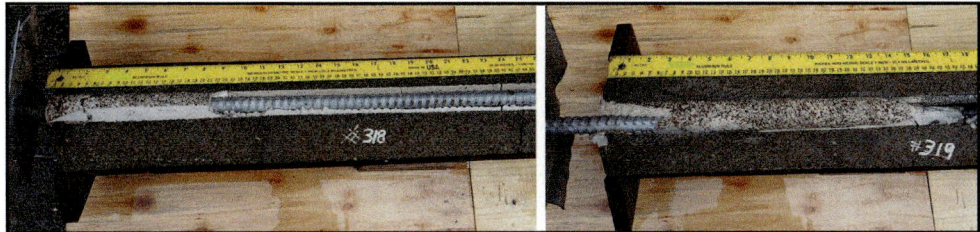

FIGURE 7.60 Dissection to expose two reinforcement systems following testing showing a fully resin encapsulated collar region.

FIGURE 7.61 Dissection to expose the reinforcement system following testing showing (a) totally decoupled collar region and (b) well-mixed resin encapsulating the toe anchor region.

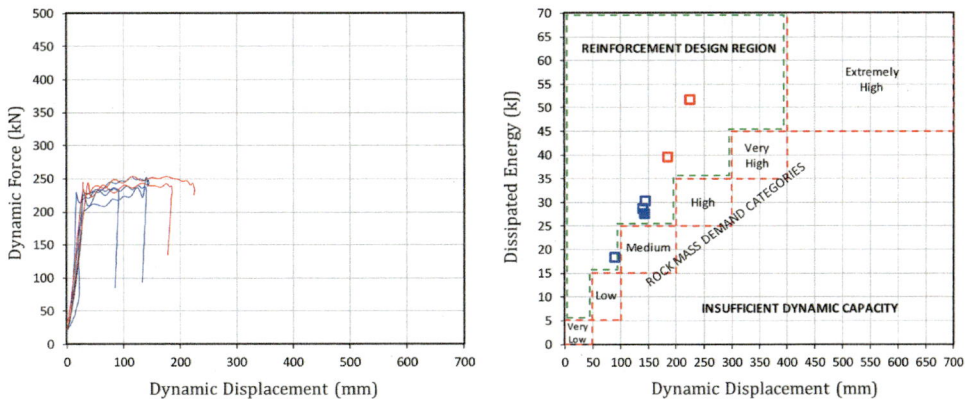

FIGURE 7.62 Combined plot of dynamic force and related energy dissipation capacity for 20-mm diameter Posimix bolts encapsulated in resin. All bolts were installed using a mining jumbo and resin cartridges. Decoupled lengths range from 1.0 to 1.8 m.

displacement are linearly related to the length of the decoupling region, with the results plotting within the design region being proposed here.

Figure 7.63 presents the combined data for 2.4 m long, 25-mm diameter decoupled Posimix bolts installed with a jumbo in 38-mm diameter holes. The bolts were installed with a range of decoupled lengths, including toe to collar, to test the bolt-plate interaction. The larger bolt diameter ensures that full resin encapsulation at the collar is achieved, as shown in Figure 7.64. The dynamic strength of the 25-mm Posimix bolts ranges from 42 to 45 kJ and with remaining capacity (i.e., the plot only shows test results from the first impact).

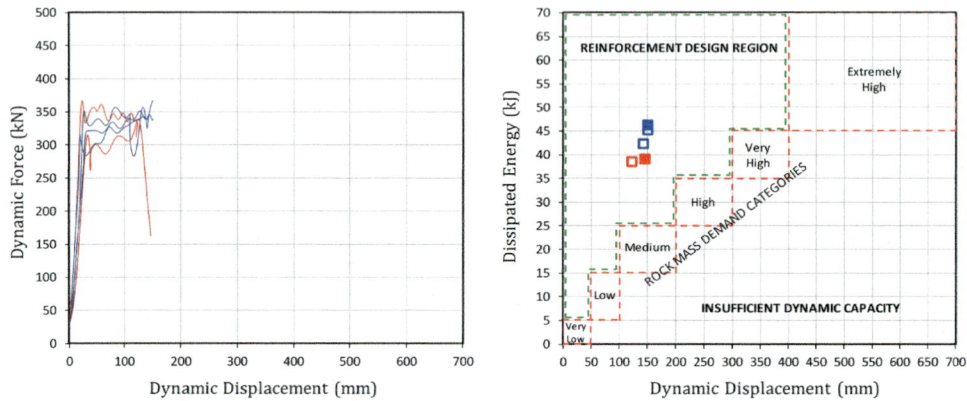

FIGURE 7.63 Combined plot of dynamic force and related energy dissipation capacity for 25-mm diameter decoupled Posimix bolts encapsulated in resin. All bolts were installed using a mining jumbo and resin cartridges. Decoupled lengths range from 1.2 to 1.6 m.

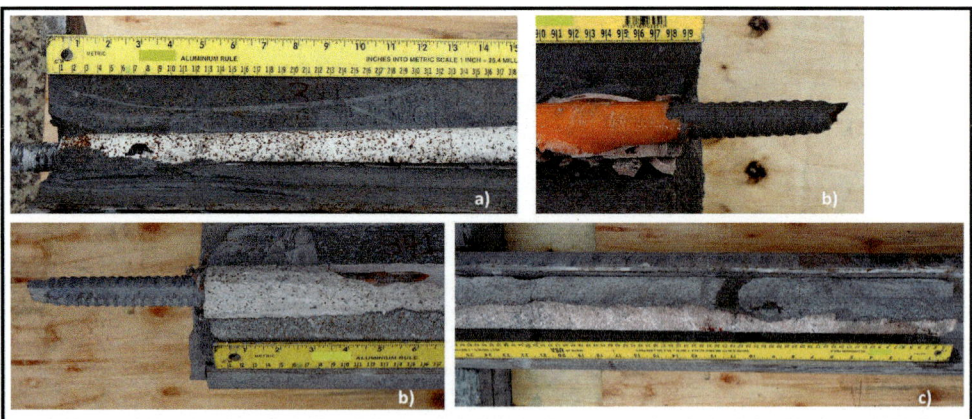

FIGURE 7.64 Dissection of reinforcement system following testing showing (a) full resin encapsulation at the collar region, (b) ruptured element within the decoupled region and (c) well-mixed resin encapsulating the toe anchor region.

7.6.3 CEMENT- AND RESIN-ENCAPSULATED D BOLT

The D bolt consists of a smooth steel bar that has a number of anchor points spaced along its bolt axis (Figure 7.65). The bolt has been designed for load transfer to occur at the anchor points and deformation of the smooth bar (Li, 2010). The anchor points are wider than the bar, which could also assist in centralizing the element within the borehole.

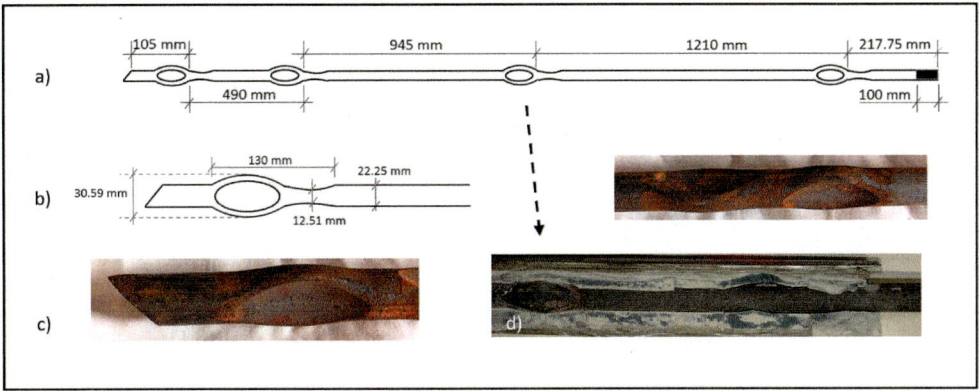

FIGURE 7.65   Long section view of a D-bolt showing (a) smooth bar and anchor positions, (b) details of anchors, (c) resin mixing end and (d) locked anchor and yielded bar following testing.

Figure 7.66 presents the results for D-bolts that are fully encapsulated in cement grout and tested under dynamic impact. The results are very consistent and indicate the yielding of the smooth bar between the anchor points. Also, the result plot within the design region is proposed here.

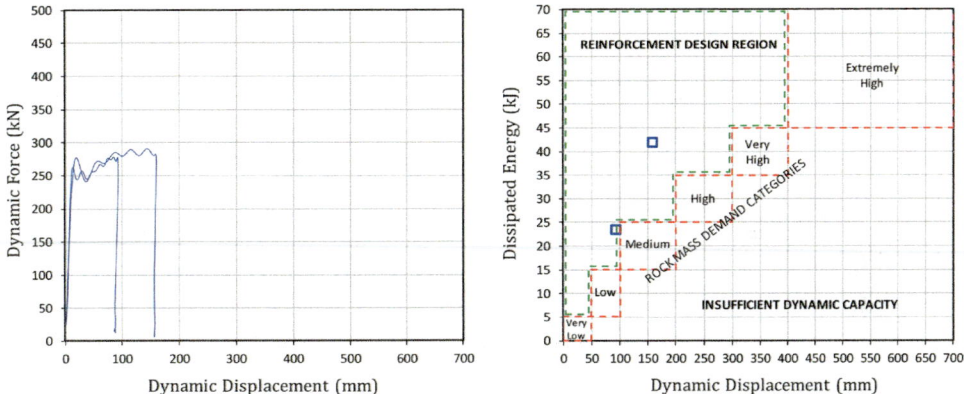

FIGURE 7.66   Dynamic force and related energy dissipation capacity for 20-mm D-bolt fully encapsulated in cement grout (unstable collar region of 1.0 m, stable toe anchor region of 1.2 m).

Figure 7.67 shows laboratory testing results for D-bolts installed using resin cartridges in a fully encapsulated configuration. The results indicate high variability linked to poor mixing of the resin. In cases where the resin was well mixed, and the D-bolts were fully encapsulated, the bolts performed well and were similar to cement-encapsulated elements. Alternatively, when the resin was poorly mixed, large displacements occurred. It is important to realize that for a similar level of demand, a different capacity can result purely due to resin mixing quality.

Resin mixing quality is controlled by the diameter of the bar, the actual diameter of the holes, the type of mixing device at the end of the bar and the bolt installation technique. If the resin is supplied by cartridges, then a finite amount of resin is available, and an enlarged hole may allow transverse displacement (Figure 7.68) or incomplete mixing, especially in the case of weak or broken rock masses.

7.6.4   CEMENT- AND RESIN-ENCAPSULATED CONE BOLT

The Cone bolt is believed to be the first rock bolt designed to use a sliding mechanism to dissipate energy. The bolt consists of a plain bar with an expanded cross-section at the toe and a thread,

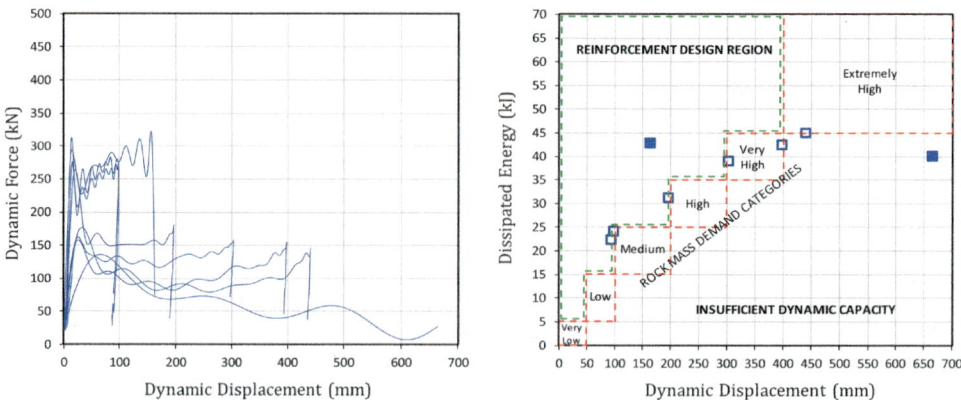

FIGURE 7.67 Dynamic force and related energy dissipation capacity for 20-mm diameter, fully coupled, resin-encapsulated D-bolts (unstable collar region of 1.0 m, stable toe anchor region of 1.2 m).

FIGURE 7.68 Example of field data showing side-way displacement of the resin cartridge by the D-Bolt resulting in a non-encapsulated collar region.

nut, washer and plate at the collar, as shown in Figure 7.69. The expanded cross-section of the bar is designed to provide resistance to pull out that is controlled by the strength and stiffness of the cement grout that encapsulates the bolt within a borehole. The shaft of the bolt is coated with saponified wax, so, ideally, there is little or no resistance to movement of the bolt relative to the encapsulating grout without affecting resistance to shear failure.

The initial prototypes were manufactured from a 16-mm diameter bar, and the majority of testing was performed on this variant. Subsequent to the development of the original Cone Bolt, demand for higher capacity elements resulted in a 22-mm diameter plain bar variant (Player, 2012).

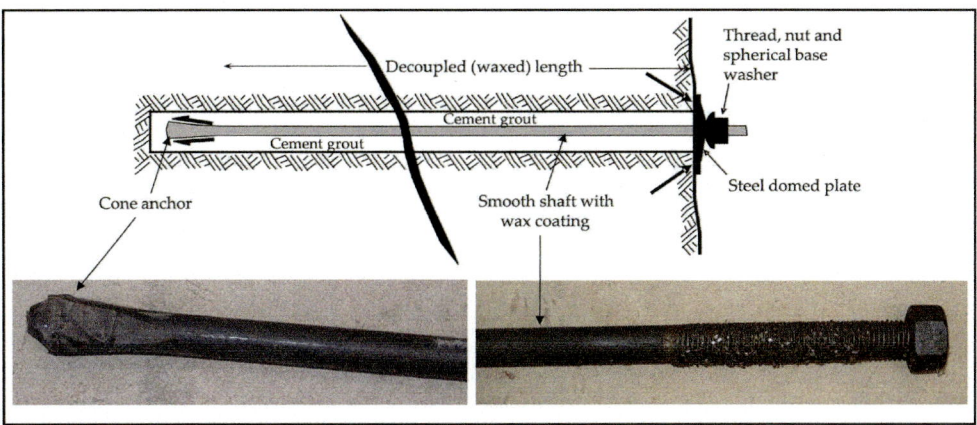

FIGURE 7.69 Cone bolt anchor designed to pull through cement grout and, therefore, increases displacement and energy dissipation capacities.

The design of the cement grout should be such that the anchor ploughs (pulls) through the grout column at a force less than the yield strength of the bolt. Both static and dynamic tests have shown that this is not the case with 'strong' grouts, and much of the measured displacement is the stretch of the bar (Player, 2012). It is, therefore, critical that the cement grout properties are designed to ensure that the cone pulls through the grout at a force less than the yield capacity of the bar (Figure 7.70). Also, the equipment and procedures used for mixing and placing the cement grout paste in a borehole result in consistent strength and stiffness of the hardened cement grout.

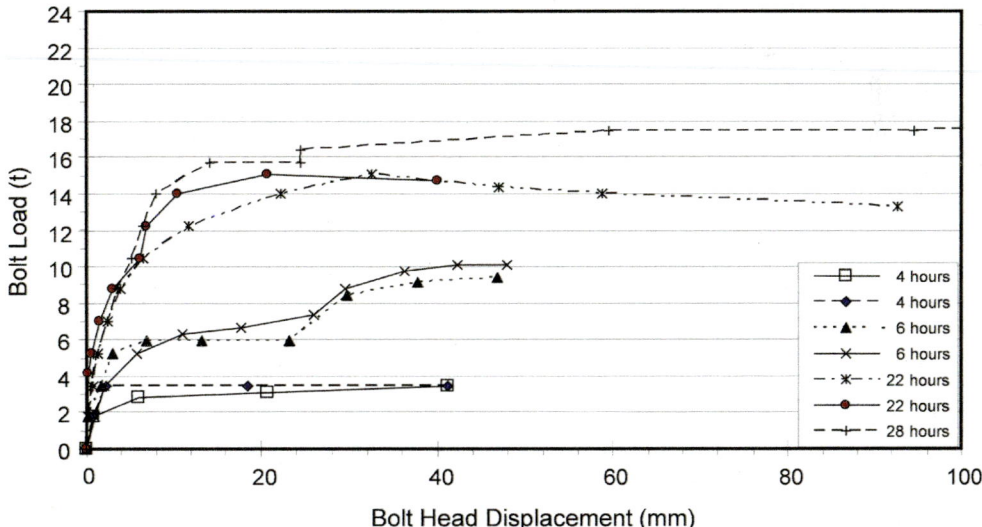

FIGURE 7.70 Cone bolt data from *in situ* static pull testing at different times after installation. A 0.35 WCR cement grout was used with the Sika HE200NN additive as the hardening agent and superplasticizer.

Figure 7.71 presents the results of dynamic testing on 22-mm diameter Cone bolts. The majority of the data did not plot within the preferred design region. However, the largest energy dissipation achieved was 59 kJ at 310 mm of displacement, indicating the potential for the bar to plough through a suitable cement grout mix (Figure 7.72). The effects of the grout quality are shown in Figure 7.73

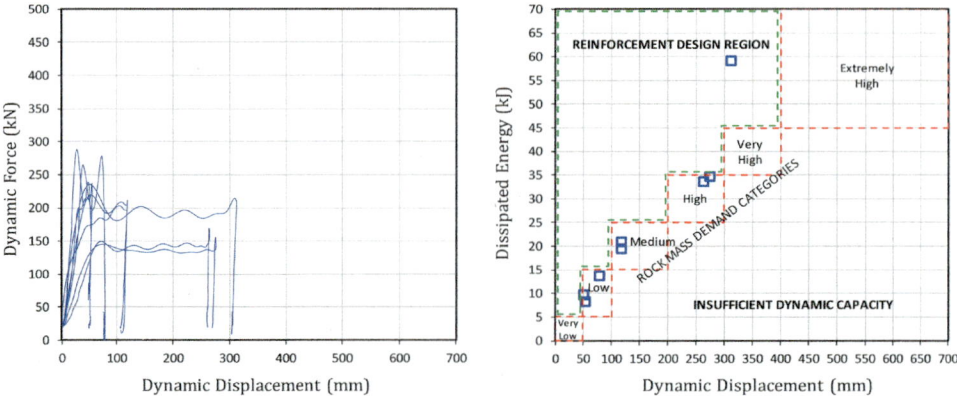

FIGURE 7.71 Dynamic force and related energy dissipation capacity for 22-mm diameter, fully coupled, cement-encapsulated Cone bolts (unstable collar region of 1.0 m, stable toe anchor region of 1.2 m).

FIGURE 7.72 Example of a Cone bolt ploughing through a cement grout within a simulated borehole under heavy confinement.

using data from the first dynamic impact. This suggests that the 22-mm Cone bolt performance is highly dependent upon the strength of the grout used (Player, 2012).

Recently a Modified Cone Bolt (MCB) has been developed in Canada. This bolt, like the original Cone bolt, has an expanded end but is designed to be encapsulated with resin grout from a two-component cartridge that is mixed during installation. The breaking of the cartridge and mixing of the resin is aided by a flat paddle attached to the expanded end (Figure 7.74). The reported results (e.g., Simser et al., 2002; Gaudreau et al., 2004) show that this bolt performs either by gross anchor displacement or element extension or a combination of both mechanisms. The fact that the bolt eventually breaks suggests that the ploughing effect is halted by strain hardening due to the densification of the grout. Figure 7.75 shows the results for an MCB, where two groups of data can be identified. One with well-mixed resin and very little cone displacement (Figure 7.76) with the energy dissipation due to bolt stretch and plate damage. The second group indicates that the resin was not well mixed because large displacements at very low loads resulted.

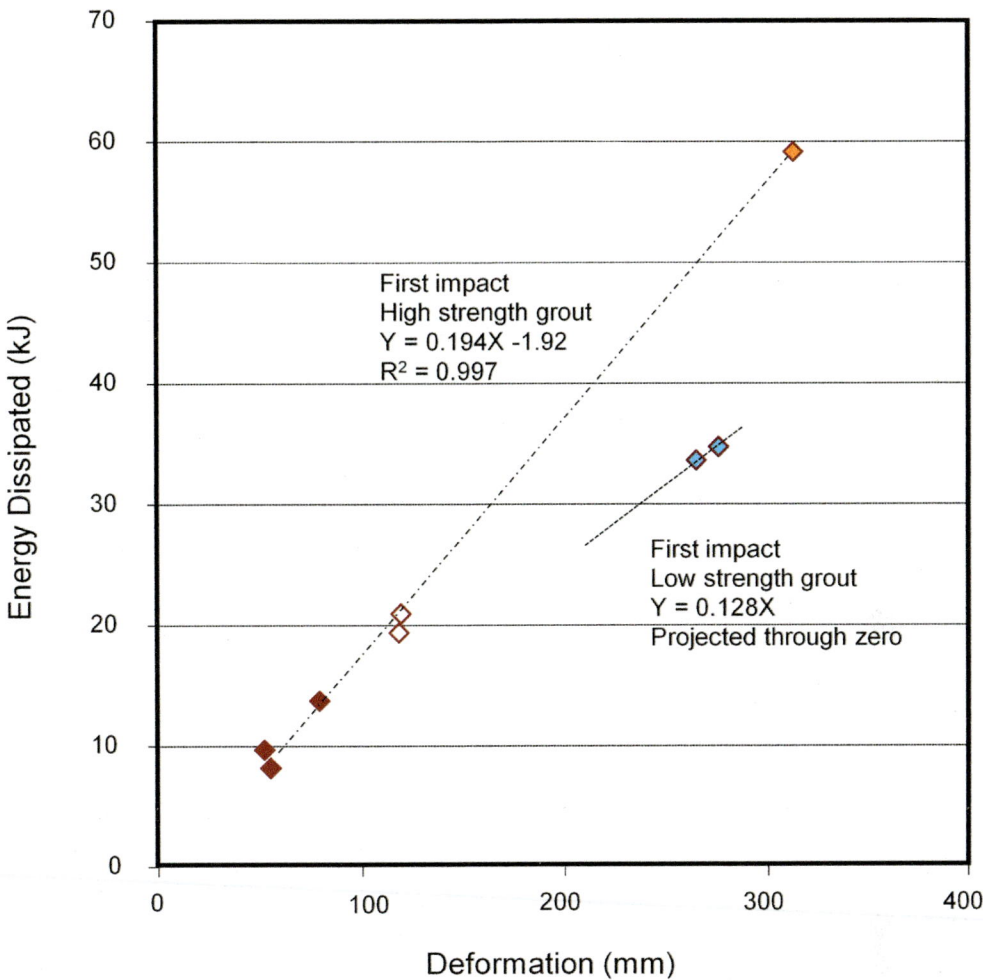

FIGURE 7.73 Difference in energy dissipation for Cone bolts ploughing through 25 MPa (low-strength) and 41.5 MPa (high-strength) cement grout.

7.6.5 CEMENT- AND RESIN-ENCAPSULATED GARFORD DYNAMIC BOLT

Figure 7.77 shows the components of the Garford Dynamic Bolt (Varden et al., 2008). The bolt consists of a 21.7-mm diameter, 5152 AVH steel grade, solid bar with a rolled M24 thread. The resin mixing device consists of a 350 mm long, 43-mm diameter coarse threaded steel sleeve crimped on the end of the bolt. The bar is covered in a polyethylene sleeve between the sliding mechanism and the thread at the collar end. This sleeve decouples the bar from the encapsulant and extends to the plate. Dynamic energy dissipation is achieved by a sliding anchor mechanism that is pressed on o the bolt below the mixing device. The mechanism comprises a piece of threaded steel with a 20-mm hole at the top and a 16-mm hole at the collar end of the bolt. When the bolt is subjected to loading at the plate, the bar is pulled through this fixed geometric constriction (Figure 7.78).

The laboratory test performance of the bolt in cement is shown in Figure 7.79, where consistent results from the yielding mechanism were determined. Sliding through the device occurred at around 150 kN, and a similar performance was achieved for repetitive testing. The bolts were also installed using a jumbo and resin cartridges to investigate if an anchor point could be secured. The results from these tests are shown in Figure 7.80. These results were not as consistent as cement

FIGURE 7.74 Details of Modified Cone Bolt anchor with its surface fixtures and a spade for resin mixing.

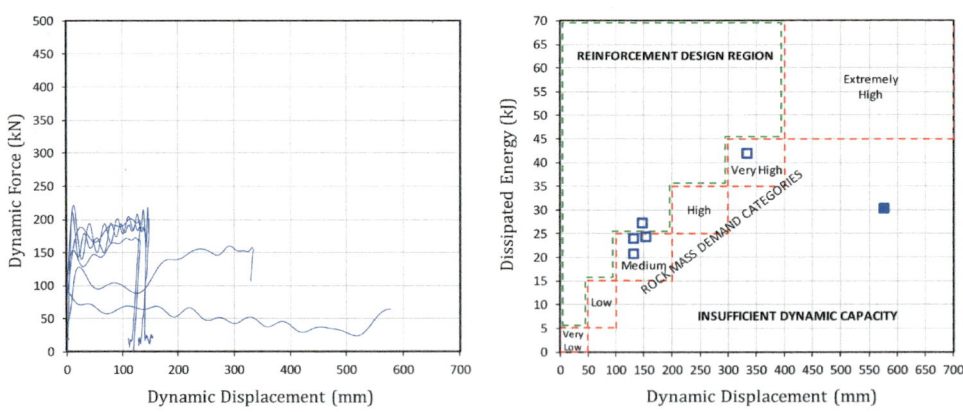

FIGURE 7.75 Dynamic force and related energy dissipation capacity for 22-mm diameter, fully coupled, resin-encapsulated Modified Cone Bolts (unstable collar region of 1.0 m, stable toe anchor region of 1.2 m).

encapsulation, and a sliding force of only 100 kN could be achieved. The peak at 400 mm displacement represents the lock-in of the sliding device at the end of the designed displacement range at the toe of the element. As with all reinforcement systems, it must be noted that all laboratory testing is undertaken without corrosion and with a fully functional plate responding to the dynamic loads. Underground observations following large seismic events have shown that the Garford Dynamic bolt does not always mix the resin cartridges or mobilize the energy dissipation mechanism, and some catastrophic failures have occurred (Figure 7.81). In some cases, bolts have pulled out with limited displacement through the sliding anchor mechanism.

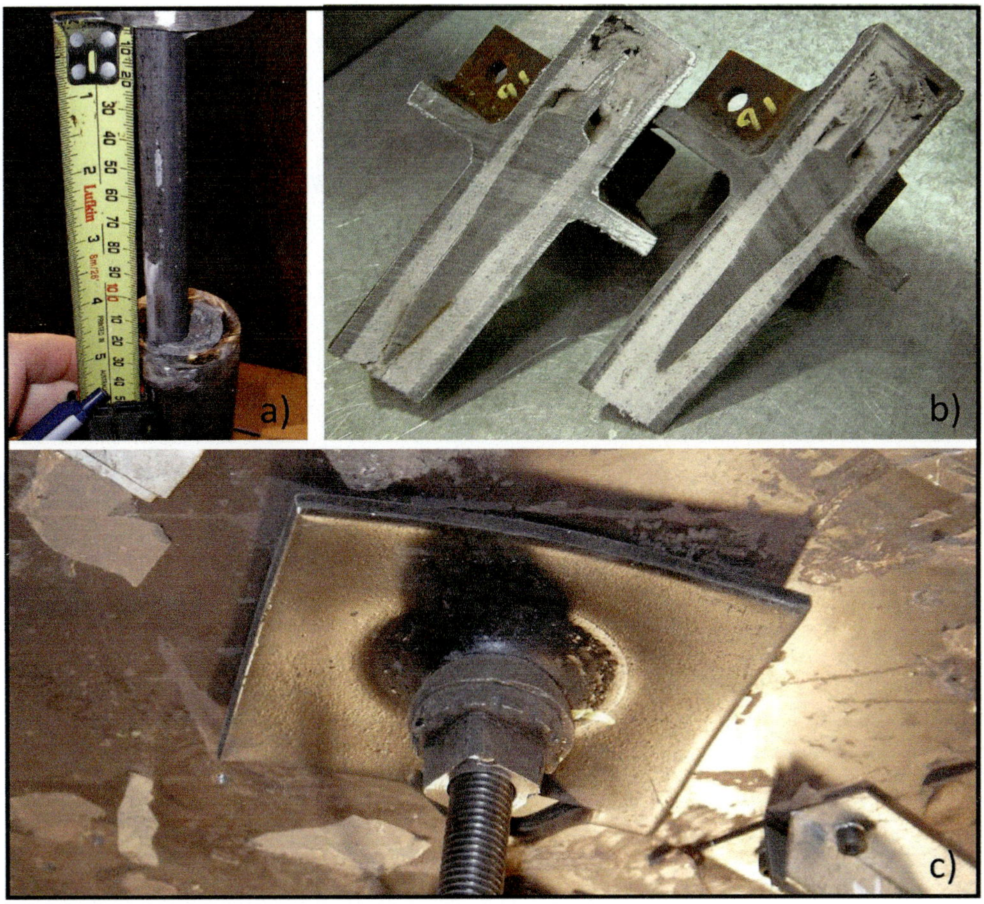

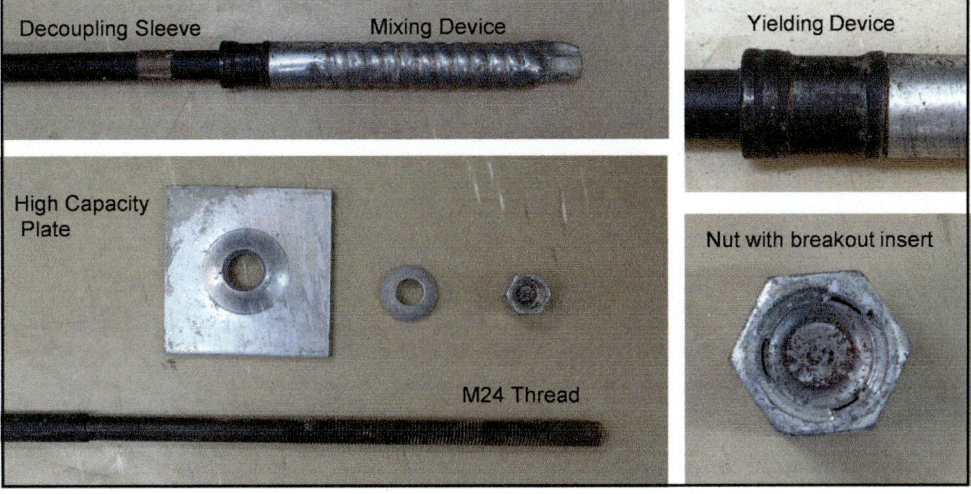

FIGURE 7.76 Details of Modified Cone bolt after dynamic testing: (a) bolt displacement at a simulated discontinuity, (b) less than 15 mm of displacement by the cone at the toe region and (c) plate deformation.

FIGURE 7.77 The components of the Garford Dynamic Bolt.

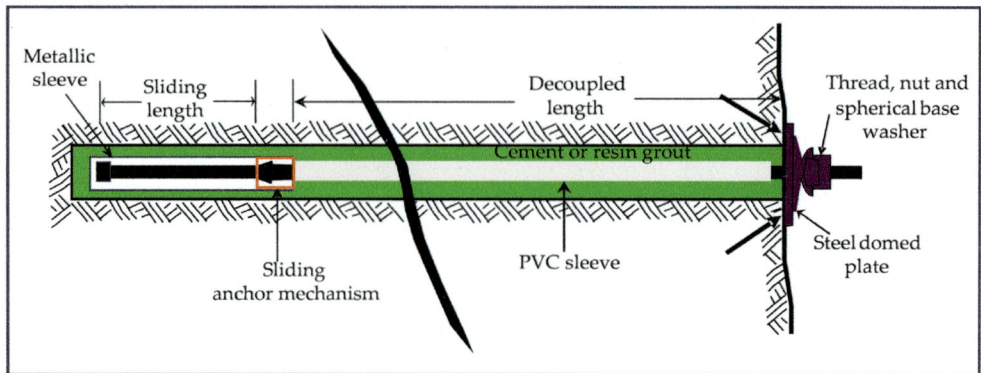

FIGURE 7.78   Schematic cross-sectional view of the point-anchored Garford Dynamic Bolt.

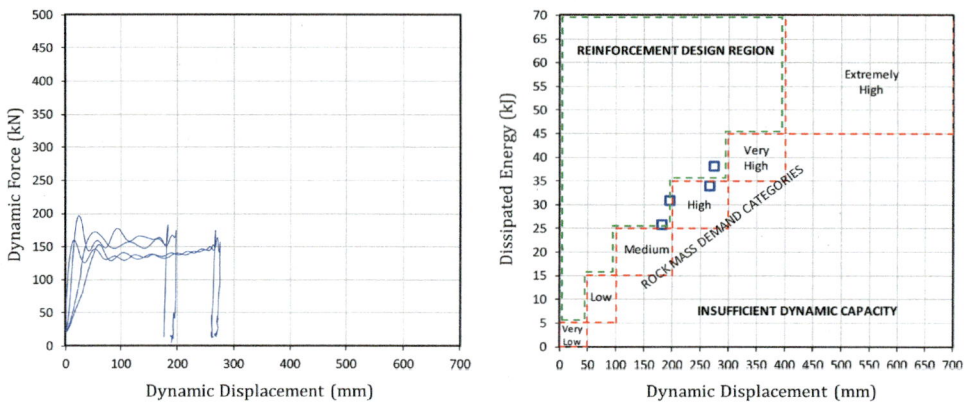

FIGURE 7.79   Dynamic force and related energy dissipation capacity for 22-mm diameter, cement-encapsulated Garford Dynamic Bolts (unstable collar region of 1.0 m, stable toe anchor region of 1.2 m).

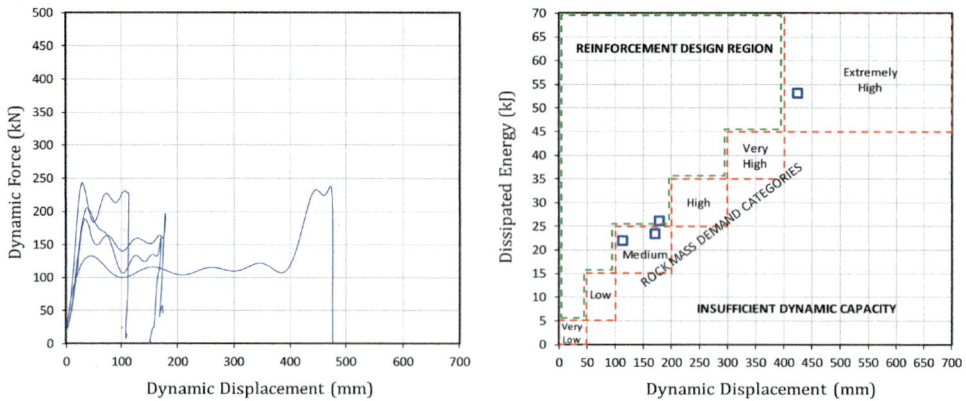

FIGURE 7.80   Dynamic force and related energy dissipation capacity for 22-mm diameter, resin-encapsulated Garford Dynamic Bolts (unstable collar region of 1.0 m, stable toe anchor region of 1.2 m).

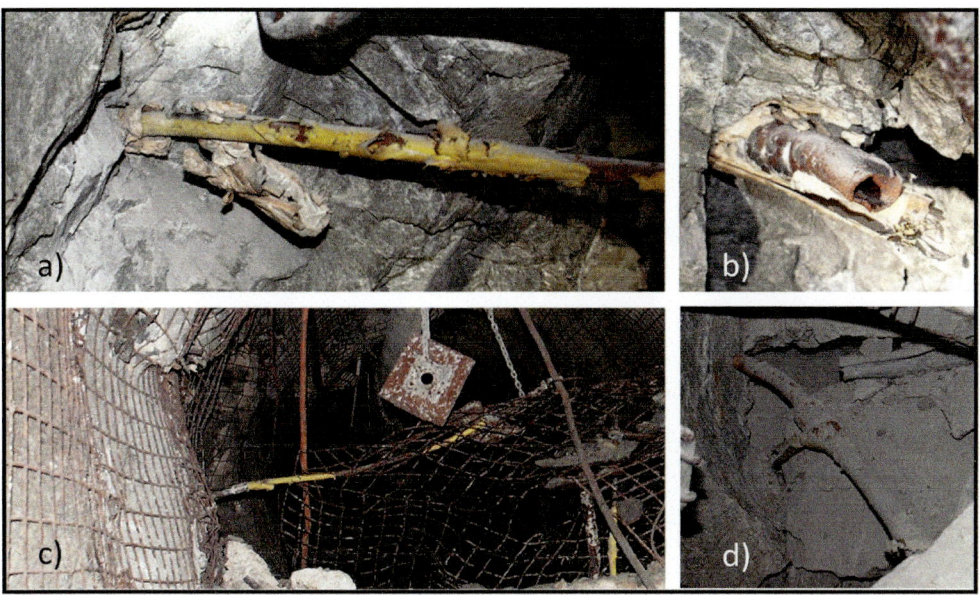

FIGURE 7.81 Field evidence following several large seismic events where catastrophic failures occurred with evidence of limited energy dissipation by the Garford yielding mechanism: (a) poor resin mixing, (b and c) bolts pulled out and (d) no yielding of the ejected bolt.

Figure 7.82 shows additional results for cement-encapsulated Garford bolts where the energy dissipation is achieved by large bolt slippage. A stick-slip mechanism of the bar being pulled through the yielding device was experienced (anchor and collar load cells showed harmonic waves transmitted along the reinforcement element). Interpretation of the instrumentation signals suggested that the end-stop device was engaged and held for a short time before failing catastrophically.

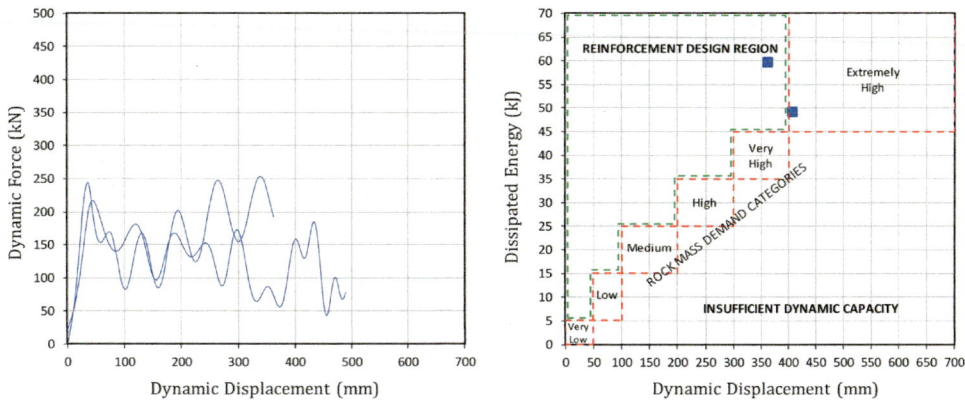

FIGURE 7.82 Catastrophic failure of two 22-mm diameter, cement-encapsulated Garford Dynamic Bolts (unstable collar region of 1.0 m, stable toe anchor region of 1.2 m).

7.6.6 CEMENT-ENCAPSULATED DURABAR BOLT

The Durabar bolt is a DMFC reinforcement system that is installed fully encapsulated in cement grout (Figure 7.83). Initial load transfer is through the plate, which then mobilizes the bolt using a

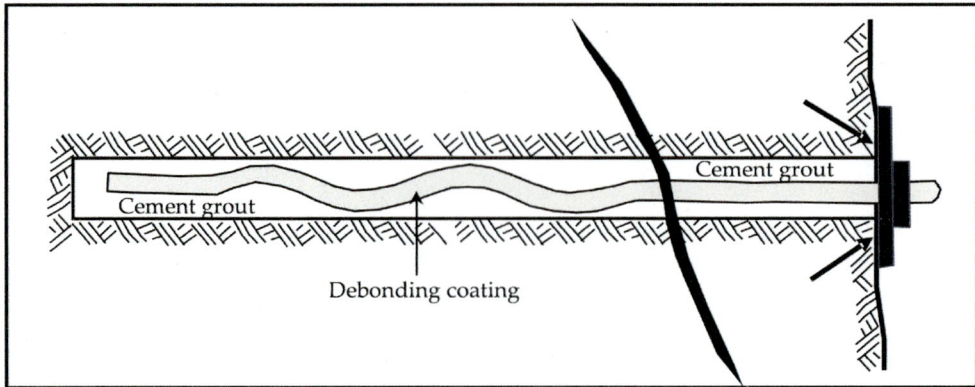

FIGURE 7.83   Schematic cross-sectional view of the Durabar Bolt.

combination of yielding and ploughing through the cement grout. As with the Cone bolt section, the strength of the cement grout plays a critical role in the performance.

Figure 7.84 shows the results from dynamic testing for 22-mm diameter Durabar fully encapsulated in cement grout. During dynamic testing, a rapid rise in anchor force is accompanied by bar yield until plate collapse occurs. This is experienced at a displacement less than the total stretch of the bar recorded from a test. After this, a combination of bar yield and bar sliding occurs, with the 'wiggles' pulling through until the energy is dissipated. The lower resistive force that occurs during the sliding phase is not fully understood given that plastic deformation of the bar is still occurring. The data shows that the division between bar yield and bar pulling/straightening mechanisms appears to be very complex.

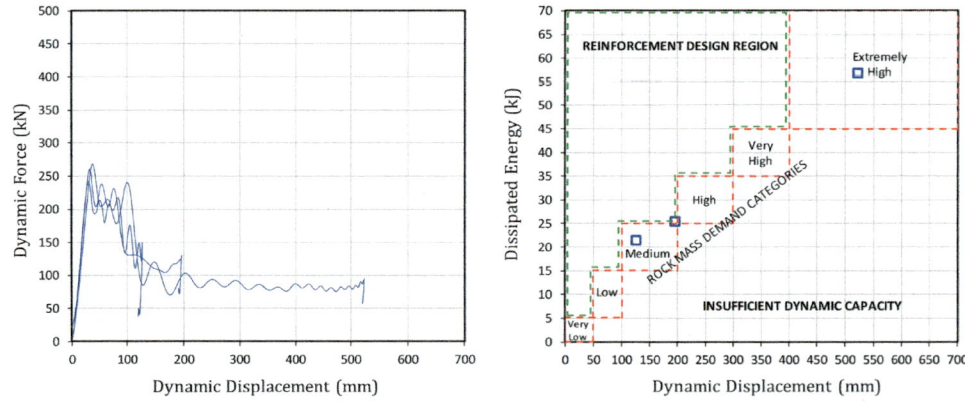

FIGURE 7.84   Dynamic force and related energy dissipation capacity for 22-mm diameter, cement-encapsulated Durabar Bolts (unstable collar region of 1.0 m, stable toe anchor region of 1.2 m).

Post-test inspection indicated a combination of bar stretch, plate collapse/inversion and displacement through the cement grout (Figure 7.85). The experimental results indicate large variability in the performance of the Durabar. Once the anchor is mobilized, the bolt is a low-capacity bolt that allows very large displacements (i.e., well in excess of 400 mm). None of these test results plot within the design region being suggested here.

7.6.7   CEMENT-ENCAPSULATED YIELD-LOK BOLT

The Yield-Lok bolt is another DMFC reinforcement system that is installed fully encapsulated in either cement or resin grout. Initial load transfer is through the plate, which then mobilizes the

FIGURE 7.85 Details of Durabar testing, including (a) bar stretch, (b) plate inversion and (c) bar ploughing through cement grout.

bolt using a combination of yielding and ploughing through a specially designed polymer coating located in the toe-end region of the element (Figure 7.86).

The element consists of a 17.3-mm round steel bar with a reported yield and ultimate tensile strength of 15 and 20 tons, respectively (Wu and Oldsen, 2010). The bolt length tested and reported here ranged from 2.4 to 3.0 m, with the last 0.8 m of the toe-end region covered with a special polymer coating. Both cement and resin-encapsulated reinforcement elements were tested. For effective

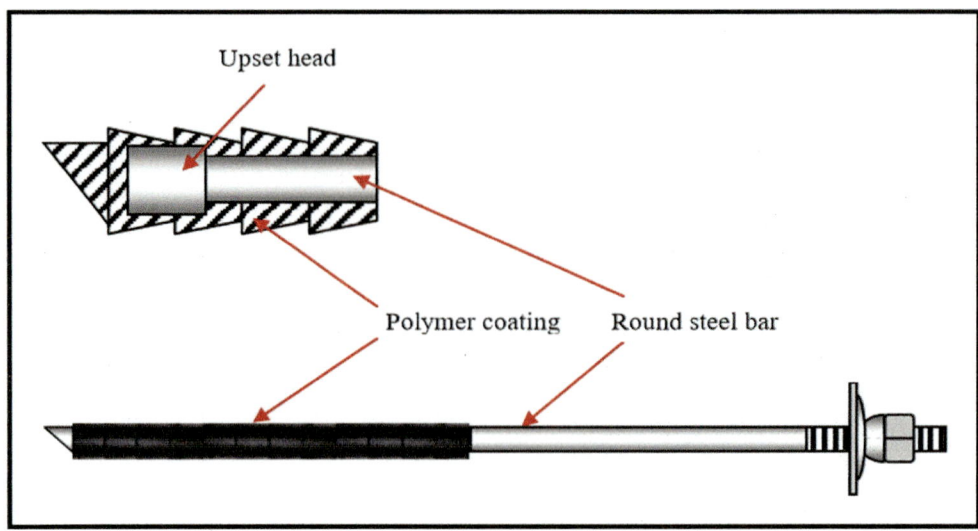

FIGURE 7.86 Long section view of the Yield-Lok reinforcement system showing the yielding mechanism (Wu and Oldsen, 2010).

installation, the saw tooth profile of the polymer coating facilitates the shredding of the resin cartridge packaging (Wu and Oldsen, 2010). The bolts possess a 123–150-mm threaded section to allow plate installation and tensioning. A relatively small plate is used compared to other commercially available energy dissipation systems.

Figure 7.87 shows the results from the dynamic testing program for bolts encapsulated in cement grout. The interpretation of the instrumentation data indicates an initial plate inversion followed by a slight stretch of the bar reaching a peak load of 200 kN for a brief period of time. The final stages of load transfer involve ploughing through the polymer coating at a constant load of approximately 100 kN.

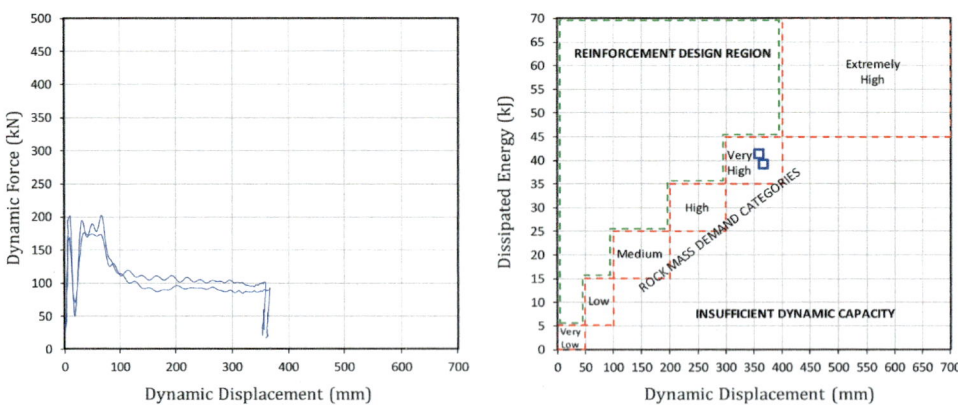

FIGURE 7.87 Dynamic force and related energy dissipation capacity for 17-mm diameter, cement-encapsulated Yield-Lok bolt (unstable collar region of 1.0 m, stable toe anchor region of 2.0 m).

Figure 7.88 shows the dynamic test results for resin-encapsulated bolts installed by mining jumbos within simulated boreholes as previously described. Because failure occurs at

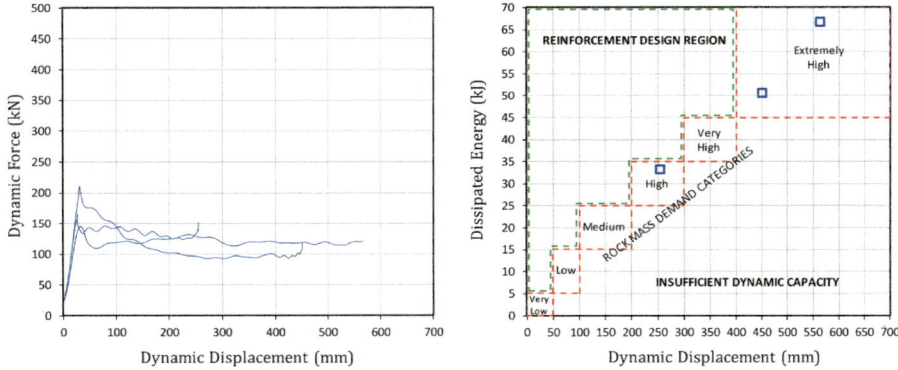

FIGURE 7.88 Dynamic force and related energy dissipation capacity for 17-mm diameter, resin-encapsulated Yield-Lok bolt (unstable collar region of 1.0 m, stable toe anchor region of 1.2 m).

the bolt-polymer coating interface, it is expected that a similar energy dissipation would be achieved with either resin or cement grout encapsulation. Figure 7.89 combines both the cement and resin-encapsulated Yield-Lok bolt test results. The data for both encapsulation systems are in agreement and show a large amount of deformation (Figure 7.90) at a very low load is possible.

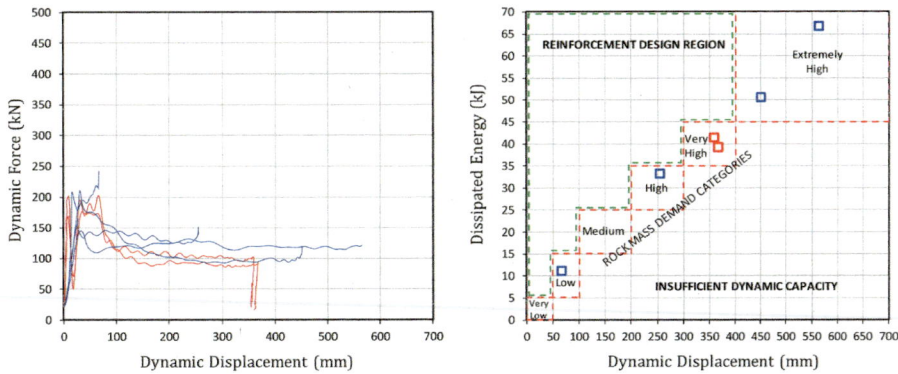

FIGURE 7.89 Dynamic force and related energy dissipation capacity for 17-mm diameter, cement- and resin-encapsulated Yield-Lok bolt (unstable collar region of 0.5–1.0 m, stable toe anchor region of 1.2–2.0 m).

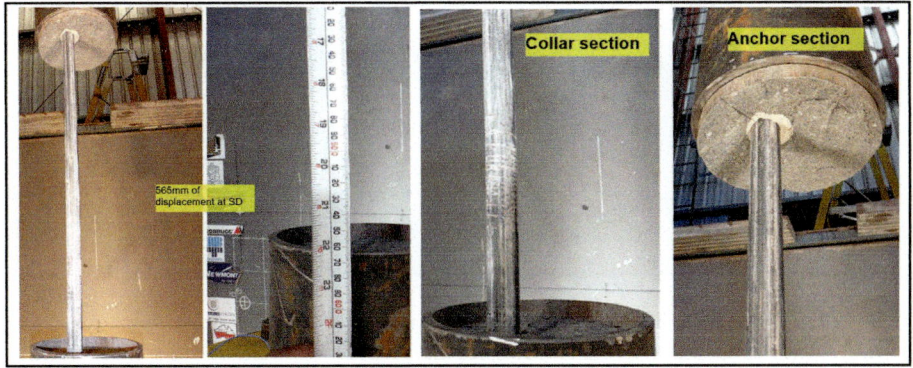

FIGURE 7.90 Example of a very large deformation following a dynamic test upon a Yield-Lok reinforcement system.

7.6.8 SELF-DRILLING ANCHOR BOLT

Self-drilling anchor bolts are hollow elements, fully encapsulated and mechanically coupled at the anchor-end and collar regions where all load transfer occurs. The elements are self-drilling (a bit is located at the anchor end) with resin injection through the element following bolt installation. The majority of the energy dissipation occurs within a 1 m long, smooth section in the central section of the bolt, where the element can elongate under dynamic load. A smooth, coarse thread is used to transfer the load at the anchor (1 m) and collar (0.6 m) regions (Figure 7.91). Because of the smooth tooth profile of the rope thread used, the nut may become loose following initial installation.

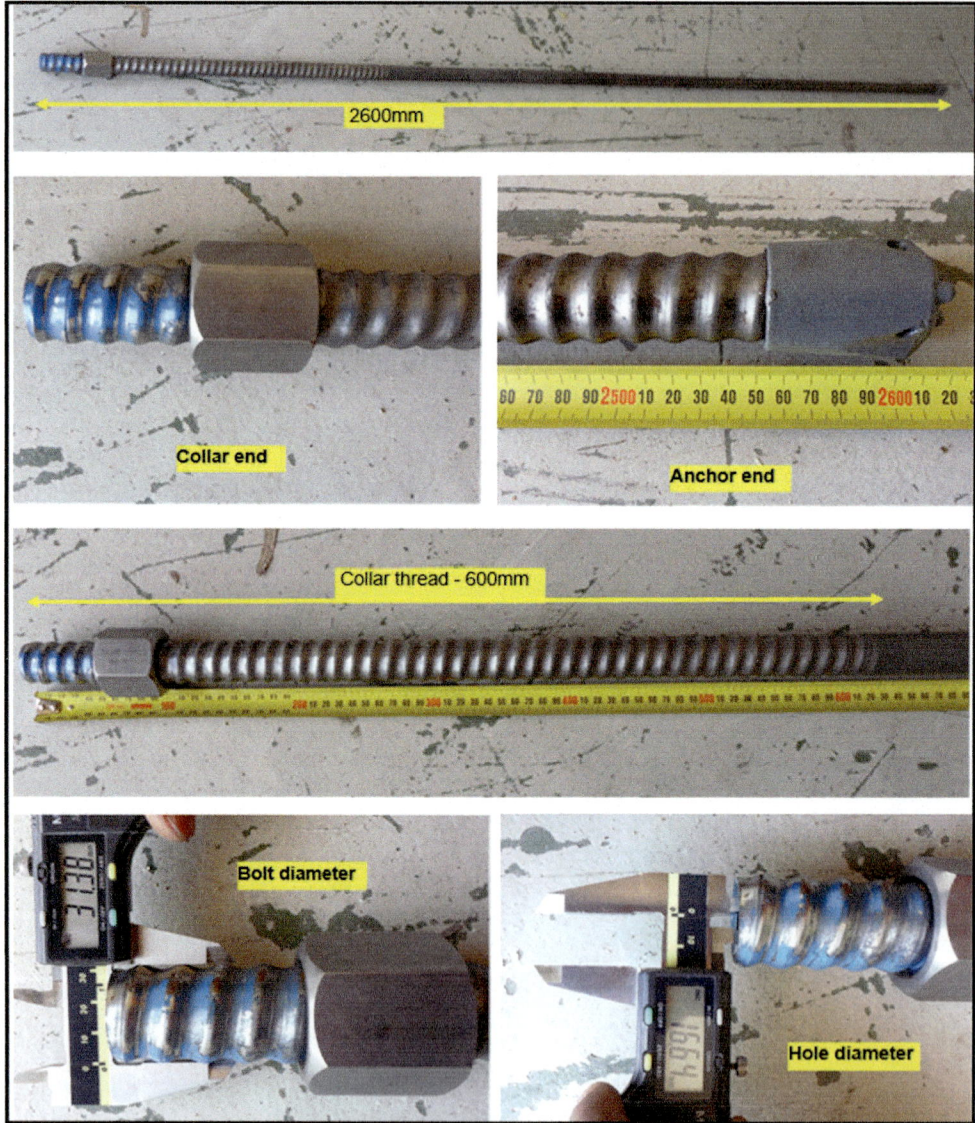

FIGURE 7.91   Geometrical details of a self-drilling, hollow rock bolt.

Figure 7.92 presents the results following dynamic loading for self-drilled, resin-encapsulated bolts. The testing was arranged so that the simulated instability occurred within the smooth middle section of the elements. That is, a simulated discontinuity was located at 1 m from the collar end, and the elements were able to dissipate 40 kJ prior to rupture at 150 mm of displacement. When loading was arranged to occur within the collar-threaded section, catastrophic failure occurred at very low energy dissipation. This suggests that the hollow bolt may be susceptible to interfacial shear failures.

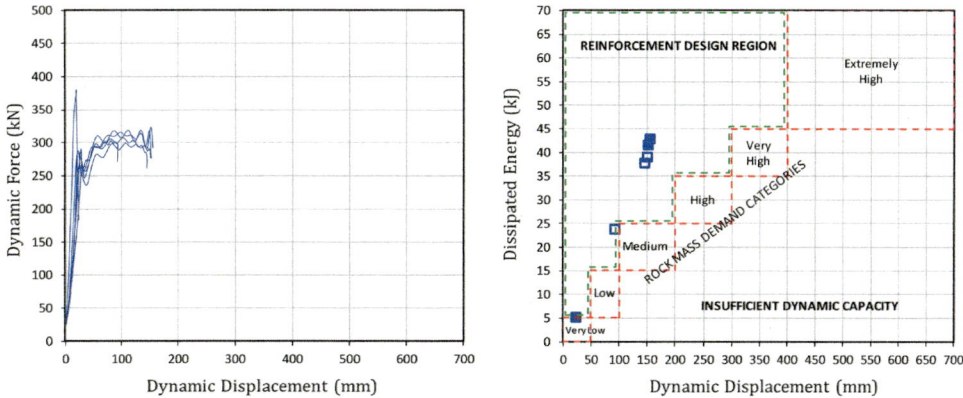

FIGURE 7.92 Dynamic force and related energy dissipation capacity for a hollow bar (OD 31.38 mm; ID 14.74 mm), resin-encapsulated SDA bolt (1.0 m deep instability, collar tail of 0.030 m, coupled collar region of 0.57 m, decoupled region of 1.0 m and coupled toe anchor region of 1.0 m).

8 Energy Dissipation of Cable Bolts

8.1 INTRODUCTION

Cable bolt reinforcement is used to stabilize wedges, large blocks or large depths of insta-bility that may form in the backs or walls of the mining development infrastructure. Cable bolts provide effective reinforcement of excavation walls where conventional rock bolts would prove geometrically inadequate due to their shorter embedment lengths or lack of capacity. In the context of deep excavations, the cable bolts provide very high retention when connected to high-capacity surface support, such as those formed in layers incorporating high tensile strength chain link mesh.

A cable bolt reinforcement system consists of four components (Windsor, 2004):

- The rock mass surrounding the borehole
- The internal fixture (cement grout)
- The element (strands)
- The external fixture (plate and barrel and wedge anchor)

The main stabilization mechanisms that operate during cable bolt reinforcement are:

- Application of compression to improve resistance against shear and tension across pre-ex-isting geological discontinuities.
- Creation of a composite beam of several layers of strata (when the cables are installed in bedded rock). The stability can be improved if individual bands can be grouped together to form a much stronger composite beam.
- Anchoring unstable zones to the stable or solid ground while providing large retention capabilities.
- Minimization of large excavation deformations arising because of violent rock mass failure and related bulking (Figure 8.1).
- Minimization of deformation related to time-dependent behaviour of weak rock masses loaded under high stress (Figure 8.2).

For high energy dissipation ground support, the stabilization process requires the implementa-tion of high-capacity surface support where rock bolts and cable bolts are installed within a staggered pattern (e.g., Arcaro et al., 2021). The reinforced excavation surface is anchored to better quality rock deeper in the rock mass by the longer cable bolts. The reinforcement length is typically taken as the potential depth of instability plus a specified length for anchorage, which usually exceeds 2 m. In practice, the typical length of a cable bolt ranges from 4 to 6 m, including the 0.5 m tail for plating. For static conditions, the cable bolt spacing is designed to provide a capacity exceeding the dead weight of the failed material. For dynamic conditions, the pattern is similar to that used for rock bolts, creating a high energy dissipation scheme is created.

DOI: 10.1201/9781003357711-8

FIGURE 8.1 Example of deformation control by cable bolts at the surface of an excavation following a sudden and violent large seismic event. Note the floor heave.

FIGURE 8.2 Example of deformation control by cable bolts at the surface of an excavation following a slow, time-dependent, deformation.

8.2 CEMENT GROUT

Cement grouting is the procedure in which a borehole (drilled into the boundary of an excavation) is filled with a cement paste which that the reinforcement element into the rock mass. This allows load transfer from a potentially unstable section of the rock mass (at the excavation boundary) back to a stable section deep in the rock mass through the reinforcing element through the load transfer concept (described in Chapter 2).

The simplest cement grouts are formed by mixing cement powder with water to form a cement paste. The physical and mechanical properties of the paste depend on the properties of the cement powder, the amount of water added and the conditions under which the cement paste is allowed to set and harden. Other factors, such as admixtures that modify the chemical reactions and physical properties of the cement grout, also affect the behaviour and properties of the cement paste. The properties also dictate the different mixing and pumping equipment that are required. Unfortunately, in many cases, the cement grout mix design is based on the ability of available equipment and not on the design requirements of a reinforcement system.

The strength of the cement grout is critical for optimizing the length of embedment needed to mobilize the ultimate capacity of the steel tendon used in a particular cable bolt system. In general, failure by slippage at the steel-grout interface will be experienced when weak grouts are used. Alternatively, rupture of the tendons can be expected when using thick, strong grouts in conjunction with stiff reinforcement systems of sufficient embedment.

8.2.1 PHYSICAL AND MECHANICAL PROPERTIES

Setting and hardening are two distinct phenomena associated with cement grouts. Setting broadly refers to the change from a fluid to a rigid state, while hardening refers to the gain in strength of a cement paste that has set. Figure 8.3 shows the composition of the cement paste schematically immediately after mixing and during the hardening phase. The cement paste initially consists of a network of capillary pores and gel pores. The pores are gradually filled with the chemical products of hydration as the cement sets and it hardens. The higher the WCR the greater the number of voids and capillaries that develop naturally in the grout as the water attempts to bleed to free surfaces. Voids may also be accidentally introduced into the grout during mixing. Voids must be minimized as they influence, amongst other things, grout strength, stiffness and permeability.

8.2.1.1 Fresh Cement Paste

For cable bolt applications, the fresh paste requires homogeneity, low air entrainment, delayed set time, appropriate pumpability characteristics and anti-bleed and anti-segregation characteristics. Cement grouts in the fluid state possess both viscosity and cohesion. As the WCR is decreased, both viscosity and cohesion increase markedly and reduce the fluidity, thereby making pumping and the flow of the grout more difficult. The cement grouts for cable bolt reinforcement must also be thixotropic. Thixotropy is a material property that enables the grout to be gel-like when stationary and to revert to fluid-like behaviour when agitated. This is an important property of cement grouts that enables them to behave as a fluid during mixing and pumping and subsequently resist segregation and flow into cracks and fissures in the rock surrounding a borehole.

The density of the fresh cement paste immediately after mixing may be used to estimate the number of voids in the grout and to indicate the viscosity. Fresh cement paste is a system comprising three components:

1. Cement particles—volume (V_c) and mass (W_c),
2. Water—volume (V_w) and mass (W_w), and
3. Air voids introduced during mixing—volume (V_{av}).

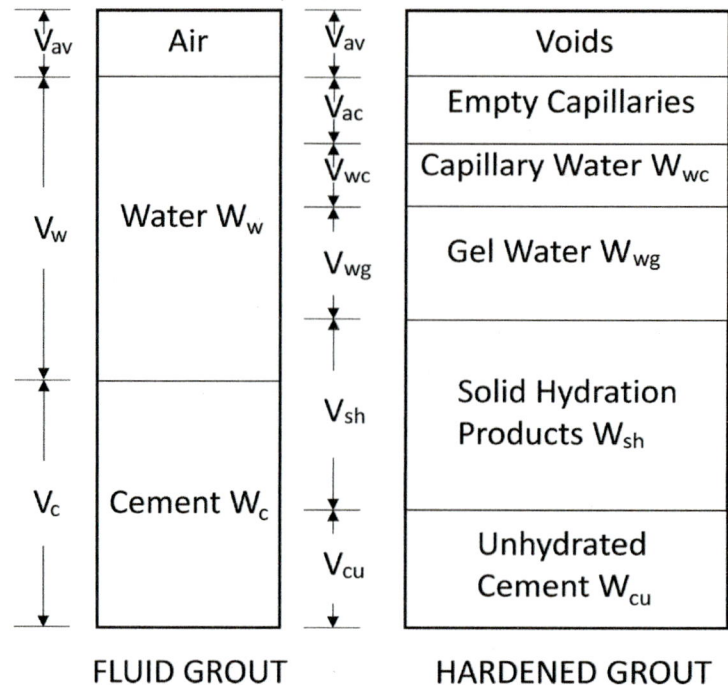

FIGURE 8.3 The components of fresh and hardened cement paste (Thompson and Windsor, 1998).

It is possible to derive the following relationship for wet bulk density (ρ_b):

$$\rho_b = \rho_w G_c \frac{WCR + 1}{(G_c WCR + 1)(1 + m)} \qquad 8.1$$

where

ρ_w is the density of water = 1.0 M_g/m^3 or 1.0 kg/litre

The WCR is defined as W_w/W_c

G_c is the specific gravity of the cement particles, usually with an assumed value of approximately 3.15

The air/void ratio 'm' is defined as $V_{av}/(V_c+V_w)$

The expression for wet bulk density described by Equation 8.1 can be plotted against WCR for various values of m and combined with the experimental results obtained by Hyett et al. (1992a) to produce Figure 8.4. The combined results suggest that the volume of air voids is increased at lower values of WCR, reflecting a greater volume of air entrainment in the cement grout during mixing due to the lower fluidity of the grout.

The rheology of the fresh grout is defined by its yield stress, which can also be used to characterize the grout viscosity during mixing and pumping. Yield stress is the stress at the limit of elastic behaviour. In other words, it is the minimum stress required to initiate grout flow at almost zero shear rate. Direct yield stress measurements can be undertaken using a method suggested by Nguyen and Boger (1985) and using a Haake VT550 viscometer.

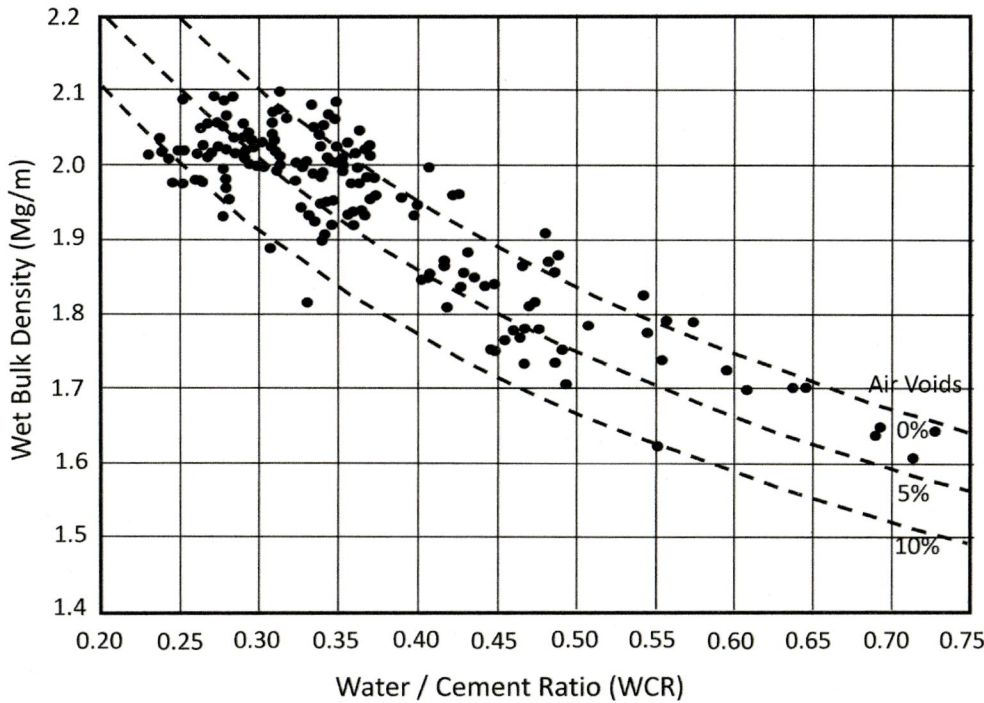

FIGURE 8.4   Variation of wet bulk density with water/cement ratio (Thompson and Windsor, 1998).

Yield stresses can be measured immediately after mixing, i.e., about 5–10 minutes after binder and water contact and at 20, 30 and 40 minutes intervals. The latter time would be the typical upper time limit for pumping and installation of a single grout batch. The viscometer uses a vane that is rotated at a shear rate of 0.5 rpm for 100 seconds, and the stress is recorded. The peak stress measured is reported as yield stress. Figure 8.5 show the relationship between the WCR and the yield stress for General Purpose (GP) Cement and other commonly used cement types, such as Low Heat Cement (LH) in Australia. The data indicates that as the WCR increases, the yield stress decreases,

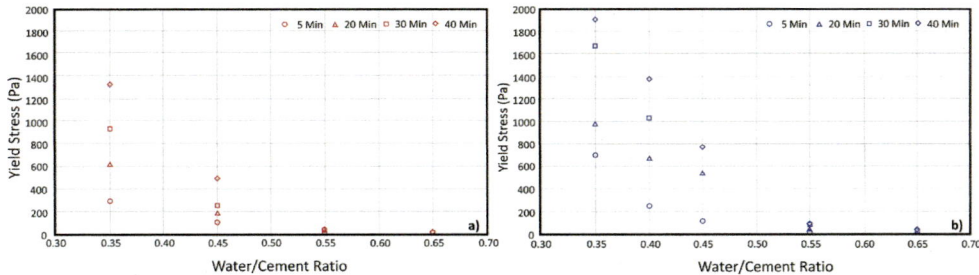

FIGURE 8.5   Yield stress of fresh grout with different water/cement ratios: (a) GP Cement and (b) Low Heat Cement.

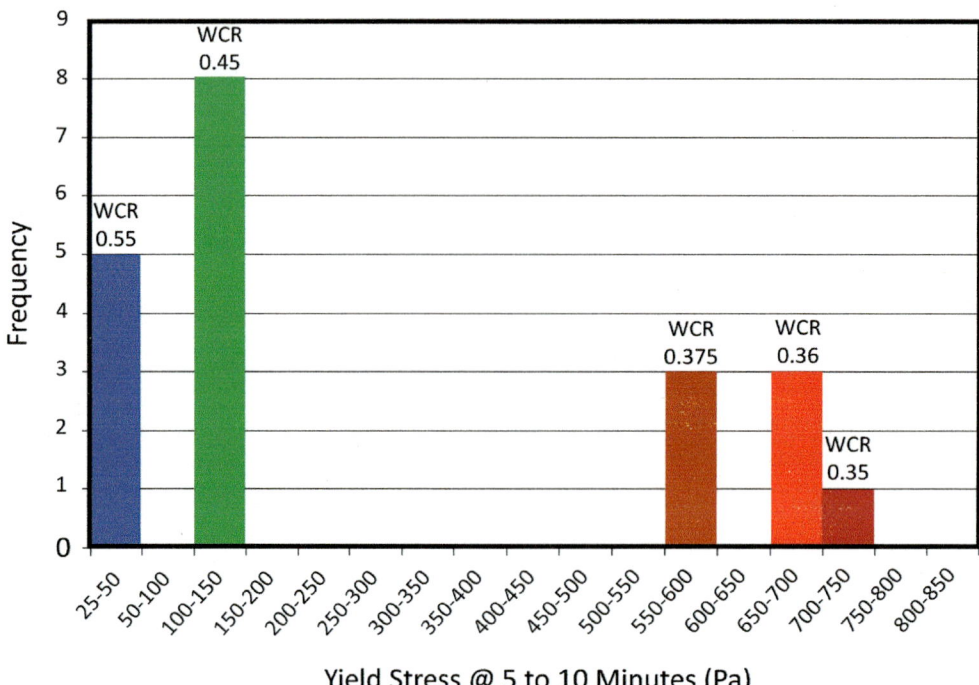

FIGURE 8.6 Typical yield stress for manual cable bolt installation (water/cement ratios of 0.55 and 0.45) and mechanized installation (water/cement ratios of 0.375–0.35). These results are for Low Heat Cement with 0.02% Methocel/weight of cement.

resulting in a lower-strength, cured grout. The yield stress (during grout mixing) for mechanized grout installations ranges from 400 to 800 Pa, while for the conventional breather tube grouting method, it ranges from 100 to 300 Pa (Figure 8.6).

8.2.1.2 Hardened Cement

The final hardened state of cement grout demands mechanical properties that allow the grout to interact with the reinforcement and chemical properties that allow it to maintain form and resistance to attack by deleterious agents (particularly sulphides) within the rock mass. The mechanical requirements include a range of compressive strength, shear strength and stiffness. The final hardened grout needs to be homogeneous, have low porosity and, for longevity considerations, low permeability and high alkalinity.

8.2.1.2.1 Hydration Process

Following placement, the cement paste will begin to hydrate, set and harden. During these processes, the paste will contain, as shown schematically in Figure 8.3, various proportions of:

1. Unhydrated cement,
2. Solid products of hydration,
3. Gel water,
4. Water-filled capillary pores,
5. Empty capillary pores and
6. Voids.

The actual proportions of cement, water and air introduced into the grout during mixing and after placement will influence whether a particular component is present and the proportion it forms of the total volume of the mixture. The overall porosity of the paste is usually in the range of 30% to 40%. The gel pores comprise about 25% of the porosity, with the capillary network forming the rest (Jastrzebski, 1976). The gel fraction of the overall cement paste can only increase whilst free water is available for adsorption by the cement. Free water is inherent in the initial mix, and hydration proceeds until all the cement particles are hydrated or the supply of free water is exhausted.

The water in a cement/water mix comprises three components:

1. Non-evaporable water;
2. Solid products of hydration; and
3. Evaporable water, comprising 'Gel water' and 'Free water'.

The gel water cannot move into the capillaries and is not available for further hydration of unhydrated cement.

Previous investigations have identified three constants that may be used to describe cement paste during the hydration process. These factors are as follows:

n_h = The porosity of the hydrated cement gel
f_n = The volume of non-evaporable water that reacts with the unhydrated cement
f_v = The volume of empty capillaries that result due to hydration
F_h = The degree of hydration

The porosity of the hydrated cement gel (n_h) is defined by:

$$n_h = \frac{V_{Wg}}{V_{Sh} + V_{Wg}}$$ 8.2

where

V_{wg} = Volume of gel water
V_{sh} = Volume of hydration solids

The volume of non-evaporable water (V_{wn}) is defined by:

$$V_{wn} = f_n f_h \frac{W_c}{\rho_W}$$ 8.3

where

f_n is the non-evaporable water constant and f_h is the degree of hydration given by

$$f_h = \frac{W_{ch}}{W_{ch} + W_{cu}}$$ 8.4

where

W_{ch} = Weight of hydrated cement.
W_{cu} = Weight of unhydrated cement

The hydration volume reduction factor (f_v) is defined as the ratio of the capillary volume increase due to the combination of unhydrated cement and water to the total volume of cement and water. f_v is used to determine V_{ac}, which is the volume of empty capillaries, in the following equation:

$$V_{ac} = f_v V_{wn} \qquad\qquad 8.5$$

Typical values for n_h, f_n and f_v can be used to explain how cement paste properties, such as compressive strength and stiffness, are affected by the WCR and curing conditions. Typical values of $n_h = 0.28$, $f_n = 0.23$ and $f_v = 0.254$ were suggested by Neville (1963) for Ordinary Portland Cement. However, the values for the constants and the resulting properties will vary somewhat with the type and particle size distribution of the cement used.

The volume of the solids (V_{sh}) that form from the reaction of unhydrated cement is given by:

$$V_{sh} = V_{ch} + V_{wn} - V_{ac} = f_h \frac{W_c}{G_c \rho_w} + f_n f_h \frac{W_c}{\rho_w}\left(1 - f_V\right) \qquad\qquad 8.6$$

The volume of gel water is then given by:

$$V_{wg} = \frac{V_{sh} n_h}{1 - n_h} = \frac{n_h}{1 - n_h} f_h \frac{W_c}{\rho_w}\left\{ \frac{1}{G_c} + K_h\left(1 - f_v\right)\right\} \qquad\qquad 8.7$$

The total volume of water (V_{wh}) required for hydration is given by:

$$V_{wh} = V_{wn} + V_{wg} = f_h \frac{W_c}{\rho_w}\left[f_n + \frac{n_h}{1 - n_h}\left\{ \frac{1}{G_c} + f_n\left(1 - f_v\right)\right\}\right] \qquad\qquad 8.8$$

This last expression may be used to calculate the minimum volume of water required to fully hydrate a given amount of cement under different curing conditions.

8.2.1.2.2 Influence of Curing Conditions

In a sealed specimen, hydration continues until the combined water is a fraction of the original water content. The actual amount of water required will vary according to the exact nature of the cement and its constituent minerals.

Using Equation 8.8, the WCR may now be given by:

$$\mathrm{WCR} = \frac{\rho_w V_{wh}}{W_c} = f_h\left[f_n + \frac{n_h}{\left(1 - n_h\right)}\left\{ \frac{1}{G_c} + f_n\left(1 - f_v\right)\right\}\right] \qquad\qquad 8.9$$

Substituting the constants defined previously and using $f_h = 1$, Equation 8.9 predicts that a WCR of 0.42 is required for full hydration under conditions where no water can either evaporate or infiltrate the grout paste during curing. Using a lower mixing WCR, less than full hydration would occur. However, additional hydration, limited by the available empty capillary space established during the setting process, may be achieved in conditions where the grout may absorb extra water during the curing process.

The extra volumes of the solids of hydration (V_{shc}) and gel water (V_{wgc}) can exceed the original volume of the additional cement hydrated (V_{chc}) by up to the volume of empty capillaries following initial hydration. This can be expressed as:

$$V_{she} + V_{wge} \leq V_{che} + V_{ac} \qquad\qquad 8.10$$

If the extra weight of hydrated cement is expressed as a fraction of the original weight of cement, then the extra hydration factor (f_e) is given by:

$$f_e = \frac{V_{ac}}{V_c} \frac{(1-n_h)}{n_h + G_c f_n (1-f_v)} \qquad\qquad 8.11$$

If f_e is set equal to $1-f_h$, then a value of f_h may be calculated to ensure that just enough empty capillaries are created in the original mix so that the extra hydration products fill this volume and any further newly created empty capillaries. The critical value of f_h is given by:

$$f_h = \frac{n_h + G_c f_n (1-f_v)}{n_h + G_c f_n (1-f_v n_h)} \qquad\qquad 8.12$$

This value of f_h can then be substituted into Equation 8.8 used previously for the WCR. Using this approach, it is found that under conditions where additional water is provided during curing, the WCR required for full hydration is 0.36 compared with the value of 0.42 when no additional water is available. Various authors (e.g., Neville, 1963, Taylor, 1997) have suggested that a minimum value of WCR of 0.38 is required for complete cement hydration for the latter curing conditions. It may be assumed that the precise value of the minimum WCR is a function of the specific cement, the method of placement and the curing conditions.

Extending the previous theoretical considerations, the density of the hardened cement paste (ρ_p) is given by:

$$\rho_p = \frac{\rho_W G_c}{1+m} \frac{1 + f_d WCR + f_h \dfrac{n_h}{1-n_h}(1-f_d)\left(1-\dfrac{1}{G_c}-f_n f_v\right)}{1+G_c WCR} \qquad\qquad 8.13$$

Here, f_d is the measure of the degree to which evaporation of free water from the capillaries has occurred. If the sample is sealed during curing, then $f_d = 1$ and the density after curing given by Equation 8.13 is the same as the density of the wet cement paste given by Equation 8.1. This theoretical finding is confirmed by the experimental results of Hyett et al. (1992), who noted little difference between the density of hardened cement paste and fresh cement paste samples.

8.2.1.2.3 Strength of Hardened Grout Paste

A simple empirical relationship for the strength of cement grouts has been established from extensive work performed on cement paste and its influence on the strength of concrete. It has been found that the strength of cement pastes is best represented as a function of the gel/space ratio (GSR), defined as:

$$GSR = \frac{V_{sh} + V_{wg}}{V_{ch} + V_w + V_{av}} \qquad\qquad 8.14$$

Empirical correlation of the observed unconfined compressive strength (S) with GSR results in the following equation:

$$S = S_0 GSR^3 \qquad\qquad 8.15$$

where S_0 is the inherent strength for a particular cement type and may be assumed as 90 MPa.

If the rate of hydration is known, then Equations 8.14 and 8.15 may be used to predict the strength at any age since the GSR changes as the hydration of cement progresses. These equations have been used to produce the three lines shown in Figure 8.7. The lines are drawn to show the variation of hardened paste strength with WCR for ideal mixing with no entrained air and for 5% and 10% of entrained air. At WCRs less than the critical value of 0.42, complete hydration can only occur if additional water is provided during the curing period. The effect of providing additional water for further hydration of cement is shown by the dotted lines. The experimental data shown in Figure 8.7 are taken from Hyett et al. (1992a). The scatter in results below the WCR of 0.42 can be attributed to the increased presence of air voids resulting from mixing and variability of the effectiveness of the curing process. Additionally, a set of qualitative regions classifying the grout strength from very low to very high are also shown in Figure 8.7. This classification forms the basis of all subsequent cable bolt energy dissipation comparisons undertaken within this chapter.

Figure 8.8 shows an experimental relationship between grout density and UCS for several cylindrical specimens tested at the WASM. The samples include a wide range of WCRs with several cement types. Curing was undertaken using a temperature and humidity-controlled curing chamber, which is shown in Figure 8.9.

8.2.1.2.4   Stiffness of Hardened Grout Paste

Similar to the relationship for strength, an empirical relationship has been derived for stiffness. Ashby and Jones (1986) have proposed:

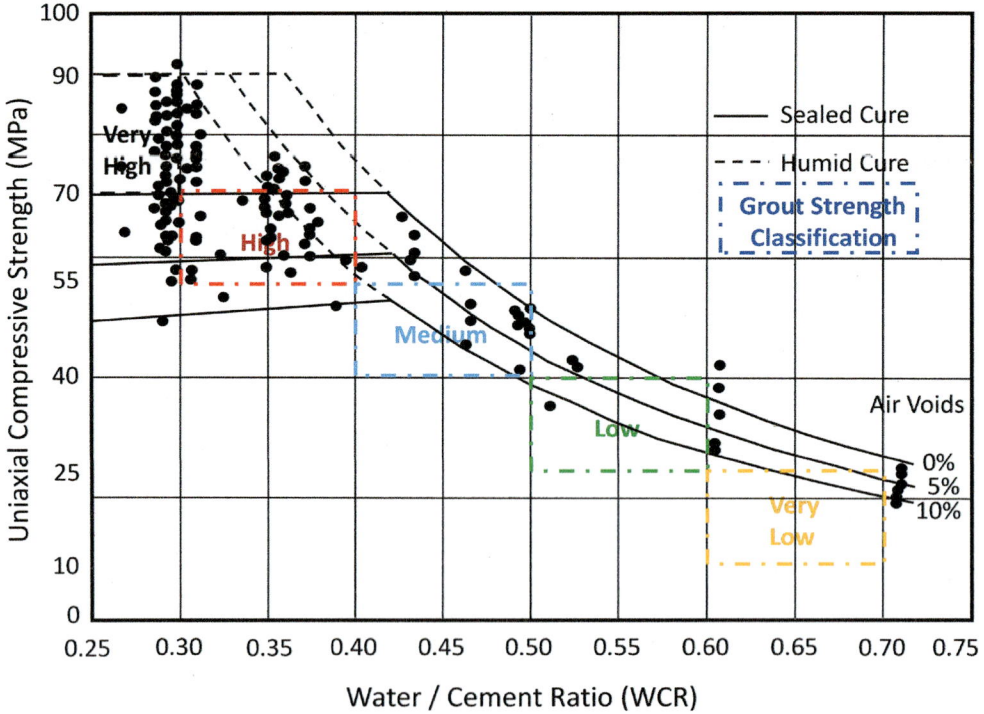

FIGURE 8.7   Classification of grout uniaxial compressive strength for different water/cement ratios (modified after Thompson and Windsor, 1998).

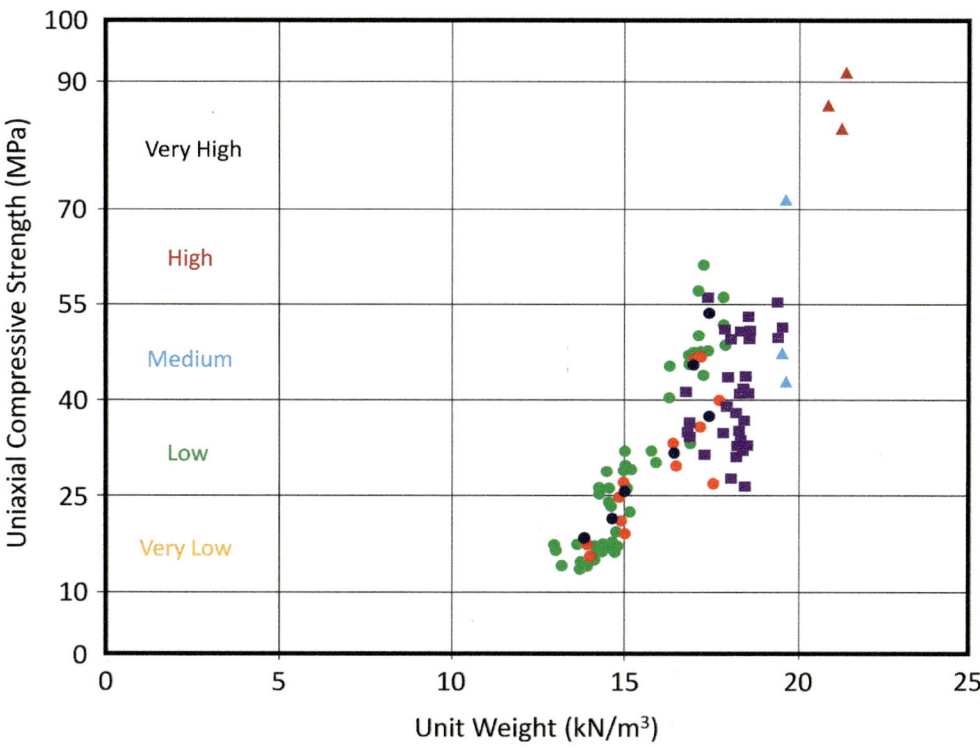

FIGURE 8.8 Relationship between grout strength and unit weight for several grout mixes.

FIGURE 8.9 Example of a temperature and humidity-controlled chamber for curing of cement grout specimens.

$$E_p = E_g \left[\rho_\pi / \rho_\gamma \right]^3 \qquad\qquad 8.16$$

where

E_p = Young's modulus of cement paste at a bulk density of ρ_p

E_g = Young's modulus of solid cement gel at a density of ρ_g

The theory of grout hydration may also be used to calculate the density of the solid cement gel. That is,

$$\rho_g = \frac{W_{sh}}{Vsh} = \rho_w G_c \frac{1 + f_n}{1 + G_c f_n \left(1 - f_v \right)} \qquad\qquad 8.17$$

Substituting the values given previously for G_c, f_n and f_v leads to a value of 2.52 mg/m³ for ρ_g.

The variation with WCR of the stiffness of hardened cement paste for different values of air voids has been estimated by firstly using Equation 8.17 to predict the solid cement gel density and then substituting the resulting value into Equation 8.16. The results from the analysis are shown in Figure 8.10, together with the grout stiffnesses obtained experimentally and reported by Hyett et al. (1992). Again, the variation in the results may be attributed to the increased likelihood of air voids being introduced during the mixing of lower WCR grouts.

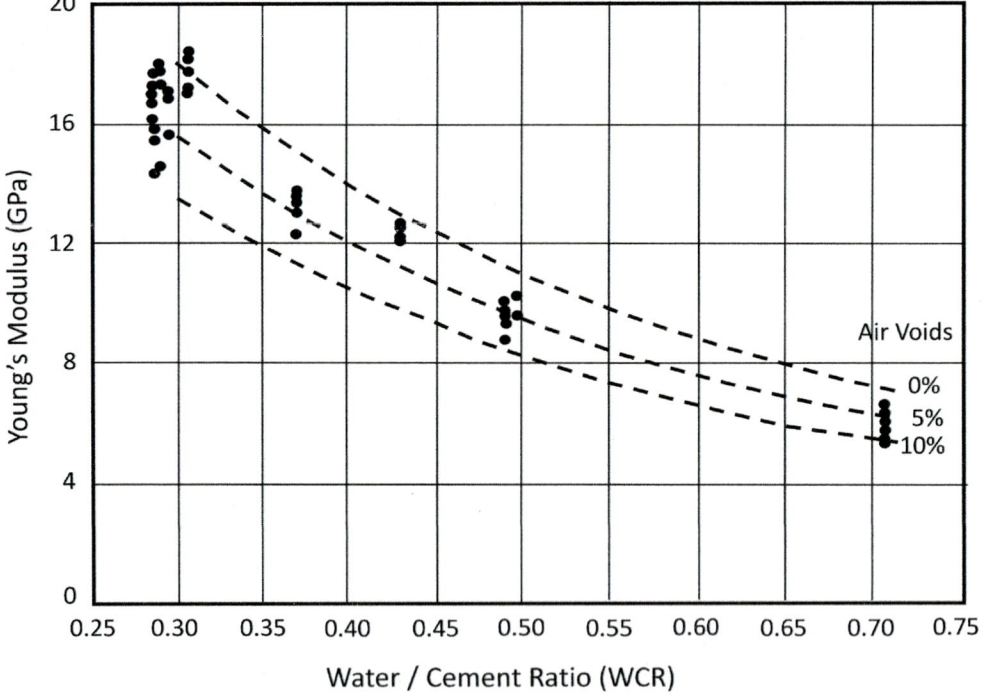

FIGURE 8.10 Variation of grout stiffness for different water/cement ratios (Thompson and Windsor, 1998).

8.2.2 GROUTING REINFORCEMENT BOREHOLES

Historically, collar-to-toe grouting methods were developed in conjunction with piston-based cement grouting pumps. The WCR for such cement grouts can range from 0.40 to 0.55. This method requires a breather tube, generally greater than 13 mm (inside diameter), to be attached to a cable bolt element before it is installed into the borehole. A permanent collar packing is placed at the hole collar as a seal to prevent grout loss. A short grouting hose of approximately 1.0 m in length is also placed permanently in the collar of the hole. The grout is pumped through the grouting hose, and when it reaches the upper end of the breather tube, it begins to flow back through this tube. When the full flow of grout emerges from the end of the breather tube, it indicates that full encapsulation of the steel tendon has been achieved (Figure 8.11).

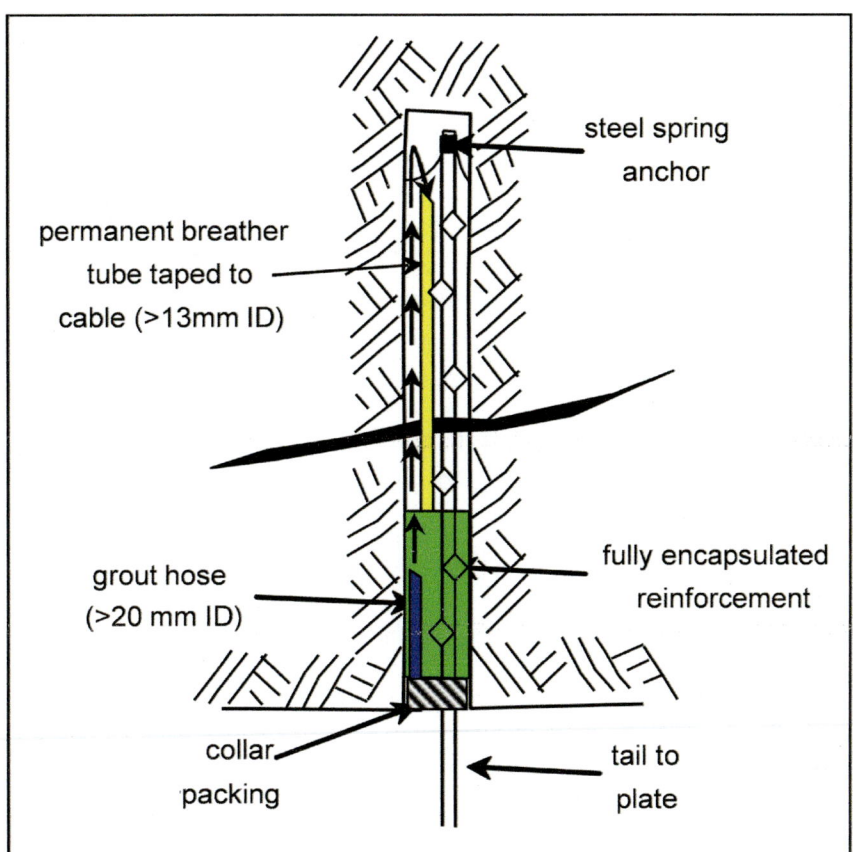

FIGURE 8.11 Example of a conventional collar-to-toe grouting method.

A typical piston-based pumping system usually requires that the mixing and pumping are carried out using the same container. This is called a one-stage grouting system. Mixing of the grout is achieved using paddles that rotate around a vertical axis (Figure 8.12). This may create a settlement of the cement particles into the area where the pumping is taking place, i.e., the bottom of the mixing tank, potentially reducing pumpability if the grout is too thick. Using a single container can lead to changes in WCRs during grouting operations. Furthermore, no accurate devices to measure the

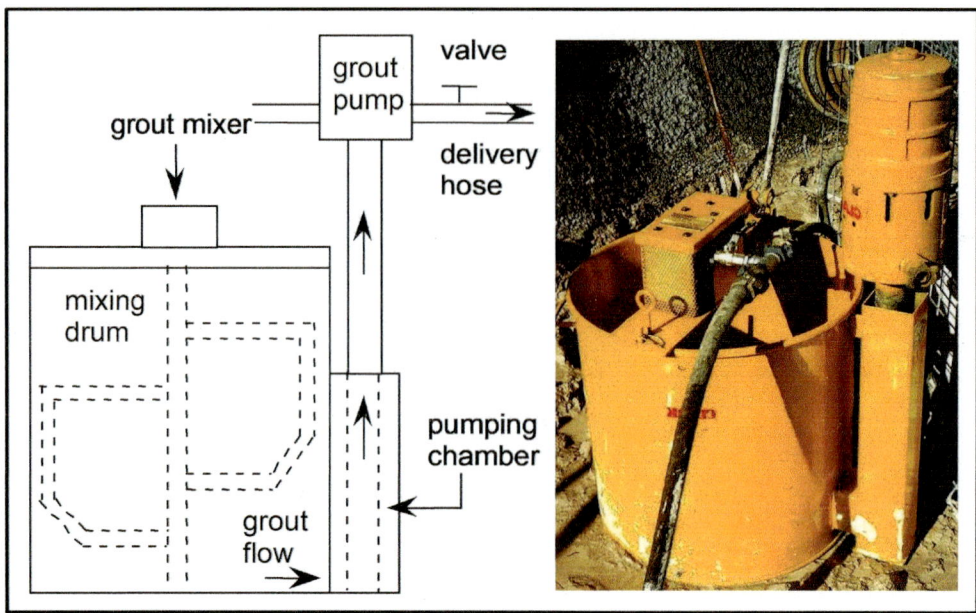

FIGURE 8.12   Example of a conventional one-stage piston-based grouting system.

amount of water being supplied to the mixing tank are rarely fitted to most conventional one-stage grouting systems. Consequently, following an initial mix design in which the WCR was probably correct, additional water could be added (while still pumping and grouting) before a corresponding amount of cement is added to the mix. This problem can be avoided if each mix is pumped separately.

8.2.2.1 Toe-to-Collar Grouting

Toe-to-collar grouting utilizes thicker pastes and involves inserting a cable bolt without a breather tube into the hole to be grouted (the need for the collar plug is eliminated), followed by subsequent grouting of the cable by means of a self-retracting grouting hose (Figure 8.13). The grout paste pushes the grouting hose out of the hole as the grouting proceeds. The optimal grouting rate is such that a self-retracting hose should be in minimal contact with an advancing grout paste inside the hole. To achieve this, the typical WCRs required usually range from 0.32 to 0.35.

This method of grouting provides several advantages, including a much faster initial cable placement and pre-grouting preparation times, savings on materials, faster rates of grouting, and a significant increase in grout strength. Strong and thicker grouts do not leak into voids and crevices encountered along the drill hole axis, thereby minimizing the wastage of cement. A disadvantage is that the increased strength of the grout may effectively decrease the deformation required to achieve the nominal steel capacity, thereby decreasing the energy dissipation capacity of the whole reinforcement system.

This technique has been implemented successfully mainly due to the use of modern two-stage grout pumps that allow a higher degree of quality control on the mixing, pumping and WCRs used. A two-stage process means that the machine has separate mixing and pumping containers that can be operated simultaneously or independently, thereby significantly increasing productivity (Figure 8.14).

The ability to mix independently of pumping allows a consistent WCR to be achieved throughout a grouting operation. This enables a high degree of quality control, as an accurate water meter allows water addition to be controlled to a precision of one-tenth of a litre. Cement, water and additives are mixed in a horizontal paddle mixer and then discharged into the lower hopper where a variable speed drive coupled to a rotor-stator pump (or mono pump) discharges the grout at the desired rate.

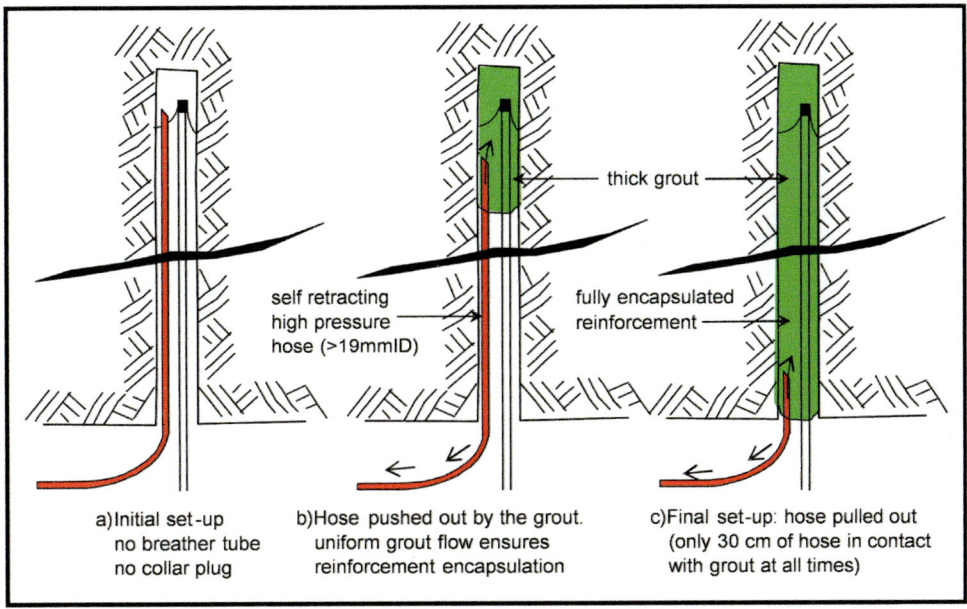

FIGURE 8.13   Toe-to-collar cement grouting method.

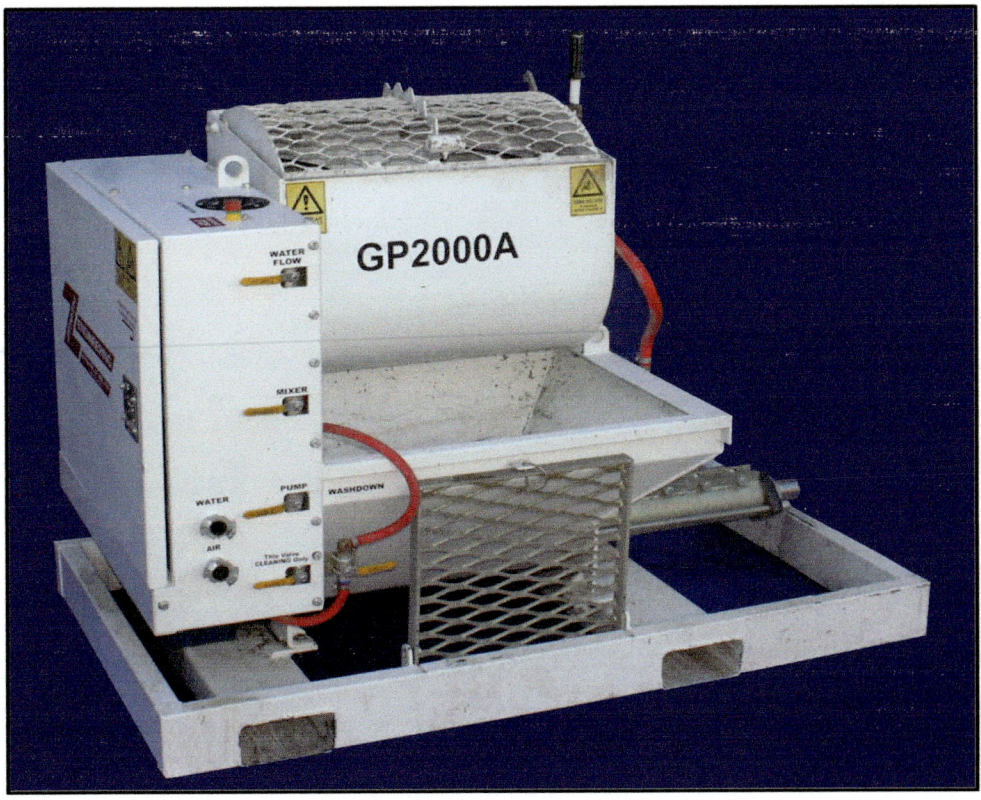

FIGURE 8.14   Example of a two-stage mono pump cement grouting machine.

A disadvantage of this method is the potential for poor cable bolt encapsulation that may result if the grouting operator retrieves the hose while grouting is underway. Consequently, in order to avoid potential encapsulation problems, the grouting house could be left in place (with no retracting) and cut when the grout reaches the collar of the hole (Figure 8.15).

FIGURE 8.15   Example of a broken cable bolt that was installed using a toe-to-collar grouting technique in which the grouting hose was left in place.

8.2.2.2   Mechanized Grouting

Toe-to-collar grouting is also carried out during the mechanized installation of cable bolts using a cable bolting machine (Figure 8.16). The process consists of hole drilling followed by cement grouting using a self-retracting hose. The rate of grouting and hose retraction is mechanized, thus ensuring that no gaps are left along the length of the holes. Once the holes are grouted, the cable bolts are inserted into the grout-filled holes. The cable bolt ends are then mechanically cut in place with a tail left for platting at a later stage.

Regardless of the grouting method used, several issues require consideration during the selection of the most appropriate grout mix design to suit a particular operation. The volume of grout that can be efficiently mixed and the ability of a machine to mix the required WCR in a reasonable amount

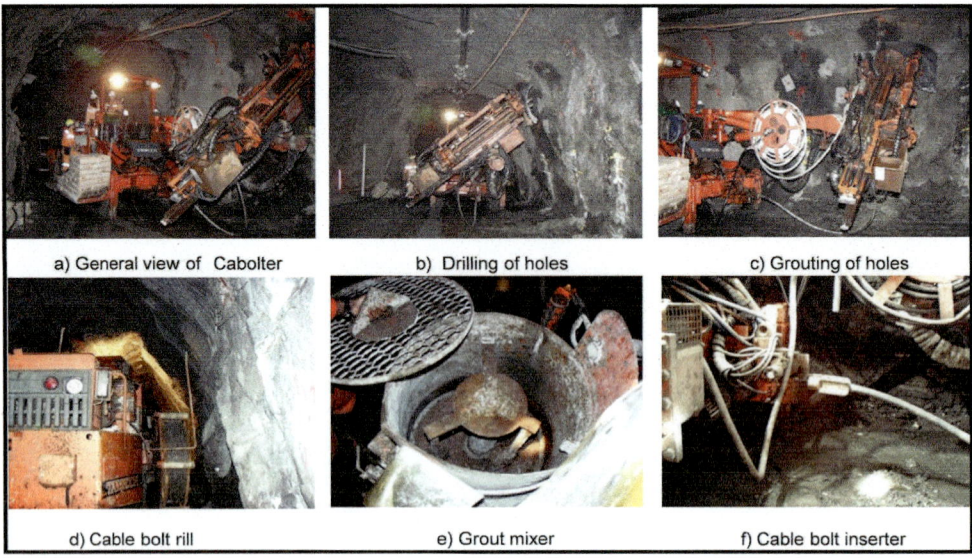

FIGURE 8.16   Mechanized toe-to-collar grouting method using the Tamrock Cabolter.

of time must be accounted for. In general, the use of additives is recommended for efficient grouting of cable bolts. Additives prevent segregation of water and cement while reducing grout shrinkage during curing. Preventing water-cement segregation at the toe end of the hole is very important to achieve the required anchorage required for load transfer. A typical additive used in Australia since the early 1980s is Methocel, which is mixed at 0.2%/weight of cement (Villaescusa et al., 1992). This prevents water and grout segregation and reduces shrinkage during curing while improving pumpability, allowing denser grouts to be used.

8.3   CABLE BOLT TYPES

The most widely used cable bolts in underground mining consist of a 7-wire, stress-relieved, high tensile steel strand with plain (round) wires. The wires are spun helically around a slightly larger diameter central (king) wire. The cables can be produced from several steel grades that provide differing yield and ultimate load capacities. The 15.2-mm diameter single strands have a minimum yield force capacity of 213 kN and a minimum breaking force of 250 kN. Alternatively, the 17.8-mm diameter cables have a minimum yield force capacity of 294 kN and a minimum breaking force of 353 kN. Figure 8.17 shows some of the typical cable bolting geometries used in the mining industry over several decades (Windsor and Thompson, 1993).

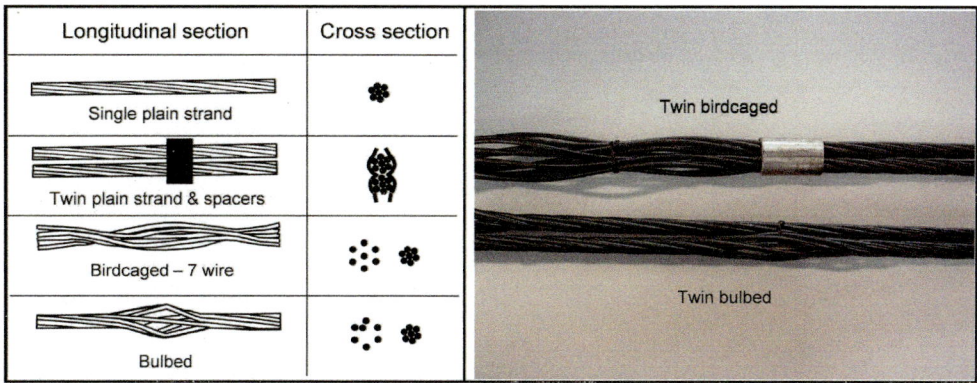

FIGURE 8.17   Typical cable bolting geometries (after Windsor and Thompson, 1993).

8.3.1   MODIFIED STRAND CABLE BOLTS

A typical plain cable bolt strand can be modified to cause a variation in cross-section along its length. These variations include bird-caged and bulbed strands, and both result in more effective load transfer between the strand and the cement grout. The more effective load transfer is reflected by the need for a shorter embedment length in which to develop the strand capacity and higher values for the force-displacement response stiffness. However, the higher values of force are likely to reach the strand capacity at much lower displacement values, thus dissipating less energy under loading. Figure 8.18

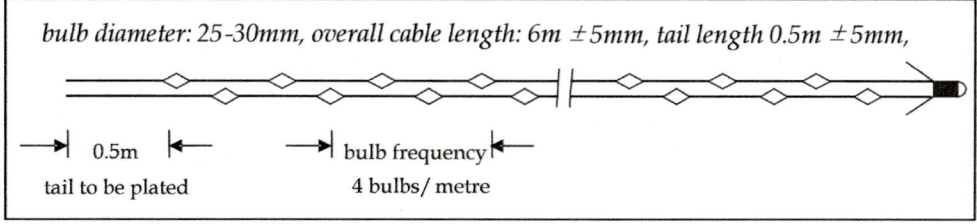

FIGURE 8.18   Schematic of twin strand bulbed cable bolt used for reinforcement in hard rock.

shows a schematic example of a twin bulbed cable geometry used for the development of excavation reinforcement in which the installed bulb frequency is 4/metre. This frequency provides a stiff reinforcement likely to minimize the displacement before strand rupture occurs. Figure 8.19 shows the results from static testing where the displacement to failure is decreased by increasing the number of bulbs per metre. The typical bulb diameter ranges from 25 to 30 mm, thereby facilitating the use of thick cement grouts that can effectively penetrate the bulbs and create anchor points along the cable bolt length.

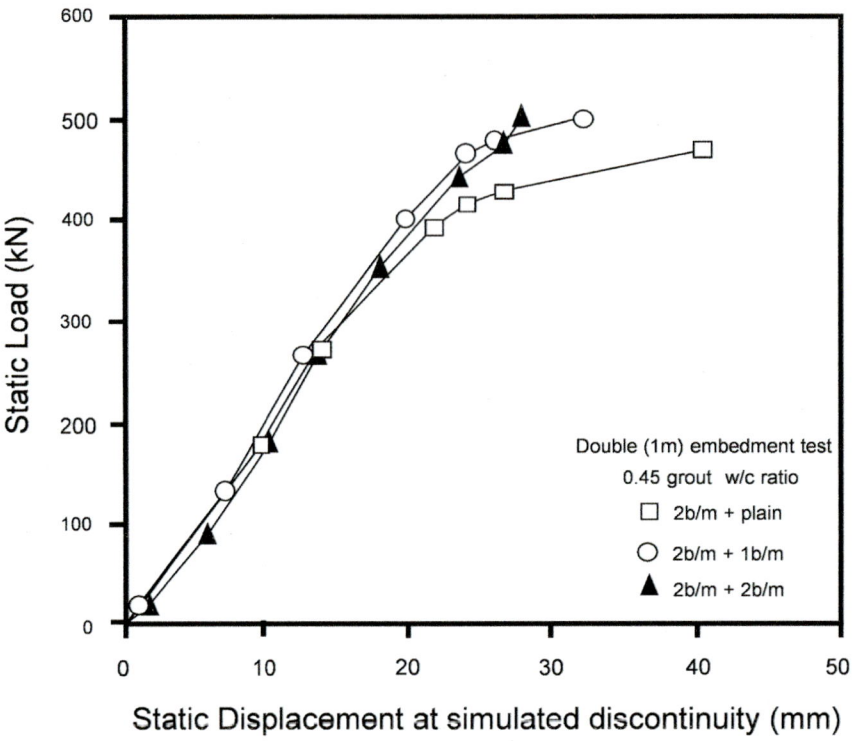

FIGURE 8.19   Split-pipe testing results from static loading showing that the displacement to failure within the simulated discontinuity increases by reducing the number of bulbs per metre.

8.3.2   PLAIN STRAND—15.2 MM DIAMETER

Plain strand cable bolts are compatible with the displacements generated at fast rates of load transfer and have been used over many years for ground support under dynamic loading conditions. The load-displacement response is a function of the WCR, as shown in Figure 8.20, where failure by slippage was experienced during the static testing experiments (Villaescusa et al., 1992).

The data indicates that for lower strength grouts, a similar displacement can be achieved, but the force does not immediately reach the strand capacity, thus potentially dissipating more energy prior to an eventual rupture. Operationally, the typical WCR used for conditions of dynamic loading range from 0.4 to 0.45, which is compatible with the breather tube grout installation method (Villaescusa, 2014). In addition, the displacement also increases when the cable bolts are not centred within a borehole, which is the norm for most inclined holes. Furthermore, plain cable bolts also

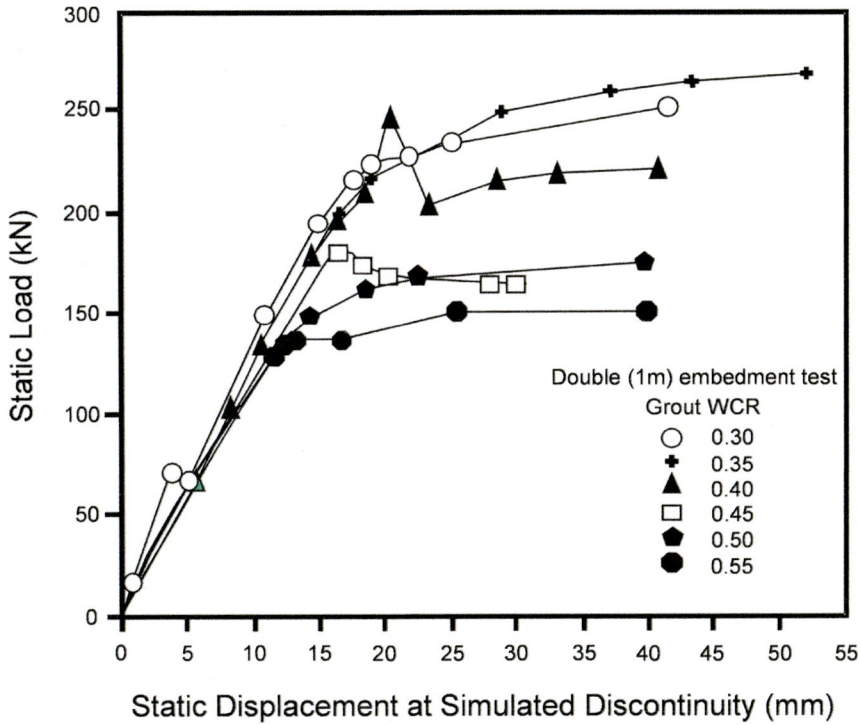

FIGURE 8.20 Static testing results from split-pipes (collar region of 1.0 m, toe anchor region of 1 m) showing the influence of cement grout on the load transfer of single, 15.2-mm diameter plain strand cables.

suffer from a significant reduction in the rate of load transfer if the borehole confinement (radial stress) reduces (Hyett et al., 1992b). Consequently, the load transfer within unstable, broken regions near the excavation boundaries is highly dependent on the external fixture (i.e., barrel and wedge) performance, as the plates are usually dynamically loaded by seismic events (Figure 8.21).

Figure 8.22 shows the results from dynamic testing of a plain, single-strand 15.2-mm diameter cable installed within split-pipes where the reinforcement element under load preferentially slips from a 1.5 m long toe anchor region. Testing was undertaken using grout with a nominal WCR of 0.4. A 1.0-m collar region was defined by a simulated discontinuity, and the cable bolt was plated. Consequently, most of the displacement occurred within the toe anchor region.

Note: In all the subsequent figures within this chapter, a solid legend symbol means a violent cable bolt rupture, while an empty symbol means a stable test result. Similarly, a legend symbol with a black outline indicates failure by slippage.

Figure 8.23 shows a similar plot for plated, plain, single-strand 15.2-mm diameter cables installed within a split-pipe with a 2.0 m long toe anchor region. The cables were grouted using a very low-strength grout (see Figure 8.7 and Figure 8.8). As the cables were plated, most of the displacement occurred within the toe anchor region. Large displacements and low residual forces resulted from the dynamic testing.

Figure 8.24 shows a comparison with similar plated cables with a slightly larger toe anchor region, which have been loaded by multiple dynamic impacts. The interpretation is that a plain, single strand cable installed in very low-strength grout is capable of displacing 200 mm and dissipating

FIGURE 8.21   Example of loaded cable bolt external fixtures following a large seismic event.

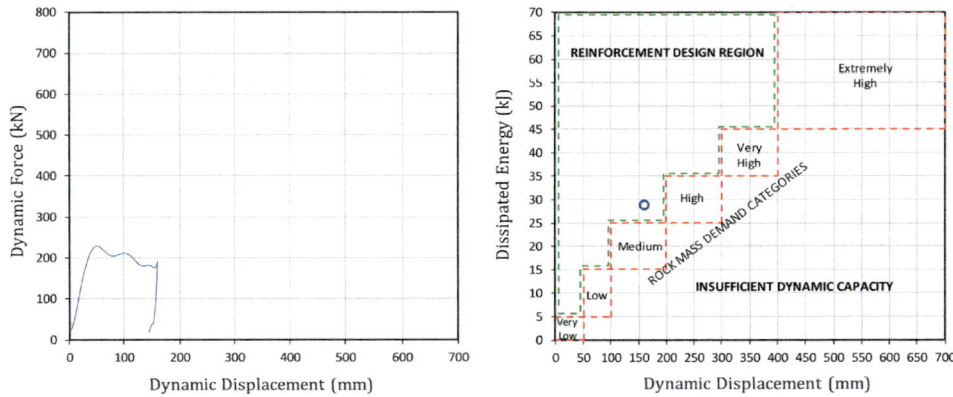

FIGURE 8.22   Dynamic force and related energy dissipation capacity for a plain, single-strand 15.2-mm diameter cable bolt showing displacement within the toe anchor region (collar region of 1.0 m, toe anchor region of 1.5 m).

up to 30 kJ. The residual force capacity remains above 100 kN, even for very large displacements that exceed 400 mm.

Figure 8.25 shows results from dynamic testing of plated, plain, single-strand 15.2-mm diameter cables encapsulated in medium-strength cement grout. The results indicate that displacements of up to 200 mm while dissipating 35 kJ are possible. The residual force capacity remains very high, exceeding 150 kN.

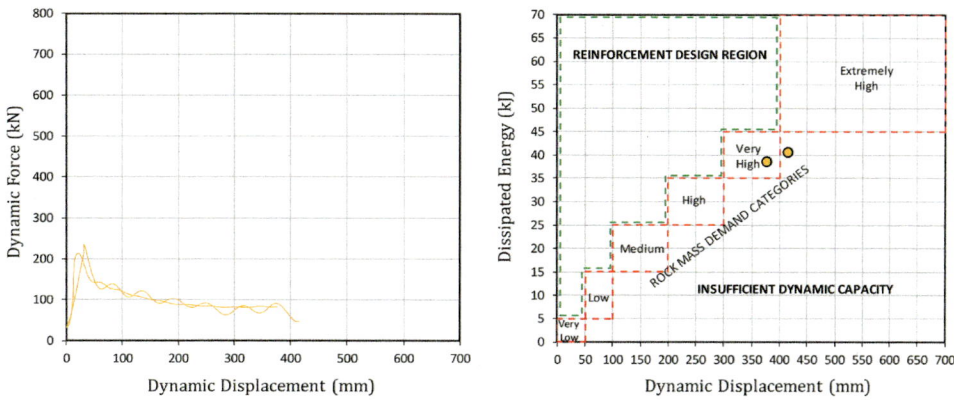

FIGURE 8.23 Dynamic force and related energy dissipation capacity for plain, single-strand 15.2-mm diameter cable bolts in very low-strength cement grout (collar region of 1.0 m, toe anchor region of 2.0 m).

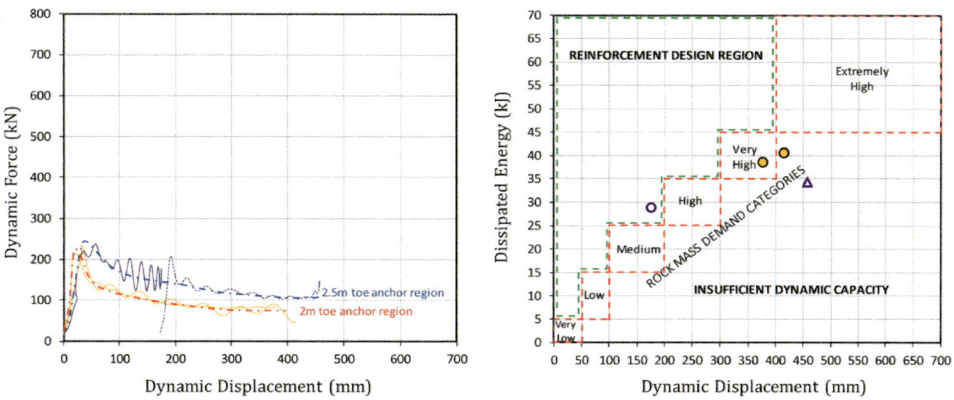

FIGURE 8.24 Dynamic force and related energy dissipation capacity for plain, single-strand 15.2-mm diameter cable bolts in very low-strength cement grout (collar region of 1.0 m, toe anchor region of 2.0–2.5 m).

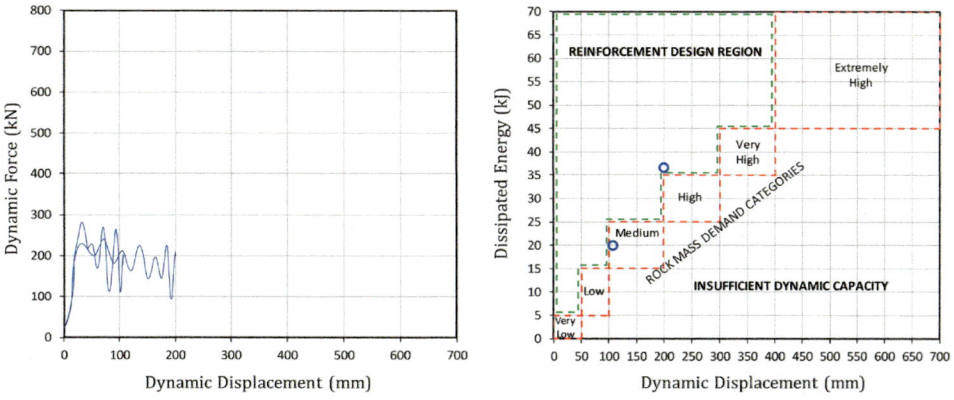

FIGURE 8.25 Dynamic force and related energy dissipation capacity for plain, single-strand 15.2-mm diameter cable bolts in medium-strength cement grout (collar region of 1.0 m, toe anchor region of 2.0 m).

Figure 8.26 shows the results for plated, plain, single-strand 15.2-mm cable bolt encapsulated in high-strength cement grout. The energy dissipation is limited to 20 kJ, and the system displacement is less than 100 mm. The cable bolts are unable to slip within the toe anchor region, and the force capacity increases until rupture occurs at a very small displacement.

Figure 8.27 presents the additional results for plated, plain, single-strand 15.2 mm cables encapsulated in high-strength grout. The double embedment split-pipe system was used with three dynamic impacts, which eventually resulted in catastrophic failure at 125 mm of total displacement. The results from the repetitive testing are shown with a triangle, with the rupture shown as a solid symbol. The results are compared with a single rupture test where the resulting energy dissipation was less than 25 kJ.

Figure 8.28 shows the results for plated, plain, single-strand 15.2-mm diameter cables encapsulated within very high strength grout. The split-pipe system ruptured at less than 50 mm of displacement, resulting in very low energy dissipation. No slippage was experienced from within the toe anchor region.

Figure 8.29 depicts a general comparison of the results for plain, single-strand 15.2-mm diameter cables encapsulated in cement grout (De Zoysa, 2022). The data shows that the energy dissipation clearly depends upon the strength of the encapsulation material. Slippage within the split-pipe

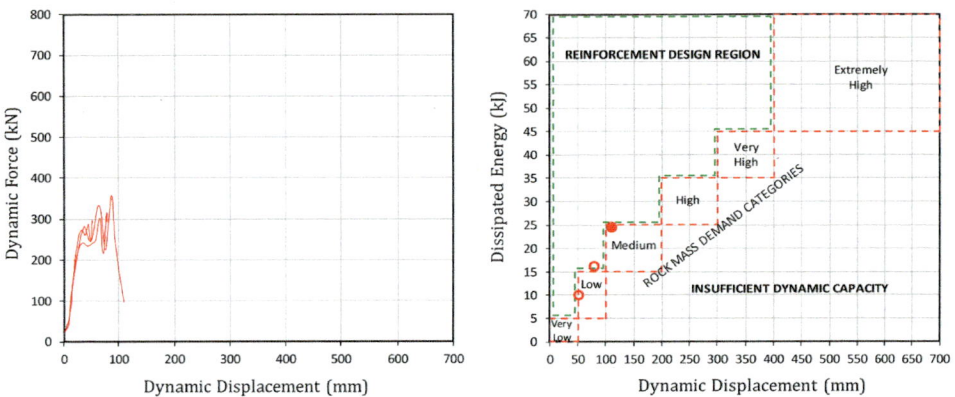

FIGURE 8.26 Dynamic force and related energy dissipation capacity for plain, single-strand 15.2-mm diameter cable bolts in high-strength cement grout (collar region of 1.0 m, toe anchor region of 2.0 m).

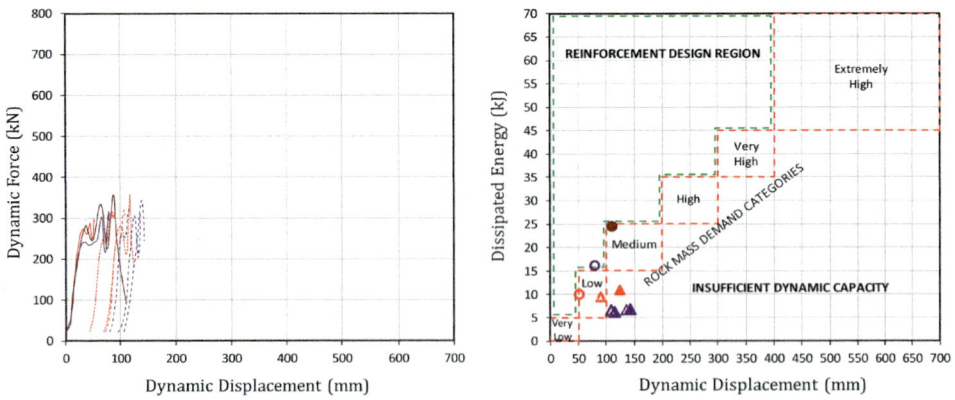

FIGURE 8.27 Dynamic force and related energy dissipation capacity for multiple dynamic impacts for plain, single-strand 15.2-mm diameter cable bolts in high-strength cement grout (collar region of 1.0 m, toe anchor region of 2.0 m).

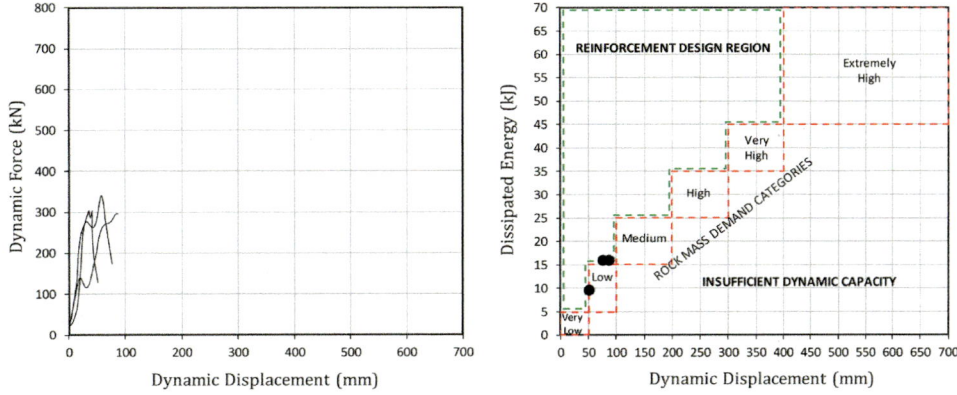

FIGURE 8.28 Dynamic force and related energy dissipation capacity for plain, single-strand 15.2-mm diameter cable bolts in very high strength cement grout (collar region of 1.0 m, toe anchor region of 2.0 m).

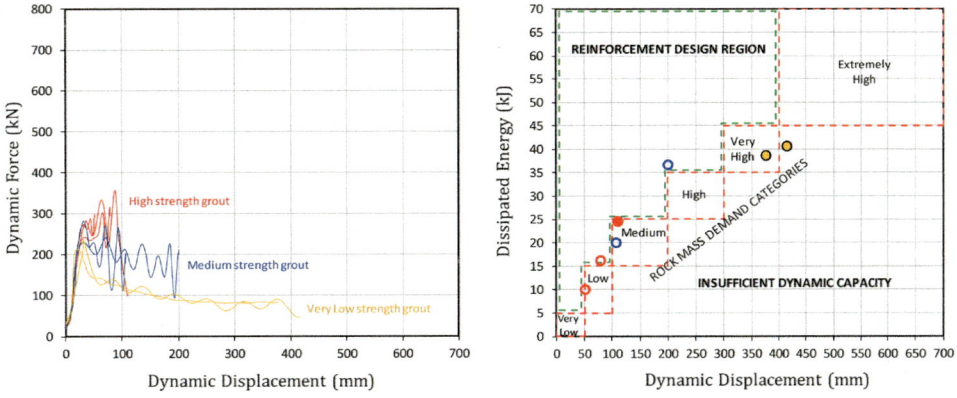

FIGURE 8.29 Dynamic force and related energy dissipation capacity for plain, single-strand 15.2-mm diameter cable bolts in cement grout (collar region of 1.0 m, toe anchor region of 2.0–2.5 m).

system only occurs for the medium- and low- to very low-strength cement grouts. High to very high strength grouts result in cable rupture at very low displacements. Figure 8.30 shows details of the ruptured cables from the WASM Dynamic Test Facility and a comparison with ruptured cables following a large seismic event in an underground mine in Western Australia. Both indicate the classic cup and cone rupture surfaces, indicating pure axial failure.

Figure 8.31 depicts a comparison between plated, plain, single and twin strand 15.2-mm diameter cables encapsulated in low-strength cement grout. Testing was undertaken using multiple dynamic loading events, with the twin strand achieving a higher force but lower displacement compatibility for similar impact forces. In the second impact, the twin strand cable dissipates 26 kJ at 130 mm of displacement, while the single strand dissipates 28 kJ at 215 mm of displacement.

It must be noted that plain, single strand cables have been used very successfully within high energy dissipation ground support schemes in Australia (e.g., Arcaro et al., 2021). This is due to their ability to displace more than 200 mm while sustaining very high residual loads exceeding 150 kN. Such capacity is sufficient to retain the broken material resulting from most seismic events. Additionally, the implementation of single strand cables results in a faster installation time. The cost saving includes a smaller hole diameter, less cement grout and fewer surface fixtures compared with the conventional twin strand cables.

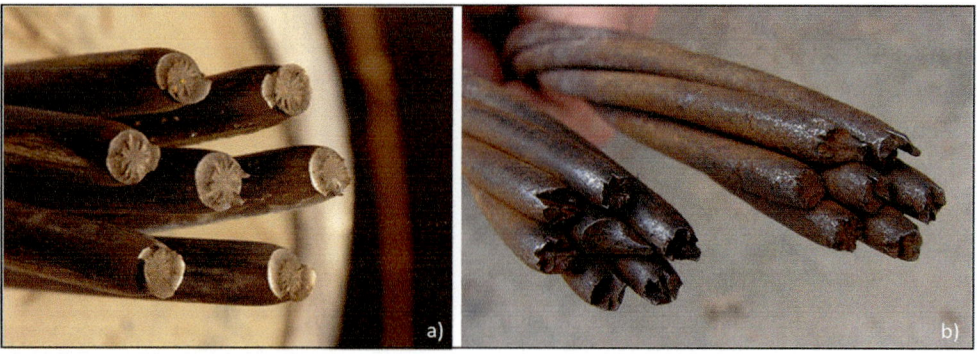

FIGURE 8.30 Example of ruptured 15.2-mm plain, single-strand cables following (a) testing at the WASM Dynamic Test Facility and (b) a large seismic event at a mine site. The consistent 'cup and cone' failure of wires indicates uniform loading.

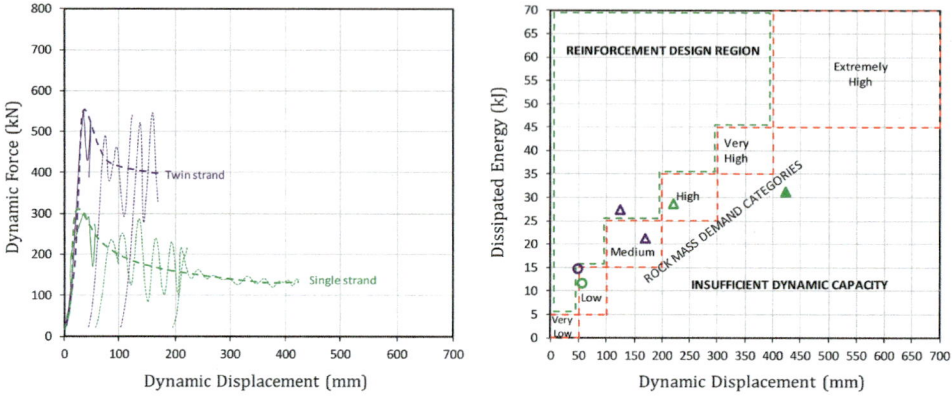

FIGURE 8.31 Dynamic force and related energy dissipation capacity for plain, single and twin strand 15.2-mm diameter cable bolts in low-strength cement grout (collar region of 1.0 m, toe anchor region of 2.5 m).

8.3.3 Plain Strand—17.8 mm Diameter

A recent trend towards replacing 15.2-mm diameter twin strand cables with single strand 17.8-mm diameter cables has resulted in a more efficient installation. Figure 8.32 shows the results for cable bolt systems tested with a simulated discontinuity very close to the plate. That is, the fully encapsulated unstable collar region was limited to 0.5 m while the toe anchor region was set to 2.5 m. Also, the elements were encapsulated using very low to very high strength cement grout. The results show that loading of short collar regions does not stretch the cables, with loading transferred to the plate where the failure of the strand at the barrel and wedge interface occurs. The maximum displacement prior to failure is limited to less than 80 mm. Therefore, when the failure geometry is located near the collar, the effect of the grout strength is only moderate, and the energy dissipation is limited to 20 kJ.

The dynamic loading at the cable bolt-plate interface can be softened with the use of yielding (collapsable) tubes or devices. Although the gain in displacement is very modest, the yielding devices appear to be significant for every grout strength, as it allows stability for a similar level of energy dissipation (Figure 8.33). The crushable device is shown with a rhomboid symbol indicating an energy dissipation of less than 20 kJ.

Figure 8.34 shows the results for plain, single strand 17.8-mm diameter cable bolts encapsulated within the medium, high and very high strength cement grout. The simulated unstable collar region

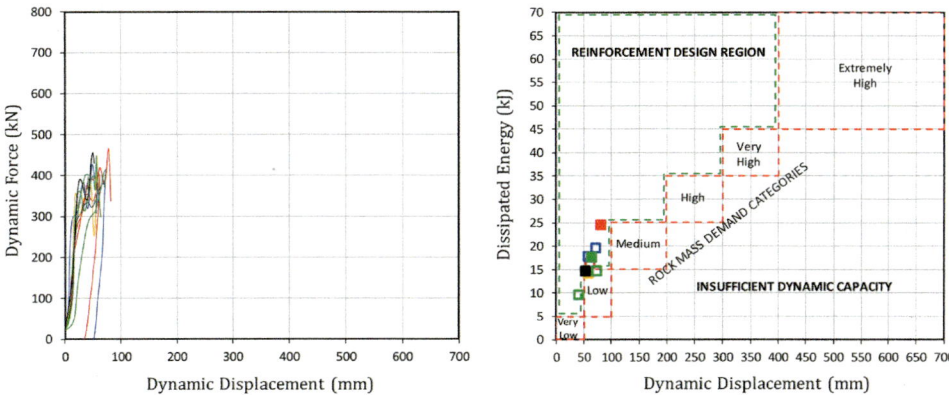

FIGURE 8.32 Dynamic force and related energy dissipation capacity for plain, single-strand 17.8-mm diameter cable bolts in very low to very high strength cement grout (collar region of 0.5 m, toe anchor region of 2.5 m).

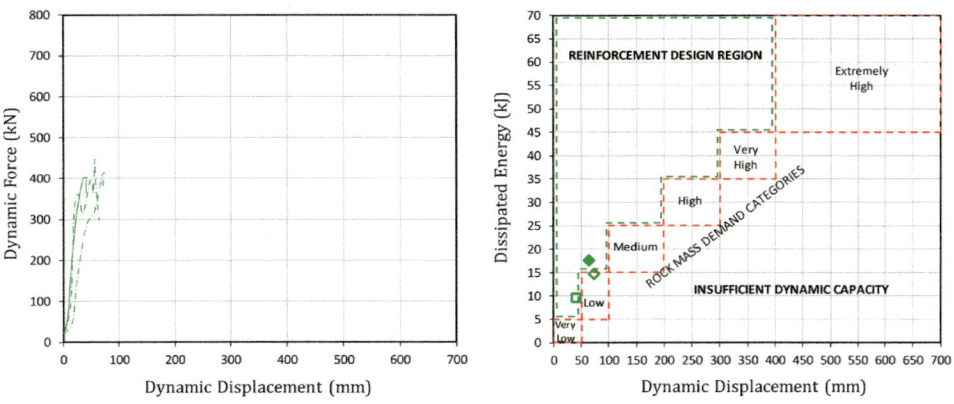

FIGURE 8.33 Dynamic force and related energy dissipation capacity for plain, single strand 17.8-mm diameter cable bolts in low-strength cement grout (collar region of 0.5 m, toe anchor region of 2.5 m).

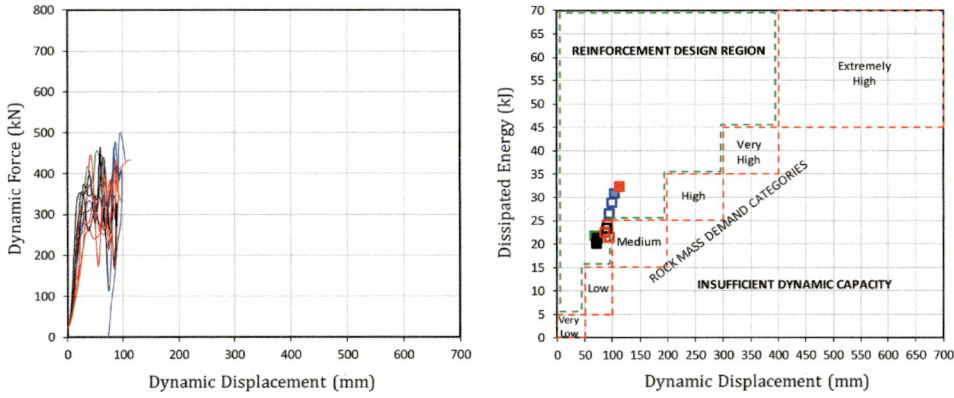

FIGURE 8.34 Dynamic force and related energy dissipation capacity for plain, single strand 17.8-mm diameter cable bolts in medium, high and very high strength cement grout (collar region of 1.0 m, toe anchor region of 2.0 m).

was set to 1.0 m and the toe anchor region to 2.0 m. The data clearly show that the energy dissipation increases with a reduction in grout strength, with the cables able to dissipate 30 kJ at 100 mm of displacement in a medium-strength grout.

Figure 8.35 shows a comparison between individual and multiple dynamic loading for 17.8 mm single plain strand cables encapsulated in medium-strength cement grout. The unstable collar region was set to 1.0 m, and the toe anchor was set at 2.5 m. The individual drop test indicates a 45 kJ energy dissipation at 225 mm of displacement. The comparison with the multiple loading test results shows that up to 35 kJ energy dissipation can be reached when multiple drop loadings reach a similar level of displacement. Both the single and multiple loading tests produce a similar overall force-displacement response (shown in red dashed line).

Figure 8.36 presents results for a single-strand 17.8-mm cable encapsulated in medium-strength grout. The cable slipped significantly in the first dynamic event, which was conducted at very high input energy. The cable bolt dissipated 45 kJ at 225 mm of displacement. In addition, two additional low-level impacts, simulating minor aftershocks, were undertaken with very little energy dissipated. The residual cable bolt force capacity remained constant at nearly 200 kN.

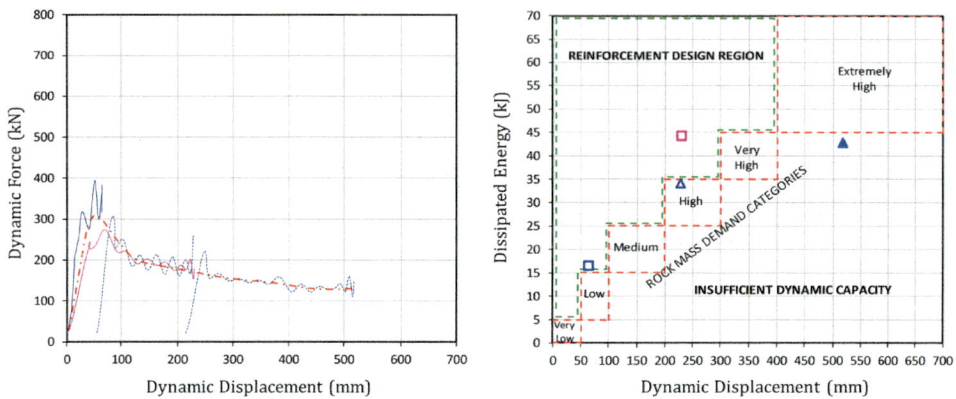

FIGURE 8.35 Dynamic force and related energy dissipation capacity for plain, single-strand 17.8-mm diameter cable bolts in medium-strength cement grout (collar region of 1.0 m, toe anchor region of 2.5 m).

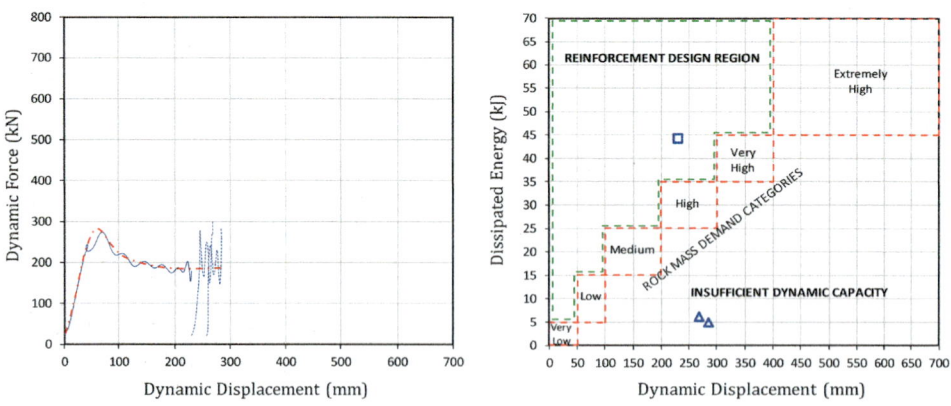

FIGURE 8.36 Dynamic force and related energy dissipation capacity for plain, single-strand 17.8-mm diameter cable bolts in medium-strength cement grout—multiple impacts (collar region of 1.0 m, toe anchor region of 2.5 m).

Figure 8.37 depicts a comparison of a single-strand 17.8 mm cable encapsulated in medium- and low-strength grouts. The results show a large energy dissipation capacity ranging from 45 to 50 kJ for 250 mm of displacement and a residual force capacity ranging from 150 to 170 kN.

Figure 8.38 shows a comparison between 17.8-mm diameter single and twin strand plain cables encapsulated in two different types of cement grout. Even though the twin strand cable was encapsulated in low-strength grout, it still exhibited a limited displacement capability. The single-strand cables were encapsulated in high-strength grout, and the displacement was greater than that achieved by the twin strand. The twin strand cable was tested using multiple loadings with failure occurring at very low energy, limited to less than 100 mm of displacement. The single-strand elements were stable at 25 kJ and 100 mm of displacement.

Figure 8.39 shows a comparison between 17.8-mm diameter single and twin strand plain cables encapsulated in medium-strength grout and similar input energies. The results show that the twin strand (rhomboid symbol) dissipates very little energy with the stable displacements limited to less than 100 mm. Results from the repeated tests for the twin strand cables are shown with a triangle

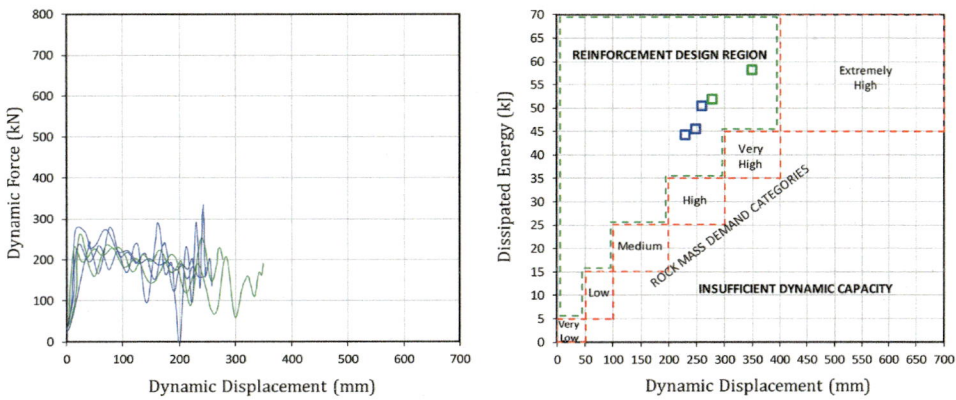

FIGURE 8.37 Dynamic force and related energy dissipation capacity for plain, single-strand 17.8-mm diameter cable bolts in medium- and low-strength cement grout (collar region of 1.0 m, toe anchor region of 2.5 m).

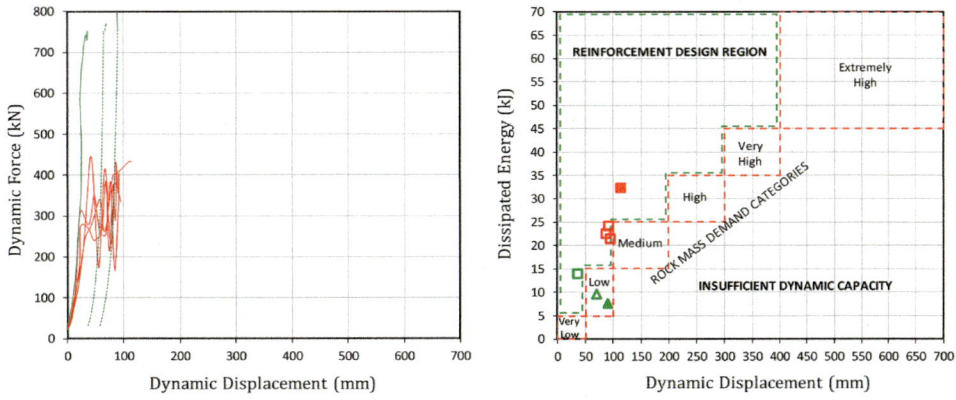

FIGURE 8.38 A comparison of dynamic force and related energy dissipation capacity for plain 17.8-mm diameter single and twin strand cable bolts encapsulated within high-strength cement grout (collar region of 1.0 m, toe anchor region of 2.0 m).

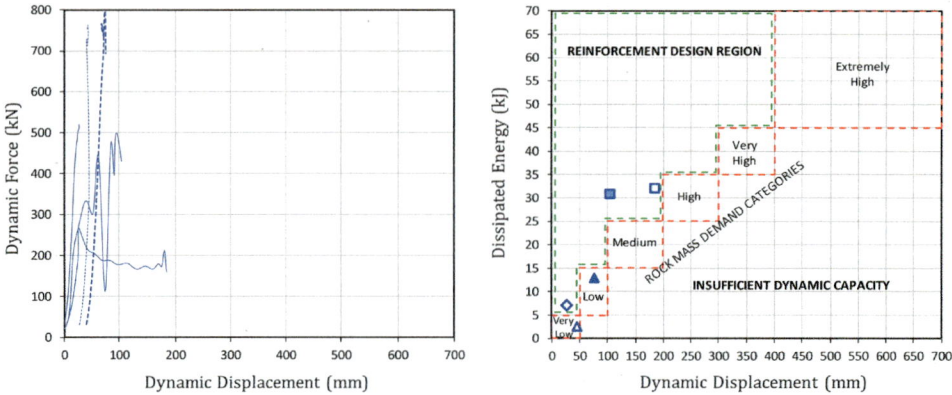

FIGURE 8.39   Dynamic force and related energy dissipation capacity for plain, single and twin strand 17.8-mm diameter cable bolts encapsulated in medium-strength cement grout (collar region of 1.0 m, toe anchor region of 2.0 m).

symbol (solid indicates rupture). In comparison, the single-strand cables can dissipate 30 kJ for the 200 mm of displacement experienced.

Figure 8.40 shows a comparison between twin strand cables encapsulated in medium- and high-strength grouts. The results indicate that although the twin strand cables can develop a large force reaction, the displacement to rupture is limited, and thus, they dissipate very little energy.

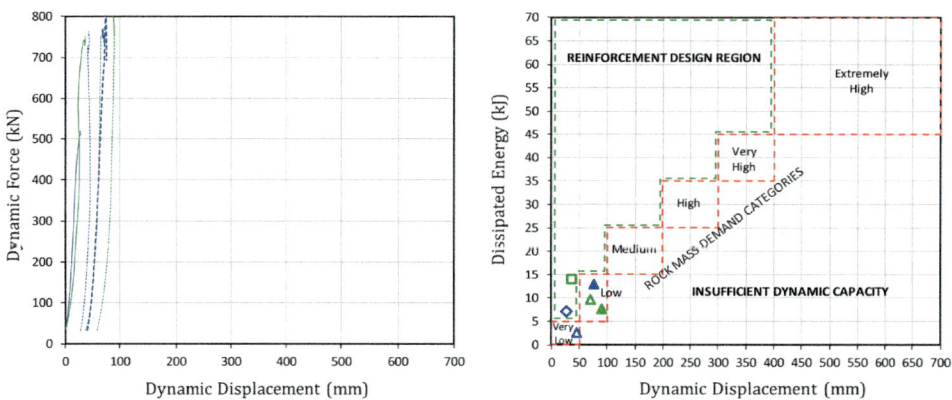

FIGURE 8.40   Dynamic force and related energy dissipation capacity for plain, twin strand 17.8-mm diameter cable bolts encapsulated in medium- and high-strength cement grout (collar region of 1.0 m, toe anchor region of 2.0 m).

Figure 8.41 presents additional results for twin strand cables with a slightly longer toe anchor region (2.5 m) and encapsulated in medium-strength grout. The results show that the twin cables can dissipate up to 25 kJ, with a greater force response to higher input loading. Nevertheless, the twin strand cables cannot sustain displacement beyond 100 mm.

Figure 8.42 depicts a comparison between 17.8-mm diameter single and twin strand plain cables encapsulated in medium-strength grout and a toe anchor region of 2.5 m. The cables were tested with similar input energies, and the results clearly indicate that the single-strand elements dissipate substantially more energy.

Figure 8.43 depicts a comparison of the energy dissipation for plain, single-strand 15.2-mm and 17.8-mm diameter cables for a range of cement grout strengths and input energies. Although the

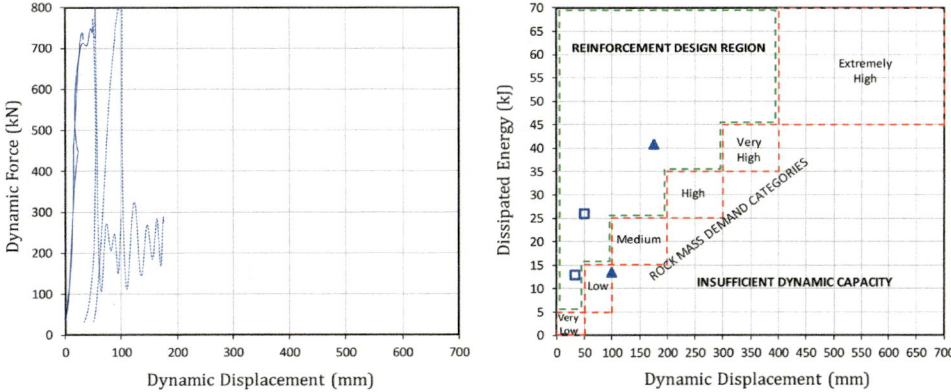

FIGURE 8.41 Dynamic force and related energy dissipation capacity for plain, twin strand 17.8-mm diameter cable bolts in medium-strength cement grout (collar region of 1.0 m, toe anchor region of 2.5 m).

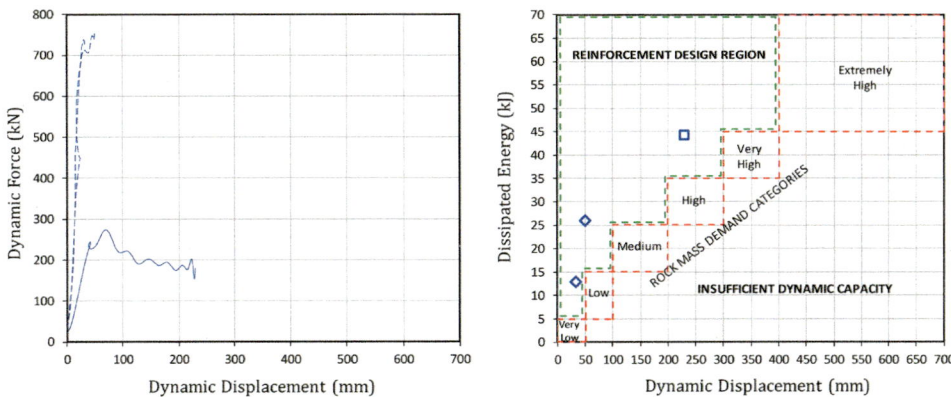

FIGURE 8.42 Dynamic force and related energy dissipation capacity for plain, single and twin strand 17.8-mm diameter cable bolts in medium-strength cement grout (collar region of 1.0 m, toe anchor region of 2.5 m).

results from both types of strands are located within the reinforcement design region, the 17.8-mm diameter cables dissipate more energy for a similar level of displacement. Regardless of the strand type, the results show that when high and very high strength grout is used, displacement is limited to less than 100 mm. In comparison, medium- and low-strength grouts allow cable bolt slippage to occur within the toe anchor regions, thus dissipating a significant amount of energy prior to cable rupture.

8.3.4 DECOUPLED STRAND

In an attempt to increase the deformation capacity, the cable bolt strands are sometimes installed with decoupling sleeves that provide regions where the element can deform according to the steel properties. In such cases, the decoupling is usually not extended to the collar and a minimum collar region is set to 0.5–1.0 m, where load transfer can occur (Figure 8.44a). This enhances retention capacity near the excavation surface, as the load transfer is not only dependent upon the performance of the surface fixtures, which can sometimes become corroded or damaged by mining equipment or blasting.

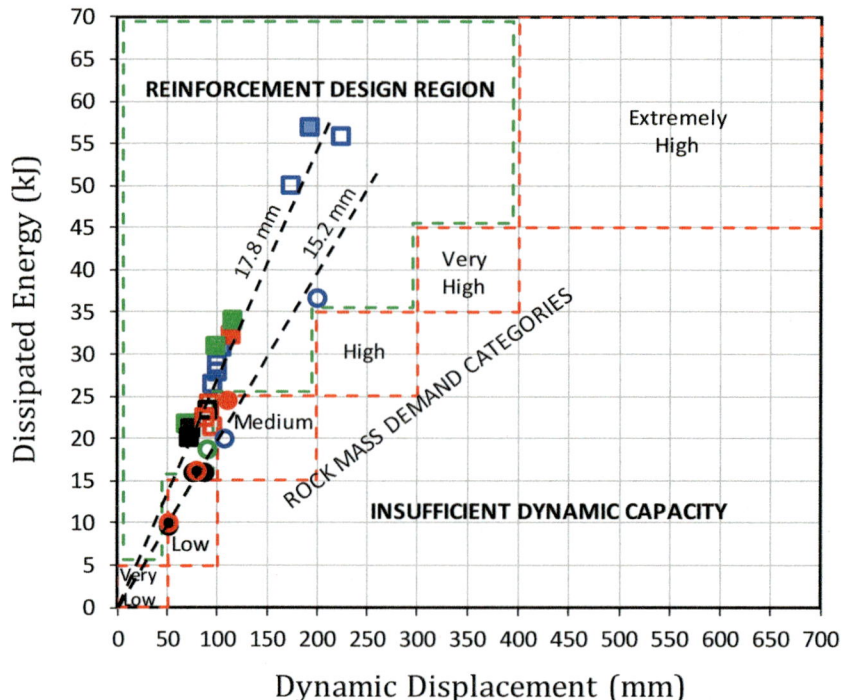

FIGURE 8.43 Summary of dynamic energy dissipation capacity for plain, single-strand 15.2-mm and 17.8-mm diameter cable bolts encapsulated in low to very high strength cement grouts (plated collar region ranging from 0.5 to 1.0 m, toe anchor region 2.0 m).

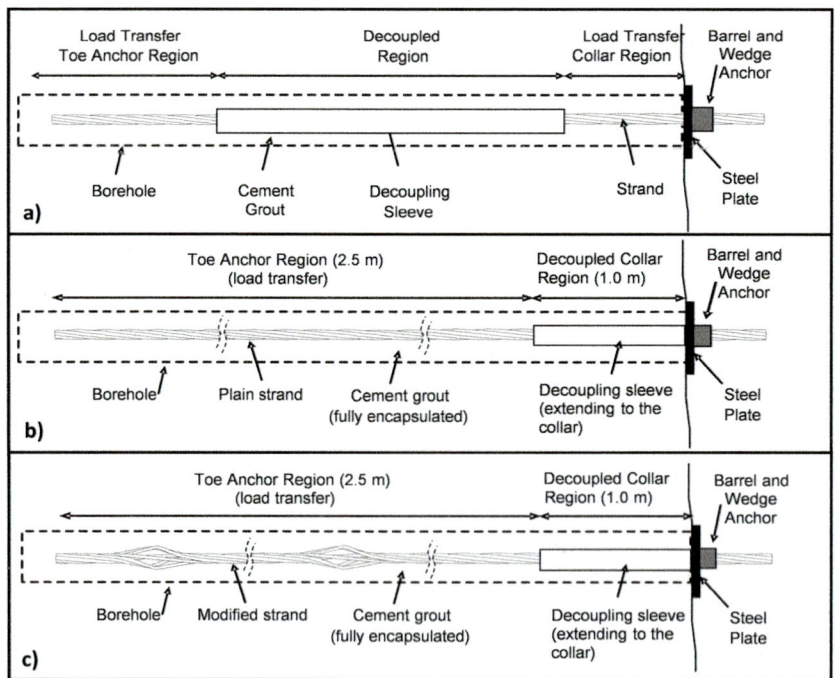

FIGURE 8.44 Schematic long section view of decoupled cable bolting reinforcement systems showing (a) generalized decoupled element with a toe and collar regions where load transfer can occur, (b) decoupling to the collar for plain strand geometry and (c) decoupling to the collar for modified strand geometry.

One disadvantage of a decoupled strand is that large, decoupled lengths may control the overall response and may result in a low reinforcement stiffness due to the extension of the free length between the anchor and collar regions. This is important when trying to match the response of a rock mass, and the provision of stiff rock bolts may be required to minimize rock mass loosening and overall displacement near the boundary of the excavation. One possible advantage of the decoupled strand is that it allows larger amounts of transverse shear displacement than fully coupled strands and solid bars. However, long-length decoupling would require a very long hole to enable a sufficient length of the toe anchor region. Also, long-length decoupling regions may allow too much displacement and increase the reliance upon load transfer within the collar region, which could be cracked and damaged, with the performance eventually being controlled by the barrel and wedge anchors.

Figure 8.45 shows the results for a 17.8-mm diameter plain, single-strand cable encapsulated in medium-strength grout. The cable was plated with a decoupled region (1.0 m) extending to the collar (to allow for an accurate cable bolt displacement measurement), as well as a fully coupled toe anchor region of 2.5 m (Figure 8.44b). The cable was repeatedly loaded, achieving over 200 mm of displacement prior to rupture, with the cable dissipating about 20 kJ during each dynamic event.

Post-testing measurements indicated that although the cable had a 1.0 m decoupling region, most of the displacement occurred at the toe anchor region. This was due to medium-strength cement grout being used for encapsulation, which is compatible with the strain of the strand at very high dynamic forces.

Figure 8.46 depicts a comparison for cables that had a plain versus a modified geometry (three bulbs) at the 2.5-m toe anchor region (Figure 8.44c). The cables were encapsulated within

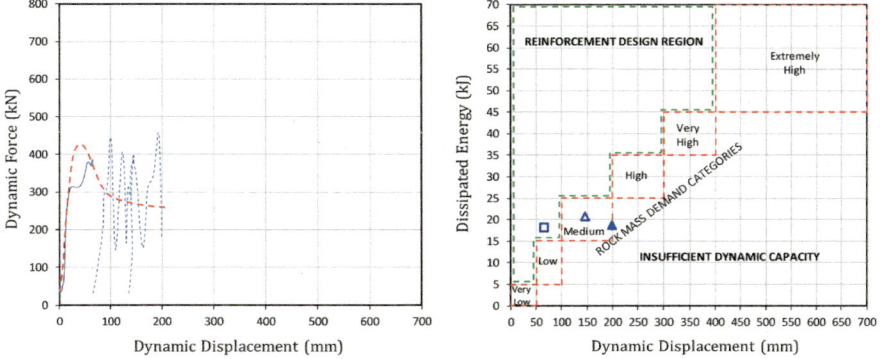

FIGURE 8.45 Dynamic force and related energy dissipation capacity from repetitive testing of plain, single-strand 17.8-mm diameter cable bolts in medium-strength cement grout (decoupled collar region of 1.0 m, toe anchor region of 2.5 m).

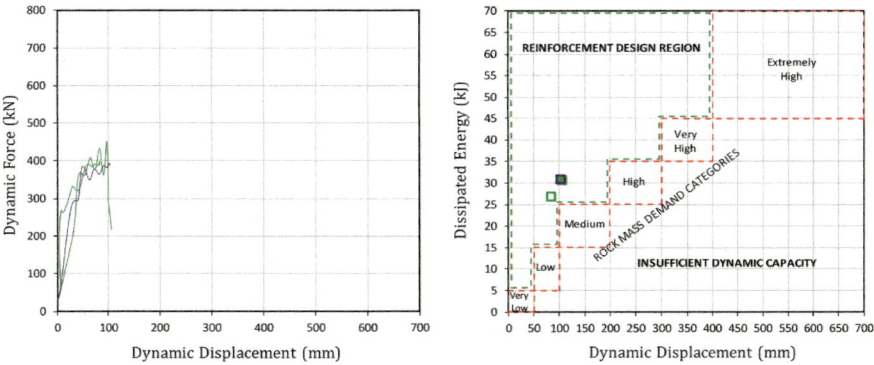

FIGURE 8.46 Dynamic force and related energy dissipation capacity for plain and modified geometry, single-strand 17.8-mm diameter cable bolts in low-strength cement grout (decoupled collar region of 1.0 m, toe anchor region of 2.5 m).

low-strength cement grout. The experiment showed that the bulbs prevented cable displacement within the toe anchor region. The decoupling region accounted for all the deformation, with rupture occurring at 100 mm of displacement. On the other hand, the plain strand cable could slide within the toe anchor region and be able to dissipate 30 kJ for similar levels of displacement.

Figure 8.47 shows a comparison of 17.8-mm diameter single-strand cable encapsulated in medium- and low-strength grout. The cables were plated and had a collar anchor region of 0.5 m, a central-decoupled region of 1.0 m and a fully coupled toe anchor region of 2.0 m (see Figure 8.44a). Three cables were installed with a modified geometry within the toe anchor region (three bulbs), while three cables had plain strand geometries. The results clearly show that the displacement occurred due to the grout failure and not due to the presence of a decoupled region. The decoupled bulbed cables were locked-in, and the displacement was limited to 100 mm (27 kJ of dissipation). In comparison, the plain cables were able to displace over 200 mm and dissipate 50 kJ of energy. Again, most of the displacement occurred within the toe anchor region. In summary, it is concluded that instead of decoupled cables, a more effective strategy for many circumstances would be to use plain strand cables fully encapsulated (from toe to collar) in medium- to low-strength cement grouts. This allows for high energy dissipation at the strand-grout interface while the residual force capacity remains very high.

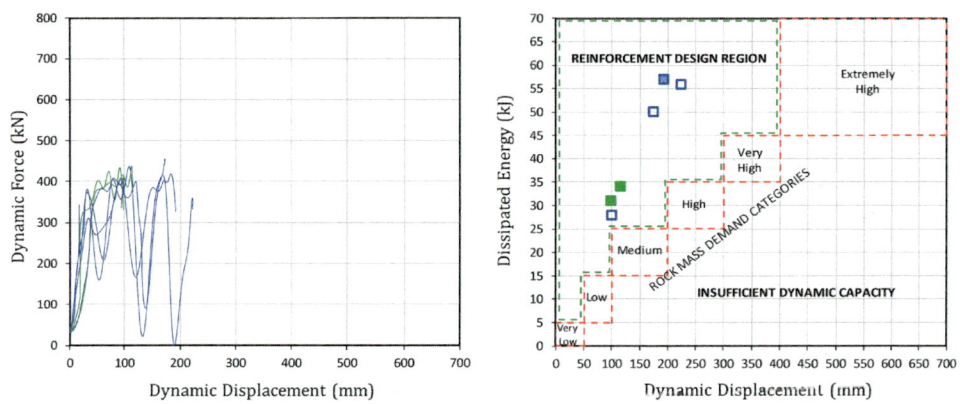

FIGURE 8.47   A comparison of dynamic force and related energy dissipation capacity for plain and modified geometry, single-strand 17.8-mm diameter cable bolts in medium- and low-strength cement grouts (Collar anchor region of 0.5 m, middle decoupled region of 1.0 m, toe anchor region of 2.0 m).

8.3.5   MULTIPLE DYNAMIC IMPACT TESTING

Figure 8.48a shows a comparison of plain, single-strand 15.2-mm diameter cable bolts encapsulated in high-strength cement grout. One cable bolt was tested to rupture using a single dynamic drop, and the other was sequentially loaded and ruptured on the third dynamic impact. Figures 8.48b show similar results, except that the sequential loading was undertaken using less input energy, with rupture occurring in the fifth dynamic impact. The results indicate that summing the energy dissipated by the individual drops does not always equal the maximum energy dissipation or resulting displacement from events that rupture the strands using a single, large dynamic impact (De Soysa, 2022). Importantly, it is suggested that both testing strategies are required to correctly characterize a cable bolt reinforcement system in order to match a particular geotechnical environment.

Combined results from multiple loading tests may be used to estimate the overall force-displacement response, as shown in Figure 8.49. The results from three dynamic impacts on a 15.2-mm diameter, single, plain cable encapsulated in low-strength cement grout were used to establish the overall reinforcement system response. Similarly, Figure 8.50 shows the estimated dynamic

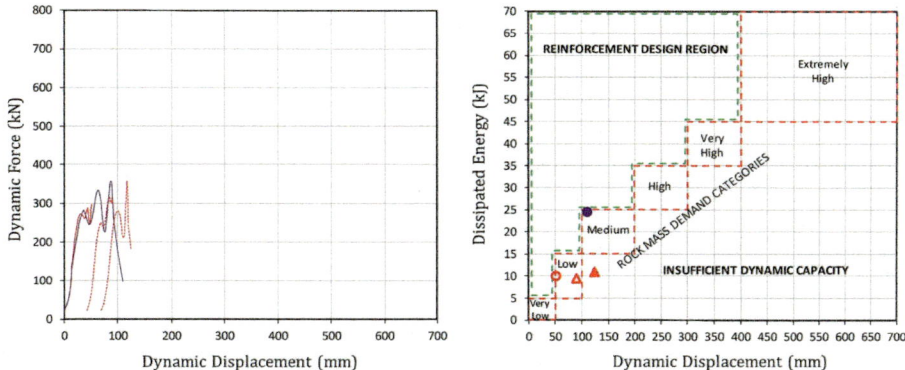

FIGURE 8.48A Comparison of dynamic energy dissipation capacity for single and multiple impacts for plain, single-strand 15.2-mm diameter cable bolts encapsulated in high-strength cement grout (plated collar region of 1.0 m, toe anchor region of 2.0 m).

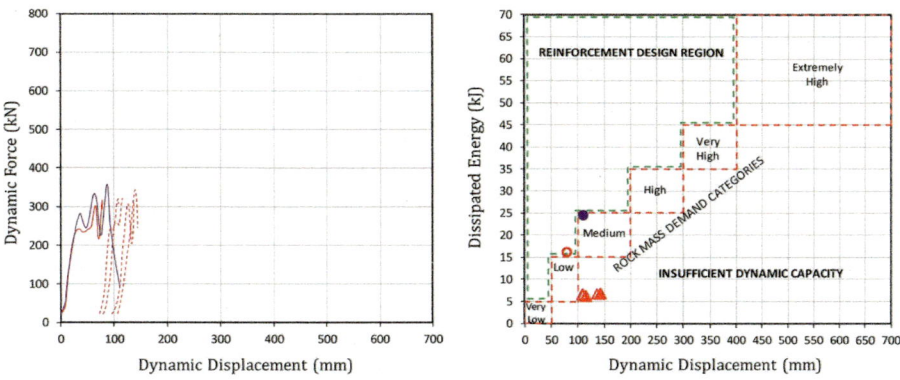

FIGURE 8.48B Comparison of dynamic energy dissipation capacity for single and multiple impacts for plain, single-strand 15.2-mm diameter cable bolts encapsulated in high-strength cement grout (plated collar region of 1.0 m, toe anchor region of 2.0 m).

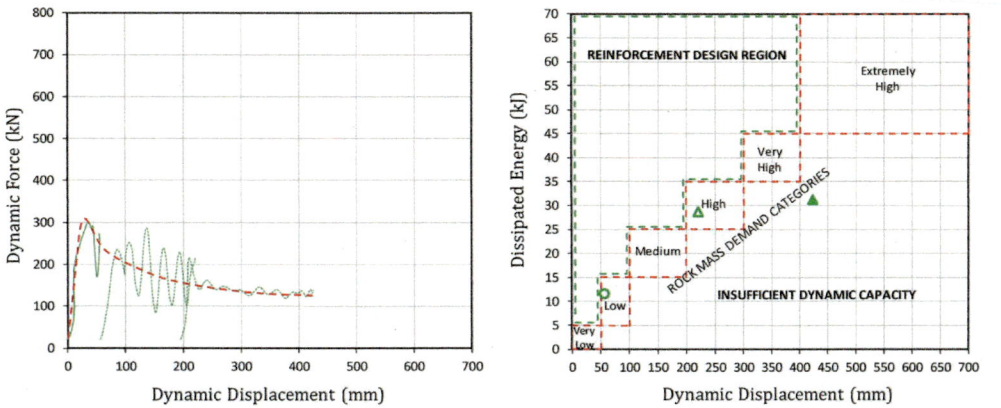

FIGURE 8.49 Example of dynamic force-displacement response and energy dissipation capacity for multiple impacts for plain, single-strand 15.2-mm diameter cable bolts encapsulated in low-strength cement grout (plated collar region of 1.0 m, toe anchor region of 2.5 m).

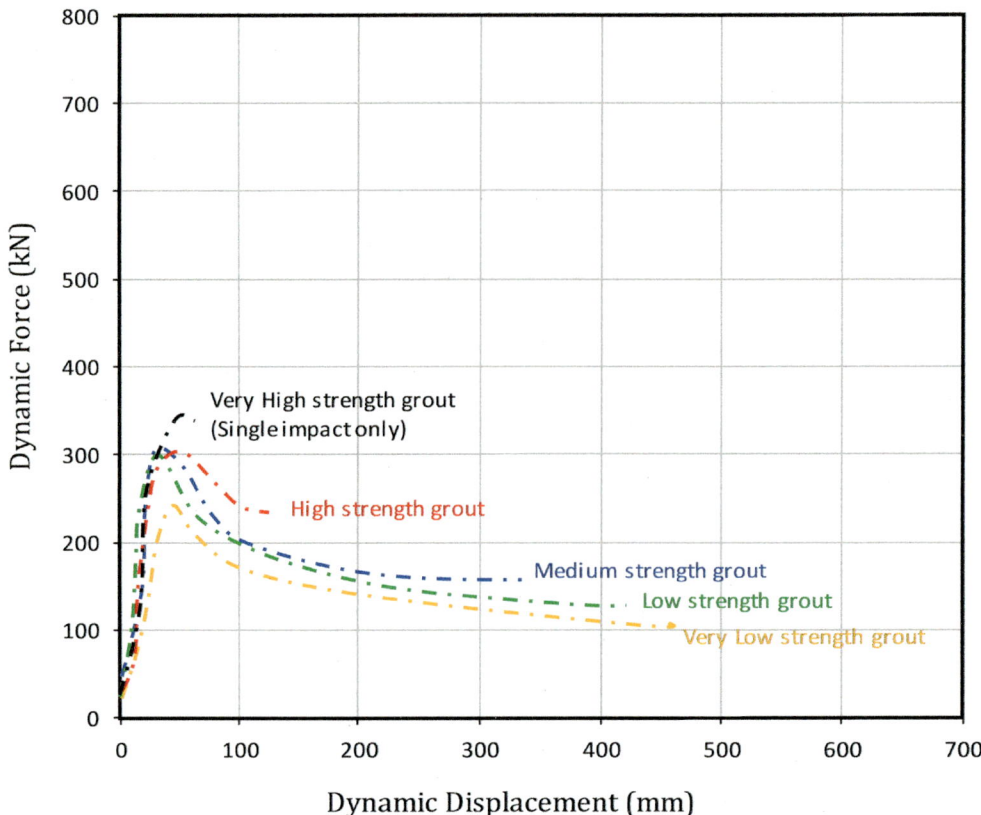

FIGURE 8.50 Dynamic force-displacement response from multiple impact testing for plain, single-strand 15.2-mm diameter cable bolts encapsulated in a wide range of cement grout strength (plated collar region of 1.0 m, toe anchor region of 2.0–2.5 m).

force-displacement response for 15.2-mm diameter plain cable bolts encapsulated in a wide range of cement grout strengths.

Like all the results presented in this book, displacement was measured at the simulated discontinuity between the collar and toe anchor regions. This is because the split-pipe simulates the entire system, including the collar region, the simulated discontinuity and the toe anchor region, which are all intersected by the reinforced element.

8.4   CABLE BOLT PLATES

Underground observations of cable bolt performance following large dynamic failures clearly show that virtually all reinforcement systems require a plate to be effective in retaining the failed rock. This is particularly true where unfavourable oriented structures are present forming slabs that can be shaken loose (Figure 8.51). Also, plates are required when stress concentration creates a region of closely spaced cracks or mobilizes closely spaced structures prior to dynamic ejection, for it is not possible to ensure sufficient load transfer to the collar region of the cables. Furthermore, the plating of cable bolts ensures that a high-capacity load-transfer sequence exists between the surface support components and all the installed cable bolt strands via the external fixtures. This mechanism of load transfer from an unstable to a stable region of an excavation is critical in minimizing the likelihood of catastrophic ground support scheme failure (see Chapter 3).

FIGURE 8.51 Example of unstable blocks falling within a reinforcement pattern for cable bolts installed without plates and without mesh connection.

8.4.1 BARREL AND WEDGE ANCHORS

The use of barrel and wedge anchors to restrain plates, straps and mesh in cable bolt reinforcing applications commenced in Australian mines in the early 1980s (Thompson, 2004). It is important to have a clear understanding of the mechanics involved and appropriate procedures to ensure the anchors are installed correctly and performed according to specifications (Thompson, 2004; Hassell et al., 2006). The typical barrel and wedge anchors are shown schematically in Figure 8.52. The wedge may be formed into two or three parts. The inner taper angle of the barrel and the outer taper angle of the wedge are approximately equal and usually can be assumed to be ~7°. The wedge is made from hardened steel and has sharp teeth formed at the inner surface, making contact with the strand.

8.4.1.1 Anchor Installation

Thompson (1992) and Thompson and Windsor (1995) have described several methods used for anchor installation. All these involve gripping and pulling on the strand and pushing on some part of the surface hardware (i.e., anchor or steel plate). These methods can be briefly summarized as follows:

- Application of tensioning force to wedge
- Application of tensioning force to both barrel and wedge (either with spring to push on the wedge or profiled nose assembly to suit the geometric properties of the barrel and wedge anchor)
- Application of tensioning force to barrel with a secondary hydraulic cylinder to push wedges home

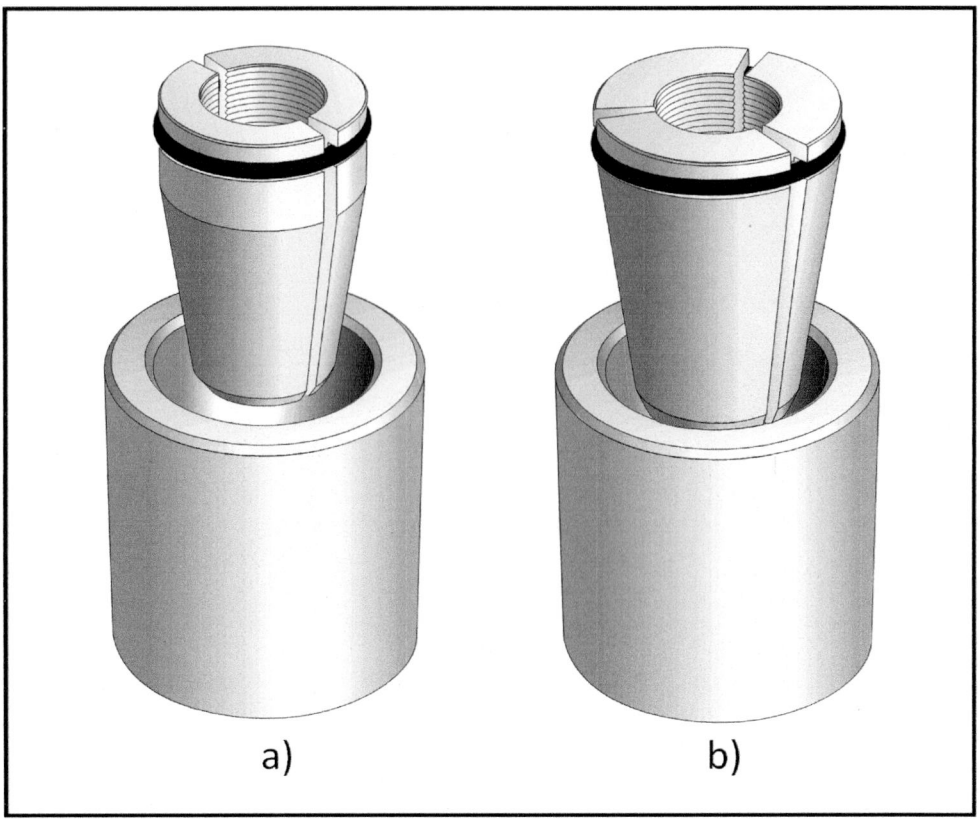

FIGURE 8.52 Schematic representation of barrel and wedge anchors with (a) two-part wedges and (b) three-part wedges.

The installation of anchors with different nose assemblies can be generalized by analysing the forces shown in Figure 8.53 (Thompson and Windsor 1995).

The strand tension (T) is given by:

$$T = KP \hspace{4cm} 8.18$$

where

P = force supplied by the hydraulic cylinder
K = Tension reduction factor, given by

$$K = 1 - \frac{P_W / P \left(1 - \tan \alpha \tan \phi_B \right) \tan \phi_C}{\tan \phi_B \left(1 - \tan \alpha \tan \phi_C \right) + \tan \alpha + \tan \phi_C} \hspace{2cm} 8.19$$

where

α = Wedge taper angle
ϕ_B = Friction angle between barrel and wedge
ϕ_C = Friction between cable and wedge

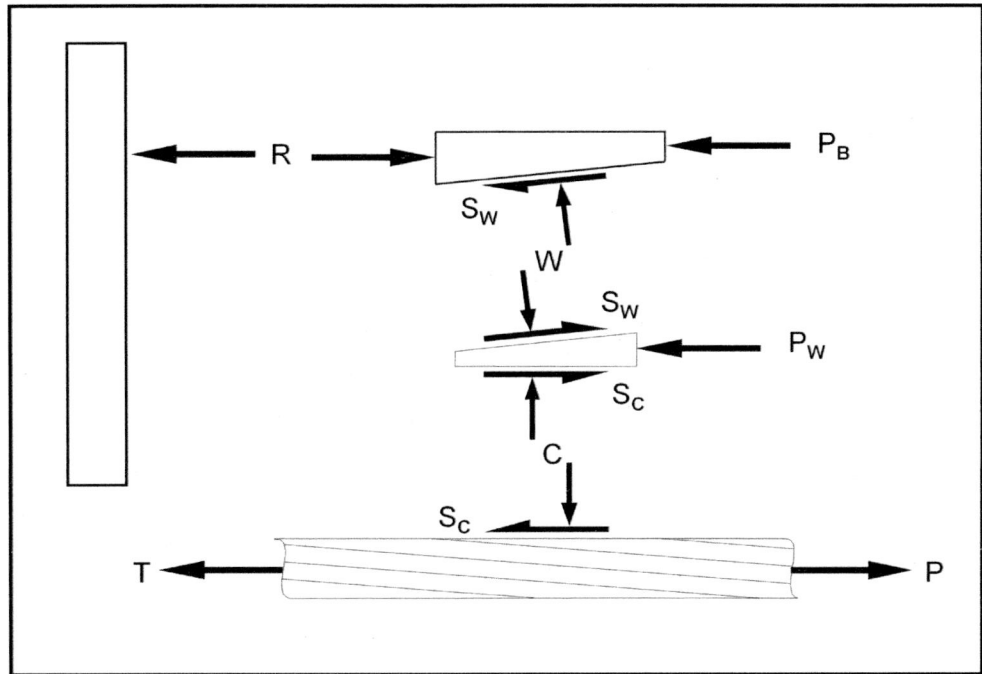

FIGURE 8.53   Forces acting on anchor during cable bolt installation (Thompson, 2004).

P_W = Force applied to the wedge
P_B = Force applied to the barrel, and given by

$$P_B = P - P_W \qquad\qquad 8.20$$

The initial residual tension developed in the strand after removal of the tensioning equipment depends on:

- Force applied by the tensioning equipment,
- Force applied to the wedge during tensioning,
- Barrel/wedge interface condition,
- Wedge/strand interface condition and
- The strand-free length.

In all cases, the strand tension behind the plate and anchor depends on the installation method and is always less than the force applied by the tensioning equipment. Figure 8.54 shows the variation of wedge outstand with strand force for a particular anchor. Figure 8.55 shows how the strand tension may reduce due to wedge draw-in after the removal of the tensioning equipment. The analysis used to predict the behaviour is based on well-established principles used in the prestressed concrete and ground anchor industry codes of practice (Thompson, 2004).

It was generally accepted that a residual tension of ~50 kN would be sufficient to ensure that the anchor was 'set' on the strand, and the anchor would subsequently be able to sustain forces up to the rupture force of the strand. However, observations from underground failures following seismic events clearly show that is not always the case (Figure 8.56).

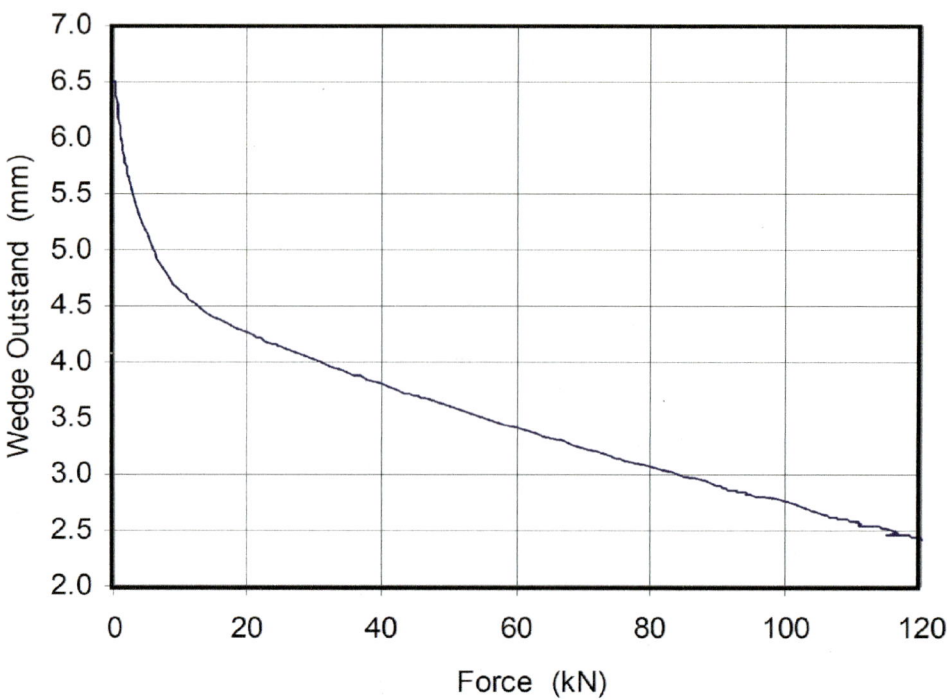

FIGURE 8.54 Typical wedge outstand from barrel versus strand force response curve (Thompson, 2004).

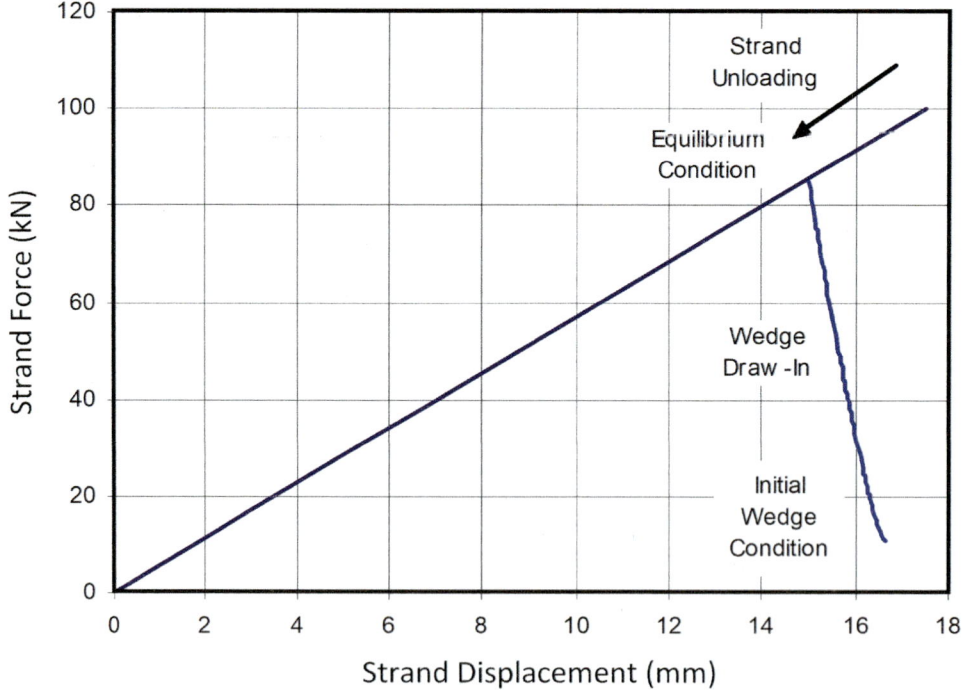

FIGURE 8.55 Theoretical prediction of strand tension loss due to wedge draw-in for 5 m free strand length initially tensioned to 100 kN, with 10 kN applied to the wedge (Thompson, 2004).

FIGURE 8.56 Examples of ejected barrels and wedges following a seismic event.

8.4.1.2 Anchor Mechanism and Performance

Barrel and wedge anchors are designed to clamp the strand and embed the teeth within the outer strand wires. In order to assess under what conditions the strand will slide relative to the barrel and wedge anchor, it is necessary to analyse the interactions between the various components of the system (e.g., Thompson, 1992, Chacos, 1993). Figure 8.57 shows the forces acting on the strand and the barrel and wedge anchor after installation and during service.

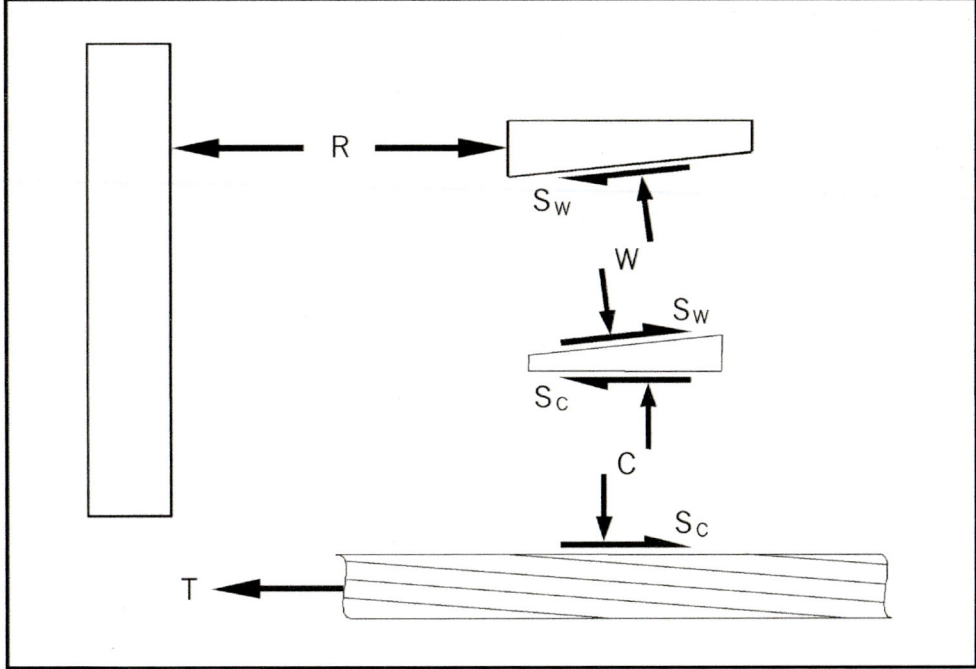

FIGURE 8.57 Component forces acting within the strand and barrel and wedge anchor system.

In this figure:

R = Force between the barrel and the plate acting against the rock/shotcrete surface
T = Tension on the strand, given by:

$$T = R \qquad\qquad 8.21$$

W = Normal force acting across the barrel-wedge interface

S_W = Shear force at the barrel-wedge interface
C = Normal force acting at the wedge-strand interface
S_C = Shear force at the wedge-strand interface

Equilibrium of forces for the barrel requires that:

$$R = SW \cos\alpha + W \sin\alpha \qquad\qquad 8.22$$

and

$$C = W \cos\alpha - S_W \sin\alpha \qquad\qquad 8.23$$

For strand sliding to occur:

$$T \geq C \tan\phi_C \qquad\qquad 8.24$$

Combining Equations (8.22), (8.23) and (8.24) results in:

$$S_W \cos\alpha + W \sin\alpha \geq \left(W \cos\alpha - S_W \sin\alpha\right)\tan\phi_C \qquad\qquad 8.25$$

The analysis assumes that sliding occurs at the barrel-wedge interface, and the sliding force S_W is given by:

$$SW \geq W \tan\phi_B \qquad\qquad 8.26$$

Substituting for S_W, eliminating W and simplifying shows that for preventing sliding, the following is required (Thompson, 2004):

$$\tan\phi_C \geq \tan\left(\phi_B + \alpha\right) \qquad\qquad 8.27$$

Thus, for sliding to occur, the following must apply (Thompson, 2004):

$$\tan\left(\phi_B + \alpha\right) \geq \tan\phi_C \qquad\qquad 8.28$$

This equation means that the barrel-wedge interface friction angle (ϕ_B) must be less than the wedge-strand interface friction angle (ϕ_C) by an amount more than the wedge taper angle.

Thompson and Windsor (1995) reported that the wedge-strand friction angle was measured experimentally to be ~45°. Given that the wedge taper angle can be assumed to be equal to 7°. This means that the barrel-wedge friction angle must be less than ~38° for the anchor to function properly. This condition is only satisfied if the interface remains new and/or lubricated (Table 8.1).

TABLE 8.1

Suggested Values of Friction Angles for Different Barrel-Wedge Surface Conditions.

Interface Condition	Chacos (1993)	Thompson (1992)
Dry, pitted, rusted, old	45°	30° to 40°
Dry, lightly, rusted, new	≈ 22°	25°
Lightly, oiled, clean,new	≈ 17°	15°
Heavily greased, clean	≈ 6°	≈10°

In order to reduce friction at the barrel-wedge interface, the application of a molybdenum-based heavy grease to the contact surface between the barrel interior and wedge exterior has proven very effective (Figure 8.58) to ensure that the wedge does not seize within the barrel for any reason, such as corrosion. The grease lubrication results in an increased normal force on the cable bolt strand during events of sudden, increased axial loading at the collar region, preventing slippage of the surface fixtures.

FIGURE 8.58 Example of the greased barrel and wedges resulting in cable bolt plate stability following dynamic loading from a large seismic event.

8.4.2 DYNAMIC TESTING RESULTS

It must be noted that all previous dynamic test results shown previously within this chapter were undertaken with plated cable bolts. Figure 8.59 shows the dynamic response of two 15.2-mm plain, strand cable bolt elements tested without an installed plate. The strands were encapsulated in cement grout with a WCR of 0.4. The results show that sliding of the unstable collar region occurred at much lower loads than the cable bolt strand capacity. The failure occurred by a complete pull-through of the strand, as shown in Figure 8.60.

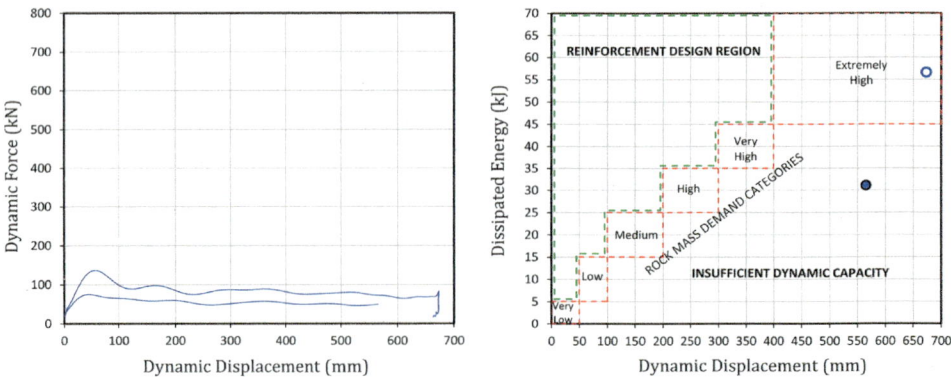

FIGURE 8.59 Dynamic force and related energy dissipation capacity for 15.2-mm plain, single-strand cable bolt showing a catastrophic pull out from a short collar region (unplated collar region of 0.6–1.0 m, toe anchor region of 1.6–1.9 m).

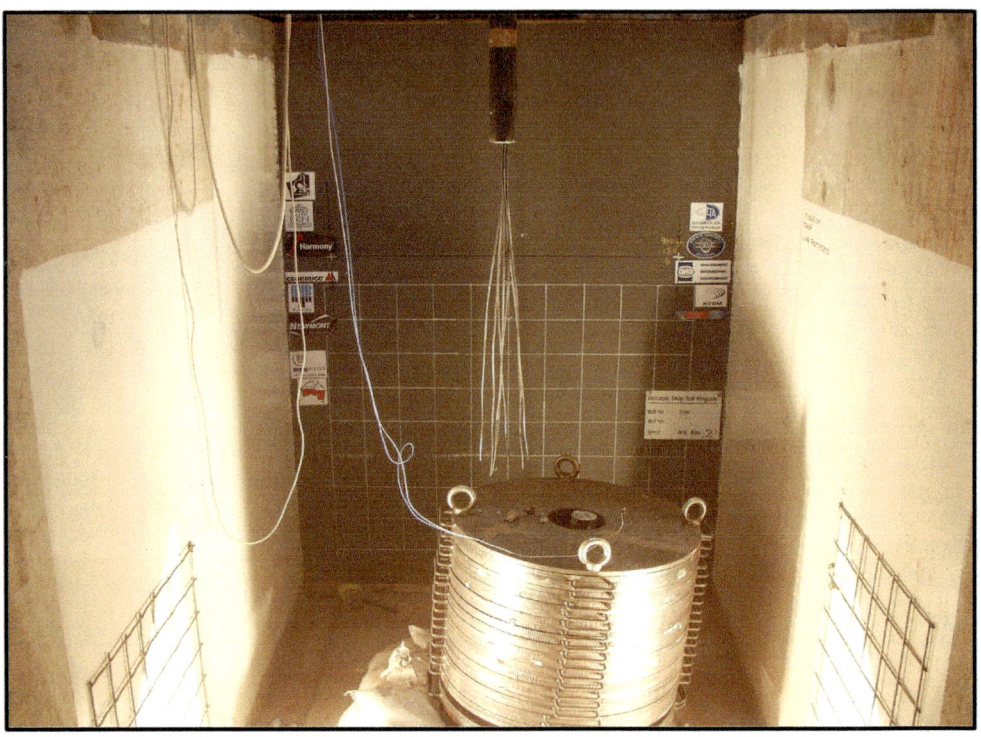

FIGURE 8.60 Catastrophic failure showing unwinding of strand because of not using a plate with barrel and wedge anchors.

9 Energy Dissipation of Mesh Support

9.1 INTRODUCTION

Steel wire mesh is a key component of the ground support schemes required to maintain the load-bearing capacity of a rock mass near the boundaries of an underground excavation. While the reinforcement elements are installed to control the overall excavation stability through keying, arching or composite beam reinforcement actions (Windsor and Thompson, 1992), a mesh is installed to connect the reinforcement elements together at the surface. In addition to retaining loose pieces of rock (or shotcrete), a mesh is required to transfer the load from areas where the reinforcement elements have been damaged to the areas where the reinforcement elements continue to have a load-bearing capacity (Figure 9.1).

Steel wire mesh for ground support is available in a variety of configurations. The most common types are welded wire mesh, comprising straight wires arranged in a rectangular or square grid and welded together, and chain link mesh, comprising regularly bent wires that are woven together and thus interconnected mechanically (Figure 9.2).

The welded wire mesh may have different wire diameters at different spacings and be supplied in various sheet sizes. The most common configurations consist of mesh with a wire diameter ranging from 3 to 5.6 mm and spaced at 100-mm centres. Sheets are generally 2.4 m wide (the maximum specified in Australia) with variable lengths, commonly 3.6–6 m. Large sheets generally cause handling and placement problems, while smaller sheets result in substantially more longitudinal and transverse overlaps during installation. Examples of mechanical handling and installation of a welded mesh using a mining jumbo are shown in Figure 9.3.

The woven (or articulated) mesh is supplied in rolls with different wire diameters and tensile strength. The wire diameters range from 3 to 5 mm, and the width and length of the roll can be manufactured to suit the spans of the supported openings. For a typical 5 m × 5 m drive, a 3.5–4.0 m wide by 17 m long roll of mesh can be installed using mining jumbos, as shown in Figure 9.4. Some woven mesh is made with high tensile steel wires, and its weight is very light in comparison with its strength and load capacity. The mesh is adaptable to irregular excavation faces where it must be installed as tight to the walls as possible in order to minimize deflections transverse to the plane of the mesh. The tight installation also facilitates shotcrete fill-in against either the rock or another shotcrete layer.

9.2 MESH LOAD TRANSFER

Thompson (2001) identified the following variables that influence the performance of mesh when it is subjected to transverse loading:

- Wire strength
- Variable wire diameters
- Variable wire spacings
- Mesh lay relative to wire loading
- Non-linear stress-strain properties for the wire

FIGURE 9.1 An example of a broken bolt with load transferred by mesh across the reinforcement pattern following a damaging seismic event.

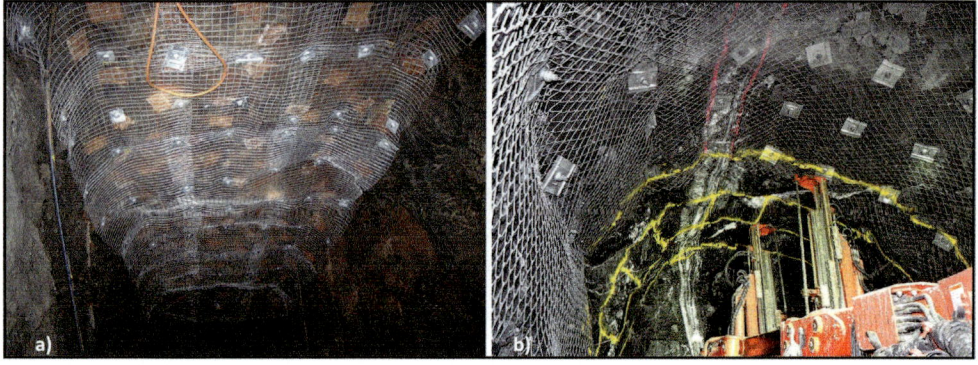

FIGURE 9.2 Different types of mesh configuration, including (a) welded wire mesh and (b) woven or chain link mesh.

FIGURE 9.3 Examples of mechanical handling and installation of welded wire mesh, including (a) surface transport, (b) decline transport and (c) jumbo installation.

FIGURE 9.4 Examples of installation of chain link (woven) mesh using a mining jumbo.

- Weld strength
- Variable bolt spacings
- Variable bolt tensions
- Slip of the mesh at the plates and bolts
- Variable mesh orientation relative to a restraint pattern
- Variable load types and areas

These requirements were identified during systematic test programs (Thompson et al., 1999). Most of these requirements are self-explanatory; however, some are not. For example, the mesh lay refers to the location of the cross-wires relative to the longitudinal wires. The relative location of the wires will influence whether the forces at a particular intersection in the mesh will produce tension or compression combined with shear in the weld connecting the wires. Variable load types and areas refer to whether the mesh is loaded at several discrete points within the mesh to simulate a large, discrete block or by distributed loading of an area of the mesh to simulate a cluster of small blocks formed in closely-spaced, jointed or highly stressed rock.

It is impossible to conduct tests to examine all the possible variables listed above. However, computer software can be used to analyse the distribution of forces in a welded mesh of any size restrained by a specified arrangement of rock bolts and subjected to an arbitrary shaped defined loading. The mesh is assumed to be comprised of cross wires and long wires connected by welds at their intersection points. The plates and rock bolts are simulated by fixed or deformable restraint at specified wire intersection points. The loading may be specified as forces and/or displacements imposed at the intersection points between wires. Two simple examples are used to demonstrate how the load is transferred through the mesh to the restraint locations:

- Square pattern of restraint with central rigid loading
- Oblique (staggered) pattern of restraint with central loading

The deformed mesh for each case is shown in Figure 9.5. The wire segments with high tensile forces have darker shades of grey. These figures clearly show the path of force transmission from the simulated rock loading to the restraint. The computer simulations, whilst able to examine all the variables, need to be complemented by actual tests on mesh samples with known boundary constraints and loadings. However, the simulations do provide guidance on what to expect, and this has assisted in designing the equipment and test configurations for both static and dynamic testing presented here.

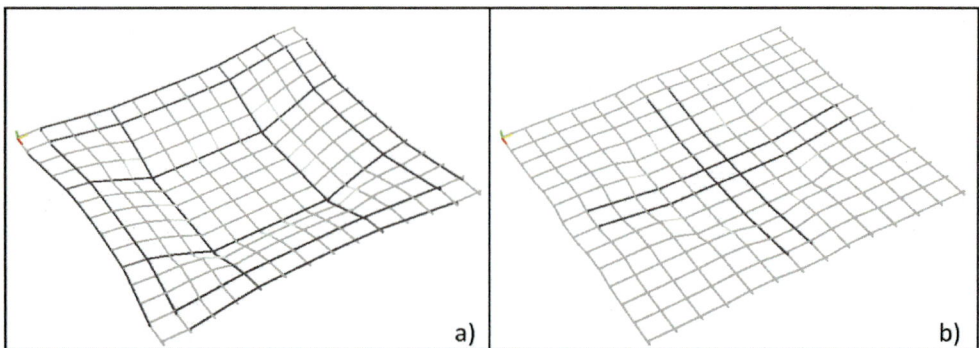

FIGURE 9.5 Deformed mesh simulation with higher axial wire forces shown in darker colours in a (a) square pattern and (b) staggered pattern.

9.3   MESH TESTING

Large-scale mesh testing for mining applications has been conducted in many countries. Test facilities have been built in South Africa (Ortlepp, 1983; Stacey and Ortlepp, 2001; Kuijpers et al., 2002), North America (Pakalnis and Ames, 1983; Tannant, 1995, 2001; Tannant et al., 1997; Dolinar, 2006), South America (Vant Sint Jan and Cavieres, 2004) and Australia (Thompson et al., 1999; Thompson, 2001; Roth et al., 2004; Potvin and Giles, 2008). It is beyond the scope of this chapter to describe, in detail, the sample configurations, test procedures, boundary conditions and results. It is sufficient to say that the different test methods each have their own advantages and disadvantages.

The results presented here have been obtained using the WASM static and dynamic testing facilities for ground support elements (Player et al., 2008; Morton, 2009). The WASM static facility comprises two steel frames: A lower frame to support the sample and an upper frame to provide a loading reaction (Figure 9.6). A mesh sample (1.3 m × 1.3 m, which is similar to the upper range of a reinforcement system pattern) is restrained within a stiff frame that rests on the support frame. A screw feed jack is mounted on the reaction frame. The screw feed jack is driven at a constant speed (4 mm per minute) and allows large displacements to be imposed on the mesh. Load is applied to the mesh through a spherical seat to a 300-mm square, 35 mm thick hardened steel plate with curvature to avoid point loading the mesh wires at the plate boundary. The force is measured using a 500 kN load cell mounted in series behind the loading point. Data acquisition is undertaken at two samples per second. Depending on the resulting deformation, testing of a sample can take up to an hour to complete.

The WASM dynamic test facility for mesh is shown in Figure 9.7. Samples are loaded using the momentum transfer concept (Player et al., 2004; Player, 2012). The mesh testing frame is bolted to a drop beam while the 1.3 m × 1.3 m mesh sample is held in place using threaded bars, shackles and eye bolts in the same configuration as the static test arrangement. A loading mass is placed into the centre of the restrained mesh. The loading consists of rounded disks, each with a mass of 50 kg. The diameter of the loading area is 750 mm. A wooden prop is placed between the loading mass and the drop beam to prevent the mass 'floating' during the initial free fall period. The drop beam and attached assembly are dropped from a specific height to generate dynamic loading on the sample. Computer software, advanced instrumentation and a high-speed camera are used to record the test data. Data acquisition is undertaken at 25,000 samples per second. Testing of a sample is completed in less than a second.

In the assessment of any ground support system, the relationship between displacement, force and energy must be assessed in relation to the expected rock mass demand. The energy dissipation

FIGURE 9.6   Details of the WASM static test facility for surface support elements.

FIGURE 9.7   Details of the WASM dynamic test facility for surface support elements.

is a function of force and displacement. Displacement is influenced, sometimes significantly, by a number of failures within a sample. For this reason, analysis of the mesh types reported here has been undertaken at rupture. Rupture may or may not correspond to the peak force achieved during a test. However, the variability of sample response once rupture has occurred means that detailed analysis producing firm conclusions cannot always be achieved.

9.3.1   Boundary Conditions

The application of appropriate boundary conditions is the most important aspect of any laboratory testing of ground support panels. The boundary conditions can either try to replicate actual loading conditions or simulate the continuation of the material beyond the sample boundary. Either way, the boundary restraint conditions applied to a sample can have a considerable impact on the resulting force-displacement reactions. Even the shape of the frame used to restrain a mesh boundary is important, with square frames favouring a square lattice of wires, such as in welded wire mesh. This is because the longitudinal and transverse axes are loaded equally, compared with, say, a rhomboidal-shaped mesh, where one axis of the mesh will be loaded preferentially when tested using a square frame. Sample clamping, cable lacing and fixed boundaries can all be implemented for boundary restraint, as shown in Figure 9.8.

FIGURE 9.8   Mesh sheet boundary restraint options: (a) sample clamping, (b) cable lacing and (d) uniformly spaced shackles.

The initial set-up investigated at WASM involved clamping the sample to the sample support frame to replicate a square bolting pattern. A square mesh sample (ranging from 1300 to 1400 mm) was placed on the sample support frame, and a boundary restraint frame was placed over the sample. Clamping posts were screwed down on each corner of the frame to hold the boundary restraint frame in place. Nevertheless, the method did not prevent slippage of the mesh and was replaced by mesh lacing using a high tensile wire rope. However, lengthy test set-up times and inconsistent data required redesign and fabrication of an attachment fixture mounted firmly against the frame to reduce the bending moments on the restraints. The chosen restraint system comprises high tensile bars, eye nuts and D shackles passing through a supplementary perimeter frame at allocated points. It allows mesh load monitoring while consistently achieving a boundary condition that simulates the continuation of the material beyond a limited sample area.

9.3.2   Loading Method

Figure 9.9 shows the 300 mm × 300 mm hardened steel plates used to load the mesh samples. The square loading flat plates are rotated 45° to be diagonal to the mesh grid to prevent slippage during testing. It must be noted that a flat base may cause point loading of the wires around the edges of the plate, particularly at large displacements. Therefore, more uniform loading conditions can be achieved with a plate having a curved base, as shown in Figure 9.10. Most of the results shown within this chapter were determined using a curved plate.

9.4   MESH FORCE AND DISPLACEMENT

The failure mechanism of welded wire mesh is really a measure of the mesh manufacturing quality. Three different welded wire mesh failure modes have been identified during laboratory testing. These can be described as a shear failure at the weld points, a failure at the heat affected zone

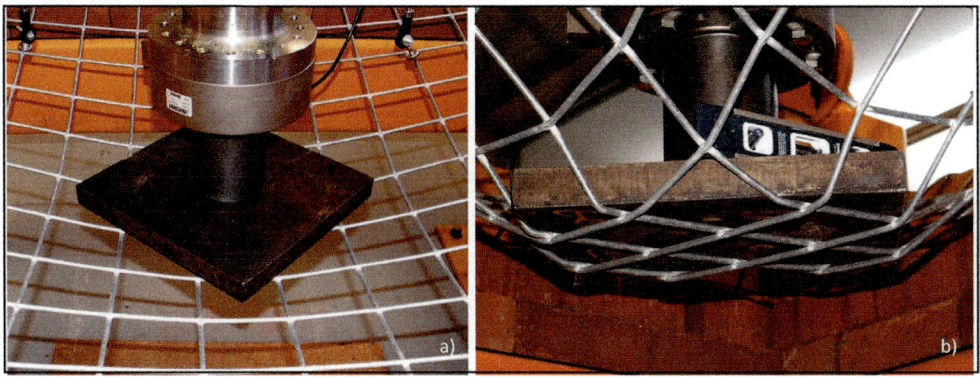

FIGURE 9.9 Square loading plate showing (a) rotation with respect to mesh and (b) point loading of wires at large displacement.

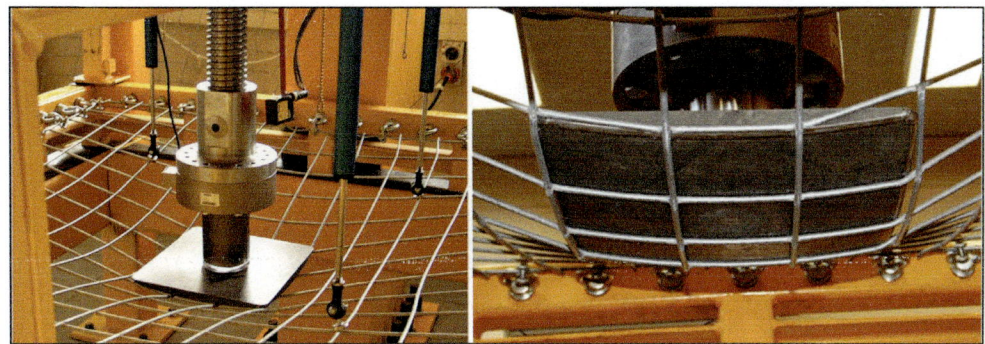

FIGURE 9.10 An example of a curved loading plate set-up, square to the mesh grid.

(HAZ) and a tensile failure of the wire (Figure 9.11). Failure at the weld is an indication of the weld technology and quality control (dirty electrodes or dirty wire) during mesh manufacturing. Failure at the HAZ is caused by the weakening of the wire during the welding process due to excessive weld head pressure and temperature, while the tensile failure of the wire is controlled by the wire manufacturing process. For ground support, the preferred mode of failure is at the HAZ or through the wire, as a mesh with poor weld strength is very susceptible to blast damage when installed very

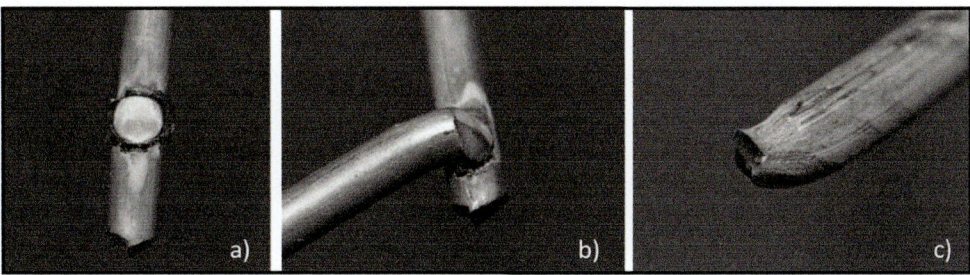

FIGURE 9.11 Welded wire mesh failure mechanisms: (a) weld failure, (b) failure of the wire at the HAZ and (c) tensile wire failure (Morton, 2009).

close to a development face (Figure 9.12). Consequently, the weld strength must be designed to have strength at least equal to that of the axial wire strength (Villaescusa, 1999).

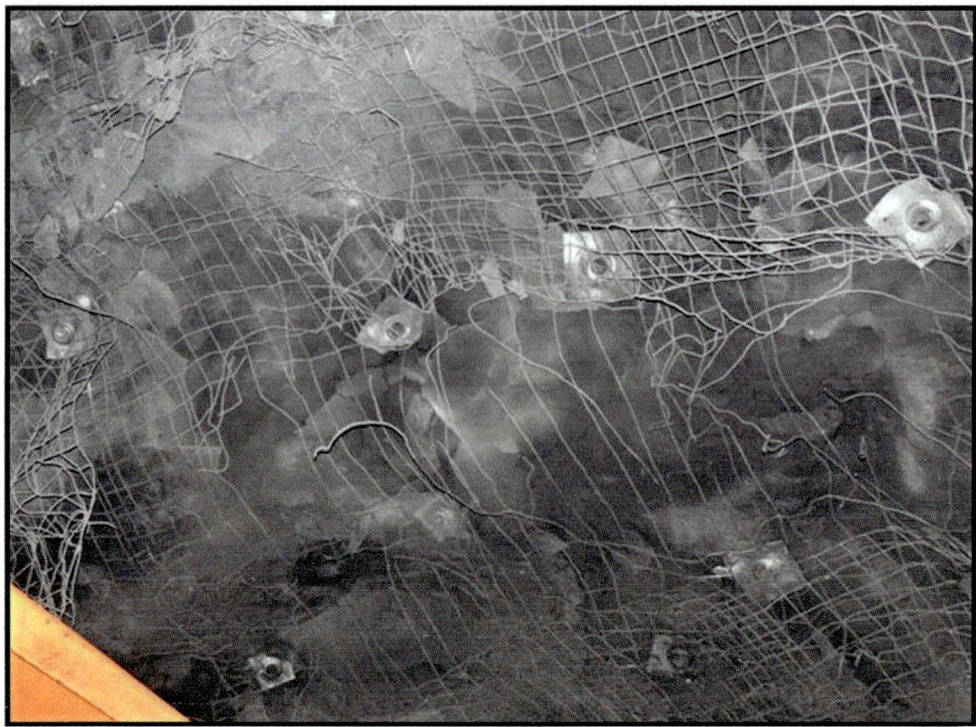

FIGURE 9.12   Example of poor weld strengths making the mesh susceptible to blast damage.

Only one failure mechanism has been observed for the woven wire mesh. The mesh fails either as a result of the edges of a load area attempting to guillotine the wires like a blade or because of the wires cutting each other at a 'link'. Generally, only one or two wires break and do not result in complete destruction or unravelling of the mesh. Underground observations indicate that a large displacement occurs when wires are accidentally cut during drilling for cable bolts (i.e., during ground support installation and prior to any rock loading). However, the mesh integrity and load transfer capacity largely remain after a dynamic event (Figure 9.13).

FIGURE 9.13   Examples of woven wire mesh failure during (a) static laboratory testing and (b) *in situ* dynamic loading following a seismic event.

9.4.1 STATIC RESULTS FOR WELDED WIRE MESH

This section presents some of the results from the WASM static test database for welded wire mesh strength, deformability and energy dissipation. The variability is due to different wire diameters and the manufacturing quality of the products tested. A uniform force, dissipated energy and displacement scale have been used to compare the results with the woven mesh. All the tests were undertaken using a 1.3 m × 1.3 m mesh panel as described in Section 9.3.

Figure 9.14 shows the results for a 3 mm wire diameter product tested on behalf of two different manufacturers. The rupture forces are less than 2 tons, and the dissipated energy is less than 1 kJ. Weld failure occurs very early in the deformation process, with an upper limit of serviceability established at 200 mm. Similar energy dissipation results were found for tests on 3-mm diameter welded wire products used at several mine sites in Mexico (Figure 9.15).

Figure 9.16 shows the results from tests on a 4-mm welded wire diameter product from two different manufacturers. The results show only a minor improvement in terms of rupture strength, deformability, and energy dissipation with respect to a 3-mm wire diameter product.

The effect of increasing the wire diameter is shown in Figure 9.17, where welded wire mesh from a single manufacturer has been tested and reported. The results clearly show an improvement for the 5.5-mm diameter wire. This welded wire (ranging from 5.5 to 5.7 mm) has become the standard

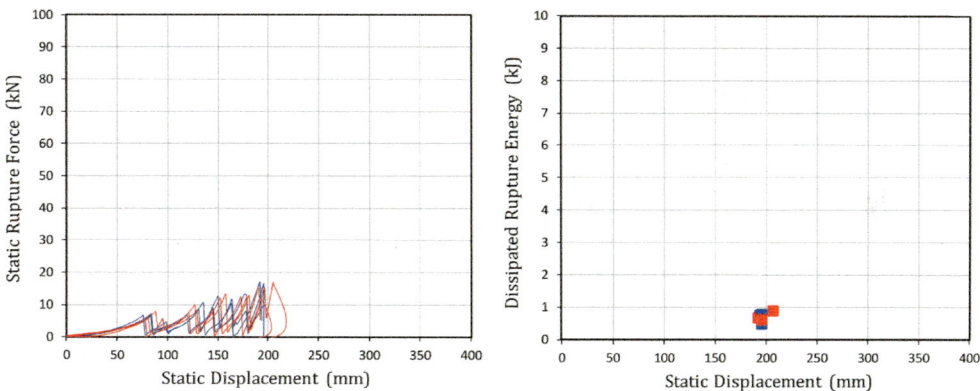

FIGURE 9.14 Rupture force and dissipated energy for welded wire mesh used in Australia (100 mm × 100 mm—3-mm diameter black wire).

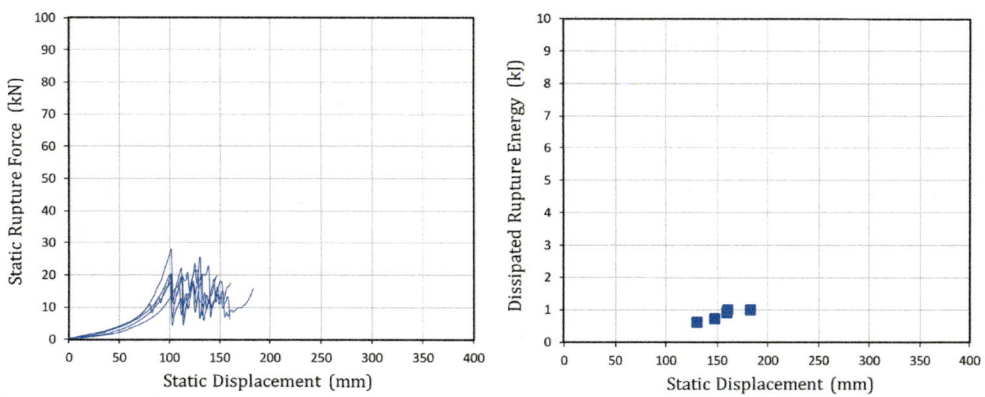

FIGURE 9.15 Rupture force and dissipated energy for welded wire mesh used in Mexico (100 mm × 100 mm—3-mm diameter black wire).

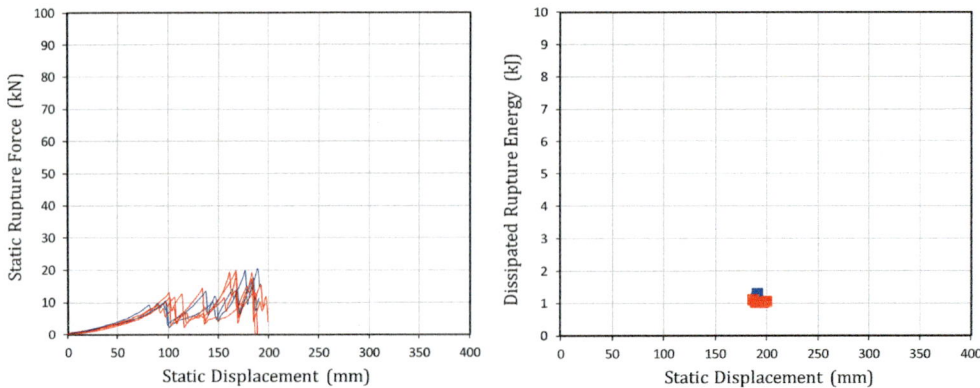

FIGURE 9.16 Rupture force and dissipated energy for welded wire mesh (100 mm × 100 mm—4-mm diameter black wire).

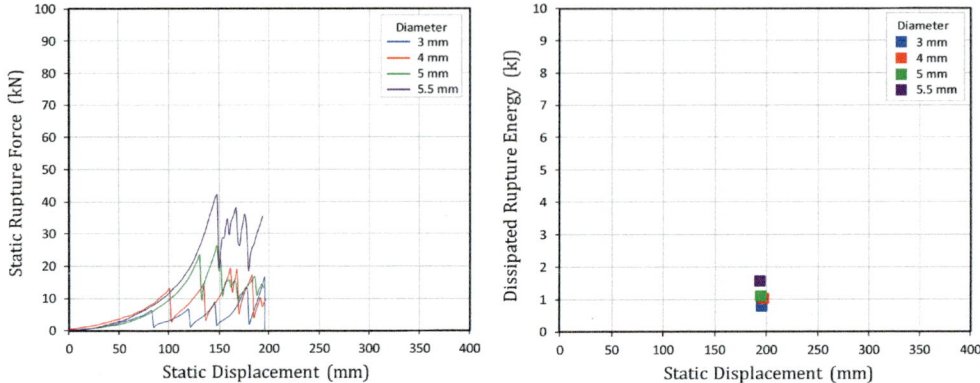

FIGURE 9.17 Rupture force and dissipated energy for welded wire mesh with a range of wire diameters ranging from 4.0 to 5.5 mm (100 mm × 100 mm).

product within the underground hard rock mining industry in Australia. Similar results are shown in Figure 9.18. The results show that the WASM test can correctly differentiate the properties resulting from very small changes in wire diameter.

Figure 9.19 shows the typical performance of welded wire mesh manufactured with 5.6 mm galvanized wire used in Australia. The results from three manufacturers have been determined using the same boundary conditions and loading rates as previously discussed. The rupture force reaches 4 tons, and the average energy approaches 2 kJ (± 1 kJ), with the resulting displacements ranging from 160 to 210 mm. The data indicate differences between different manufacturers and within sample batches from the same factory.

Figure 9.20 shows detailed results for 5.6-mm diameter galvanized welded wire mesh where corrosion effects are shown to significantly influence the results. The corrosion is characterized by a reduction in original wire diameter with different levels of corrosion: light (5.0–5.6 mm), moderate (4.5–5.0 mm), high (4.0–4.5 mm) and severe (less than 4.0 mm) Hassell et al. (2010).

9.4.2 Static Results for Woven Mesh

Two types of woven link mesh have been tested, including a mild steel mesh from Chile with a square grid and a high tensile mesh from Australia with a rhomboid grid. The static test results for

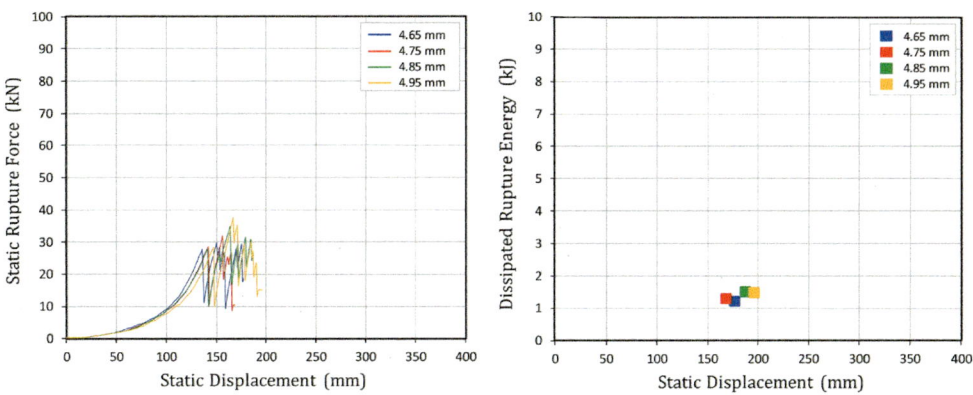

FIGURE 9.18 Rupture force and dissipated energy for welded wire mesh with a range of wire diameters ranging from 4.65 to 4.95 mm (100 mm × 100 mm).

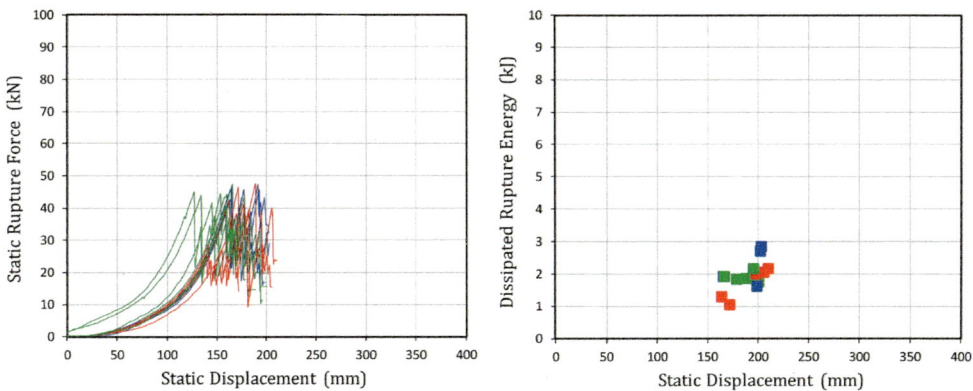

FIGURE 9.19 Rupture force and dissipated energy for welded wire mesh (100 mm × 100 mm—5.6-mm diameter galvanized wire).

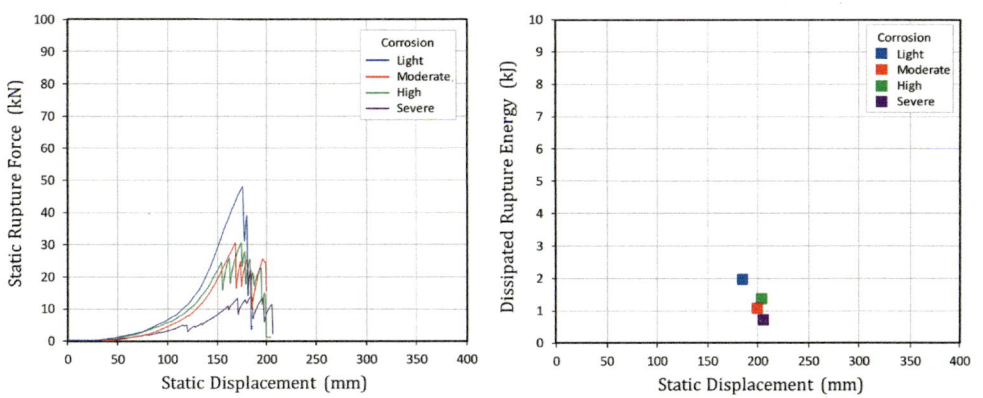

FIGURE 9.20 Rupture force and dissipated energy for corroded welded wire mesh (100 mm × 100 mm—5.6-mm diameter galvanized wire).

the square grid woven mesh with 4- and 5-mm diameter mild steel wires are shown in Figure 9.21. The results indicate an increased capacity compared with the welded wire mesh of a similar diameter. It must be noted that the 4-mm diameter product had a higher tensile strength compared with the 5-mm wire mesh. Additionally, the product is lighter and, thus, more suitable for manual installation.

Figure 9.22 indicates similar results for woven mesh (with a rhomboidal grid) manufactured with high-strength steel wires. Results for 3-, 4-, 4.6- and 5-mm diameter wire variants are shown. A large force and deformation capacity in comparison to the conventional 5.7-mm diameter welded wire mesh can be achieved by using woven high tensile wire mesh.

Figure 9.23 shows a comparison of 4-mm diameter wire for several commercially available products. Testing was undertaken using a large diameter plate, which shows that load area and shape also control mesh response. Nevertheless, the advantages of using high-capacity woven wire mesh in terms of higher forces and smaller displacement are clearly evident.

Figure 9.24 shows a comparison of force, displacement, and energy dissipation for several commercially available non-steel mesh products. Testing was undertaken using the same procedure described earlier for the welded wire steel products. The results show a large variability in terms of displacement to failure. Due to their light weight and ease of installation, this type of mesh has

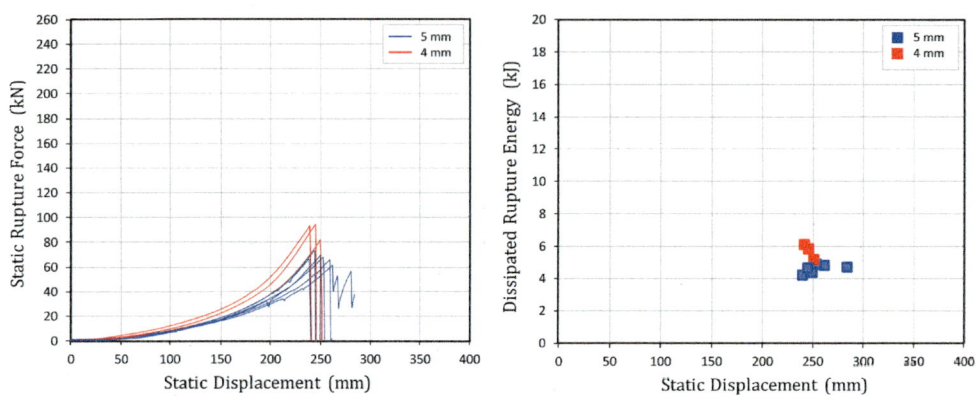

FIGURE 9.21   Rupture force and dissipated energy for mild steel woven (chain link) mesh (110 mm × 110 mm, galvanized wire product).

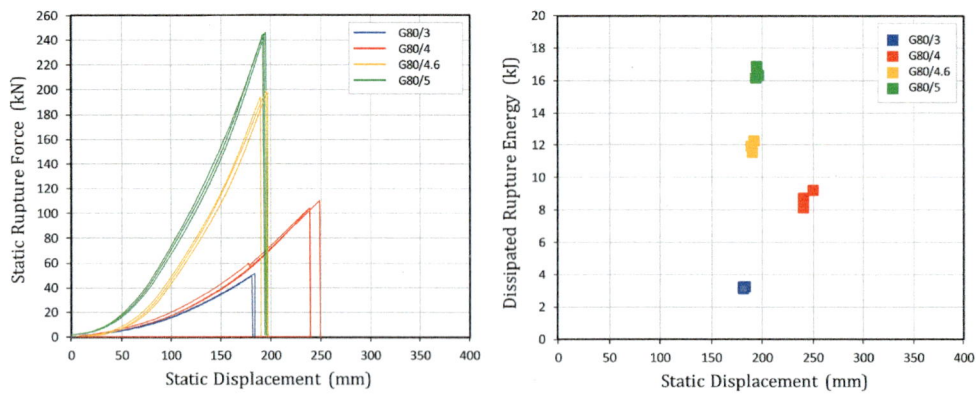

FIGURE 9.22   Static rupture force and dissipated energy for high tensile steel woven (chain link) mesh (80 mm diamond shape, galvanized wire).

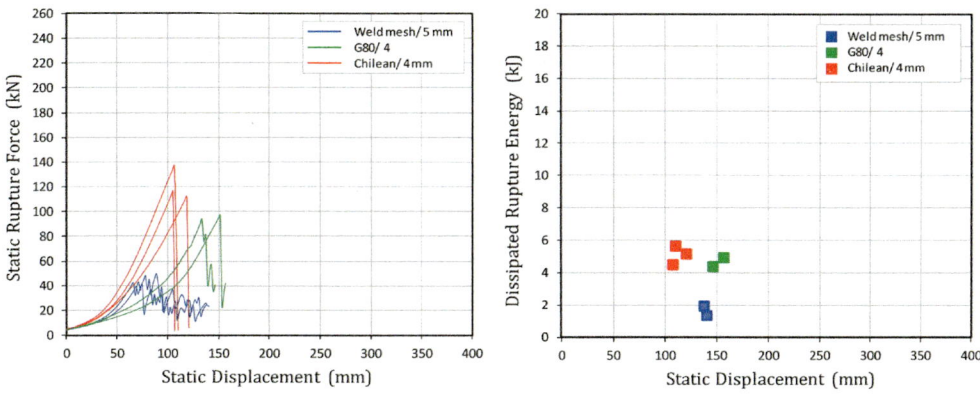

FIGURE 9.23 A comparison of rupture force and dissipated energy for woven (chain link) and welded wire mesh loaded with a large diameter plate.

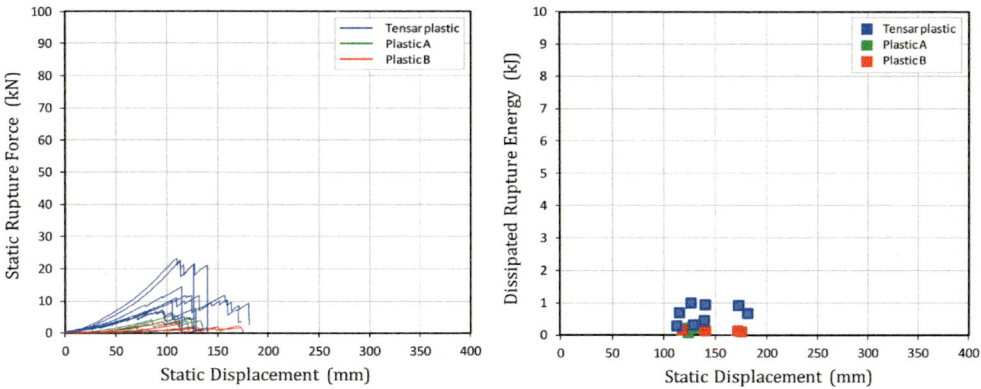

FIGURE 9.24 A comparison of rupture force and dissipated energy for three types of plastic mesh.

been suggested for face support. However, in all cases, the energy dissipation was less than 1 kJ (indeed, for some products, the capacity was extremely low), making them unsuitable for any level of dynamic loading.

9.4.3 Dissipated Static Energy

The energy dissipation for mesh products is best determined using a cumulative plot of energy dissipation versus displacement, as shown in Figure 9.25. The results for woven mesh made of mild steel show that large deformations can be expected before the mesh can reach a similar capacity to that of the conventional (5.6-mm diameter) welded wire mesh. The data show that the woven mesh products have a larger displacement capacity compared with the welded wire mesh products. Also, their ultimate energy dissipation capacity exceeds the values of welded wire mesh, even for smaller-diameter wire variants.

Figure 9.26 provides a similar comparison for high tensile strength wire-woven mesh products. The results indicate that the mesh can respond faster and with a much higher force capacity than the conventional welded wire mesh. Also, the high tensile mesh has a large displacement capacity while dissipating a significant amount of energy compared with conventional welded wire mesh products.

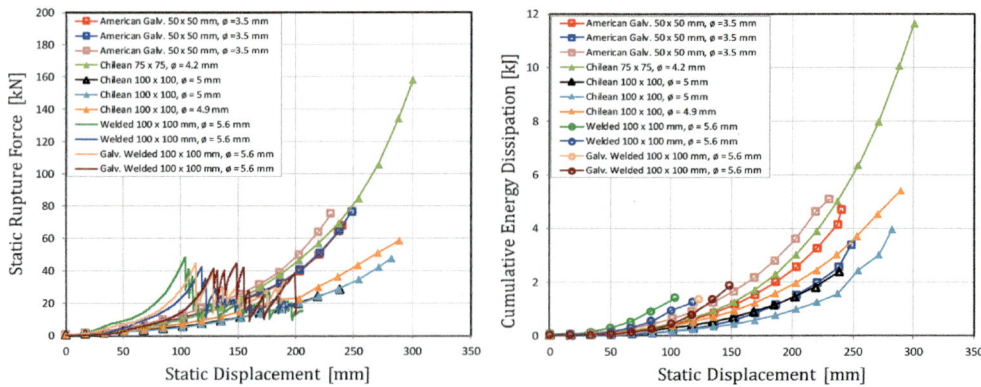

FIGURE 9.25 A comparison of static force displacement and dissipated energy for two types of mild steel woven mesh products versus the conventional galvanized welded wire mesh.

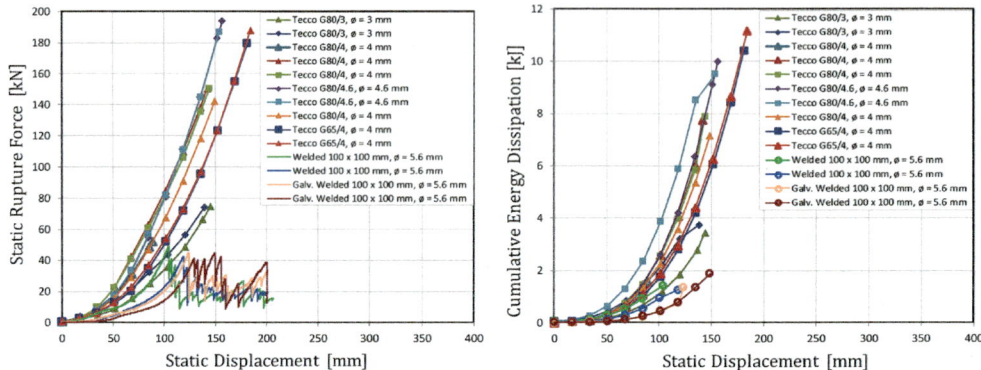

FIGURE 9.26 A comparison of static force displacement and dissipated energy for high tensile woven products versus the conventional galvanized welded wire mesh.

9.4.4 Dynamic Results for Welded Wire Mesh

Figure 9.27 shows the results from dynamic testing of welded wire mesh panels. The results for conventional galvanized welded wire mesh (5.6 mm diameter) products from four different manufacturers are shown. The data indicate an energy dissipation range of 1–3 kJ with an average dynamic displacement of 200 mm.

9.4.5 Dynamic Results for Woven Mesh

Figure 9.28 shows the dynamic test results for mild steel woven mesh manufactured in Chile. The mesh comprised panels made of squared woven wires with different wire diameters. The force-displacement graph shows the initial static displacement prior to dynamic loading, indicating susceptibility to localized static loading. Hence, the recommendation of shotcrete infill (see Chapters 4 and 10). Also, it must be noted that a solid symbol indicates rupture.

Figure 9.29 shows the results for medium tensile (≈800 MPa) mesh being manufactured in Chile. A large initial static displacement was recorded that exceeded 250 mm, with total displacement ranging from 400 to 450 mm. Very large dynamic capacities ranging from 14 to 19 kJ were attained. However, the large displacements experienced would probably dictate extensive rehabilitation

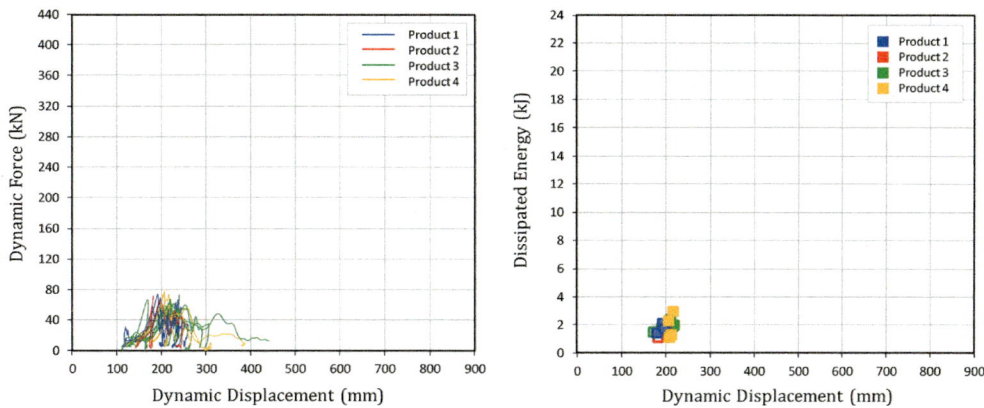

FIGURE 9.27 Dynamic force displacement and dissipated energy for welded wire mesh (100 mm × 100 mm—5.6-mm diameter galvanized wire).

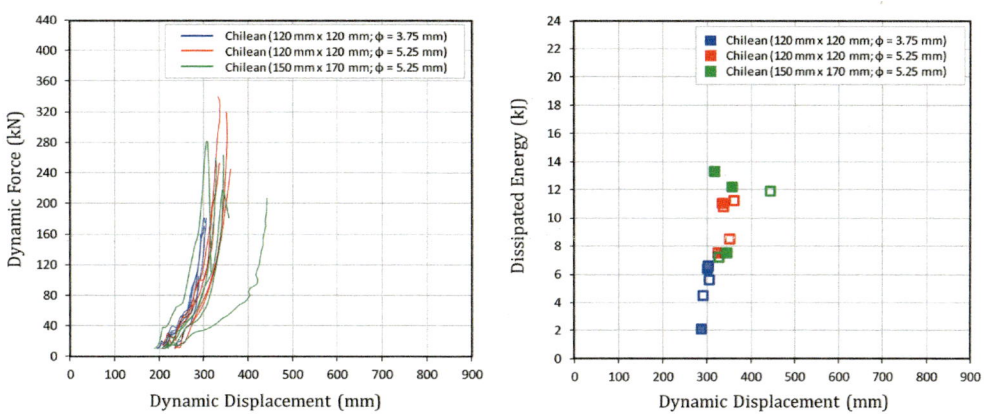

FIGURE 9.28 Dynamic rupture force and dissipated energy for mild steel woven mesh manufactured in Chile.

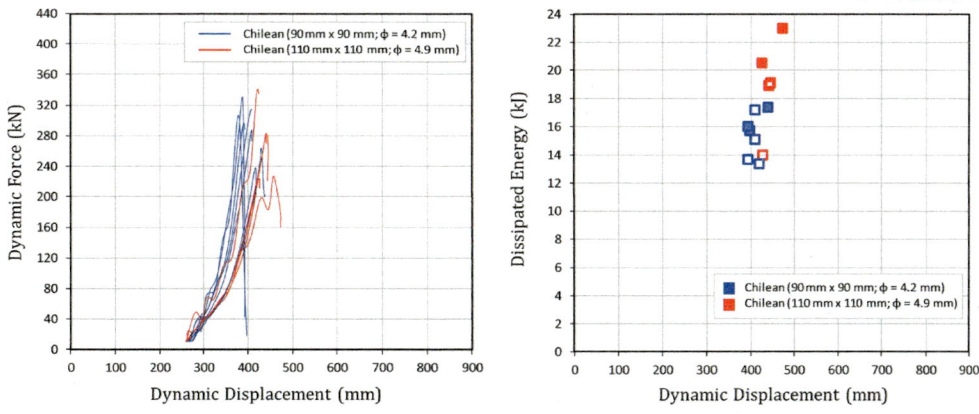

FIGURE 9.29 Dynamic rupture force and dissipated energy for medium tensile woven mesh manufactured in Chile.

following the dynamic loading events. Hence, the need to embed the mesh within the shotcrete layer using the shotcrete fill-in process described in Chapter 10.

Figure 9.30 presents the dynamic test results for high tensile (≈1800 MPa) mesh manufactured in Australia. The woven mesh has a rhomboidal shape (Figure 9.31), which is not entirely suitable for testing within square testing facilities, such as the one built at the WA School of Mines (Figure 9.6). The rhomboidal mesh is manufactured with a different number of longitudinal and transverse wires leading to uneven loading at the mesh boundaries, resulting in an underestimation of strength. A range of Tecco G80 products was tested, engaging longitudinal and transverse wires ranging from 13×7 to 13×13, with a clear influence on the dynamic response and related energy dissipation. Also, for the same products, as more wires were engaged, the deformation from static loading increased. The data show that the square frame boundary conditions clearly influenced the experimental results.

Irrespective of the different boundary conditions established high-tensile diamond mesh does not unravel once a wire fails (Roth et al., 2004). This was also observed with mild steel chain-link mesh, as reported by Stacey and Ortlepp (2001) and Van Sint Jan and Cavieres (2004). Due to the high-tensile steel, the weight of the mesh is very light compared to its strength and load capacity. It

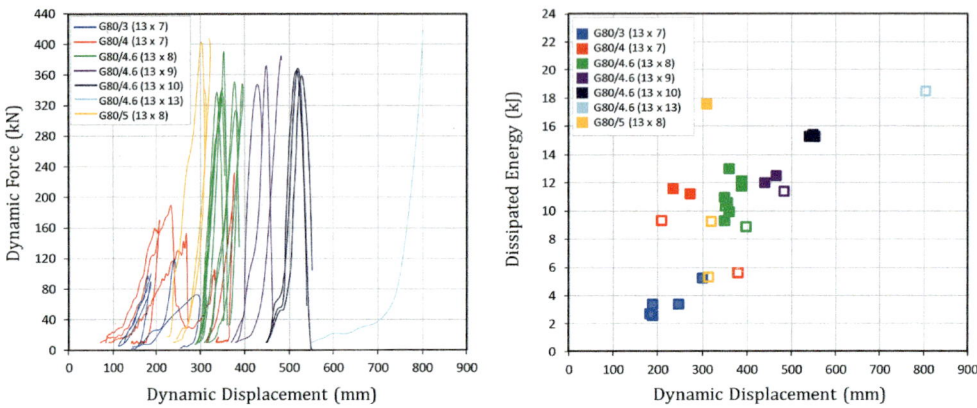

FIGURE 9.30 Dynamic rupture force and dissipated energy for high tensile woven mesh manufactured in Australia.

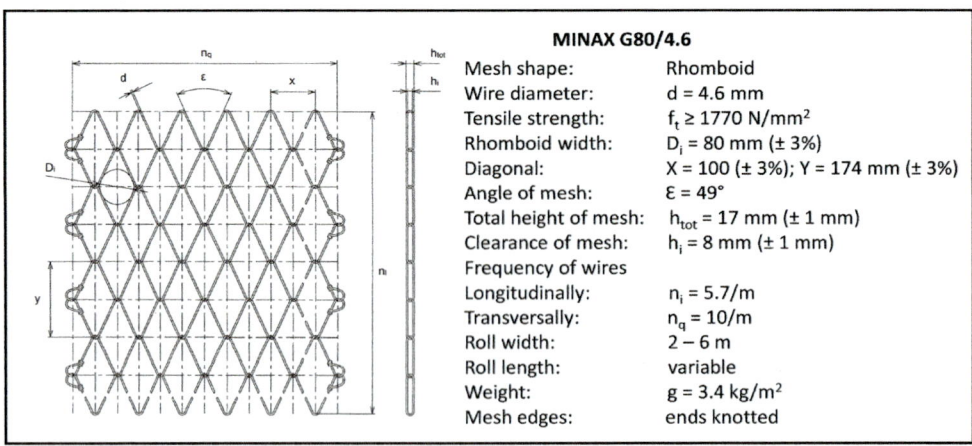

FIGURE 9.31 Geometrical and technical details for Tecco G80/4.6 high tensile woven mesh.

is adaptable to an irregular face and should be installed as tight to the face as possible to facilitate thin layers of shotcrete fill-in (see Chapter 10).

9.4.6 Dissipated Dynamic Energy

The energy dissipated during dynamic testing is best determined using plots of cumulative energy dissipated as a function of dynamic displacement. Figure 9.32 shows the cumulative energy dissipation results for a range of Geobrugg (G80) products in which the rupture point is shown as a solid star. The results are shown with respect to the total displacement, which includes the initial static loading prior to dynamic loading. As indicated earlier, the rhomboidal mesh is not suitable for testing within a square frame. The data clearly show that for the same product, the results depend upon the number of longitudinal and transverse wires being mobilized by the loading area during testing.

A comparison of the dynamic energy dissipation among multiple products is facilitated by translating the data to the origin and tracking the dynamic displacement, as shown in Figure 9.33. The data indicate that when the mesh is constrained, it can dissipate a significant amount of energy with minimal deformation. It is achieved when the mesh is embedded in shotcrete and provides energy dissipation exceeding 15 kJ for ranges of displacement similar to the rupture deformation of most commercially available rock bolts.

Figures 9.34 and 9.35 show the data for the Geobrugg G 80/4.6 woven mesh product, which has been shown widely within this book. A result for the conventional (5.6 mm diameter) galvanized welded wire mesh is also presented for comparison. The larger static displacement experienced with the woven mesh is clearly evident. Therefore, to minimize the need for rehabilitation (due to excessive deformation) after a dynamic event, the mesh should be installed with shotcrete fill-in (see Chapter 10). The data also show the stiff response of the mesh, with the full response achieved in less than 100 mm, thus neatly matching the displacement response of most reinforcement systems.

Figures 9.36 and 9.37 indicate similar responses for woven mesh manufactured in Chile. The mesh is manufactured with square and woven apertures and is thus more suitable for testing using square frames, as is the case at the WASM testing facility. Similar to the static case, the data show that the dynamic displacement occurs over a larger range of displacement compared with the Geobrugg G80 products.

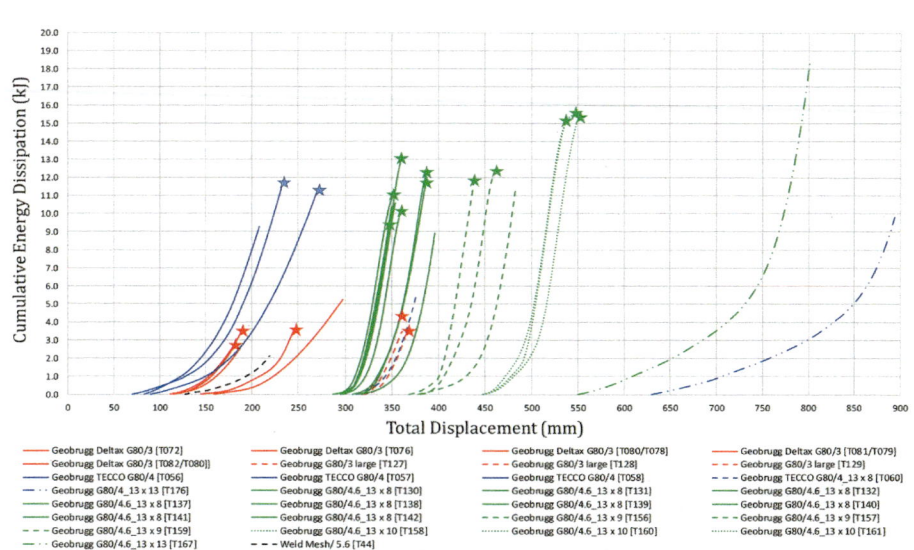

FIGURE 9.32 A comparison of dissipated energy versus total displacement (static and dynamic) for high tensile G80 woven mesh products (test number in parenthesis).

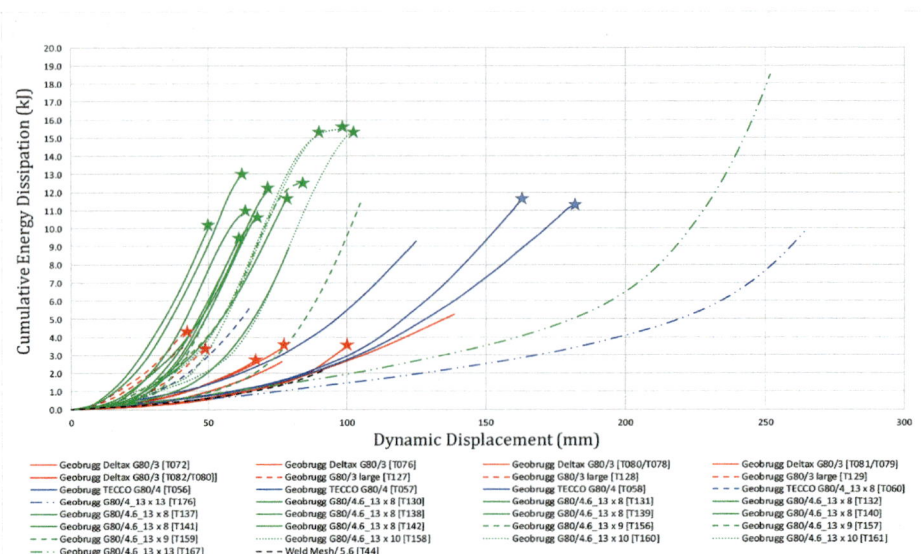

FIGURE 9.33   A comparison of dissipated energy versus dynamic displacement for high tensile G80 woven mesh products.

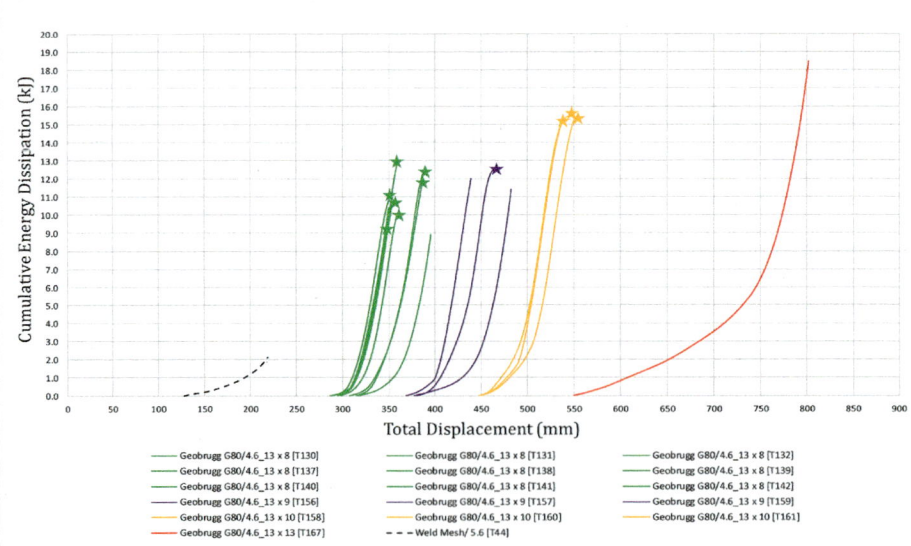

FIGURE 9.34   A comparison of dissipated energy versus total displacement (static and dynamic) for high tensile G80/4.6 woven mesh products and conventional galvanized (5.6 mm diameter) welded wire mesh.

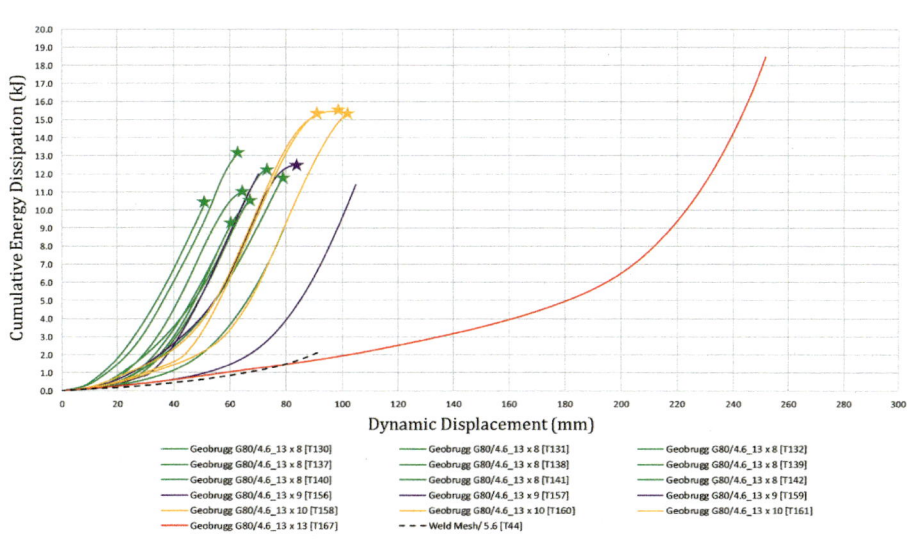

FIGURE 9.35 A comparison of dissipated energy versus dynamic displacement for high tensile G80/4.6 woven mesh and conventional galvanized (5.6 mm diameter) welded wire mesh.

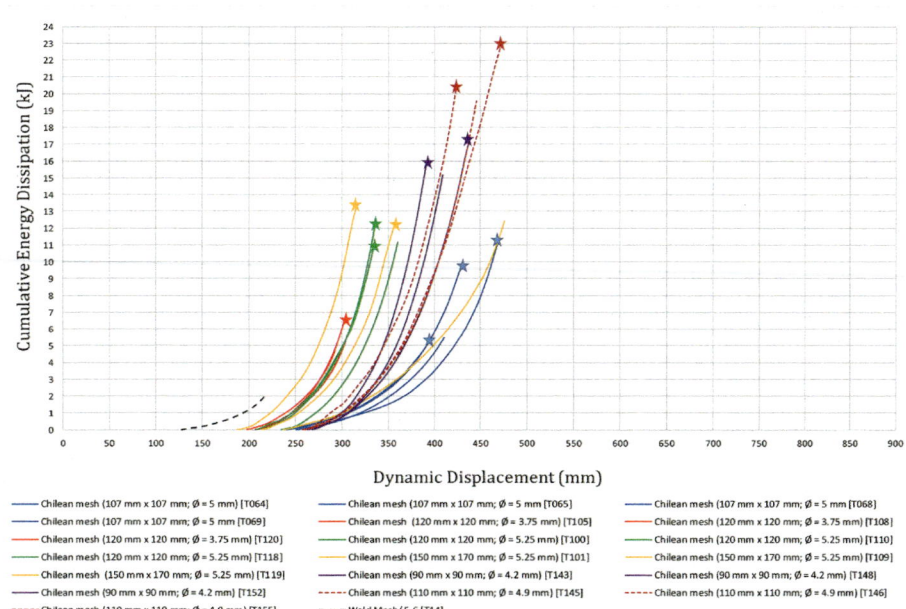

FIGURE 9.36 A comparison of dissipated energy versus total displacement (static and dynamic) for several Chilean woven mesh products.

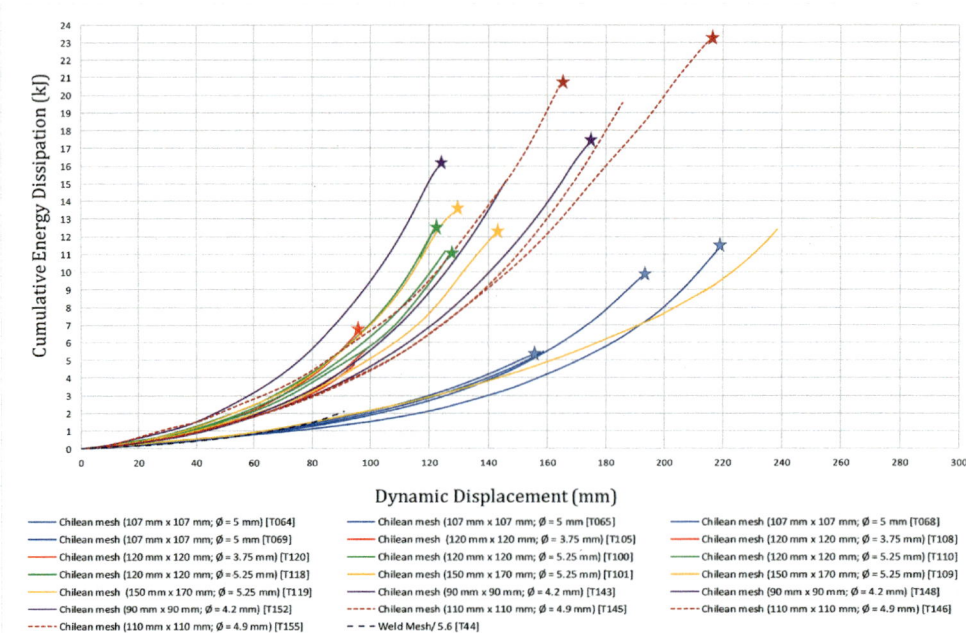

FIGURE 9.37 A comparison of dissipated energy versus dynamic displacement for several Chilean woven mesh products.

10 Energy Dissipation of Shotcrete Support

10.1 INTRODUCTION

Shotcreting is a surface support technique in which a specially designed concrete mix is sprayed at high speed onto rock excavation surfaces to improve rock mass integrity and assist load-carrying capacity at the rock surface. The benefits of using shotcrete compared with other ground support techniques have been demonstrated, particularly where the rock mass is of poor quality and is easily disturbed when attempting to scale or drill boreholes for the installation of reinforcement. Mesh reinforced shotcrete is usually required in hard strength rock masses where time-dependent or violent fracturing due to induced stress concentration can progressively or suddenly occur.

Wet mix shotcrete is now widely accepted in mines throughout the world, especially those prone to violent rock failure due to induced stress changes. In Australian mines, the use of shotcrete has continued to grow since the late 1980s due to its success in stabilizing excavations. The future use of shotcrete will continue and is likely to increase further as mines attempt development within the higher stress regimes and more difficult conditions that generally accompany mining at depth.

10.2 SHOTCRETE MIX DESIGN

A shotcrete mix comprises cement, supplementary cementing minerals, fine and coarse aggregates, chemical admixtures, water and usually fibres (Table 10.1). It begins as a separated heterogeneous mixture of solids and fluids, which are combined, mixed, pumped and sprayed to form a heterogeneous but uniform layer of solids and fluids that harden into a solid concrete layer. The mechanical properties of shotcrete change as the material hydrates from a fluid to a paste and then hardens to a solid. The physical, chemical and mechanical properties of each of these constituent materials affect the mechanical properties of the gelling, setting and hardening of the final shotcrete.

The two most widely used applications use the dry-mix or wet-mix processes. The dry-mix process comprises:

1. Cement, coarse and fine aggregates, mineral additives and (with or without) fibres, which are thoroughly mixed and fed into the delivery equipment and carried by compressed air through the delivery hose to a nozzle; and
2. The nozzle, which is fitted with an internal water ring, through which water is introduced under pressure and mixed with the other ingredients and sprayed at high velocity onto a rock mass surface.

The equipment used for dry-mix comprises two distinct types: Pressure vessels and rotary or continuous-feed guns. The detailed arrangements of dry-mix equipment are described in 'Guide to Shotcrete' (ACI 506.5R-09). Today, dry-mix shotcrete is mainly used in projects with relatively small volumes where high flexibility is required (Melbye and Garshol, 1997). The wet-mix process comprises cement, coarse and fine aggregates, mineral additives, (with or without) fibres and mixing water, which are thoroughly mixed in similar to conventional concrete. This mixture is fed into delivery equipment or an agitation truck, conveyed by compressed air to a nozzle and sprayed at high velocity onto a rock mass surface.

TABLE 10.1

Typical Range of Constituent Materials in a Shotcrete Mix (Saw, 2015).

Constituents		Dry mix	Wet mix	Unit
Cement		400–420	400–450	kg/m³
Fine aggregate (fine to coarse sand)		1380–1510	1110–1180	
Coarse aggregate (maximum 9–12.5 mm)		235–400	500–535	
Water		150–200	175–200	
Mineral additive (supplementary cementing materials) [1]	Silica fume		20–50	
	Fly ash		20–75	
	Granulated blast-furnace slag		20–50	
Chemical admixture [1 & 2]	Air-entraining		0.3–0.5	litre/m³
	Accelerator		20–22	
	Hydration stabilizer	–	1–2	
	Water reducer	–	1–2	
	Superplasticizer	–	3–7	
Steel fibres [3]			30–40	kg/m³
Synthetic fibres [4]			5–8	kg/m³

(1) Different types of mineral additives may be used in a particular mix

(1 & 2) Chemical admixture dosage should follow the manufacturer's guidelines

(3) For steel fibre reinforced shotcrete

(4) For synthetic fibre reinforced shotcrete

10.3   MATERIAL PROPERTIES

10.3.1   CEMENT

Generally, cement exhibits adhesive and cohesive properties that make it capable of bonding mineral fragments into a compacted mass. All types of cement are mainly composed of four compounds, which are tricalcium silicate (C_3S), dicalcium silicate (C_2S), tricalcium aluminate (C_3A) and tetracalcium aluminoferrite (C_4AF).

10.3.2   MECHANISM OF HYDRATION OF CEMENT

When cement is mixed with water, a series of simultaneous and consecutive reactions occur between water and the cement constituents. These reactions are covered by the term hydration, which comprises two stages, namely setting and hardening. Generally, setting refers to the change from a fluid to a rigid state, and hardening refers to the gain in strength of a cement paste that has been set.

Tricalcium silicate (C_3S) constitutes about 50% to 70% of cement by weight and tends to dominate the early hydration period that comprises setting and early strength development. In addition, it is the component most responsible for the formation of the calcium silicate hydrate gel (CSH), the main product of hydration (Bullard et al., 2010). Figure 10.1 shows a typical curve relating the fraction of C_3S consumed with time as determined by quantitative X-ray diffraction analysis. Taylor et al. (1984) suggested two stages of reactions that can be divided into four periods. The first stage reaction occurs primarily in the pre-induction period and continues very slowly into the induction period. The second stage reaction begins at the end of the induction period and continues through an acceleratory period, in which the main reaction occurs rapidly and is followed by post-acceleratory periods, wherein slow and continuing chemical reactions occur. The beginning and end of these periods are difficult to pinpoint precisely.

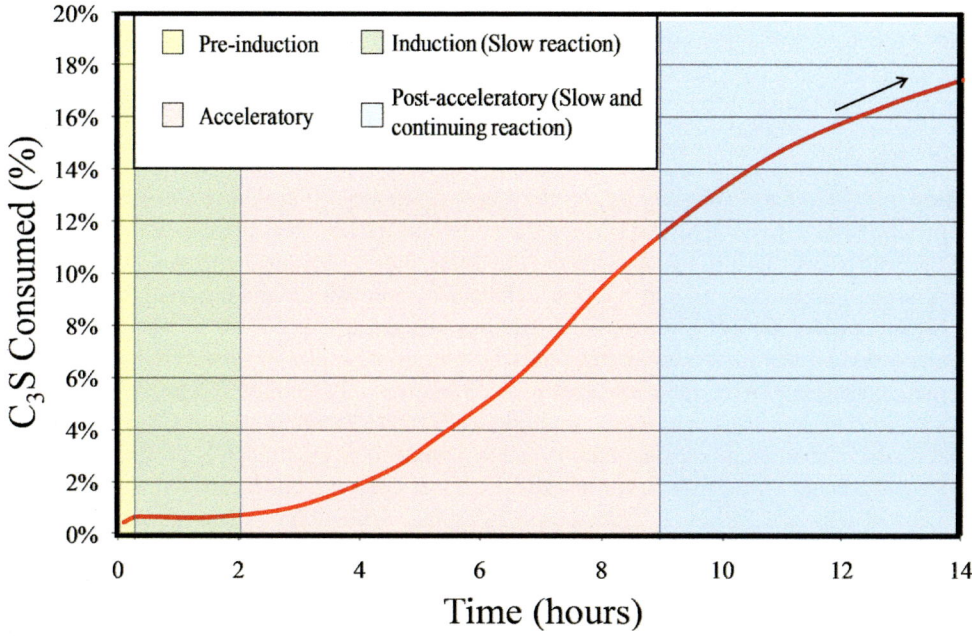

FIGURE 10.1 Typical curve of C_3S hydration as a function of time (modified after Taylor et al., 1984).

The mechanism of C_2S hydration is similar to that for C_3S but with a much lower rate of reaction. The products are also similar, apart from a much smaller content of CH (Taylor, 1997).

The reaction of C_3A hydration is very fast. There is no period of slow reaction, and the setting is almost instantaneous. This quick setting is undesirable in concrete or cement grout. Therefore, calcium sulphate (Gypsum—$CaSO_4 \cdot 2H_2O$) is added to the cement to delay the reaction of C_3A. This initial period of the rapid reaction quickly gives way to a period of low heat output. The duration depends on the quantity of calcium sulphate in the system. When the added calcium sulphate has been consumed, the rate of reaction rapidly increases again.

Generally, the hydration mechanism of C_4AF is similar to that of C_3A. However, differences in hydration behaviour may occur depending on the ability of the hydrate to accommodate ferric ions in their structure (RILEM, 1986). The hydrous ferric hydroxide is produced as a separate phase (Fukuhara et al., 1981).

10.3.3 SUPPLEMENTARY CEMENTING MATERIALS

10.3.3.1 Silica Fume

Silica fume, as defined in ASTM C1240–10a, is a 'very fine pozzolanic material, composed mostly of amorphous silica produced by electric arc furnaces as a by-product of the production of elemental silicon or ferrosilicon alloys'. It may be added directly to concrete (shotcrete) as an individual material or in a blend of Portland cement and silica fume. The silica fume generally contains more than 90% silicon dioxide (SiO_2).

Using silica fume in high-strength and high-performance concrete has become an increasingly accepted practice. Silica fume enhances the properties of concrete by both physical and chemical mechanisms. The physical mechanisms include reduced water bleed, provision of chemical nucleation sites and more efficient packing of the solid particles. The reduced bleeding increases cohesiveness and thus reduces segregation (Sellevold, 1987). Silica fume accelerates the hydration of cement during the early stages by providing nucleation sites where the products of cement hydration can more readily precipitate from the solution (Sellevold and Nielsen, 1987). The ACI 506.5R-09 'Guide for specifying underground shotcrete' recommended using silica fume in shotcrete because it can improve the shotcrete performance with respect to rebound and maximum achievable build-up. The ACI 234R-06 provides a comprehensive guide for the use of silica fume in concrete.

10.3.3.2 Fly Ash

According to ASTM C618–08a, fly ash is 'the finely divided residue that results from the combustion of ground or powdered coal and is transported by flue gasses'. It is primarily the inorganic portion of the source coal in a particulate form (Popovics, 1992). According to ASTM C618–08a, fly ash is classified into classes 'F' and 'C' by its chemical composition and physical properties.

Fly ash comprises 60–90% amorphous (glassy) spheres and a 10–40% crystalline phase. Its reaction depends largely on the breakdown and dissolution of a 'glassy' structure by hydroxide ions and the heat generated during the early hydration of the hydraulic cement fraction (ACI 232.2R-03). The reaction of the fly ash continues to consume $Ca(OH)_2$ to form additional CSH. The amount of heat that evolves from chemical reactions is usually reduced when fly ash is proportioned together with Portland cement in the mix. If the Portland cement content is reduced, fly ash can extend the setting time of concrete. When fly ash is used to replace a portion of cement in a unit volume of concrete, the amount of paste increases. Lane (1983) reported that an increase in paste volume produces fresh concrete with greater plasticity and better cohesiveness. The compressive strength and the rate of strength gain in hardened concrete are affected by the chemical and physical properties of a particular type of fly ash and cement. A Class 'F' fly ash can develop lower strength in 7 days or less when the concrete is cured at room temperature. When concrete is cured in a moist condition, the continued pozzolanic activity of fly ash provides strength development at later ages (Abdun-Nur, 1961). Class 'C' fly ashes often show a higher rate of reaction at early ages than Class 'F' fly ashes. However, certain Class C fly ashes may not show the later stage strength gain. The ACI 232.2R-03 provides a comprehensive guide for the use of fly ash in concrete.

10.3.3.3 Slag Cement

According to ASTM C219–07a, slag cement is defined as 'hydraulic cement consisting predominantly of ground granulated blast-furnace slag'. The granulated blast-furnace slag (GBFS) is a 'glassy' granular material formed when molten blast-furnace slag is rapidly chilled by immersion in water. The blast furnace slag is a non-metallic product developed in a molten condition simultaneously with iron in the blast furnace. The composition of blast-furnace slag is determined by that of the ores, fluxing stone and impurities in the coke charged into the blast furnace. The major oxides of silica, alumina, lime and magnesia constitute 95% or more of the total oxides (ACI 232.2R.03). ASTM C989–10 provides three grades of Slag cement, depending on their respective mortar strengths when blended with an equal mass of Portland cement.

The hydration of slag cement in combination with Portland cement at normal temperature is a two-stage reaction (Regourd, 1980; Roy and Idorn, 1982). Initially and during the early hydration, the predominant reaction is with alkali hydroxide, but the subsequent reaction is predominantly with calcium hydroxide. The principal hydration product is the same as when Portland cement hydrates, that is, calcium-silicate hydrate (Smolczyk, 1978). Wood (1981) reported that the workability of concrete containing slag cement was improved when compared with concrete without slag cement. The compressive strength and rate of strength gain in concrete containing slag cement vary depending upon the slag activity index of particular slag cement. Hogan and Meusel (1981) reported that, compared with Portland cement concrete, using Grade 120 slag cement typically results in reduced strength at 1–3 days and increased strength beyond seven days of curing. Using Grade 100 results in lower strength at one to 21 days but equal or greater strength after 21 days. Grade 80 typically gives reduced strength at all ages. ACI 506.5R-9 'Guide for specifying underground shotcrete' suggested that the use of slag cement is acceptable only if all shotcrete performance requirements can be demonstrated during preconstruction testing. This is especially relevant to early strength gain.

10.3.4 MIXING WATER

The mixing water is the free water present in freshly mixed shotcrete. It has three main functions: (1) It reacts with the cement powder, thus producing hydration; (2) It acts as a lubricant, contributing to the workability of the fresh mixture; and (3) It secures the necessary space in the cement paste for the development of hydration products (Popovics, 1992). Two important issues here are quantitative and qualitative. The quantitative issue concern is how much water should be added to the batch. It is indicated by the WCR. The WCR is given by the mass of free water relative to the mass of cement powder comprising the shotcrete mix. The WCR is known to affect most of the chemical and mechanical properties of the fresh and hardened shotcrete (Thompson and Windsor, 1998).

An important qualitative issue concerns the question of what quantities of impurities render the water unsuitable for shotcrete mixes. The ACI 506.5R-9 'Guide for specifying underground shotcrete' suggested that mixing water used for underground shotcrete shall comply with the ASTM C1602/C1602M. In general, potable water is used without testing for conformance with the ASTM standard. Non-potable water is permitted for use in any proportions to the limits that meet the required performance and chemical limits set by the ASTM standard.

10.3.5 AGGREGATE

Aggregate is a granular material, such as sand, gravel, crushed stone, or iron blast-furnace slag, used with a cementing medium to form hydraulic-cement concrete or shotcrete (ASTM C125–10a). It occupies at least three-quarters of the total volume of a given shotcrete mix, and thus, its properties greatly affect the properties of the fluid and hardened concrete or shotcrete. Generally, aggregates can be classified in several ways, such as according to their petrography, specific gravity, particle size distribution (PSD), natural or manufactured, crushed or naturally processed and chemically inert or reactive (Popovics, 1992). Among these, PSD is the most frequently used classification. Based on PSD analysis, the aggregate can be divided into coarse and fine. Generally, aggregate retained on the 4.75-mm sieve is defined as coarse aggregate and that passing the 4.75-mm sieve and retained on the 75-μm sieve is defined as fine aggregate. Aggregates comprising both coarse and fine particles are called combined mixed aggregates (ASTM: C125–10a). ASTM: C125 does not define the upper limit of coarse aggregate. Based on Australian standard AS 1289.3.6.1: 'Methods of testing soils for engineering purposes—Soil classification tests—Determination of the particle size distribution of a soil—Standard method of analysis by sieving' fine aggregate ranges from fine sand to fine gravel, and coarse aggregate ranges from medium gravel to cobles. Figure 10.2 shows coarse and fine aggregate with different combined grading limits to produce typical shotcrete for underground application. According to ACI 506.5R-09, grade No. 1 aggregates are sometimes used in finishing coats and other thin layers, and grade No. 2 aggregates are normally preferred for ground support and linings. The American Concrete Institute has suggested that coarse aggregate should have a maximum size of 9.5 mm for underground shotcrete application (ACI 506.5R-09). This limitation is based on the pumping and/or spraying equipment in order to avoid too much rebound (Melbye and Garshol, 1997). According to Armelin and Banthia (1998), the kinetic energy of the incoming particle and the mechanical properties of the substrate and its cohesion are all key parameters to predict if a particle will either be trapped in the wet shotcrete layer or rebound from it.

Industrial research on conventional cast-in-place concrete suggested that extending the grading of aggregate to a larger maximum size lowers the water requirement of the mix so that the WCR can be lowered as a consequence of an increase in strength (Higginson et al., 1963). Reny and Jolin (2011) suggested that the use of coarse aggregate in a shotcrete mixture will provide a wide range of benefits to both plastic and hardened properties. The question is: 'What is the optimum maximum aggregate size?' Based on the knowledge from the design of 'Cyclopean concrete', where large stones or 'Plums' are used, the maximum aggregate size should not be greater than one-third of the thickness of the concrete (Neville, 1995). A typical thickness of shotcrete used for underground

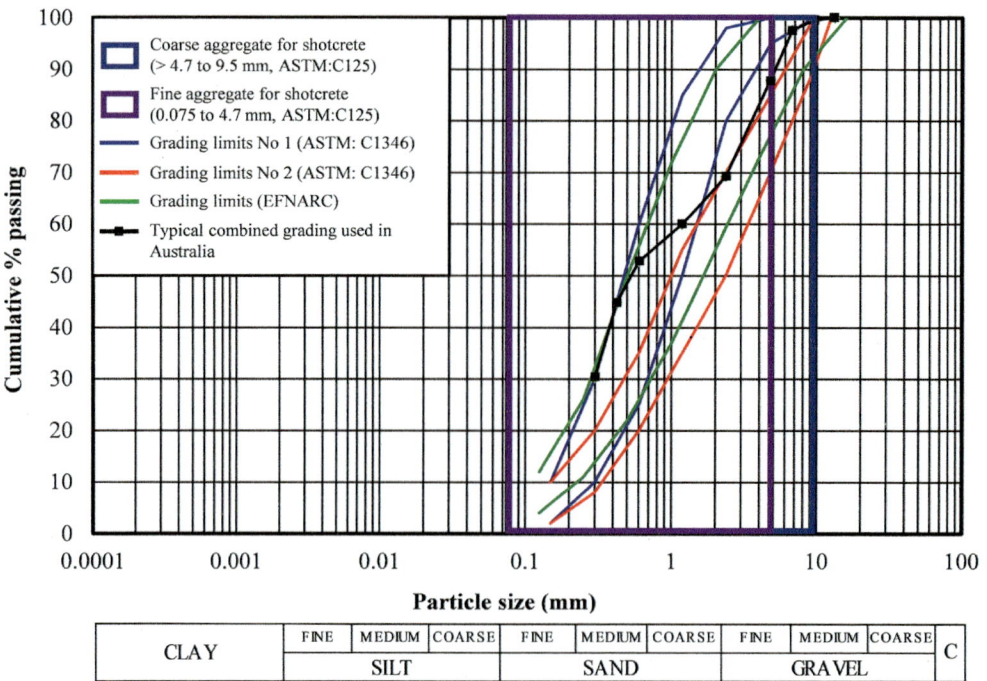

FIGURE 10.2 Grading limits for combined aggregate for shotcrete (ASTM: C125, C1436, ACI 506.5R-09 and AS1289.3.6.1).

mine ground support usually ranges from 50 to 75 mm. Therefore, the maximum aggregate size could approach 17 mm.

In addition to a PSD, the particle shape and texture, specific gravity, bulk density, moisture content, mineral and chemical properties and the strength of aggregate also influence the properties of shotcrete.

10.3.6 Fibres

Fibres are slender filaments, which may be discrete or in the form of bundles, networks, or strands of natural or manufactured materials, which are required to be distributed uniformly throughout the fresh cementitious mixture (ASTM C125–10a). Since ancient times, fibres have been used to reinforce brittle materials. For example, straw was used to reinforce bricks and horsehair was used to reinforce mortar and plaster. In the early 1960s, the first major investigation was made to evaluate the potential of steel fibres as a reinforcement for concrete (Romualdi and Batson, 1963). Since then, a substantial amount of research, developmental experimentation and the industrial application of fibre-reinforced concrete has emerged. There are numerous fibre types available for commercial and experimental use. The basic categories are steel, glass, synthetic and natural fibre materials, though ACI 506–5R-09 suggested that only steel and synthetic fibre should be used for underground shotcrete applications. The physical and mechanical properties of fibre reinforced shotcrete (FRS) are influenced by the fibre types and dimension (length to diameter ratio), geometry, dosage, distribution and bond characteristics between the fibre and shotcrete matrix.

Figure 10.3 shows that fibres with various orientations to a reinforced crack actually respond in different failure modes (Windsor, 1996). The responses are predominantly shear, tensile and compression in nature. The most common response is a combination of these modes. The force-displacement relationship of the individual fibres can be described with a characteristic diagram, as shown in Figure 10.4 (Saw et al., 2009).

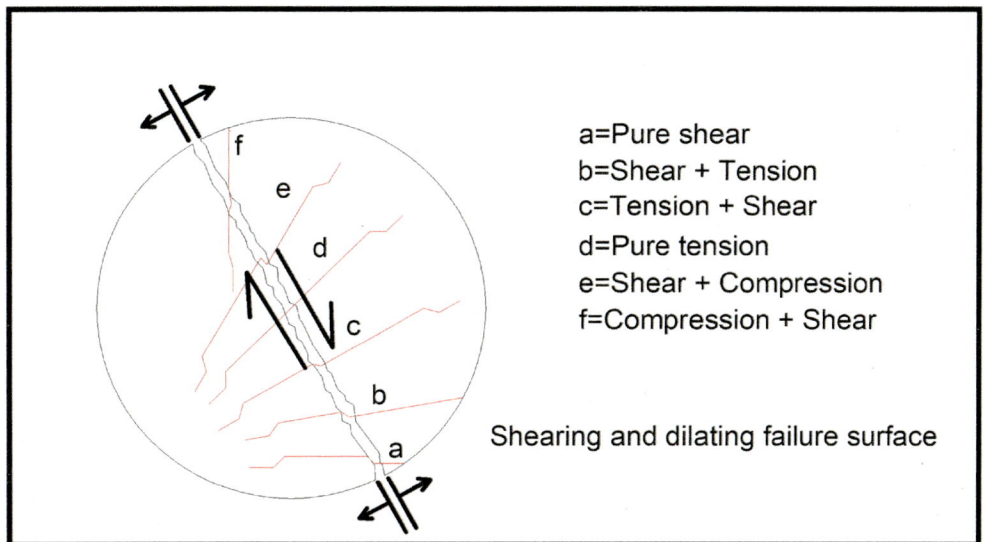

FIGURE 10.3 Fibre with various response modes for different fibre orientations to a reinforced crack (Saw, 2015).

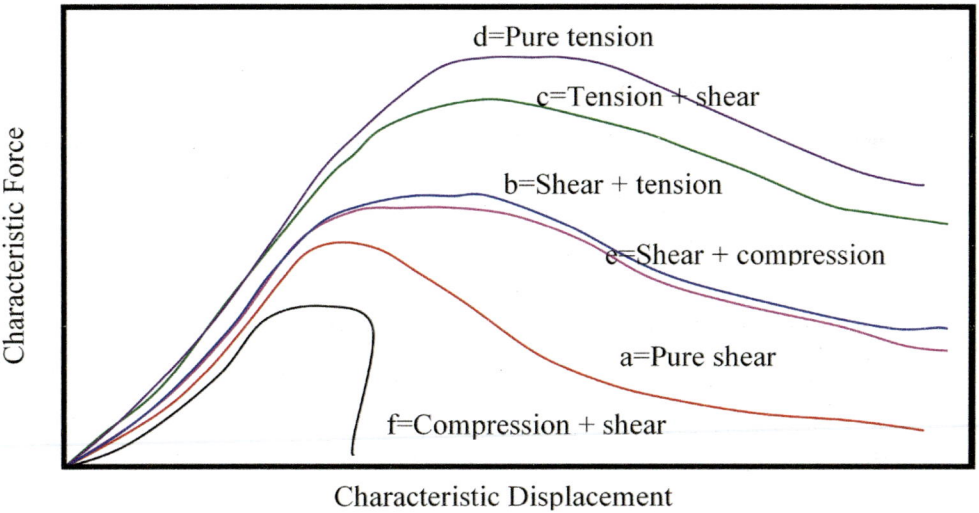

FIGURE 10.4 Force-displacement characteristic diagram for different fibre orientations to a crack (Saw, 2015).

Windsor (2009) suggested that the adhesive, frictional and mechanical bond strength between the fibre and the cementitious matrix controls the mechanical performance of the failure surface (or cracked shotcrete), not the strength of the fibre. Figure 10.5 (Windsor, 2009) illustrated the effect with a crack system in a synthetic fibre reinforced shotcrete surface on a mine excavation wall shown in 'A' with magnifications at crack points 'B' and 'C'. Collectively, views 'A', 'B' and 'C' show that:

- The fibres lie in the plane of the layer, oriented approximately normal to crack.
- The orientations of fibres across the crack indicate the geometry of crack displacement.
- Some fibres remain bonded in each crack face depending on crack displacement (A).

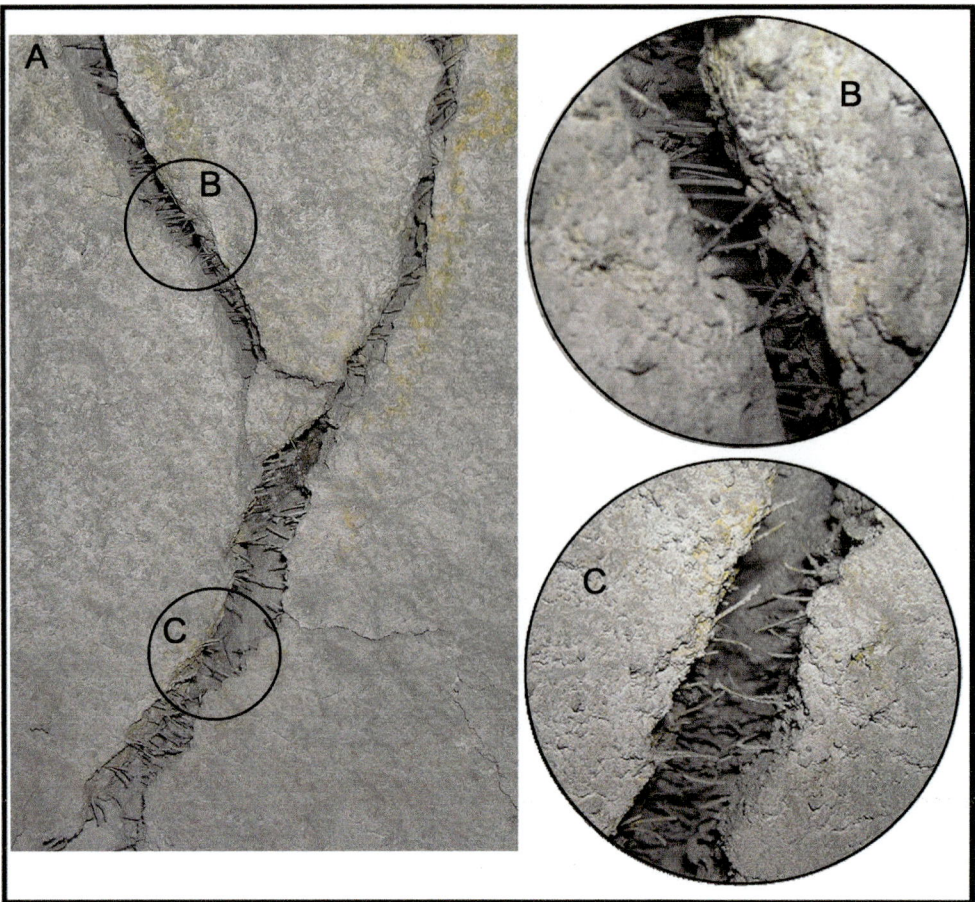

FIGURE 10.5 Cracked surfaces in a synthetic fibre reinforced shotcrete surface on a drive wall shown in 'A' with magnifications at crack points 'B' and 'C' (Windsor, 2009).

- Some fibres have pulled out of each crack face depending on crack displacement (B).
- Very few, if any, broken fibres can be found.
- Little, if any, of the fibres actually project from the layer surface.

10.3.6.1 Steel Fibres

Steel fibres are defined as short, discrete lengths of steel which are sufficiently small to be randomly dispersed in an unhardened concrete mixture (ACI 544.1R-96). Basically, five general types of steel fibres are available, and they are used based on the manufacturing process. These include cold-drawn wire (Type—I), cut sheet (Type—II), melt-extracted (Type—III), mill cut (Type—IV) and modified cold-drawn wire (Type—V). Steel fibres can be either straight or deformed (ASTM A820/A 820M). The ASTM A820 specified the standard dosage for steel fibres in reinforced concrete or shotcrete as between 30 and 40 kg/m³ of shotcrete.

10.3.6.2 Synthetic Fibres

Synthetic fibres are made from polymers such as acrylic, aramid, carbon, nylon, polyester, polyethene and polypropylene. Polypropylene base synthetic fibres are currently the most widely used fibres for shotcrete application. The fibre length, width and thickness of macro synthetic fibre are typically

48, 1.28 and 0.5 mm, respectively. A typical macro synthetic fibre dosage is between 5 and 8 kg/m³ of shotcrete. A typical length of micro synthetic fibre is 6–20 mm and a diameter of 0.015–0.03 mm.

10.3.7   ADMIXTURES

An admixture is defined as 'A material other than water, aggregate, hydraulic cement and fibre reinforcement used as an ingredient of concrete or mortar and added to the batch immediately before or during its mixing' (ASTM C125). It includes mineral and chemical admixtures. Mineral admixtures, such as silica fume, fly ash and slag, were described earlier in Section 10.3.3 as supplementary cementing materials. Admixtures in this section refer to chemical admixtures. The admixtures may remain in a free state as solids or in solution, which may interact at the surface or be chemically combined with the constituents of cement or cement paste. The type and extent of interaction might influence the physical-chemical and mechanical properties of concrete, such as water demand, hydration, the composition of the products, setting times, microstructure, strength and durability (Ramachandran, 1995). ASTM C1141/C1141M provides a standard specification for admixtures for shotcrete. Admixtures used in underground shotcrete applications mainly fall into four categories: (a) accelerator, (b) superplasticizer or high range water reducer, (c) hydration stabilizer and (d) air-entraining admixture. The following sections describe the effects of those admixtures and the chemistry underlying those effects.

10.3.7.1   Accelerator

Most of the accelerators used for shotcrete in underground civil and mining applications contain aluminium sulphate in a liquid state (Myrdal, 2007). The accelerator is added to a shotcrete mix as a separate component at the nozzle during spraying. This type of accelerator is widely known as an alkali-free accelerator. The addition of alkali-free accelerators changes the chemical reactions, especially in the early stages. Bürge (2001) found that aluminium sulphate reacts with calcium hydroxide ($Ca(OH)_2$) and produce additional ettringite crystals and aluminium hydroxide.

The aluminium sulphate of these accelerators promotes the crystallization of ettringite at a very early stage. Within 4 hours of hydration, these crystals grow and fill most of the capillary pores. Usually, the formation of ettringite within the first 30 minutes is largely sufficient to set a shotcrete mix (Paglia et al., 2001).

The addition of an alkali-free accelerator is known to reduce the shotcrete durability with respect to chloride resistance, permeability, sulphate attack, and freezing and thawing action (ACI 506–5R). Most accelerators result in lower 28-day strength when compared with non-accelerated shotcrete. Complex factors that may cause low strength are the formation of C-S-H with a higher C/S ratio, retardation of C3S hydration and rapid initial setting followed by large heat development producing a more porous structure (Ramachandran, 1995).

10.3.7.2   Superplasticizer

Superplasticizers are used in shotcrete for three different purposes or their combined purpose: (1) To increase workability (pumpability and shootability) at a given mix composition, (2) To reduce the mixing water, at a given cement content and workability, in order to reduce the WCR, (3) To reduce both water and cement quantity (Collepardi et al., 1999). Superplasticizers belong to a class of water reducers that are capable of reducing water content by about 30–40%. According to ASTM C494/C494M, most of the commercial superplasticizers can be classified as 'type F—high range water reducing admixtures'.

10.3.7.3   Air-Entraining

An air-entraining admixture is 'an admixture that causes the development of a system of microscopic air bubbles in concrete, mortar, or cement paste during mixing' generally used to increase workability and resistance to freezing and thawing (ACI 212.3R-10). An air-entraining admixture does not generate air voids. It stabilizes the air voids trapped in the concrete during the mixing process (Powers, 1968). As the air-entraining admixture molecules are inserted between adjacent water molecules at the water surface, the mutual attraction between the separated water molecules is reduced. Lowering the surface tension stabilizes the bubbles against mechanical deformation and rupture, making it easier for bubbles to be formed. Without the presence of an air-entraining agent, the smaller bubbles, which have higher internal pressure, coalesce to form larger bubbles that have a greater tendency to escape to the surface and burst (Nagi et al., 2007).

Air-entraining admixtures have no appreciable effect on the rate of hydration of cement or on the heat evolved during that process. Apparently, they do not affect the chemical composition of the hydration products (Dolch, 1984). Beaupre (1994) suggested that the workability of a fresh shotcrete mix is increased to meet the pumpability requirement by introducing a large amount of entrained air bubbles (10–30%) into the fresh mix. The initial large amount of air will be lost during pumping and shooting due to the compaction process. In addition, it can improve accelerator distribution within the shotcrete. It results in uniformly higher early strength achievement (Hauck and Kristiansen, 2010).

10.3.7.4 Hydration Stabilizer

Hydration stabilizers are used to limit pre-hydration of shotcrete and to extend the retention time of batched shotcrete when the handling and logistical constraint typically associated with underground construction causes prolonged transport and discharge time (ACI 506.5R-9). It is also commonly referred to as hydration controlling admixture or retarder (ASTM C494). The stabilizer can suspend cement hydration for up to 72 hours. Usually, an activator or accelerator is added to the shotcrete mix at the nozzle as it is sprayed.

The stabilizer is adsorbed onto the surfaces of growing particles of hydration products and complex calcium ions (Ca^{2+}) in an aqueous solution (Taylor, 1997). The adsorption causes the precipitation of a low-permeability film coating the cement particles, slowing down further hydration (Edmeades and Hewlett, 2008).

10.4 SHOTCRETE SUPPORT SYSTEM

The structural design of a shotcrete support system involving three main types of shotcrete systems, such as plain shotcrete, fibre reinforced shotcrete and mesh reinforced shotcrete, has been suggested by Windsor (1999). A shotcrete support system comprises a system of four principal components: (1) the rock, (2) the cementitious matrix, (3) the aggregate and (4) the reinforcement. Each component of the system is involved in load transfer interactions. The load transfer in a shotcrete support system occurs at the surface on the boundary of the rock mass. The 'capacity' of a support system can be defined as the force-displacement response by the integrated behaviour of these components and their multiple interactions to a given arrangement of loadings and boundary configurations.

Two forms of capacity can be identified:

- Nominal capacity—Associated with contrived testing arrangements of loading and boundary configurations
- *In situ* capacity—Associated with the *in situ* arrangement of loading and boundary configurations

The force-displacement relations may be characterized by a number of important features, such as stiffness, yield, ultimate, residual and failure. The support system capacity can be categorized into (1) Force capacity (F_c), which is the maximum force that the system is capable of achieving with an associated displacement; and (2) Displacement capacity (δ_c), which is the maximum displacement the system is capable of achieving with an associated force.

The force-displacement response needed to assist a rock mass collapse mechanism to achieve equilibrium is defined as the 'demand' of the support system. The collapse mechanism is determined by the interaction of:

- The excavation geometry and orientation;
- The intact rock geometry and its mechanical properties;
- The rock structure geometry and its mechanical properties; and
- The rock mass environment (e.g., stress regime, hydrogeological regime).

Shotcrete collapse mechanism may be controlled by the structures or the stress acting on the rock mass structural or stress regime. The concepts of demand and capacity are illustrated in Figure 10.6a and b. In Figure 10.6a, the demand characteristic is independent of displacement (as in gravitational collapse mechanisms), and in Figure 10.6b, it is initially displacement-dependent (common in stress-driven collapse mechanisms). In the first case, if force capacity exceeds demand, then the equilibrium is achieved at a displacement depending on the system stiffness. In the second case, the force capacity, displacement capacity and stiffness of the system collectively determine if equilibrium can be achieved.

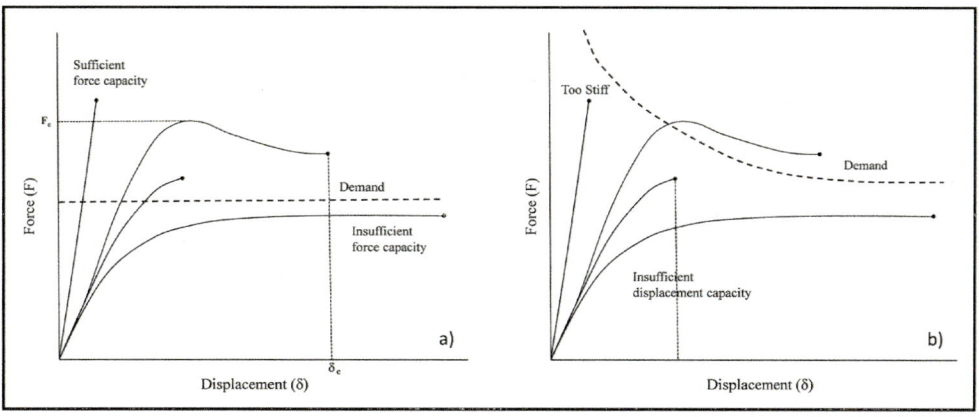

FIGURE 10.6   The concept of demand and capacity is shown in terms of force displacement: (a) displacement-independent demand and (b) displacement-dependent demand (Windsor, 1999).

10.4.1   ROCK SURFACE AND SHOTCRETE PROFILES

The interaction between the rock surface and shotcrete is critical for the proper design of a shotcrete support system. Figure 10.7 shows a horseshoe-shaped tunnel excavated in a massive rock, stratified rock and jointed rock mass. Figure 10.7a shows the typical results when smooth wall blasting techniques are employed in massive rock. Figures 10.7a, b and c show typical results when conventional drill and blast excavation are employed. It is clear that the surface of an excavation has a roughness, which is dependent on the excavation method and the rock structure.

The shotcrete profile is important in the structural engineering calculations for the shotcrete support system. There are five possible shotcrete profiles dependent upon the roughness of the rock surface and the shotcrete spray technique. These are:

- Coated (C) profile—Coat the rough rock excavation surface with a minimum thickness to produce a rough-coated surface;

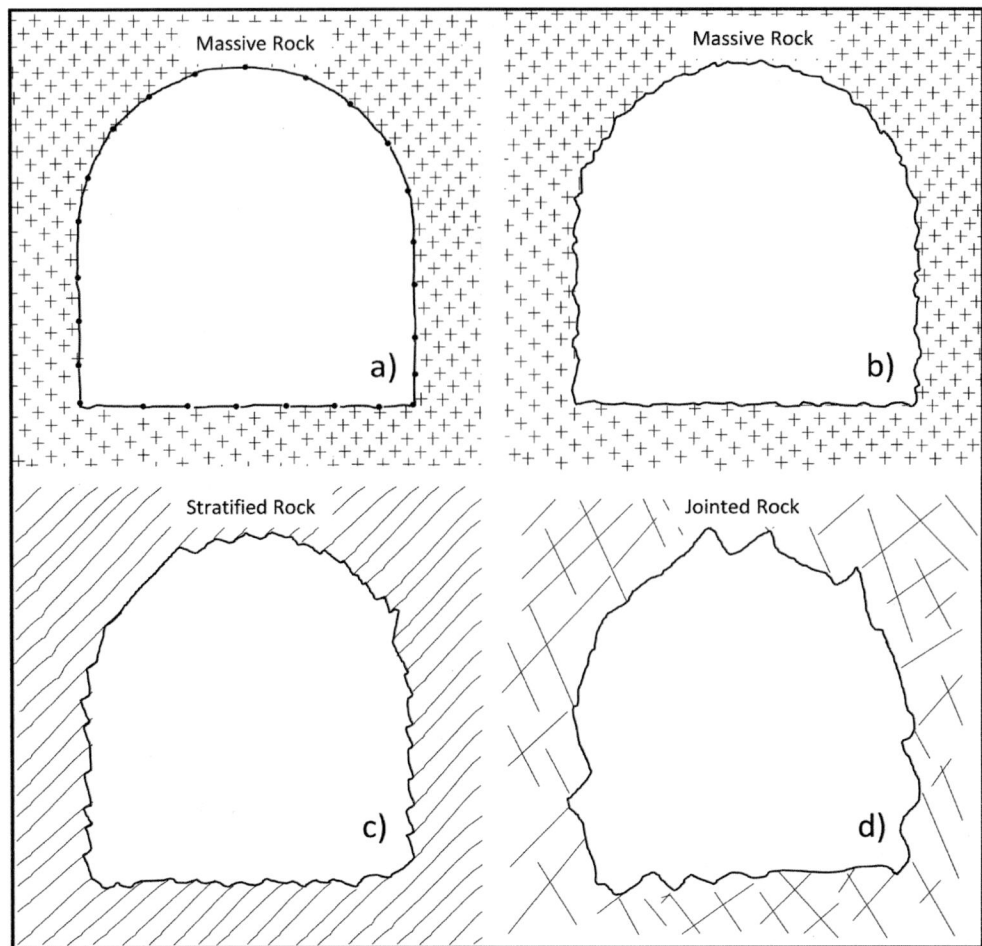

FIGURE 10.7 Surface irregularities in a horseshoe-shaped tunnel, including (a) machine or smooth wall excavation in massive rock, (b) conventional drill and blast excavation in massive rock, (c) conventional drill and blast excavation in stratified rock and (d) conventional drill and blast excavation in jointed rock (Windsor, 1999).

- Filled (F) profile—Fill the rough rock surface to a smooth surface defined by the 'teeth' or 'tips' of the rock projections to produce a relatively smooth filled surface;
- Coated filled (CF) profile—Coat the rough rock surface with an even minimum thickness and partially backfill the irregularities or 'notches' to produce a rough-coated and filled surface;
- Filled covered (FC) profile—Fill the irregularities to an 'F' profile and then apply a minimum thickness cover over the tips to produce a smooth, filled and covered surface; and
- Specified excavation (SE) profile—Fill all overbreak and irregularities and continue with covering until a given excavation design surface is achieved to a specified excavation geometry.

Figures 10.8–10.10 illustrate the five possible shotcrete support profiles for a horseshoe-shaped tunnel produced by a drill and blast excavation method.

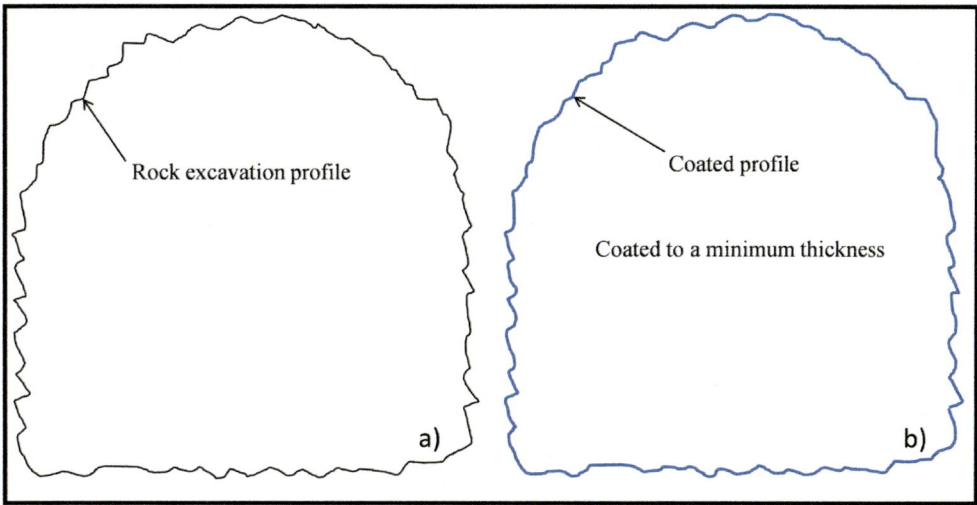

FIGURE 10.8   (a) Drill and blast rock excavation profile and (b) excavation profile coated to a minimum shotcrete thickness (Windsor, 1999).

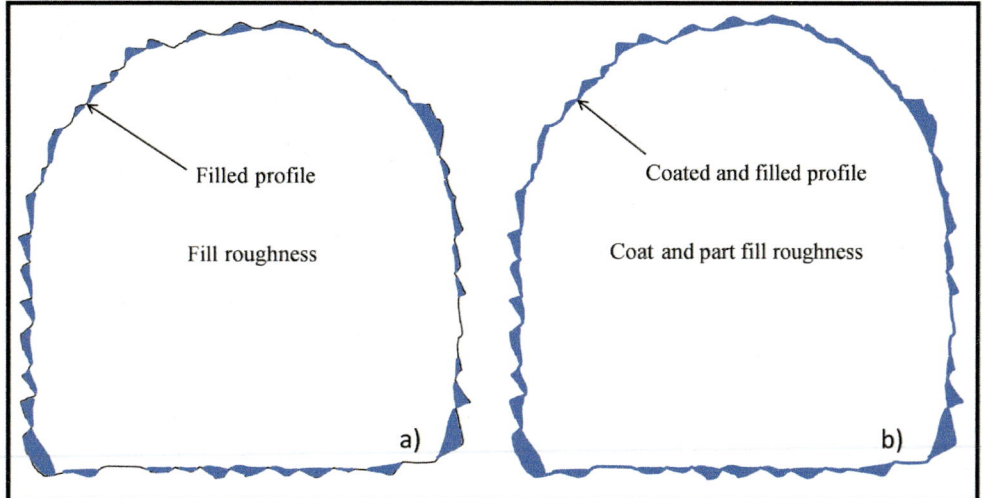

FIGURE 10.9   Excavation profile (a) filled to a smooth surface and (b) coated and partially filled excavation irregularities (Windsor, 1999).

10.4.2   DEFORMATION MECHANISMS

An infinite variety of possible loadings and equally large numbers of possible boundary configurations exist. However, two main mechanisms are considered: Firstly, a stress-controlled deformation mechanism, and secondly, a structurally-controlled deformation mechanism. The total response and total capacity can be a combined response mode comprising a limited number of fundamental responses, including:

- Compression,
- Tension,

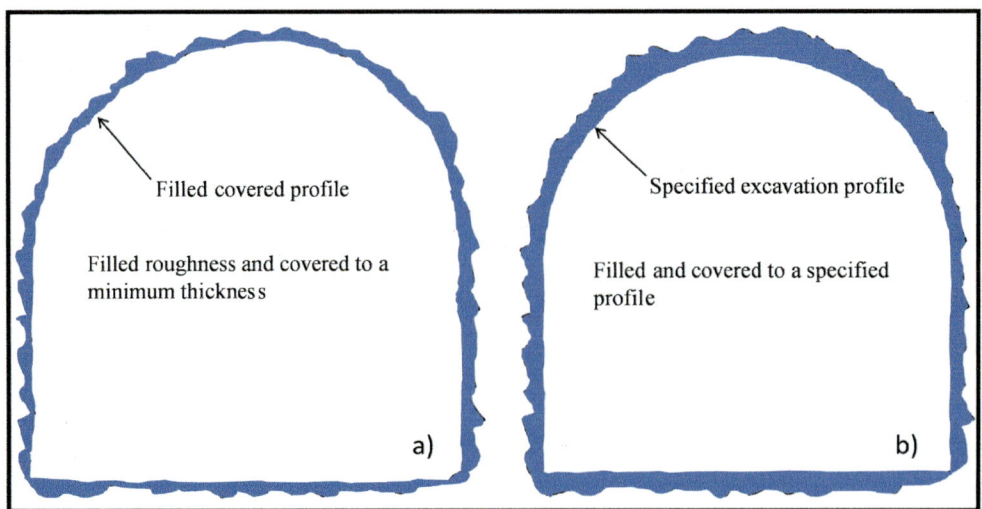

FIGURE 10.10 Excavation profile (a) filled to a minimum thickness and (b) filled to a specified profile (Windsor, 1999).

- Shear,
- Torsion,
- Flexural and
- Combinations.

The typical response modes are illustrated in Figure 10.11. Each fundamental response mode has an associated capacity (in terms of force and displacement) or mechanical properties, such as strength (in terms of compressive, tension, shear and flexure) in elastic and plastic phases. The mechanical properties can be determined from different laboratory test arrangements.

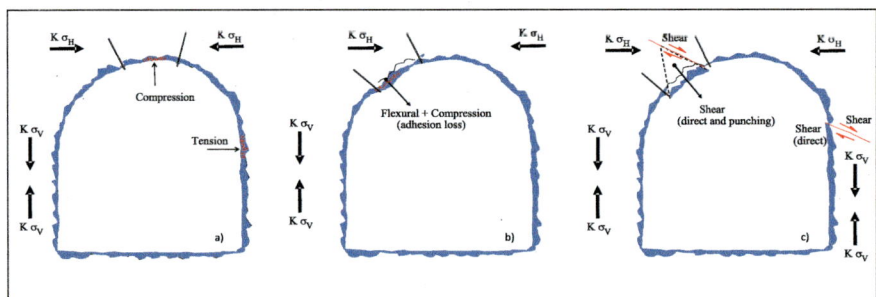

FIGURE 10.11 Modes of response for excavation with filled roughness and covered to a minimum shotcrete thickness: (a) compression and tension, (b) flexural and compression and (c) shear modes of failure (modified after Windsor, 1999).

10.5   STATIC PERFORMANCE OF FRESHLY SPRAYED SHOTCRETE

The performance of freshly sprayed shotcrete is particularly critical for In-Cycle Shotcrete (ICS). ICS, in a true engineering sense, involves precise mark-up and drilling of face holes (using modern drilling jumbos), blasting (using careful blasting techniques), water jet scaling (using specialized hydro-jetted machines) and high-quality shotcrete (using modern shotcrete machines). Its purpose is to set a timed re-entry into the workplace, followed by rock bolting and meshing.

Figure 10.12 shows a typical underground mining process incorporating shotcrete in a cycle which takes approximately 9 hours to complete a 5.3 m × 5.5 m × 3 m (Width × Height × Advance) tunnel (O'Shea, 2005). It can be seen that the re-entry time is a significant component of the total cycle time. Thus, re-entry time sometimes presents a conflict between increased productivity and workplace safety. Figure 10.13 shows an example of freshly sprayed shotcrete detached and fallen due to its weight. Figure 10.14 shows an unstable block that is punching through a fresh shotcrete layer due to gravity loading. These conditions need to be considered when determining the safe re-entry time below a freshly sprayed concrete layer. A review of the current literature does not sufficiently address the issue of re-entry time under freshly sprayed shotcrete. Currently, re-entry times range from 2 to 4 hours based on the time when the UCS of shotcrete (based on penetration resistance) reaches 1.0 MPa (Clements, 2004).

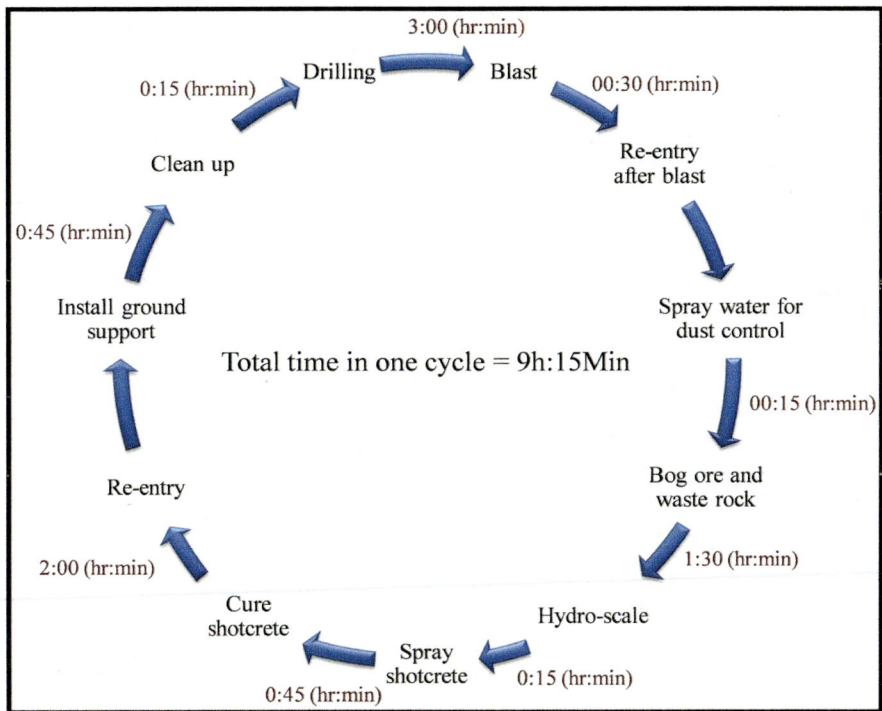

FIGURE 10.12   Underground mining process using ICS for a mining excavation having a width of 5.3 m, a height of 5.5 m, a typical advance length of 3 m and a shotcrete layer thickness of 50 mm (modified after O'Shea, 2005).

10.5.1   REVIEW OF SHOTCRETE EARLY STRENGTH

Considerable early strength data on shotcrete (collected according to ASTM C403) has been published (e.g., Jolin et al., 1999, Rispin et al., 2003, Clements, 2004, Knight et al., 2006, O'Toole and Pope, 2006 and Bernard, 2008). Figure 10.15 shows the UCS of shotcrete at various curing times (determined by the penetration resistance method) published by different authors. The implication

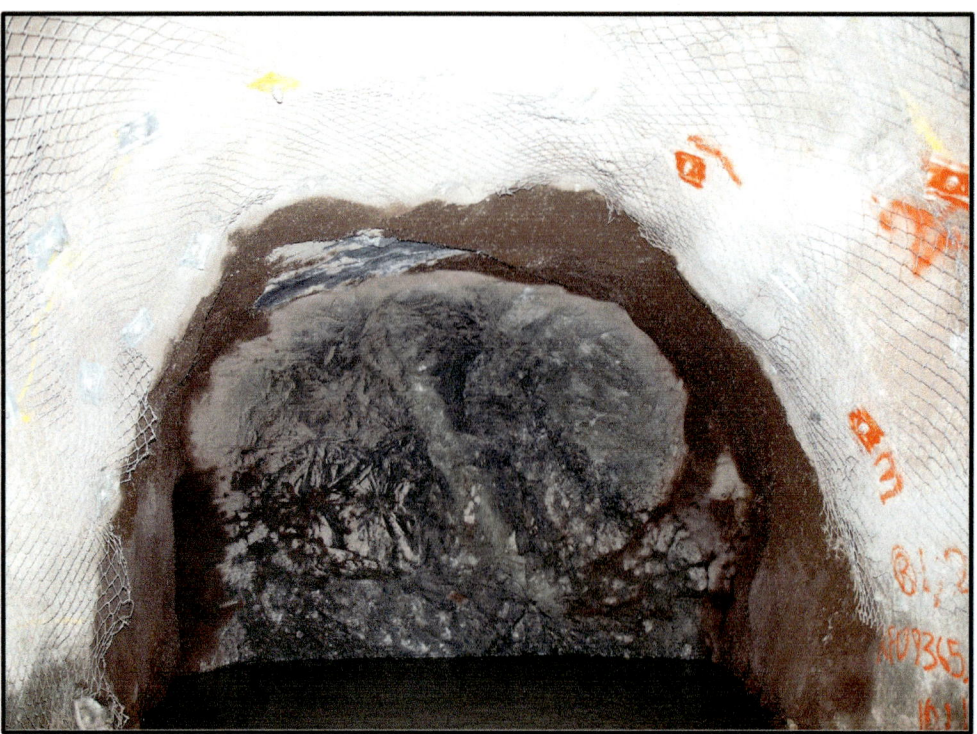

FIGURE 10.13 Example of a freshly sprayed shotcrete layer becoming unstable due to its own weight.

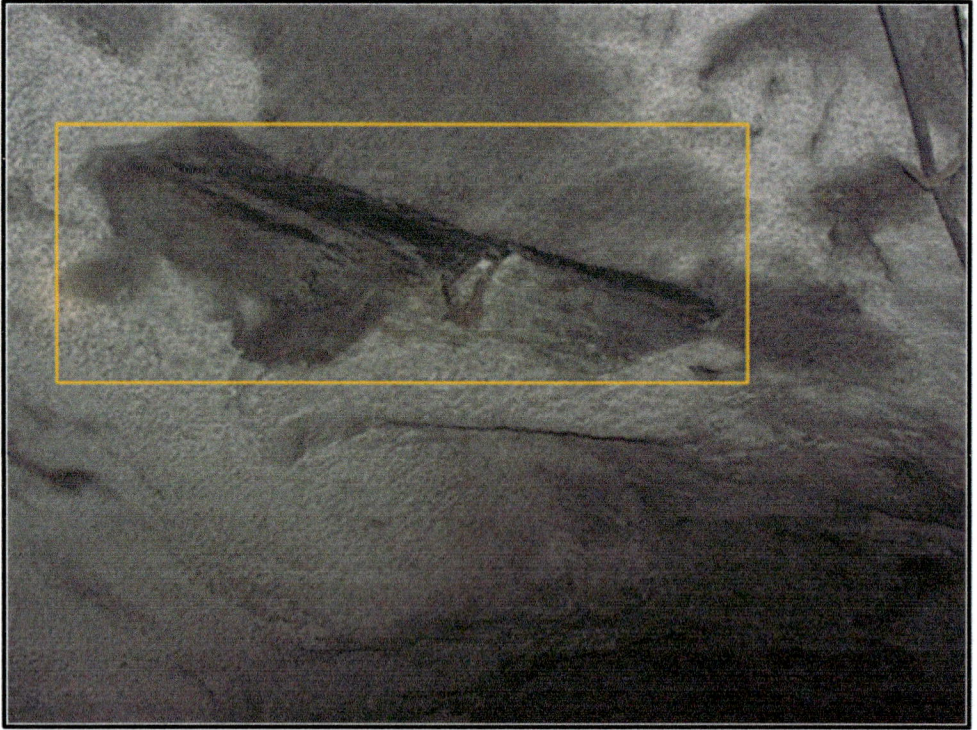

FIGURE 10.14 Example of a gravity-driven block failure through a freshly sprayed shotcrete layer.

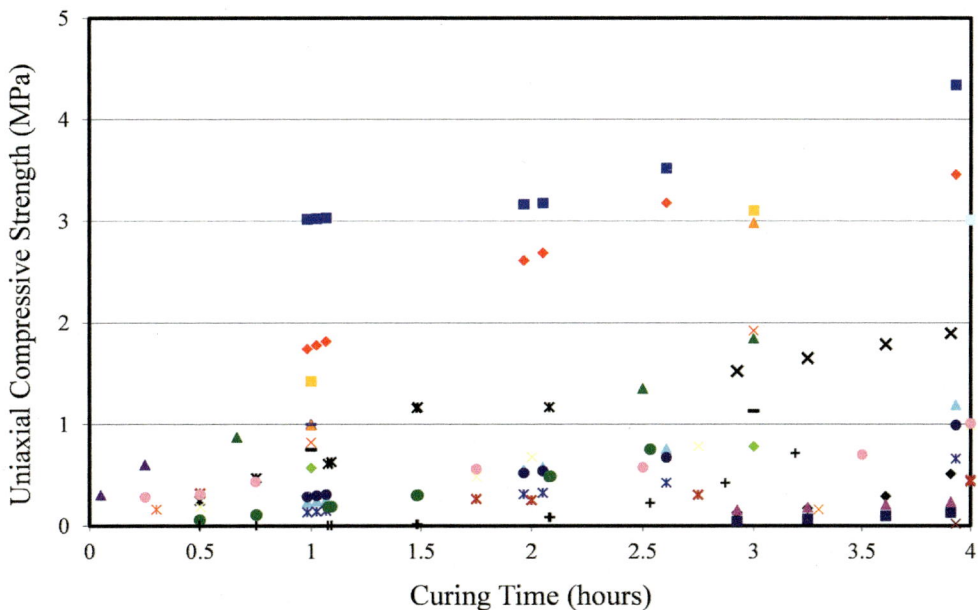

FIGURE 10.15 Typical published UCS results for freshly sprayed shotcrete determined by penetration resistance (Saw, 2015).

of UCS reaching 1.0 MPa as a benchmark for re-entry time is somewhat confusing, as most sites do not seem to reach this strength even four hours after spraying. In addition, penetration resistance can be increased by the presence of aggregates within the shotcrete mix, potentially leading to misleading indications of strength. The shear strength of a fresh (immediately after it has been sprayed) layer is more critical than its compressive strength.

10.5.2 SHEAR STRENGTH OF FRESHLY SPRAYED SHOTCRETE

Two conflicting requirements can be identified for a shotcrete mix. Firstly, it must have the rheological properties of a fluid in order to be pumped and sprayed. Secondly, it must have the mechanical properties of a solid to create a stabilizing structural layer. The rheology of the mix depends on the fluid/solid constituents and their particle size distribution, which, in turn, affect the mechanical properties of the *in situ* paste, its hardening and the mechanical properties of the final hardened layer.

The physical properties of the hardened layer that are important are its density, void ratio and permeability. However, the mechanical properties of the shotcrete layer that are most significant in rock support action include its strength (compressive, tensile, shear and adhesion) and its stiffness (flexural, biaxial and shear). These mechanical properties are dictated by the rheology of the paste, the hardening mechanism of hydration and the underground environmental conditions (i.e., temperature and moisture) during hydration and curing.

Prior to hardening, the mechanical properties of the paste are dictated by the cementitious matrix, comprising the cement, mineral additives, chemical admixtures and water. After hydration, the shotcrete should possess the mechanical properties of the hardened matrix plus some additional strength due to the presence of coarse aggregate particles, mesh reinforcement and fibres. It is important to note that these mechanical properties improve with hydration from the wet paste, to the stiff paste, to the hardened paste and finally to the fully hardened and cured shotcrete. Consequently, the mechanical properties of freshly sprayed shotcrete are those associated with the cementitious matrix, hereafter referred to as 'shotcrete paste'. Therefore, it is predominantly the changes with time of the curing shotcrete paste that will indicate the mechanical response of freshly sprayed shotcrete.

The shear strength of shotcrete pastes without a chemical accelerator can be determined using a standard vane shear test apparatus, such as a viscometer (VT550), 0–3 hours after curing. During this period, the hydration products in the shotcrete pastes are in the fluid gel state. After 3 hours, the hydration products start transforming into a solid gel state, and its shear strength cannot be determined with the standard vane shear test apparatus. After about 4 hours, the shear strength of shotcrete pastes can usually be determined using the conventional triaxial compression test method. The shear strength is estimated using the Mohr-Coulomb shear failure criterion, that is:

$$\tau = c + \sigma_n \tan \phi \qquad\qquad 10.1$$

where τ and σ_n represent shear stress and normal stress, respectively. The parameters c and φ are assumed to be constants representing the cohesion and the angle of internal friction, respectively (Hoek and Brown, 1980). In reality, 'c' and 'φ' change with the stress levels. In the case of the shotcrete at the excavation boundary, the normal stress (σ_n) is considered to be insignificant. Therefore, the minimum shear strength is essentially cohesion.

10.5.2.1   Development of Shear Strength of Shotcrete Paste with and without the Influence of a Chemical Admixture

A summary of the shear strength development of shotcrete pastes with and without the influence of chemical admixtures for the first four hours of hydration is presented in Figure 10.16. The chemical admixtures were added at the dosage recommended by the manufacturer of the particular admixture, as detailed previously in Table 10.1.

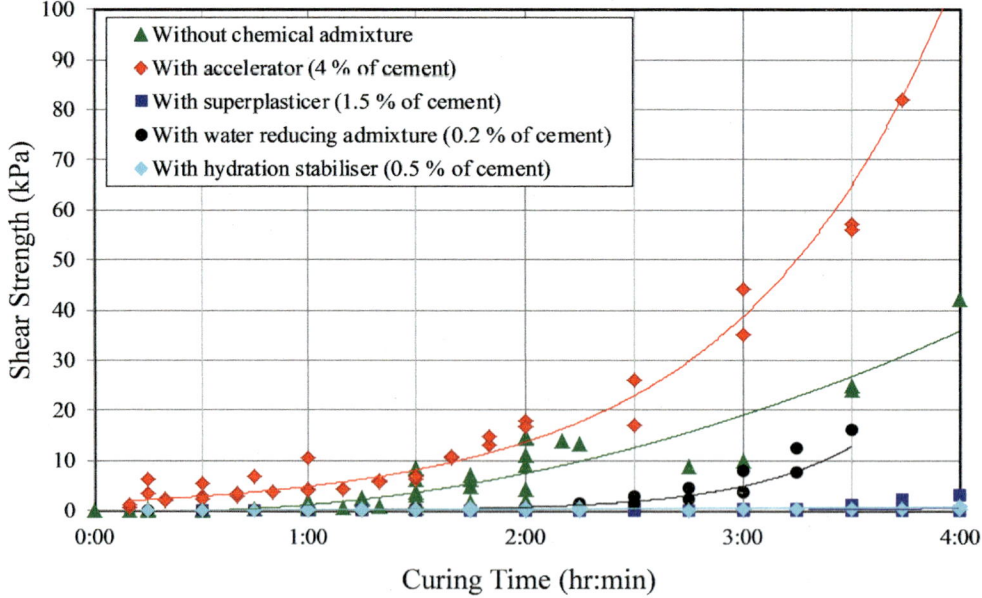

FIGURE 10.16   Developments of shotcrete paste shear strength (with and without the influence of chemical admixtures) during the first four hours of hydration (Saw, 2015).

The results show that, generally, the shear strength increases rapidly during the first four hours of curing in all shotcrete paste mixes except for the shotcrete paste mixed with a hydration stabilizer.

10.5.2.2 Development of Shear Strength of Shotcrete Paste with Various Combinations of Mixed Components

Figure 10.17 summarizes the shear strength of shotcrete paste with an accelerator, shotcrete paste reinforced with synthetic fibres and accelerator and shotcrete paste reinforced with synthetic fibres, aggregates and accelerator for the first four hours of curing. The specifications of these components were provided previously in Table 10.1. The shear strength of the shotcrete paste, reinforced with synthetic fibre, aggregates and accelerator, was slightly higher than those of shotcrete pastes mixed with synthetic fibres and accelerator. The test results suggest that the addition of aggregates (sand and coarse aggregate) increased the cohesiveness of the paste and resulted in an increased shotcrete shear strength at an early age.

10.5.3 STRUCTURAL REQUIREMENTS FOR A FRESHLY SPRAYED SHOTCRETE LAYER

Two structural requirements can be identified for a freshly sprayed shotcrete layer. Firstly, it must support its own mass within minutes of being applied to the surface and, secondly, it must support the superimposed mass of an estimated unstable volume of rock. In the first instance, the shotcrete supports its own mass by the development of an adhesive bond (comprising adhesion and mechanical interlock) between itself and the substrate and by the development of intrinsic shear strength.

Consider a one-metre square block of rock that could be formed within a typical rock bolt pattern used in the mining industry supported by a shotcrete layer having a thickness (t_s) and unit weight of (γ_s), as shown in Figure 10.18.

10.5.3.1 Requirements for Self-Support

In the case of zero shotcrete shear strength, the minimum bond strength (σ_{vs}) required for a shotcrete layer is equal to the vertical stress due to its own weight. This may be calculated as

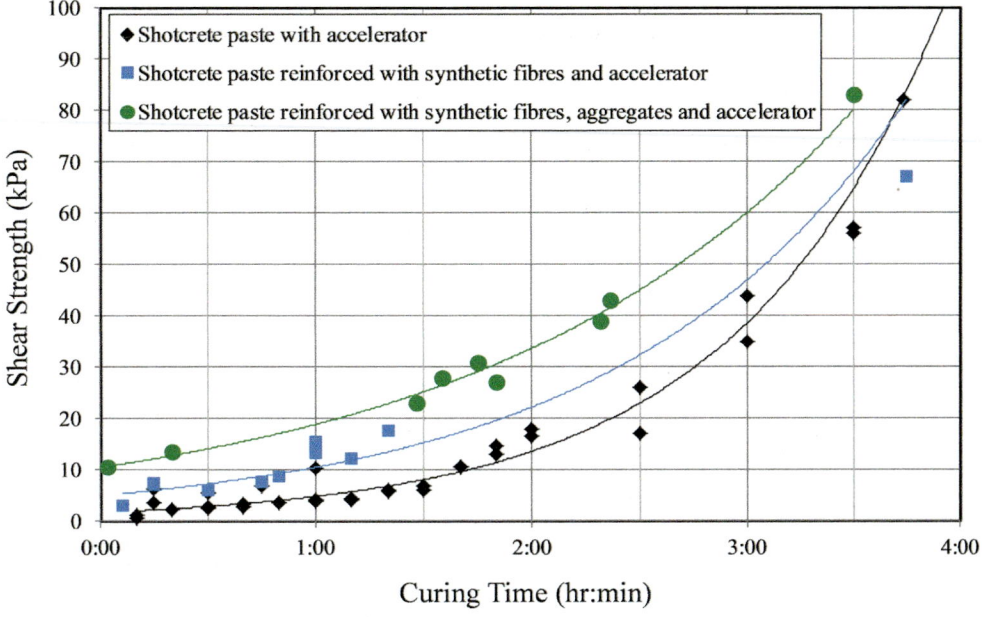

FIGURE 10.17 Development of shear strength in shotcrete with different components (Saw, 2015).

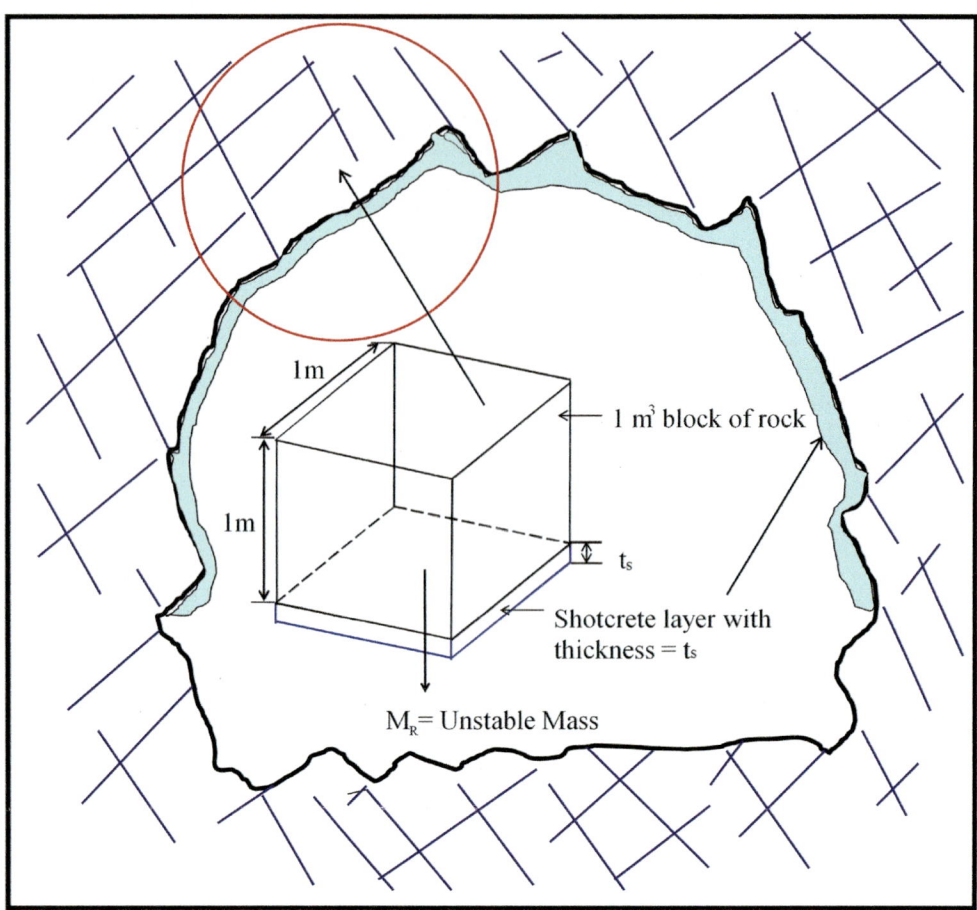

FIGURE 10.18   A conceptual 1.0 m³ block of rock interacting with a shotcrete layer of thickness t_s (Saw, 2015).

$$\sigma_{VS} = \gamma_S t_S$$ 10.2

The typical unit weight for synthetic fibre-reinforced shotcrete is 23 kN/m³. Therefore, the minimum bond strength required for a shotcrete layer thickness, measured in mm, is about $\gamma_s\, t_s/1000$ kPa (or about 0.023 t_s kPa).

Figure 10.19 shows a shotcrete layer and a potential shear failure plane within the layer. In the case of zero bond strength, the minimum shear strength required for the shotcrete to support its own weight is calculated from

$$\tau_S = \frac{F_S}{A_S}$$ 10.3

where

τ_S = Shear strength of shotcrete (kPa)
F_S = Force (kN) due to the self-weight of the shotcrete layer
A_S = Cross-sectional area (m²) of the shear plane

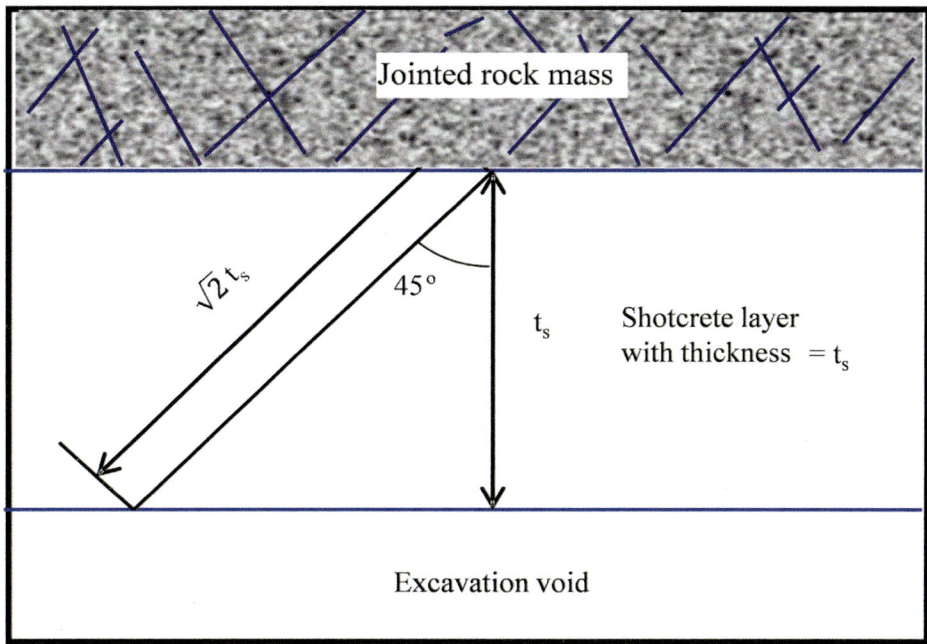

FIGURE 10.19 Long sectional view showing thickness of shotcrete layer along an expected punching shear plane (Saw, 2015).

For calculation purposes, the cross-sectional area is given by

$$A_s = L_s \sqrt{2\, t_s}$$

10.4

where L_S is the average length of the shear surface.

In general, the equation for the force due to gravity is

$$F_s = V_s \gamma_s$$

10.5

where V_S is the volume of shotcrete (m^3).

Therefore, Equation 10.3 may be rewritten, and the shear strength can be calculated from

$$\tau_S = \frac{V_S \gamma_S}{L_S \sqrt{2} t_S}$$

10.6

The volume of shotcrete (V_S) supporting a 1 m^3 block of rock is given by

$$V_S = 1 \times 1 \times t_S$$

10.7

Substituting Equation 10.7 into Equation 10.6, we obtain

$$\tau_S = \frac{1^2 t_{SS} \gamma_S}{L_S \sqrt{2} t_S}$$

10.8

By eliminating 't_s' from Equation 10.8, the required shear strength of shotcrete is now given by

$$\tau_S = \frac{\gamma_S}{L_S \sqrt{2}} \qquad\qquad 10.9$$

For a 1.0-m square face, $L_S = 4.0$ m. Therefore, the minimum required shear strength is

$$\tau_S = \frac{23}{4\sqrt{2}} = 4 \text{ kPa} \qquad\qquad 10.10$$

That is, the minimum shear strength required for shotcrete to support its own weight is typically about 4 kPa.

In almost all cases where bond and shear strength develop simultaneously after spraying, laboratory investigations and *in situ* experience have shown that the required strength levels for shotcrete to support itself are easily achieved.

10.5.3.2   Required Shotcrete Shear Strength for Self-Weight and Block Support

During service, the shotcrete must be capable of supporting the mass of loose rock blocks that may become unstable and represent a risk to personnel that enter an excavation. The volume of loose rock that may become unstable is naturally minimized during blasting by waves that vibrate the excavation surfaces and by subsequent hydro-scaling procedures that clean the excavation surfaces.

The specific arrangement of excavation span, stress and structural geology associated with each excavation will be different, and the specification of a single or standard unstable volume of rock is not possible. However, example calculations may be used to show that, within a few hours of spraying, shotcrete is quite capable of supporting a significant volume of unstable rock and that this volume or mass of rock may be calculated as a function of layer thickness and time after spraying.

Two scenarios will be considered: Firstly, the one-cubic metre block shown in Figure 10.18 and, secondly, the equilateral tetrahedral block of rock with one-metre side lengths as shown in Figure 10.20. If the latter block existed in the roof of an underground excavation, its face triangle (i.e., the expression of its shape visible in the roof when viewed from beneath) would be an equilateral triangle with a side length of 1.0 metre.

The minimum shear strength required for shotcrete to support each of these two blocks with a 1.0-metre side length can be calculated from

$$\tau_S = \frac{F_T}{L_S \sqrt{2} t_S} \qquad\qquad 10.11$$

where F_T is the total force to be resisted by the shotcrete in shear and given by

$$F_T = F_s + F_R \qquad\qquad 10.12$$

and

$$F_R = M_R g \qquad\qquad 10.13$$

where M_R is the mass of unstable rock and g is the gravitational acceleration (9.81 m/s^2). L_s is 4.0 m for a 1.0 m^3 block, and L_s is 3.0 m for an equilateral tetrahedral block of rock with 1-metre side lengths.

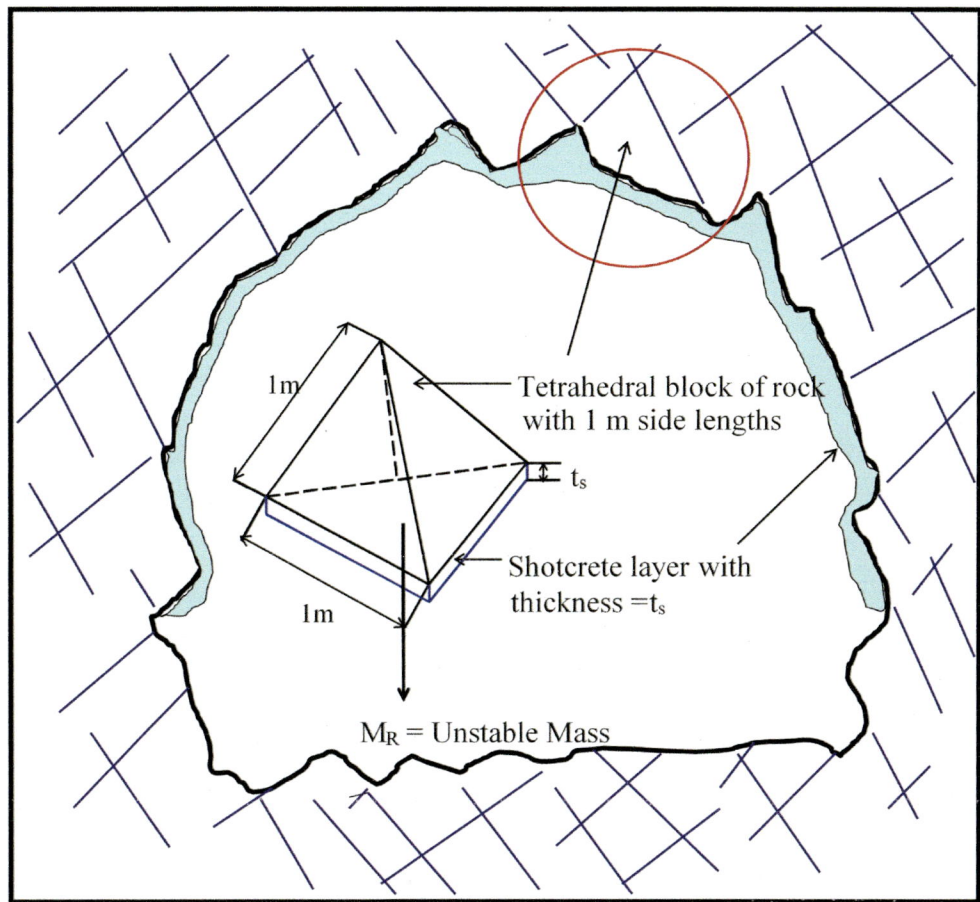

FIGURE 10.20   An equilateral tetrahedral block of rock loading a shotcrete layer, having a thickness t_s (Saw, 2015).

The minimum shear strengths required for shotcrete of different thicknesses to support each of the two different shaped blocks are shown in Figure 10.21.

10.5.4   SAFE RE-ENTRY TIME

The ICS safe re-entry time for underground mine excavations at a given shotcrete thickness can be calculated from the relation between the minimum shear strength required to develop in freshly sprayed shotcrete with the development of shear strength of shotcrete paste with curing time, as described in Section 10.5.2.

Figure 10.22 shows the shear strength required to be developed for different thickness layers of shotcrete to support the layer and a one-metre cube block of rock. It was computed for a rock with a common unit weight of 27 kN/m^3, resulting in a total block mass of 2700 kg. The graphs show that a shotcrete layer of about 50-mm thickness needs to develop a shear strength of about 97 kPa. This level of shear strength will develop within about three hours and 50 minutes after spraying for a shotcrete mix with an accelerator, synthetic fibres and aggregates.

Figure 10.23 shows the shear strength required to be developed for different thickness layers of shotcrete to support an equilateral tetrahedral block of rock with 1.0-metre side lengths in addition to supporting its own mass. Again, this is computed for a rock unit weight of 27 kN/m^3 resulting in a total block mass of 300 kg. The graph shows that an average shotcrete layer of about 50 mm

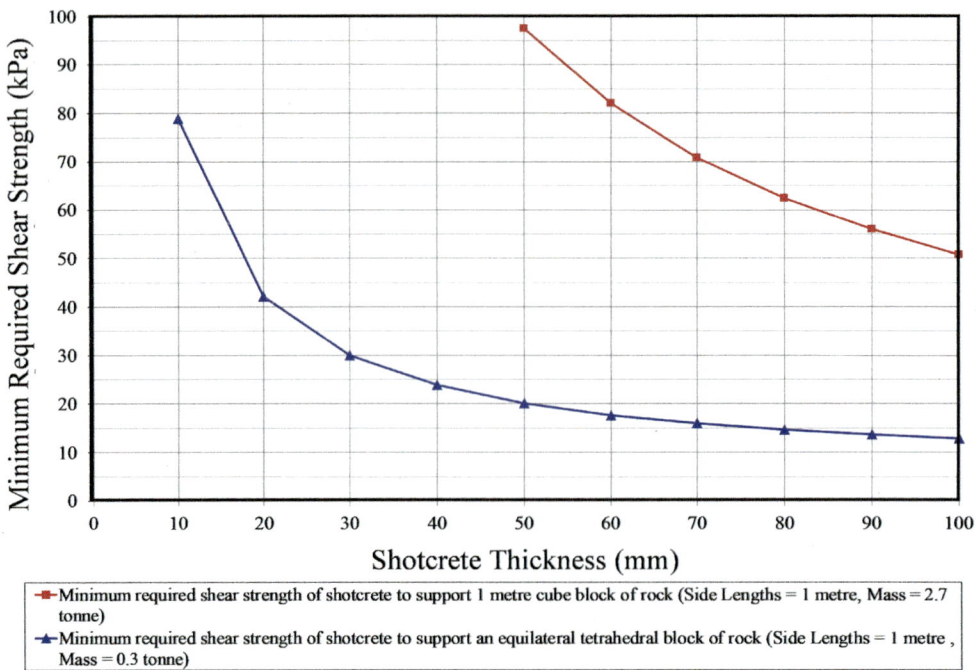

FIGURE 10.21   Minimum shear strength required for shotcrete with different thicknesses to support a cube block of rock and an equilateral tetrahedral block of rock both with 1-metre side lengths (Saw, 2015).

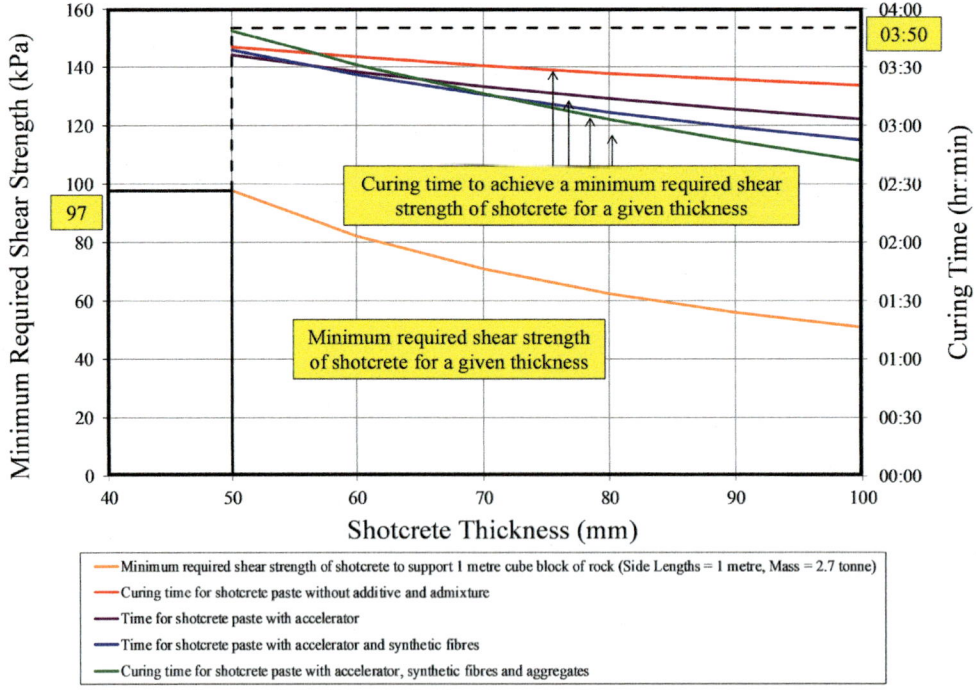

FIGURE 10.22   Minimum shear strength required to develop in a shotcrete layer to support itself and a block of rock with one cubic metre volume (Saw, 2015).

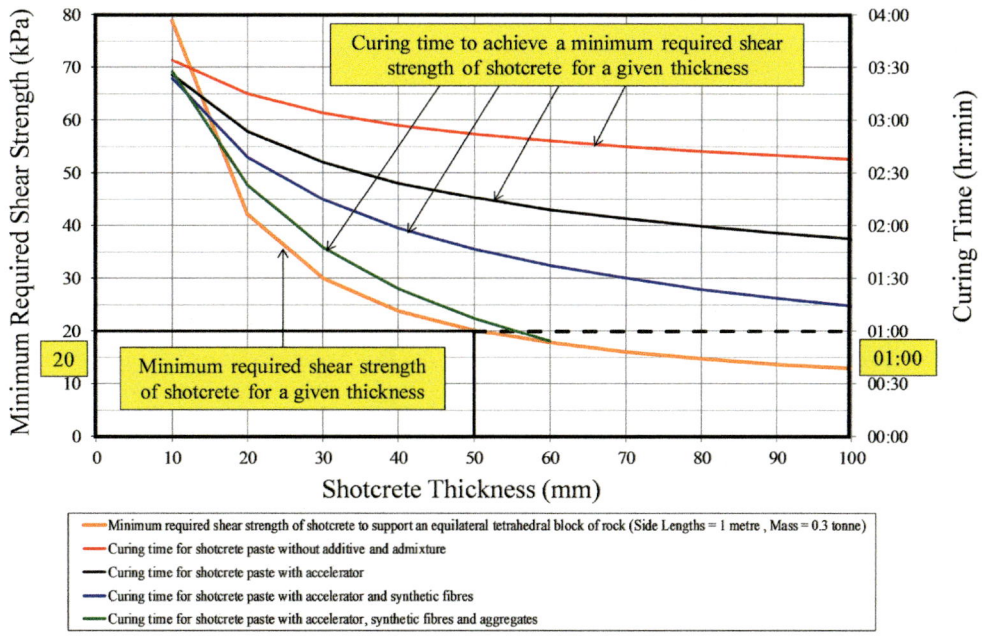

FIGURE 10.23 Minimum shear strength required to develop in a shotcrete layer to support itself and an equilateral tetrahedral block of rock with 1-metre side lengths (Saw, 2015).

thickness needs to develop a shear strength of about 20 kPa. This level of shear strength will develop within about one hour after spraying for a shotcrete mix with an accelerator, synthetic fibres and aggregates. Clearly, a greater curing time will be required for thinner layers of shotcrete prior to safe re-entry.

10.6 STATIC PERFORMANCE OF CURED SHOTCRETE

10.6.1 Uniaxial Compressive Strength (UCS)

The UCS test is the most common test method to determine the capacity of shotcrete. The UCS test should be conducted according to the test method suggested by ASTM 'Standard test method for compressive strength of cylindrical concrete specimens', or (C39/C39 M-09a). Tejchman and Kozicki (2010) reviewed results from different researchers on steel fibre-reinforced concrete and summarized the most important physical and mechanical properties of steel fibre-reinforced concrete/shotcrete. The UCS of cementitious materials is highly influenced by the mix design, curing time, curing environment and spray method and spray technology.

Figures 10.24 and 10.25 show the typical UCS development of synthetic fibre-reinforced shotcrete with curing time established from the WASM Rock Mechanics laboratory database (2010–2014). The data suggest a UCS range from 7 to 25 MPa after 1 day of curing and from 15 to 55 MPa after 28 days of curing.

Figure 10.26 presents typical stress versus strain curves from UCS tests conducted on steel fibre-reinforced shotcrete samples (Saw et al., 2013). The yield region displayed in the curves increases with increasing UCS. After yield, non-linear strain hardening can be observed until a peak occurs. After a peak, localized damage develops and strain softening and/or a 'snap-back' begins. The 'snap-back' implies that the material failed in a brittle mode. Globally, the steel fibre reinforced-shotcrete continued to deform in shear associated with dilation and the load taken by the steel fibres.

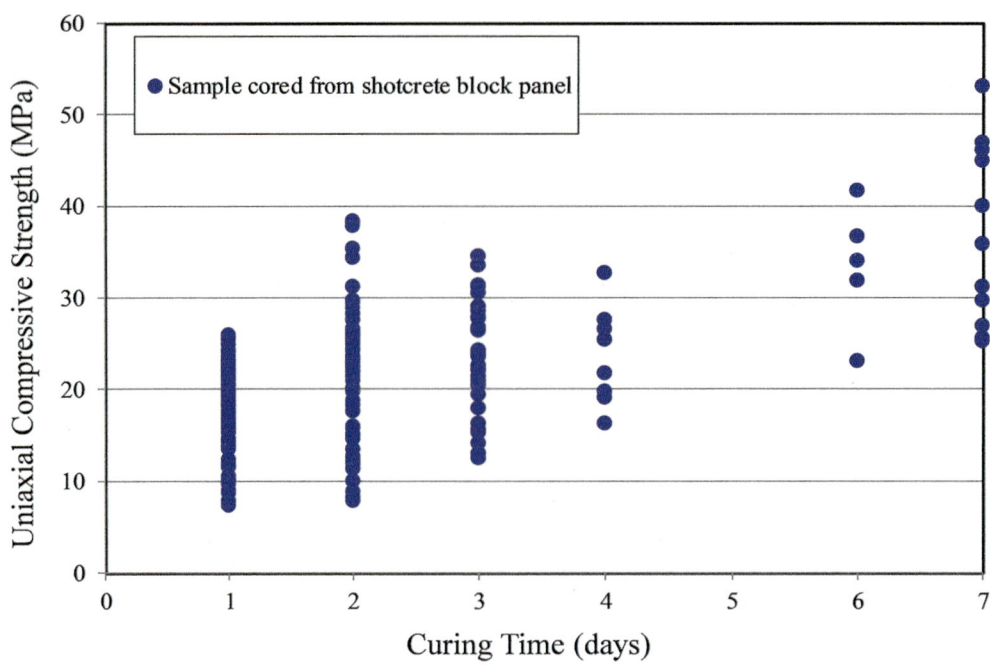

FIGURE 10.24 A typical UCS development of synthetic fibre-reinforced shotcrete with a curing time of 1–7 days.

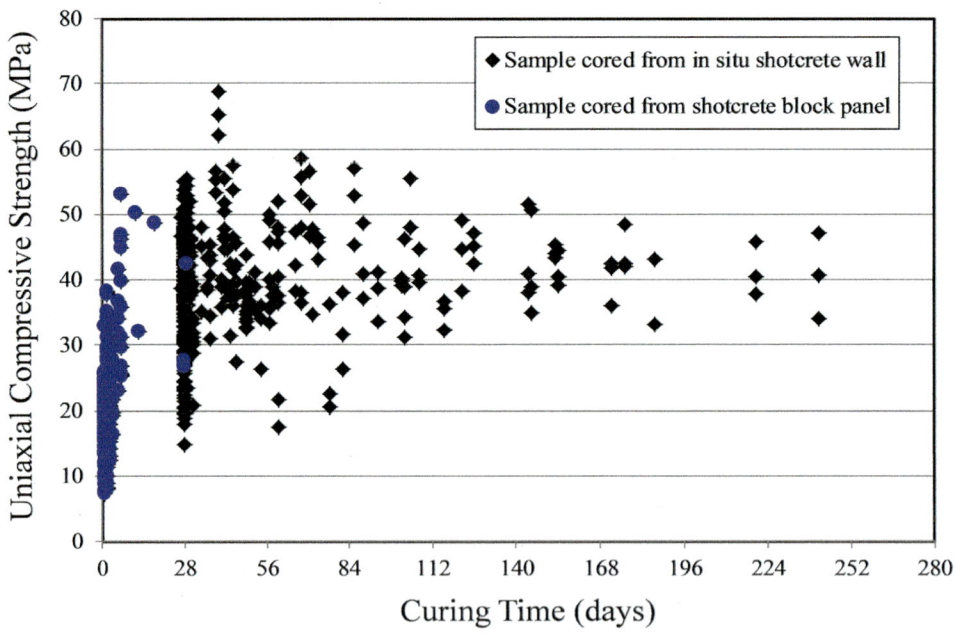

FIGURE 10.25 UCS development of synthetic fibre reinforced shotcrete with a curing time of 1–241 days.

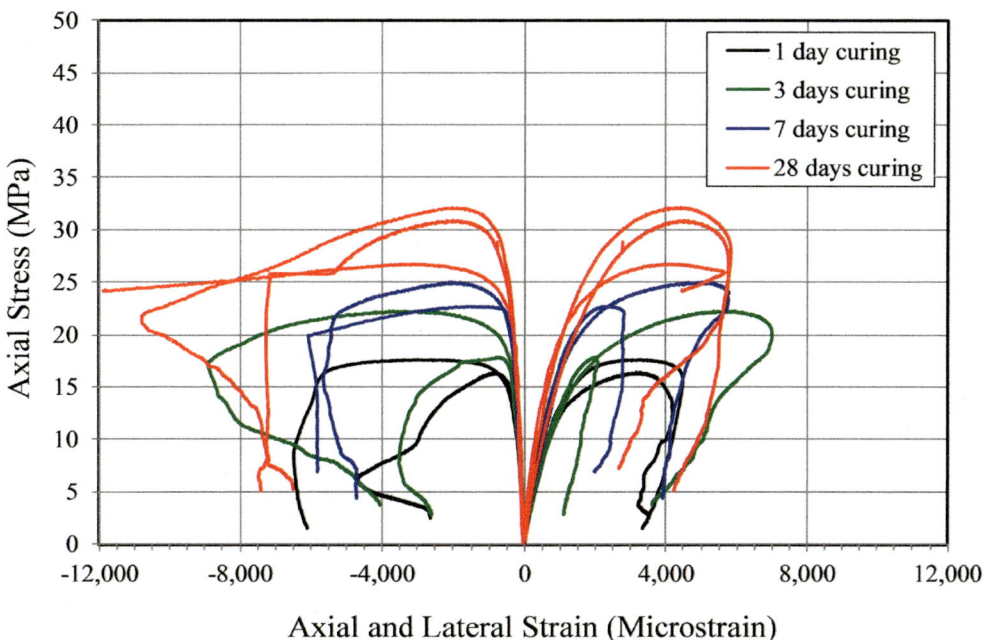

FIGURE 10.26 A typical stress versus strain curves from the UCS test conducted on steel fibre-reinforced shotcrete with a curing time of 1–28 days (Saw, 2015).

10.6.2 TENSILE STRENGTH

The determination of the tensile strength of shotcrete is a complex experimental procedure due to the stability of the load control system, gripping mechanisms and the eccentricity of the specimen (Velasco et al., 2008). Therefore, the tensile strength of shotcrete is usually determined by the indirect test method suggested by the International Society of Rock Mechanics (ISRM, 1978). The typical tensile strength development in synthetic fibre-reinforced shotcrete with curing time is shown in Figure 10.27 Internal results from the WASM Rock Mechanics Laboratory database (2010–2014) suggests a tensile strength range from 2 to 5.5 MPa after 28 days of curing.

Typical load-displacement curves from uniaxial tensile strength determined by the indirect Brazilian test are shown in Figure 10.28. It suggests that after the first crack, the load is taken by fibres, and the ultimate tensile strength depends on the number and orientation of the effectively embedded fibres.

10.6.3 TENSILE BOND STRENGTH OF ROCK—SHOTCRETE INTERFACE

The tensile bond strength of the rock-shotcrete interface depends upon a number of factors, such as surface preparation (Kuchta, 2002; Malmgren et al., 2005), compaction between shotcrete and the rock substrate (Brennan, 2005), the substrate rock type (Hahn and Holmgren, 1979), the excavated surface profile (Windsor, 1999) and the spray methodology. It is usually determined by a pull-out test. In a pull-out test, a direct tensile load is applied to a core drilled through the shotcrete and the rock substrate. The tensile failure load at the rock-shotcrete interface is used to calculate the tensile bond strength (ACI 506.4R-04, ASTM C1583, SS 13 72 43). Results of *in situ* tensile bond strength for a rock-shotcrete interface are presented in Table 10.2 and suggest a range from 0.59 to 4.77 MPa.

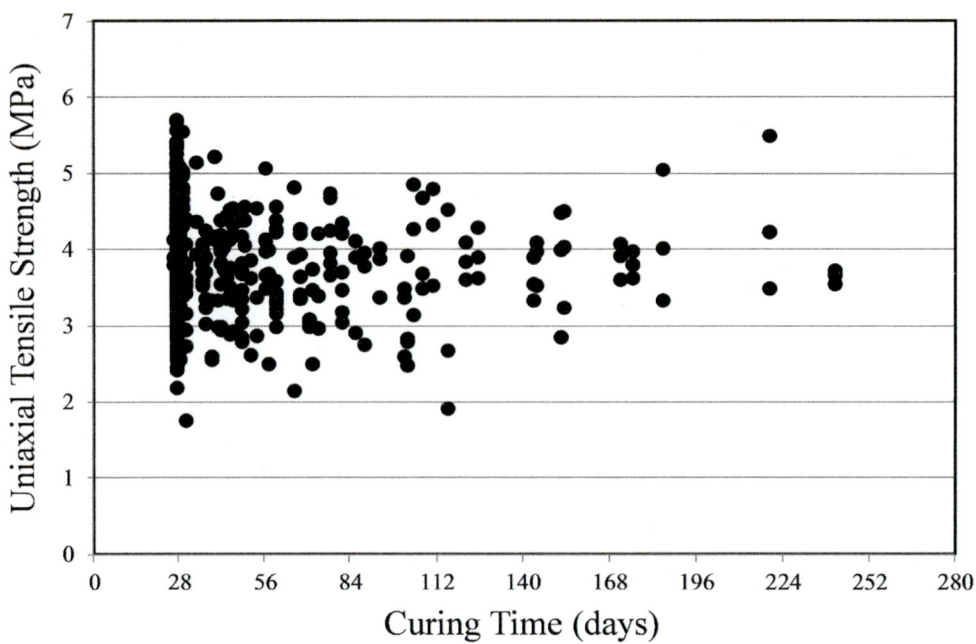

FIGURE 10.27 Typical tensile strength development rates of synthetic fibre-reinforced shotcrete with a curing time of 28–241 days.

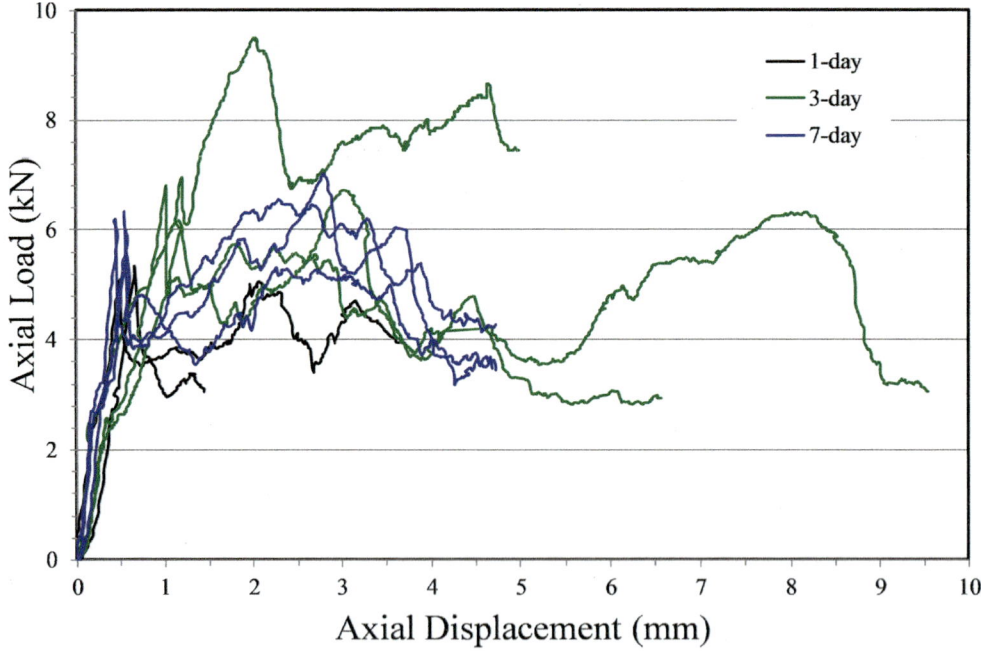

FIGURE 10.28 Load-displacement curves from uniaxial tensile strength tests at different times after sample preparation (Saw, 2015).

TABLE 10.2

In Situ Tensile Bond Strength of Rock-Shotcrete Interface (Jenkins et al., 2004).

Shotcrete thickness (mm)	Curing (Days)	Tensile bond strength (MPa)	Surface preparation
–	–	0.59	**Hydro-scaling**
–	–	0.68	
70	150	3.98	
75	150	3.18	
30	150	2.70	
70	150	2.83	
80	150	4.27	**Jumbo Scaling**
35	150	4.77	
45	150	4.37	

10.6.4 SHEAR STRENGTH

Shear strength is the intrinsic strength of a material or component to resist failure or yield where the material or component is subjected to a shear load. An excess shear force tends to produce a sliding failure in a material along a plane parallel to the direction of the force (Gere and Timoshenko, 1990). Therefore, the shear strength of shotcrete is the shear force required to produce sliding or yielding.

Typical shear strength development of steel fibre-reinforced shotcrete with curing time is shown in Table 10.3 (Saw et al., 2009). The shear strength parameters were calculated based on Coulomb's failure theory. Alternatively, the peak and residual strength envelopes plotted on the 'p-q' plane are also shown in Figures 10.29 and 10.30, respectively. Generally, while the shear strength increases with curing time, the friction and dilation angles do not change significantly. The residual strength

TABLE 10.3

Typical Shear Strength Development in Steel Fibre-Reinforced Shotcrete with Curing Time (Saw et al., 2009).

Batch No.	Curing (Days)	Shear strength			
		Peak		Residual	
		c (MPa)	$\Phi°$	c (MPa)	$\Phi°$
1	3	5.4	40	3.4	42
	7	7.7	35	4.8	35
	28	7.9	38	6.8	18
2	1	3.9	45	2.1	45
	3	4.3	40	2.6	41
	7	5.4	40	4.1	41
3	1	4.7	36	4.9	32
	3	5.7	38	–	–
	7	5.8	40	5.2	38
	28	7.6	38	–	–

is influenced by confining pressure. The main cause of an increase in shear strength is an increase in cohesion, with the slopes of the relation indicating similar friction angles.

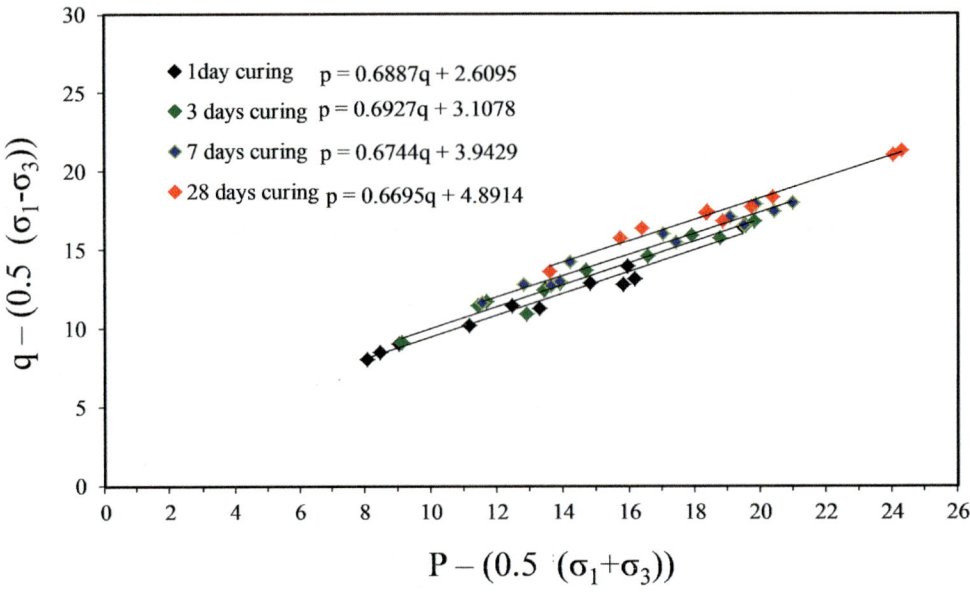

FIGURE 10.29   Peak shear strength envelopes plotted on the p-q plane at various curing times (Saw, 2015).

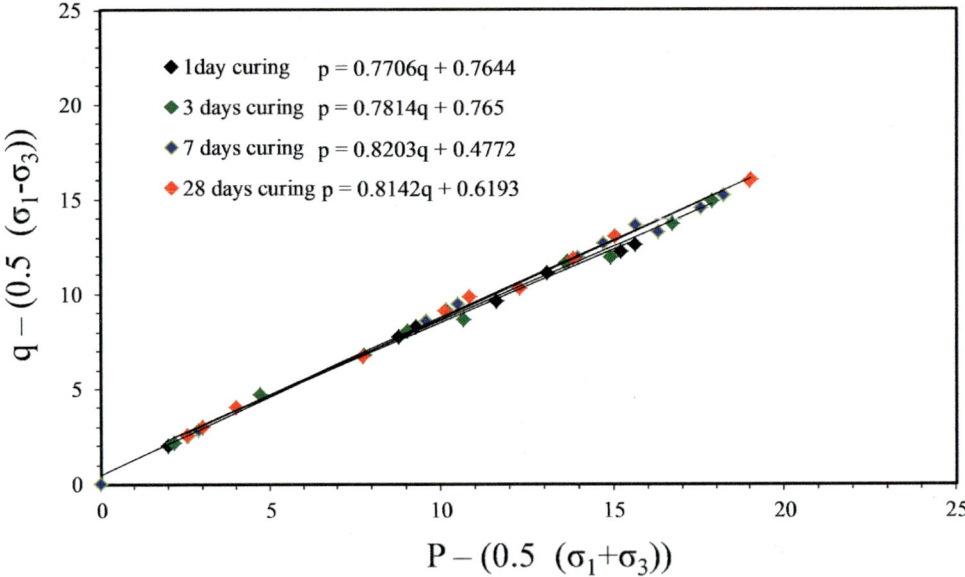

FIGURE 10.30   Residual shear strength envelopes plotted on the p-q plane (Saw, 2015).

10.6.5 TOUGHNESS

The toughness of a shotcrete support system relates to its ability to dissipate energy and deform plastically before failure. In other words, it is the amount of energy dissipated by the shotcrete support system response when it is subjected to static or dynamic loading. It is used to assess the support system where the transient forces immediately after loading would be sufficient to cause support failure (given that the system did not deform until the force demand reduced to an acceptable level). The energy dissipation can be determined by calculating the area under the force-displacement curve.

A test method for flexural toughness of fibre-reinforced concrete (using a centrally loaded round panel), also known as the Round Determinate Panel (RDP) test, has been suggested to determine the energy dissipation capabilities of a shotcrete panel (ASTM C1550–05). In this test, a circular panel of fibre-reinforced shotcrete cast is subjected to a central point load while supported on three symmetrically arranged support pivots. The load is applied through a hemispherical-ended steel piston advanced at a prescribed rate of displacement. Load and deflection are recorded simultaneously up to a specified central deflection. The energy dissipated by the panel at 5, 10, 20, or 40 mm deflections is defined as toughness at these displacements of the fibre-reinforced shotcrete (ASTM 1550–05). Figure 10.31 shows the toughness of synthetic fibre-reinforced shotcrete development with curing time as per the ASTM C1550 test method (Slade and Kuganathan, 2004 and Bernard et al., 2006). However, determining energy at an arbitrary displacement is not indicative of the energy capacity of a shotcrete support system. In order to effectively assess the energy dissipation capacity of a shotcrete support system, the cumulative energy dissipation variation with central displacement should be considered.

10.7   SHOTCRETE FAILURE MECHANISMS

The basic shotcrete support theory proposed by Deere et al. (1969) has gained the most recognition; however, the complexity of the interaction between the shotcrete and the rock mass, and the difficulty in measuring this reaction, means further development of the support mechanism has not occurred. Studies by Holmgren (1976, 2001) and Fernandez-Delgado et al. (1976) show that adhesion loss and flexure are the primary modes of shotcrete failure. A further review conducted by Barrett and McCreath (1995) identified that shotcrete capacity in the blocky ground, under static

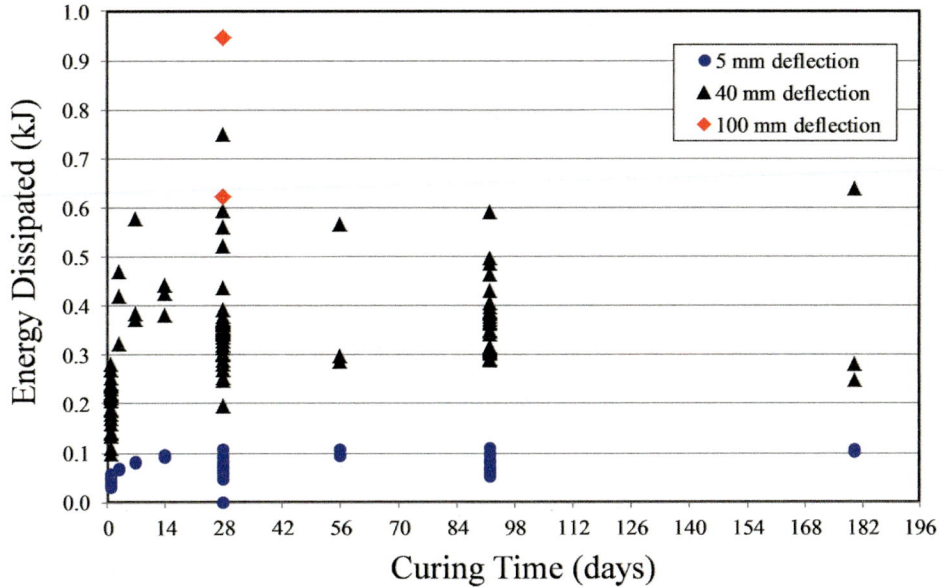

FIGURE 10.31   Toughness of synthetic fibre reinforced shotcrete development with curing time as per the ASTM C1550 test method. (Compilation of data published by Slade and Kuganathan, 2004, Bernard et al., 2006.)

conditions, is governed by six mechanisms: Adhesion loss, direct shear, flexural failure, punching shear, compressive failure and tensile failure. Adhesion loss occurs when the bond between the shotcrete and the rock is lost, often due to poor surface preparation prior to spraying or shrinkage of the shotcrete during curing. It may also be that certain rock types contain minerals that cannot sustain adhesion forces. Flexural failure is the bending failure of the shotcrete and can only occur after the adhesion is broken. For flexural failure to occur, the shear strength of the material must be higher than the flexural strength. The authors disagree with the definitions provided by Barrett and McCreath (1995) for 'direct shear' and 'punching shear'. Direct shear (or shear failure) occurs over a single planar interface, typically represented in two dimensions as a line. Punching shear, described as direct shear by Barrett and McCreath (1995), is the shear that occurs over a complex three-dimensional surface. Punching shear failure is a combined mechanism of flexural failure and shear failure. The modified failure mechanisms are illustrated in Figure 10.32. These failure mechanisms are not well understood, and further research is required to understand the complexities of the rock/shotcrete interaction, particularly under the influence of dynamic loading.

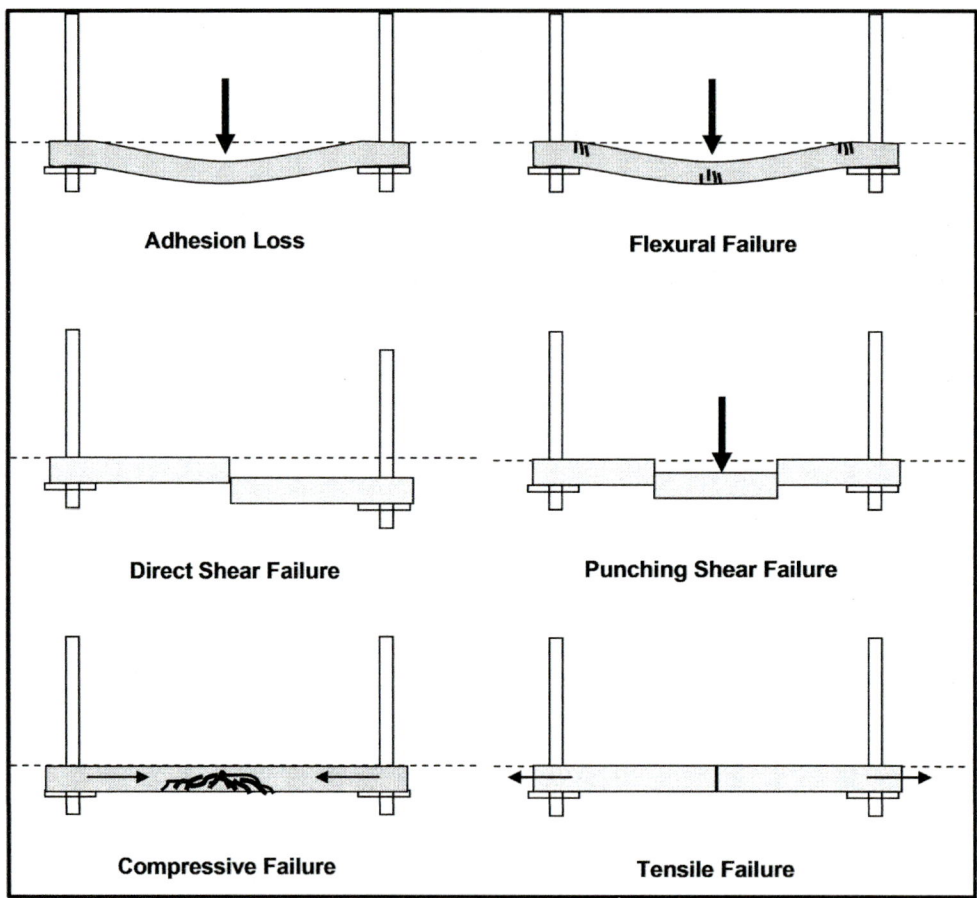

FIGURE 10.32   Modified shotcrete failure mechanisms (after Barrett and McCreath, 1995).

10.7.1   SHOTCRETE LOAD TRANSFER

In general, the reaction of a shotcrete layer in response to loading (w) is provided by a complex combination of in-layer forces (P), bending moments (M) and shear forces (V), as shown schematically in Figure 10.33. Note that P may be compressive as shown or tensile. In the case of very thin layers (i.e., Thin Spray on Liners), very little resistance can be expected to develop for concave surfaces without excessive displacement. It is also possible that the shotcrete layer may be flat or convex. The latter two shapes are less efficient in providing support action, particularly when the layer is brittle and may crack due to flexure.

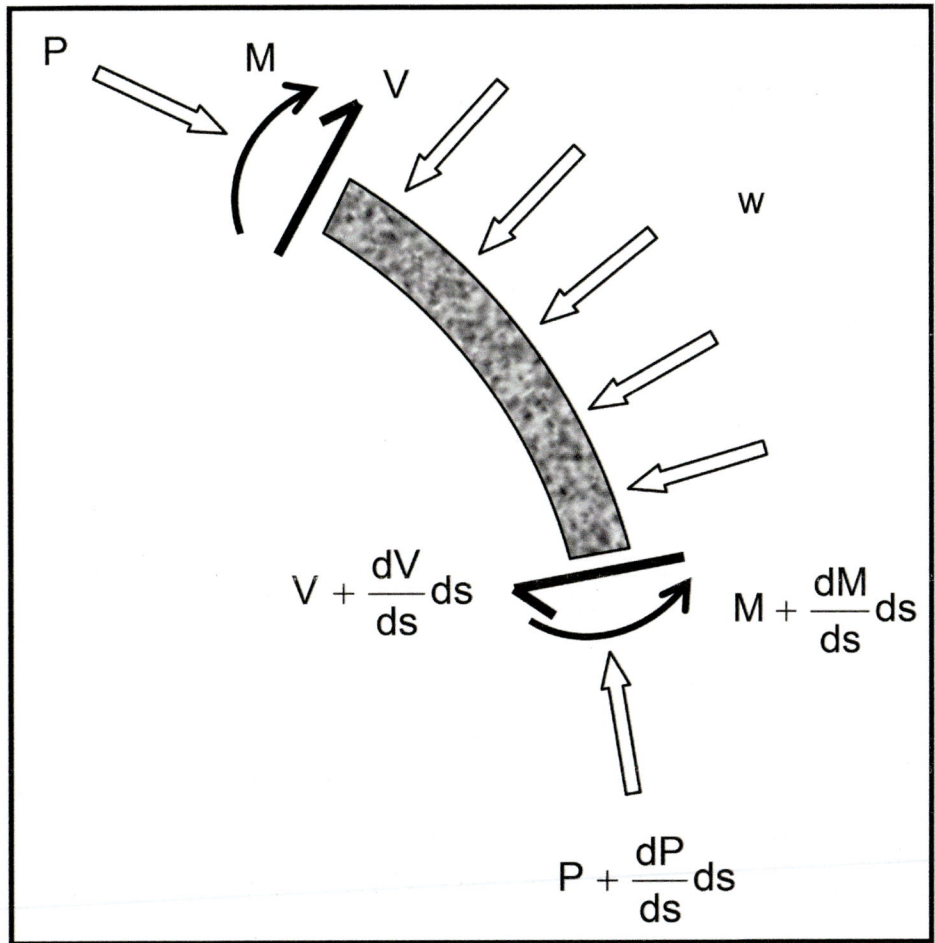

FIGURE 10.33   Forces acting on a segment of a curved arch section.

10.8   LARGE SCALE STATIC TESTING

In 2004, the WASM Rock Mechanics research group designed and built a large-scale static testing facility to evaluate shotcrete and mesh support failure mechanisms under static load conditions for samples resembling actual bolt spacings. A large-scale punch shear test method similar to that used by Holmgren (1976) and Fernandez Delgado et al. (1976) was adopted to test the energy dissipation capability of shotcrete support layers (Morton et al., 2009, Villaescusa et al., 2010, 2015).

Unstable rock blocks are simulated loading a rough, planar surface of rock substrate drilled to create a centrally dislocated region. 1.5 m × 1.5 m sandstone and granite slabs were selected as substrate materials due to their ready availability in various uniform thicknesses (from 20 to 70 mm).

The natural rough surface texture ensures adhesion of the shotcrete to the substrate, similar to most hard rocks encountered at mine sites. To prepare the substrate, the centre of each slab is marked, and then a 500-mm diameter disk is formed by drilling through the slab with a large diameter coring bit (Figure 10.34). Plastic foam is inserted between the central disk and the granite substrate to prevent shotcrete penetration, while shotcrete is sprayed with a high-pressure nozzle.

FIGURE 10.34   The creation of a dislocated centre within a rough granite substrate.

The substrate and the dislocated disk are fixed to a metal frame to facilitate safe transportation to the mine sites for spraying purposes. The external boundary of the shotcrete panels is enclosed by a wooden form. The purpose of the wood form is to control the thickness while spraying on an almost vertical surface (Figure 10.35). The panels are sprayed by mining operators using conventional, remote control underground equipment and shotcrete mixes (using standard accelerator dosages). The sample area of 1.5 m × 1.5 m provides a realistic spraying surface area, where the suitability of the equipment and the skills of the operators can be quantified. Following shotcreting, the samples are cured at the WASM in controlled conditions of temperature and humidity that replicate a particular underground environment.

FIGURE 10.35   Example of an operator spraying a shotcrete panel and a detailed view of the panel after spraying an initial layer of shotcrete.

For mesh-reinforced shotcrete, a second layer is sprayed with the thickness kept as thin as possible to fill in the gaps between the mesh and the first layer (see Chapter 4). The spraying of the second shotcrete layer is illustrated in Figure 10.36, with the outline of the mesh structure still clearly observable after spraying.

The sprayed panels are left on the mine sites for one day to prevent any damage or changes to the sprayed shotcrete layers during the early curing stages. The panels are then transported to the

FIGURE 10.36 Example of second layer spraying (*shotcrete fill-in*) and the plastic wrap for the first day of curing at the mine site.

temperature and humidity-controlled chambers at WASM Rock Mechanics Laboratory for proper curing. The plastic wraps are removed, and the samples are left to cure for 7–28 days (± 1 day) under specific humidity and temperature conditions resembling particular underground mining conditions. The samples are not water cured as this could increase the hydration of the cement, resulting in a strength increase that would not normally be associated with mining applications.

After the specified curing period, the wooden form is removed, and the panels are transported to the WASM Large Scale Test Facility. The panels are marked with a reference grid for monitoring purposes during testing. A forklift is used to rotate the panels and to place them into the static testing facility, as shown in Figure 10.37. A vertical load is applied to the dislocated disk using a spherical seat, as shown in Figure 10.38. Manual reference measurements are taken before testing, with load and displacement measurements continuously collected during testing. A video camera is used to record the shotcrete behaviour and to reconcile critical events, such as loss of adhesion to the substrate, initial shotcrete rupture and start of mesh loading.

FIGURE 10.37 Details of mesh-reinforced shotcrete panel ready for static testing.

FIGURE 10.38 Details of the applied loading with subsequent cracking of the mesh-reinforced shotcrete layers.

Figure 10.39 displays the force-displacement response for several shotcrete layers. The data clearly indicate that the shotcrete material itself ruptures at very small displacements (limited to less than 50 mm). The rupture is controlled by the shotcrete layer thickness, as shown in Figure 10.40. In comparison, rupture of the combined layer is controlled by mesh performance. Figure 10.41 shows a comparison of reinforced layer performance for welded wire and high tensile woven mesh. Prior to catastrophic failure, the displacements for welded wire mesh reach up to 80 mm, while for high tensile woven mesh, they exceed 300 mm without layer rupture (Figure 10.42).

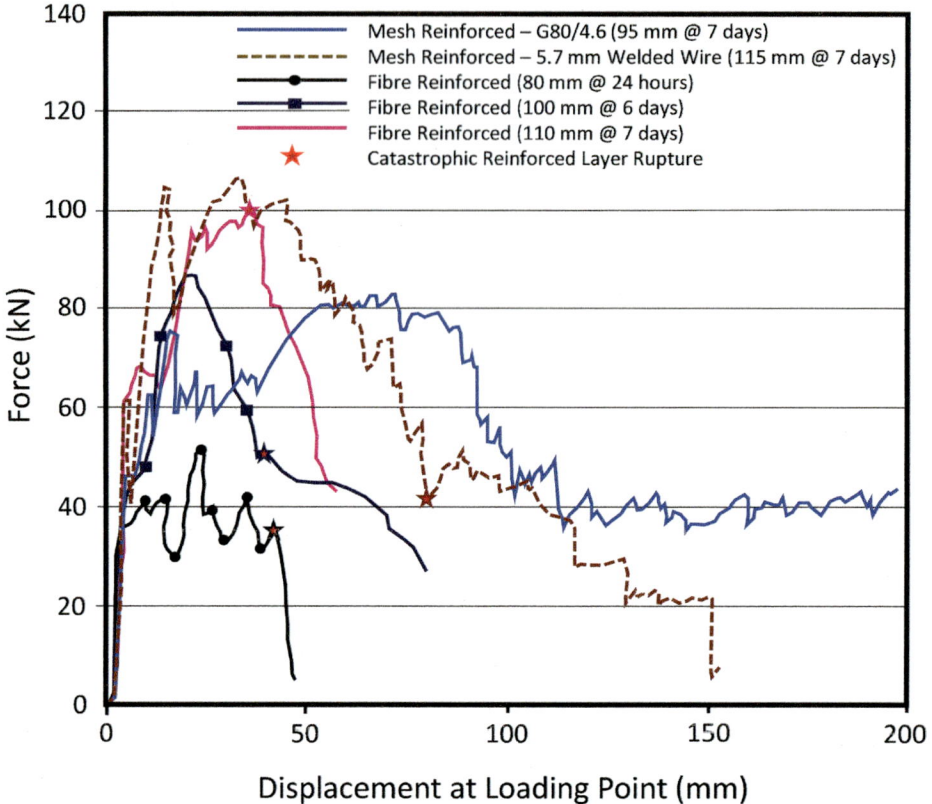

FIGURE 10.39 A comparison of force versus displacement curves for several shotcrete layers.

10.9 LARGE SCALE DYNAMIC TESTING

10.9.1 SHOTCRETE TEST SET-UP

A loading frame, sample configuration and methodology have been developed at the WASM to enable dynamic testing of large-scale shotcrete panels. These panels are of similar configurations to those described earlier for static testing. The testing has been designed to investigate the combined effects of adhesion and flexural strength. The samples consist of a rough rock substrate (Figure 10.43) with a 500-mm diameter disk created at the centre (and isolated from the surrounding substrate) over which a shotcrete layer is sprayed. The test geometry produces a large version of the well-known 'punch test' used to cause bending and shear loads in the shotcrete layer and separation forces at the interface between the shotcrete and the rock substrate.

The WASM shotcrete dynamic test facility comprises two steel frames: a lower frame to support the sample and an upper frame to provide the load reaction. The sprayed rock substrates are

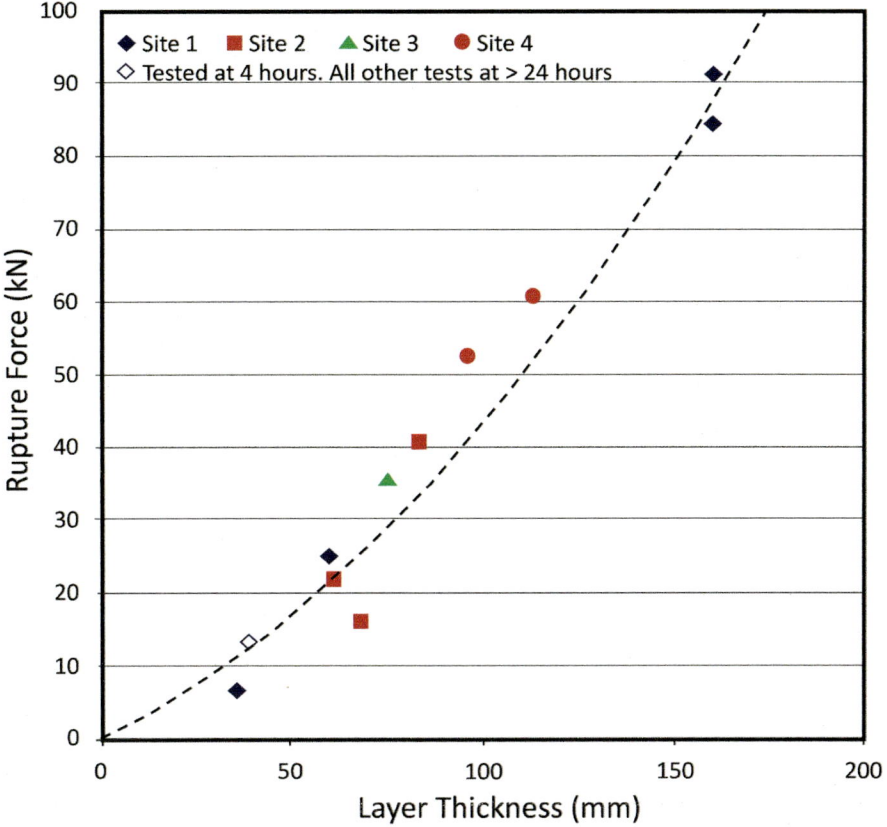

FIGURE 10.40 A comparison of rupture force with shotcrete layer thickness.

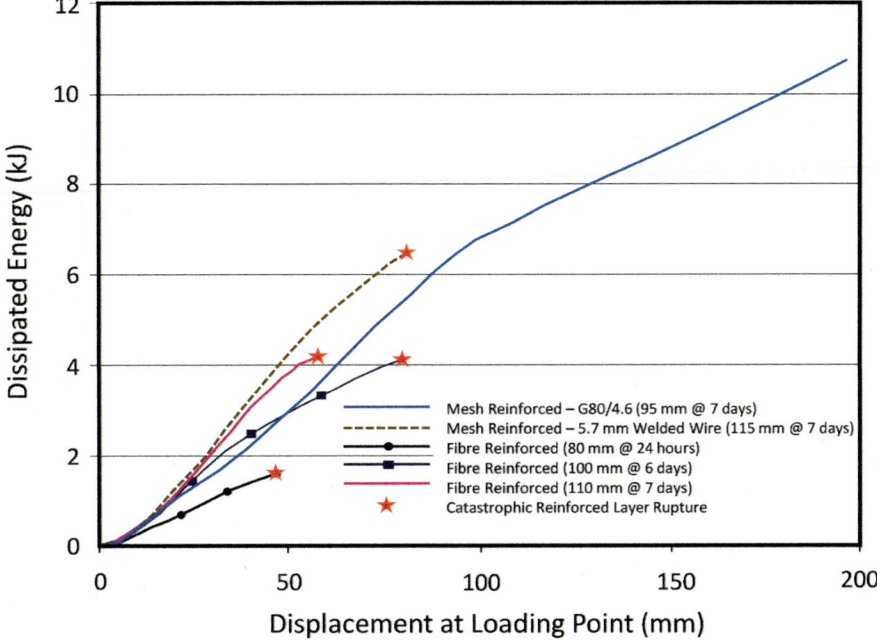

FIGURE 10.41 Dissipated static energy for a number of shotcrete layer configurations.

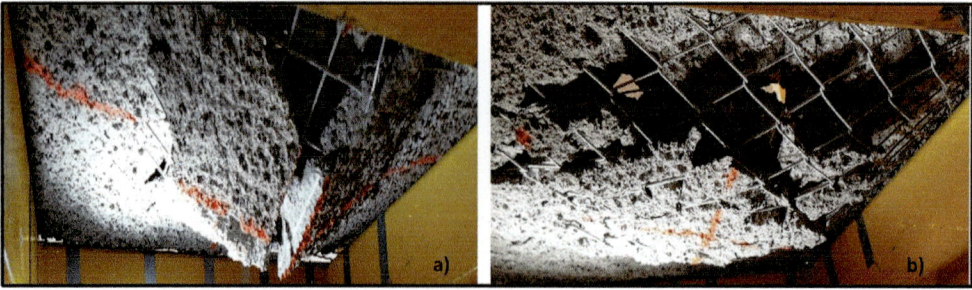

FIGURE 10.42 Example of mesh-reinforced shotcrete layer performance under large displacement with (a) welded wire and (b) high tensile woven wire.

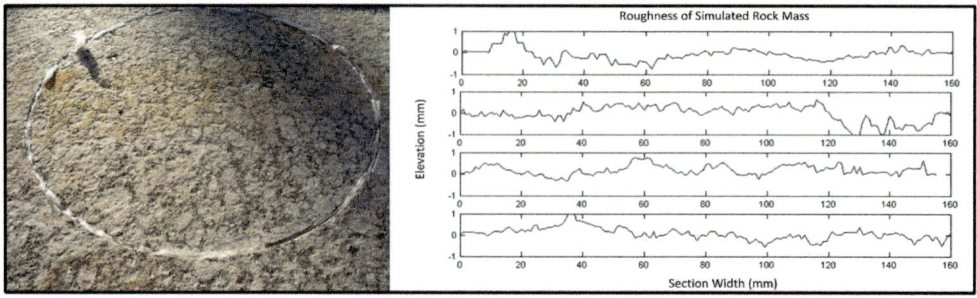

FIGURE 10.43 Details of surface roughness measurements for several granite substrates.

clamped in place to provide continuous edge support like that used in large-scale static testing. In addition, the shotcrete layer its restrained by rock bolts near the edges in one of three configurations: four bolts located near the corners (square pattern), four located at the centres of the edge spans (oblique pattern) and at all eight locations. A loading mass is placed into the centre of the shotcrete panel (Figure 10.44), and the samples are loaded using the momentum transfer concept (Player et al., 2004; Player, 2012). The drop beam with the attached assembly is dropped from a specific height to generate dynamic loading on the sample. Computer software, advanced instrumentation and a high-speed camera are used to record the test data. High-speed data acquisition is undertaken at 25,000 samples per second.

FIGURE 10.44 Details of shotcrete panel ready for dynamic testing.

10.9.2 SHOTCRETE FAILURE MECHANISM

The force and displacement resulting from a dynamic shotcrete test are similar to those that occur from dynamic testing of mesh (Chapter 9). The major difference is the self-weight of the shotcrete panel, which must be included in the analysis of loading and energy calculations. For this case, the free body diagram is shown in Figure 10.45, and the assumed displacement mechanism is shown in Figure 10.46.

Yield line theory was chosen to describe the mechanisms of failure for flat slabs of shotcrete. Yield line theory is based on the premise that a layer, loaded to failure, will develop 'yield lines' in the most highly stressed areas. Yield lines are continuous 'plastic hinges' associated with the failure mechanism. It was found that, for the purposes of design, the patterns conform to a few simple rules:

1. Yield lines end at a layer boundary.
2. Yield lines are straight.
3. Axes of rotation generally lie along the lines of continuous support (e.g., a square slab supported on four sides as in the EFNARC (1996) test) and occur adjacent to pivot support (e.g., round panel test described by ASTM (2002) and Bernard (2003)).
4. Yield lines between adjacent rigid regions pass through the point of intersection of the axes of rotation for those regions.

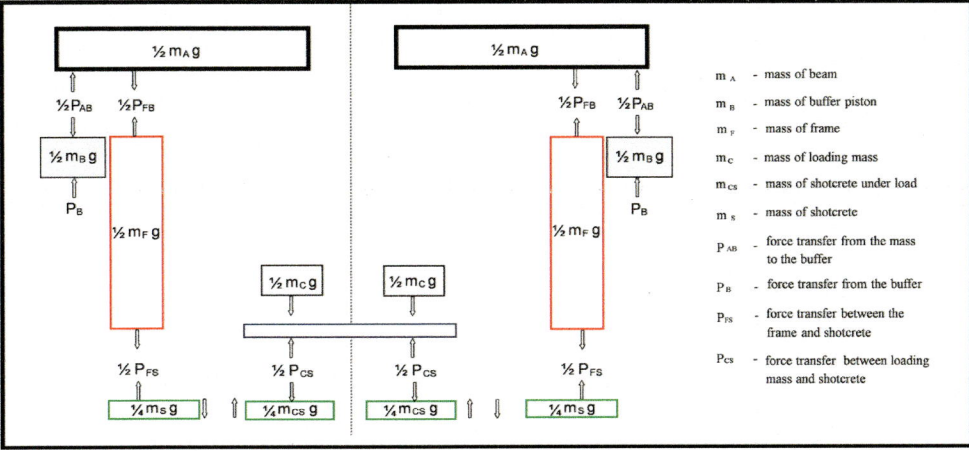

FIGURE 10.45 Free-body diagram showing momentum transfer mechanisms within a shotcrete panel test at the WASM Dynamic Test Facility.

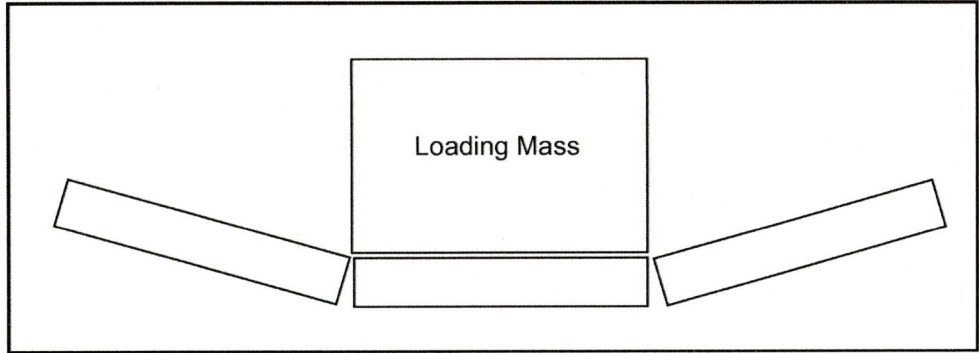

FIGURE 10.46 The assumed crack and displacement mechanism within a shotcrete panel test.

Once the yield line pattern is defined, then relationships can be developed between the applied forces and the internal resisting moments. In yield line analysis, it is usual to define resisting moment (M) in units of force times distance per unit length (e.g., Nm/m). The shotcrete moment-rotation relationship (M,θ) depends on several factors, such as:

- The thickness of the layer,
- The type and distribution of any internal reinforcement and
- The mechanical properties of the materials.

The simple rules listed above can be used to define possible yield line patterns for complex loadings and support geometries. For example, yield line patterns for shotcrete panel testing are given in Figure 10.47 (e.g., Tran et al., 2001, 2005). Laboratory observations at WASM have shown that theoretical relationships based on yield line theory correlate well with the results from static testing of shotcrete using different samples and loading configurations that produce a flexural response.

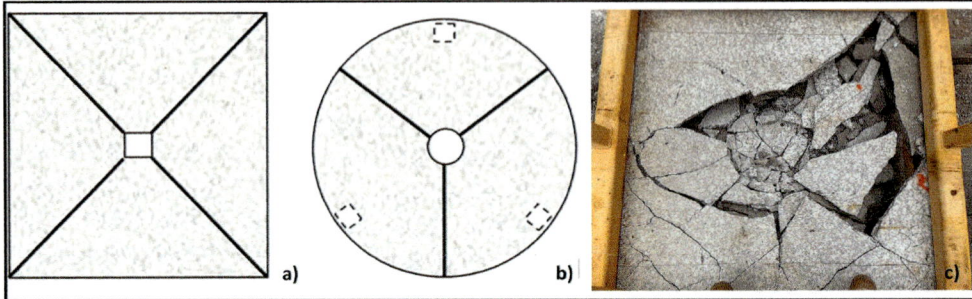

FIGURE 10.47 Yield line pattern for (a) centrally loaded square slab supported at the edges, (b) centrally loaded circular slab supported at the three equally spaced pivots and (c) the collapse pattern following a large dynamic loading.

10.9.3 SHOTCRETE ENERGY DISSIPATION

10.9.3.1 Fibre-Reinforced Shotcrete

Dynamic testing was first conducted on fibre-reinforced shotcrete. The details for the fibre-reinforced layer and test sample specifications are summarized in Table 10.4 and Table 10.5, respectively. The dynamic testing program involved four tests on the same concrete mix design with two different types of plastic fibres.

TABLE 10.4

Shotcrete Layer Specifications for Fibre-Reinforced Shotcrete Testing.

Sample number (Flat slab)	Slab size (W × L) (m)	Shotcrete thickness (mm)	Fibre type	Restraint
1	1.4 × 1.5	90	Shogun	3 sides
2	1.4 × 1.5	102	Shogun	3 sides
3	1.4 × 1.5	77	Reoco Hookshot	4 sides
4	1.4 × 1.5	79	Reoco Hookshot	4 sides

A summary of dynamic test results is given in Table 10.6. Sample 1 was subjected to two separate dynamic loadings. For Samples 1 and 2, although the shotcrete layer cracked and deformed, it did not fail completely. However, the calculations show that the 90-mm and 102-mm thick fibre-reinforced shotcrete samples can only dissipate 1.8 and 3.5 kJ of energy, respectively. Samples 3 and 4 had a smaller thickness, and the dynamic test violently punched through the fibre-reinforced layer at levels of input energy similar to those used with Sample 1.

Details of the typical crack patterns created by testing and their comparison with field observations are given in Figure 10.48 and Figure 10.49, respectively. The crack patterns have some similarities to those expected from yield line theory for centrally loaded slabs that are simply supported on the edges. Figure 10.50 shows a comparison of performance for Samples 1, 2 and 4 after dynamic loading.

TABLE 10.5
Test Specifications for Fibre-Reinforced Shotcrete Sample Dynamic Testing.

Sample number	Slab size (W × L) (m)	Loading mass (kg)	Impact velocity (m/s)	Input energy (kJ)
1A	1.4 × 1.5	446	3.30	2.4
1B	1.4 × 1.5	446	4.45	4.4
2	1.4 × 1.5	446	5.69	7.2
3	1.4 × 1.5	446	3.30	2.4
4	1.4 × 1.5	446	3.30	2.4

TABLE 10.6
A Summary of Performance for Fibre-Reinforced Shotcrete Sample Dynamic Testing.

Test number	Curing age (days)	Peak force (kN)	Dissipated energy (kJ)	Mode of failure
KB-1A	25	17.5	0.01	Bending/cracking
KB-1B	25	93	1.8	Bending/cracking
KB-2	32	96	3.5	Bending/cracking
KB-3	11	–	–	Punch through
KB-4	29	–	–	Punch through

FIGURE 10.48 Crack pattern underneath the shotcrete panel after testing of Sample 2 and its comparison with field damage.

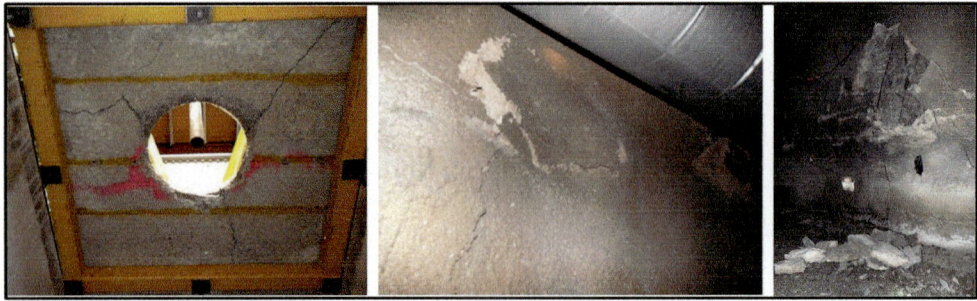

FIGURE 10.49   Punch through failure after testing of Sample 4 and its comparison with field damage.

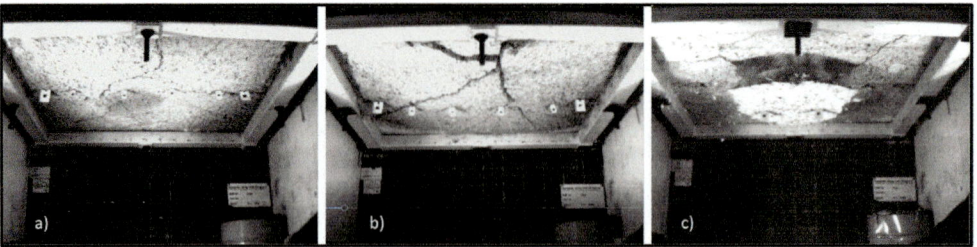

FIGURE 10.50   Dynamic performance comparison: (a) Sample 1 crack failure, (b) Sample 2 crack failure and (c) Sample 4 punching failure.

A second testing campaign of fibre-reinforced shotcrete was undertaken using the same material specifications, spraying equipment and operators. Table 10.7 lists the conventional laboratory strength properties of core samples collected from the sprayed panels. Figure 10.51 shows the contours of the sprayed thickness, where up to a 20 mm over-spray is evident. The mix was dosed with 7.0 kg/m^3 of plastic fibres.

Table 10.8 shows a summary of the dynamic performance for three of the tests undertaken. One of the sprayed panels (KB-8) was able to resist multiple impacts and was tested twice with permanent cracks and resulting plastic deformation, as shown in Figure 10.52. Figure 10.53 shows contours of cumulative deformation. The panel could resist a total deformation of 20 mm, with adhesion loss between the fibrecrete layer and the granite substrate detected after the sample was disassembled. One of the tests was designed to achieve catastrophic failure (input energy of 15 kJ) with the dynamic load penetrating the layer for energy dissipation of 6.3 kJ at the first crack.

Figure 10.54 shows the dynamic force-deformation response for the three dynamic tests. The cracked but stable fibre-reinforced shotcrete panels were able to resist multiple loading events while resisting large loads ranging from 160 to 180 kN. However, the dynamic central displacement was limited to less than 35 mm. In comparison, when the input energy was increased to a moderate level, catastrophic failure was experienced through the relatively thick layer (Figure 10.55).

Tables 10.9 and 10.10 show the fibre density across the cracks following static laboratory testing and after the dynamic rupture of a shotcrete layer, respectively (Figure 10.56). The data shows that fewer fibres are detected following a dynamic impact compared to that detected after conventional static testing. Nevertheless, very few fibres span across the cracks and only about 20% of them rupture during dynamic testing.

The overall interpretation is that shotcrete is characterized by an initial stiff response up to the initiation of cracking, followed by a steady softening response with dynamic loading after the initiation of failure. Even though a shotcrete layer can provide immediate support to a rock mass and

TABLE 10.7

Conventional Laboratory Strength of Fibre-Reinforced Shotcrete Samples.

Sample number	Curing days (30°, 90% humidity)	UTS (MPa)	Point loading (MPa)	UCS (MPa)
KB 6–12	41	5.27	–	–
KB 6–13	41	7.78	–	–
KB 6–20	41	8.36	–	–
KB 6–21	41	5.44	–	–
KB 8–1	13	8.22	–	–
KB 8–3	13	8.75	–	–
KB 8–6	13	5.69	–	–
KB 6–11	41	–	2.78	–
KB 6–14	41	–	2.74	–
KB 6–19	41	–	3.36	–
KB 8–2	13	–	2.72	–
KB 8–4	13	–	2.64	–
KB 8–5	13	–	2.39	–
KB 6–1	41	–	–	28.90
KB 6–2	41	–	–	42.59
KB 6–7	41	–	–	25.90
KB 8–25	13	–	–	28.00
KB 8–27	13	–	–	22.35
KB 8–28	13	–	–	32.64
Average (MPa)		**8.2**	**2.8**	**30**

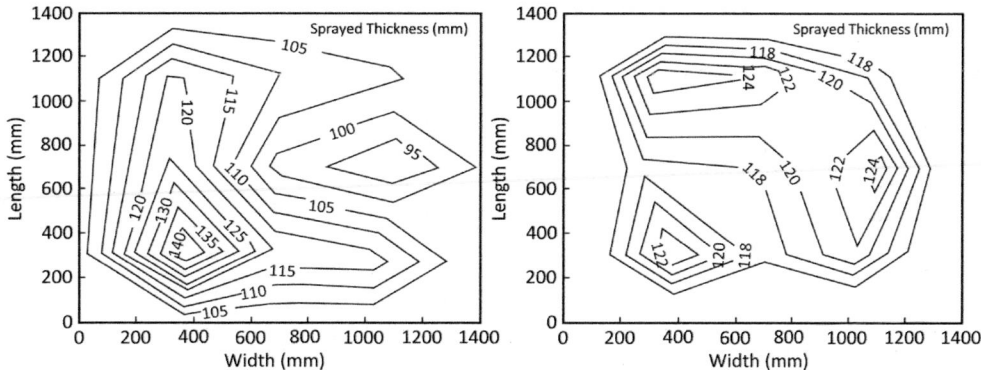

FIGURE 10.51 Measured contours of sprayed shotcrete produced when aiming for a 100-mm target thickness.

inhibit loosening, the fibre-reinforced shotcrete is not at all compatible with the large displacements likely to occur in highly stressed underground excavations. In high deformation environments (either slow, time-dependent or sudden violent failures), the limited displacement capacity of the fibre-reinforced shotcrete systems will result in a loss of surface restraint. In these circumstances, shotcrete support must be complemented with mesh that has a much higher displacement capacity. Furthermore, the displacement compatibility of a mesh-reinforced shotcrete layer will greatly depend upon the type of mesh reinforcement used.

TABLE 10.8

Test Specifications and Dynamic Performance for Two Nominal 100 mm Thick Fibre-Reinforced Layers.

Test #	Loading mass (kg)	Thickness range (mm)	Impact velocity (m/s)	Input energy (kJ)	Max disp (mm)	Max force (kN)	Energy dissipated (kJ)
KB-8A	457	116–128	5.84	7.8	17.4	183.1	2.26
KB-8B	457	116–128	6.09	–	32.3	167.5	2.90
KB-7	2327	90–140	3.64	15.4	260	164.3	6.32

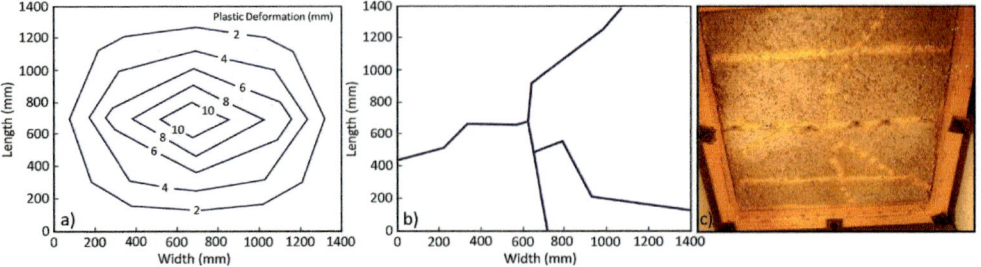

FIGURE 10.52 Contours of plastic deformation and details of the permanent cracks resulting from an initial dynamic test.

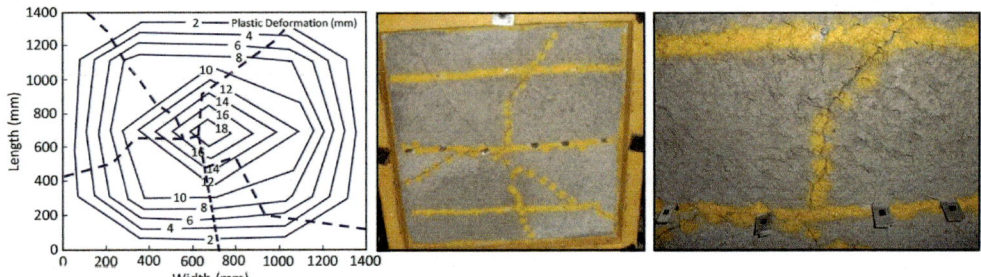

FIGURE 10.53 Contours of cumulative plastic deformation and details of the permanent crack pattern and width resulting from two dynamic tests.

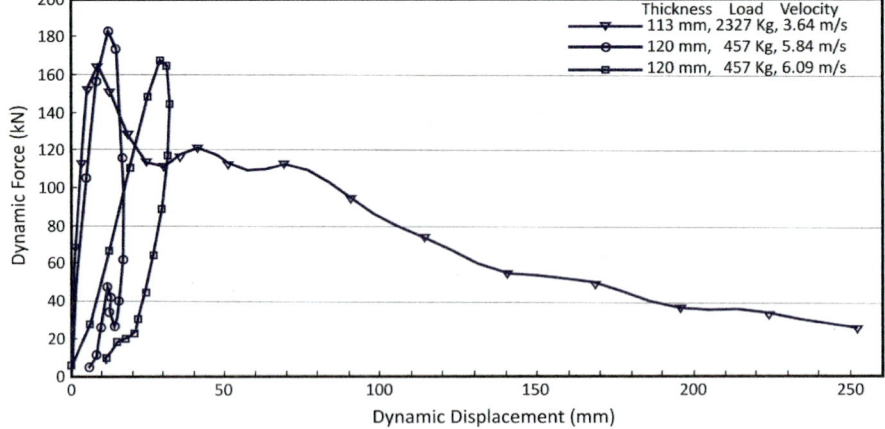

FIGURE 10.54 Dynamic force-displacement response for stable and unstable fibre-reinforced shotcrete layers.

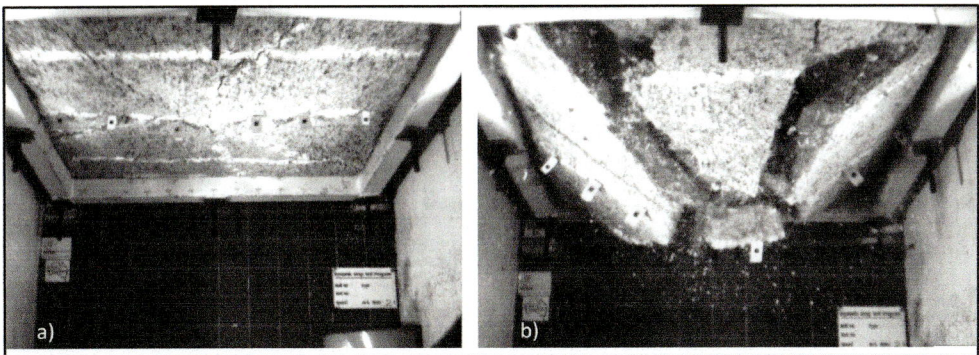

FIGURE 10.55 Dynamic performance comparison showing (a) a stable but cracked layer and (b) combined punching and flexural failure.

TABLE 10.9
Density of Synthetic Fibres across a Crack for 7 kg/m³.

Sample number	Test type	Number of fibres	Crack area (cm²)	Density fibres (fibres/cm²)	Average density
KB 6–12	UTS	8	13.60	0.59	**0.53**
KB 6–13	UTS	9	13.63	0.66	**fibres/cm²**
KB 6–20	UTS	10	13.31	0.75	
KB 6–21	UTS	6	13.19	0.45	
KB 8–1	UTS	6	12.90	0.47	
KB 8–3	UTS	6	12.71	0.47	
KB 8.6	UTS	4	12.62	0.32	
KB 6–11	Point Load	6	12.93	0.46	**0.50**
KB 6–14	Point Load	13	13.46	0.97	**fibres/cm²**
KB 6–19	Point Load	7	12.97	0.54	
KB 8–2	Point Load	5	13.70	0.36	
KB 8–4	Point Load	5	12.07	0.41	
KB 8.5	Point Load	3	12.29	0.24	
Average fibre density (fibres/cm²)					**0.52**

TABLE 10.10
Density of Synthetic Fibres after Dynamic Testing of Specimen KB-7.

Lavyer thickness (mm)	Width (mm)	Total fibres	Fibres broken	Area (cm²)	Fibre density (fibres/cm²)	Broken fibres (fibres/cm²)
140	100	29	9	140	0.207	0.064
105	100	31	4	105	0.295	0.038
110	100	26	2	110	0.236	0.018
100	100	28	4	100	0.280	0.040

(Continued)

TABLE 10.10

Continued

Lavyer thickness (mm)	Width (mm)	Total fibres	Fibres broken	Area (cm^2)	Fibre density (fibres/cm^2)	Broken fibres (fibres/cm^2)
100	100	44	6	100	0.440	0.060
90	100	24	6	90	0.267	0.067
113	100	25	7	113	0.221	0.062
126	100	36	10	126	0.286	0.079
100	100	29	4	100	0.290	0.040
120	100	39	3	120	0.325	0.025
130	100	30	11	130	0.231	0.085
120	100	39	4	120	0.325	0.033
Average fibre density (fibres/cm^2)					**0.284**	**0.049**

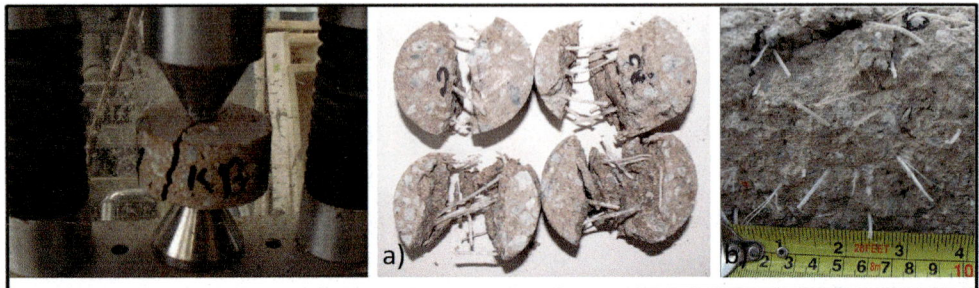

FIGURE 10.56   Density of synthetic fibres across a crack following (a) conventional laboratory testing and (b) dynamic rupture.

10.9.3.2   Mesh-Reinforced Shotcrete

10.9.3.2.1   Welded Wire Mesh

Figure 10.57 presents the pre- and post-test results for a dynamic test of a shotcrete layer reinforced with 5.7-mm diameter welded wire mesh. The measured average layer thickness was 110 mm, with the shotcrete sprayed in two passes in order to embed the mesh within the ground support layer. Testing was initially undertaken using very low input energy, which resulted in minimal damage to the support layer. However, an additional test resulted in catastrophic results, with the layer unable to arrest the failure (Figure 10.58).

10.9.3.2.2   Woven Wire Mesh

Figure 10.59 shows the pre- and post-test results for a shotcrete layer reinforced with G80/4.6 high tensile woven mesh. The measured average layer thickness was 89 mm, with the shotcrete sprayed in two passes in order to embed the mesh within the ground support layer. Dynamic testing was undertaken using higher input energy compared with the results shown earlier in Figure 10.60. The test results demonstrated that the layer was able to arrest the failure, and the mass did not penetrate the mesh-reinforced shotcrete layer, which dissipated 22 kJ at 225 mm of displacement (Figure 10.58). Similar to underground observations following a seismic event, a significant amount of shotcrete ejection was experienced at the dynamic impact point.

FIGURE 10.57   Details of dynamic testing of a shotcrete layer reinforced with a 5.7-mm diameter welded wire mesh (a) prior to testing and (b) catastrophic results after testing.

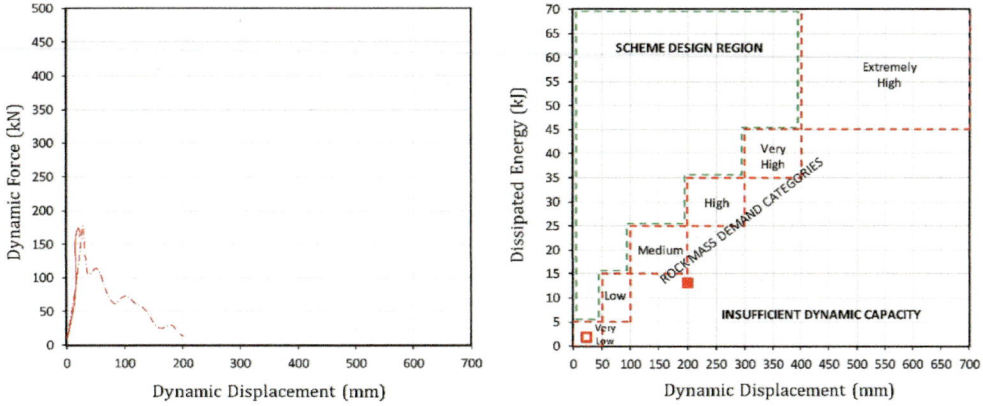

FIGURE 10.58   Dynamic force displacement and related energy dissipation prior to catastrophic punch failure for a 110 mm thick shotcrete layer reinforced with a 5.7-mm diameter welded wire mesh.

FIGURE 10.59   Details of dynamic testing of a shotcrete layer reinforced with a G80/4.6 high tensile mesh (a) prior to testing and (b) failure arrested showing shotcrete ejection.

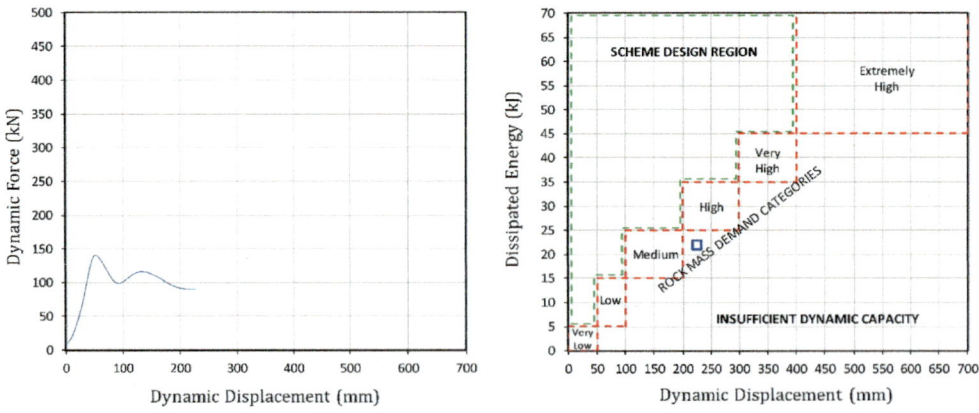

FIGURE 10.60   Dynamic force displacement and related energy dissipation prior to catastrophic punch failure for an 89 mm thick shotcrete layer reinforced with a high tensile G80/4.6 mesh.

11 Dynamic Performance of Ground Support Schemes

11.1 INTRODUCTION

Ground support schemes consist of combined rock reinforcement and surface support systems. As described in Chapters 7 and 8, a rock reinforcement system comprises elements fixed within a borehole, drilled into the rock mass and having an exterior face plate with an external fixture. A pattern of reinforcement systems is often used to support a large instability and prevent large-scale ground deterioration. The type of element and the spacing of each system depend on the materials available locally and the prevailing geological conditions, namely the geometry of the potentially unstable blocks, the loading conditions, and the resulting failure mechanisms. As described in Chapters 9 and 10, a surface support system is used to connect the reinforcement systems to prevent the unravelling of the rock mass within and across a reinforcement pattern. Importantly, the surface support systems allow load transfer from unstable regions, where reinforcement systems may have failed, to stable regions, where reinforcement systems still have load-bearing and displacement capacities. Hence, the surface support systems are restrained by the face plates used with the rock reinforcement systems to form one integrated ground support scheme.

11.2 LOAD TRANSFER

In order to properly interpret and utilize the results from dynamic testing facilities, it is necessary to understand the *in situ* interaction between ground support schemes and a rock mass. An important aspect of a ground support scheme response is the amount of rock mass deformation and the rate at which it occurs. Should there be an ejection of rock, a reinforcement system should dissipate the momentum generated by an unstable region. The momentum is equal to the mass of instability multiplied by its initial ejection velocity. The momentum of the ejected rock is resisted by the transfer of load to the stable region of the rock mass. The load developed in the reinforcement causes deceleration and a reduction in the unstable mass velocity.

To simulate this, a laboratory test must be a double embedment configuration with collar and toe anchor lengths and a collar geometry that allows tension to be applied to the surface restraint (Chapter 6). This generates the correct load transfer to the reinforcement system and allows for failure to occur at whichever is the weakest location, that is, the surface hardware, the collar region of the reinforcement element, the reinforcement element at the simulated discontinuity or displacement within the toe anchor region. It is also possible that a volume of rock may fail violently, immediately adjacent to the surface support, as shown in Figure 11.1 and Figure 11.2. That is, failure may occur between reinforcement systems or in a volume of rock surrounding a borehole in which reinforcement has been installed. In the former case, the support is required to transfer the dynamic load through the surface support into the reinforcement through the external fixture. In the latter case, the support and reinforcement respond and combine in an attempt to sustain the dynamic loading. In the former case, it is possible for the surface support to fail and not transfer the load to the reinforcement system. This is often observed where rock simply pushes through low-strength mesh or punches through a layer of shotcrete. In the latter case, the proportion of load and energy dissipated by the reinforcement and surface support will be complex. The complexities are related to

DOI: 10.1201/9781003357711-11

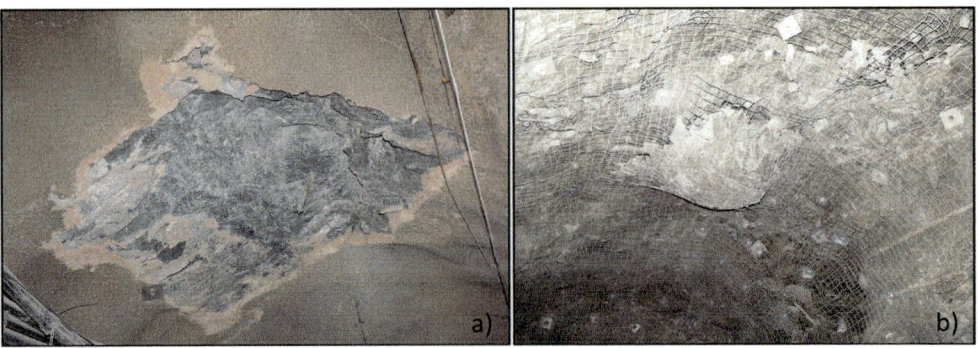

FIGURE 11.1 Examples of rock failure, loading mostly the surface support between reinforcements: (a) shotcrete with no load transfer and (b) welded wire mesh with load transfer.

FIGURE 11.2 Example of rock failure, loading both surface support and reinforcement.

the relative stiffnesses of all the system components in the scheme, with the reinforcement elements often dissipating most of the energy.

It is possible for the reinforcement system to fail, either internally or at the collar, and then for the surface support to transfer the load subsequently to adjacent reinforcement systems within the ground support scheme. The WASM Dynamic Test Facility was developed with both mechanisms in mind, that is, a panel of surface support system restrained either only at the edges to simulate a continuous sheet or at the edges with a single reinforcement at the centre of the panel.

In both cases of initiated dynamic loading, a susceptible excavation may involve a detachment process in which a single block or fragments of rock may attempt to eject from the surrounding rock mass into the excavation and load the ground support scheme. This type of ground support loading is unlikely to be instantaneous but rather takes a finite time. For a remote event, this is related to the seismic wave velocity, amplitude, the acceleration pulse of the dominant frequency and fracture velocity within the rock mass. For a local event, the excitation will be a 'pulse' of loading that will result in a mass (m) of rock moving at a particular velocity (v). That is, the failure volume will have a change in momentum that is related to a force (Ft) acting over a short time (dt) as defined by the impulse equation:

$$\Delta Mv = \int F_t dt \qquad\qquad\qquad 11.1$$

In these models of ground support loading, it is clear that prior to a seismic event, the rock mass and ground support are stationary. After the event, any disconnected mass is accelerated to a certain velocity in a short but finite time. Hence, in a test facility, it is not appropriate to apply load instantaneously, rather, it should be applied quickly, and the rate of application should be measurable and similar to that considered to cause damage to underground mine ground support. It is not appropriate to simply apply the load to the exposed collar of a reinforcement system (Carlton et al., 2013).

11.3   FREE BODY DIAGRAMS

Chapter 6 explained that loading within the WASM Dynamic Test Facility is provided by the relative velocity between a loading mass and a drop beam, as the momentum of the loading mass is transferred through the reinforcement and support systems to the drop beam (Thompson et al., 2004). The reinforcement system will yield, slide or break in the process of dissipating the dynamic load. The process will depend on the design and material properties of the reinforcement element, the encapsulation medium, and the interaction with the borehole. This is best done by applying the input energy via impact by dropping the modelled rock mass and reinforcing system and the rock mass that remains afterwards onto an engineered impact surface with a given stiffness (i.e., by dropping the test assembly onto buffers). The energy dissipation by the impact surface must be measurable and reported with all other parameters from the test.

The equilibrium of each component in a test configuration forms the basis for the engineering analysis of a test and results in the reinforcement or surface support system performance being expressed in terms of a force-displacement response and energy dissipation.

Comprehensive free body diagrams are required to identify the force interactions between all the components involved during a dynamic test, including combinations of reinforcement systems, mesh and shotcrete panels, as shown Figure 11.3 (see also Chapter 6, Figure 6.28). The free body force diagram and associated component displacements are used to refine the requirements for the types of instrumentation and their locations on the various components. These diagrams are key in the design and development of data analysis software together with the following assumptions:

- The drop beam and support testing frames behave as rigid bodies. That is, their deformations are considered infinitesimally small compared with the reinforcement and support displacements.
- The beam displacement, velocity and deceleration after impact are considered to be equal to the corresponding response from the buffers. This means that parameters must be measured for either the beam or the buffer piston.
- Vertical movements alone are considered in the analysis, so only vertical components from the accelerometers are used, and the rotational and horizontal components of acceleration (and displacement) are ignored.

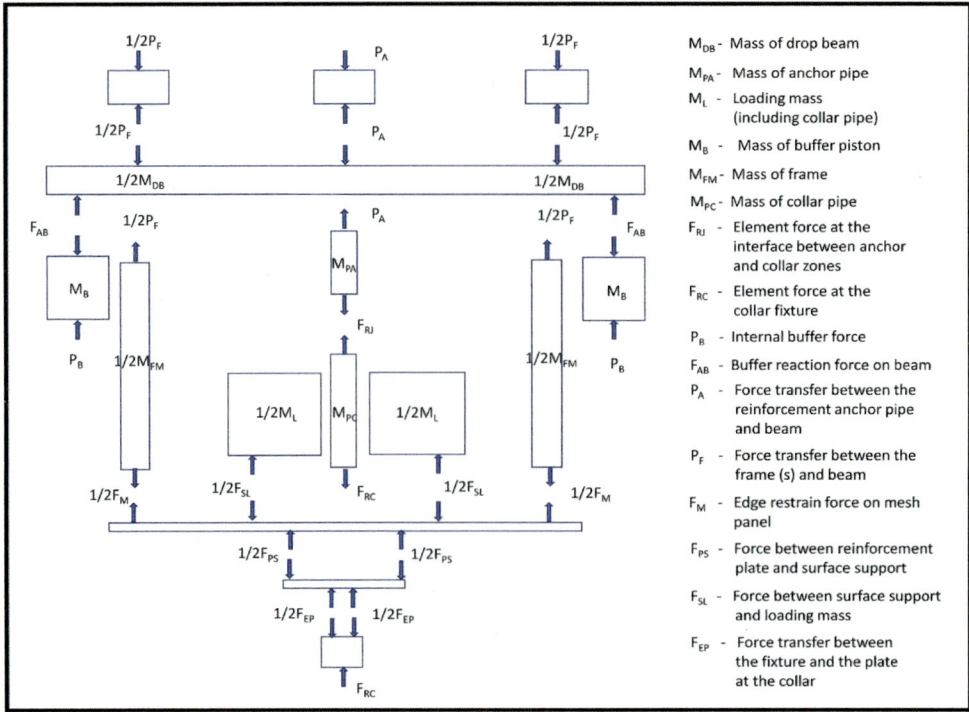

FIGURE 11.3 Free body force diagram for dynamic testing of combined reinforcement and mesh panels (Villaescusa et al., 2010).

- Filtering data does not influence the overall outcomes from the results or the relative performance of the classes of reinforcement system classifications but will affect some details in the graphs.
- The zeroing of the filtered data to a common impact time that coincides with a start time from the digital video does not significantly influence the overall result; however, it does change the initial shape of the response graphs.
- The number of buffers in a test does not significantly affect the results. This was validated by testing samples of the same reinforcement system impacting either two or four buffers and comparing the resultant dynamic force-displacement responses.

11.4   COMBINED REINFORCEMENT AND MESH SCHEMES

The results from twelve tests are presented here to describe the performance of a ground support scheme. The combined reinforcement and mesh scheme test program involved three different types of reinforcement system, namely:

- Fully coupled, cement-encapsulated, 20-mm diameter threaded bar (2.4 m long);
- A decoupled (1000 mm), cement-encapsulated, 20-mm diameter threaded bar (2.4 m long); and
- A decoupled (1400 mm), cement-encapsulated, 20-mm diameter threaded bar (3.0 m long).

The reinforcement systems were combined with four different types of mesh support systems, namely:

- TECCO G80 high tensile woven mesh, 4 mm wire diameter (G80/4);
- Mild steel, woven mesh, 4 mm and 5 mm wire diameters; and
- Conventional, galvanized welded wire mesh, 5.6 mm wire diameter.

11.4.1 Sample Preparation and Testing

Figure 11.4 shows a number of simulated boreholes as well as details of the cement encapsulation for the threaded bar used in the experiments. All the elements were centralized and grout with a WCR of 0.4 was used with dynamic testing undertaken after 28 days of curing.

The mesh was fixed onto shackles attached by eye bolts to the test frame (Figure 11.5). The number of restraints for each side depended upon the type of mesh (woven mesh or welded wire mesh) and the size and shape of a mesh loop (typically a diamond shape for woven mesh and a square or rectangle for welded wire mesh). Targets with black squares on a white background were positioned

FIGURE 11.4 Long and plan section views of cement-encapsulated threaded bar.

FIGURE 11.5   Details of mesh testing frame and attachments of mesh to the frame.

securely to the mesh to measure the initial mesh deflection and to track the mesh deformation with a high-speed video camera.

The simulated boreholes (the steel pipe configuration) were positioned through the drop beam, and a loading mass was used to simulate an unstable region of a rock mass. A curved plate was fixed between the load mass and the mesh with conventional surface hardware (plate) fixed to the reinforcement element using a nut. The total system (drop beam + mesh frame + reinforcement system + mass) was lifted with a crane and positioned above the drop pit. The entire test configuration is shown in Figure 11.6.

11.4.2   Data Analysis

The objective of data analysis is to determine the variations of accelerations, velocities and displacements with time for each component in a test. Specialized software is used to calculate the forces based on the mass of the component and the corresponding accelerations/decelerations. The masses and their velocities are used to calculate the kinetic energies of the components. In addition, the loss of potential energy during the test after impact is calculated. The analysis involves filtering and processing of the data from accelerometers, load cells and potentiometers, and consists of three basic stages:

- Reviewing and selecting data for analysis,
- Filtering of the selected data and
- Analysis of the filtered data over the required time interval.

Basically, the analysis process can be divided into two parts: First, video data analysis tracked from a high-speed camera (calculated displacement), and second, sensor data analysis from accelerometers, load cells and linear potentiometers. The raw data is disturbed by mechanical and electrical

FIGURE 11.6 Details of a ground support scheme (a) prior to and (b) following dynamic testing.

noise due to the vibration and metal-to-metal contact during the impacts. Filtering the raw data is essential to interpreting the results in a meaningful way (Figure 11.7).

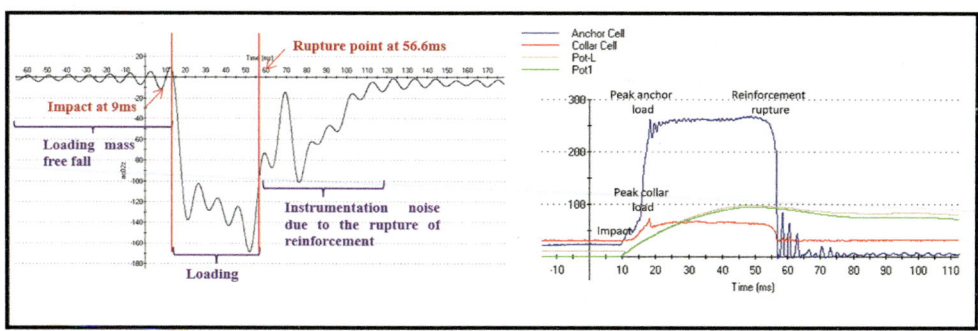

FIGURE 11.7 Example of analysis showing filtering of raw data using (a) Fast Fourier Transform filter and (b) Butterworth filter.

11.4.3 DATA

The results from two ground support scheme test programs will be presented here. Importantly, the individual contributions of reinforcement and support systems towards the overall performance of the combined schemes are calculated. Table 11.1 and Table 11.2 detail the sample specifications for the two campaigns of testing reported in this section.

Table 11.3 details the combined test specifications of impact velocity, total loading mass and the nominal input energy for the two test programs. For Test Program 1, the input velocities varied from 5.4 to 6.8 m/s, while the respective input energies varied from 27 to 53.2 kJ. The second

TABLE 11.1

Sample Specifications—Ground Support Scheme Test Program 1.

Test No	Test ID	Reinforcement system	Bar length (mm)	Bar diameter (mm)	Decoupled length (mm)	Support system	Wire diameter (mm)
1	195	Fully encapsulated threaded bar	2400	20	0	TECCO 80	4
2	196	Fully encapsulated threaded bar	2400	20	0	TECCO 80	4
3	197	Fully encapsulated threaded bar	2400	20	0	Welded Wire Mesh	5.6
4	198	Fully encapsulated threaded bar	2400	20	0	Welded Wire Mesh	5.6
5	199	Decoupled Posimix	2400	20	1000	Welded Wire Mesh	5.6
6	200	Decoupled Posimix	2400	20	1000	TECCO 80	4
7	201	Decoupled Posimix	2400	20	1000	TECCO 80	4
8	202	Decoupled Posimix	2400	20	1000	Welded Wire Mesh	5.6

TABLE 11.2

Sample Specifications—Ground Support Scheme Test Program 2.

Test No	Test ID	Reinforcement system	Bar length (mm)	Bar diameter (mm)	Decoupled length (mm)	Support system	Wire diameter (mm)
1	231	Decoupled Posimix	3000	20	1400	CODELCO	4
4	234	Decoupled Posimix	3000	20	1400	CODELCO	5
5	235	Decoupled Posimix	3000	20	1400	TECCO 80	4
6	236	Decoupled Posimix	3000	20	1400	TECCO 80	4

test program involved higher impacts, where the velocities of the loading mass varied from 7.0 to 7.3 m/s, with respective input energies ranging from 52.8 to 57.5 kJ.

11.4.3.1   Cement-Encapsulated Rebar and G80/4 Mesh

The fully cement encapsulated threaded bar and G80/4 mesh scheme were implemented widely by the main research sponsor at the time of the experiments. Figure 11.8 shows the #195 ground support scheme set-up prior to testing at the WASM Dynamic Test Facility. The scheme was tested with 36.6 kJ input energy. The force versus time and displacement responses for sample #195 are represented in Figure 11.9. The reinforcement system ruptured at 47.6 ms after impact, with a maximum displacement at the simulated discontinuity of 90 mm. Following the rupture of the bolt, the mesh force increased, and the loading mass was brought to a rest.

Figure 11.10 shows the conditions after the dynamic impact on the scheme. The threaded bar ruptured with a maximum displacement at the discontinuity of 90 mm, while the woven mesh survived with a maximum displacement of 115 mm.

TABLE 11.3

Ground Support Scheme Test Sample Specifications.

Program No	Test ID	Impact Velocity (m/s)	Loding Mass (m/s)	Intial Input Energy (KJ)
1	195	5.8	2158	36.6
1	196	6.0	1778	32.0
1	197	5.4	1869	27.0
1	198	5.8	2248	38.7
1	199	6.8	2248	53.2
1	200	6.1	2158	40.9
1	201	6.8	1778	41.6
1	202	5.8	2158	37.0
2	231	7.2	2158	56.0
2	234	7.1	2158	54.9
2	235	7.3	2158	57.5
2	236	7.0	2158	52.8

FIGURE 11.8 Details of pre-test set-up for ground support scheme #195 consisting of one fully encapsulated threaded bar and G80/4 woven mesh.

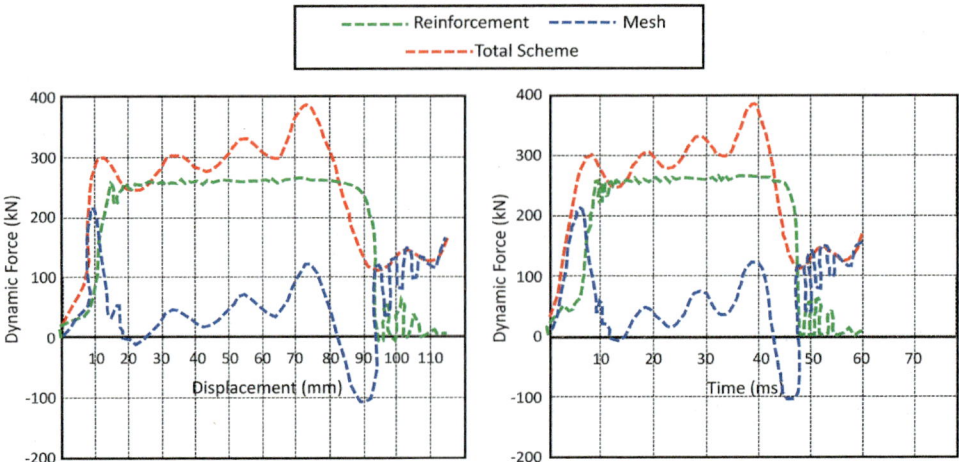

FIGURE 11.9 Reinforcement, mesh and ground support scheme dynamic response for sample #195 (fully cement encapsulated threaded bar and G80/4 mesh which achieved stability).

FIGURE 11.10 Details of post-test conditions for ground support scheme #195 showing broken reinforcement element, with loading mass stabilized and no mesh damage.

The energy balance is defined as the sum of dissipated energy (the sum of energy dissipated by the reinforcement system, support system, buffers and residual kinetic energy of the loading mass, the beam, and the mesh frame) divided by the input energy (kinetic energy of the entire system at impact plus the change in potential energy). The energy balance for a test is calculated at the failure/rupture point or when the relative velocity between the loading mass and the drop beam becomes zero. The endpoint for test #195 was defined at 47.6 ms when the reinforcement ruptured (Figure 11.11). This ground support scheme dissipated 17.9 kJ, and the energy balance was 96.5%, which is an excellent reconciliation.

Figure 11.12 shows the anchor and collar load cell responses for sample #196. The drop beam impacted the buffers at 0.011 seconds, with the anchor load reaching 265 kN. The mesh was installed very tight against the loading mass and failed at 0.043 s, followed quickly by the fully encapsulated bar, which ruptured at 0.052 seconds. The combined scheme failed catastrophically, as shown in Figure 11.13.

Figure 11.14 shows the loading mass velocity and the reinforcement loading velocity across the simulated discontinuity, while Figure 11.15 shows the force versus time and displacement response

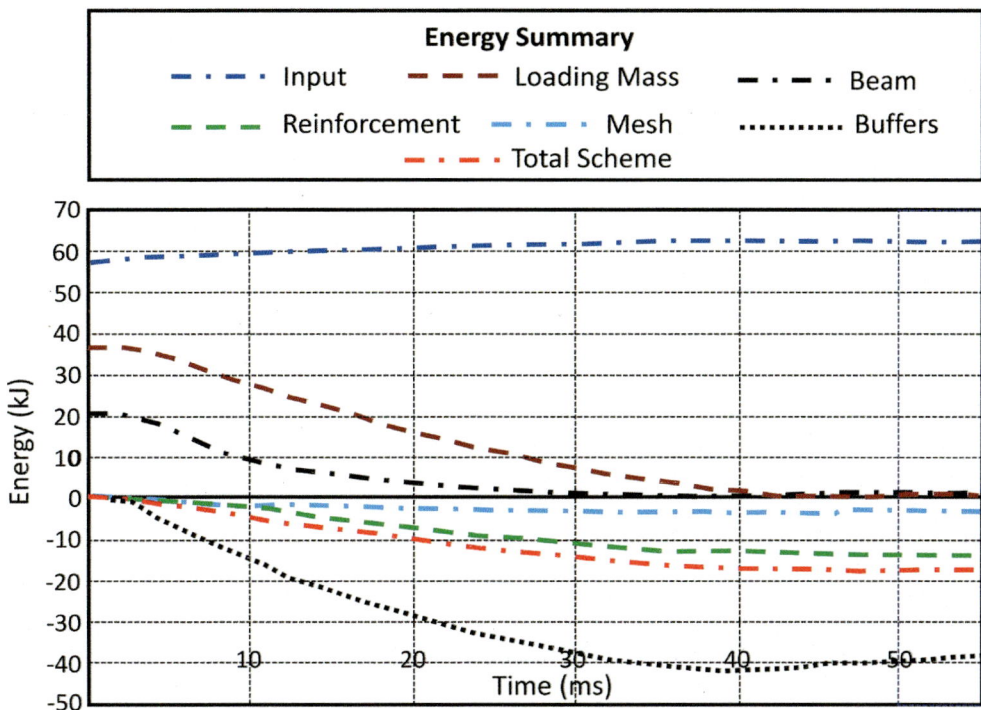

FIGURE 11.11   Energy balance for a stable ground support scheme (fully cement encapsulated threaded bar and G80/4 mesh—scheme #195).

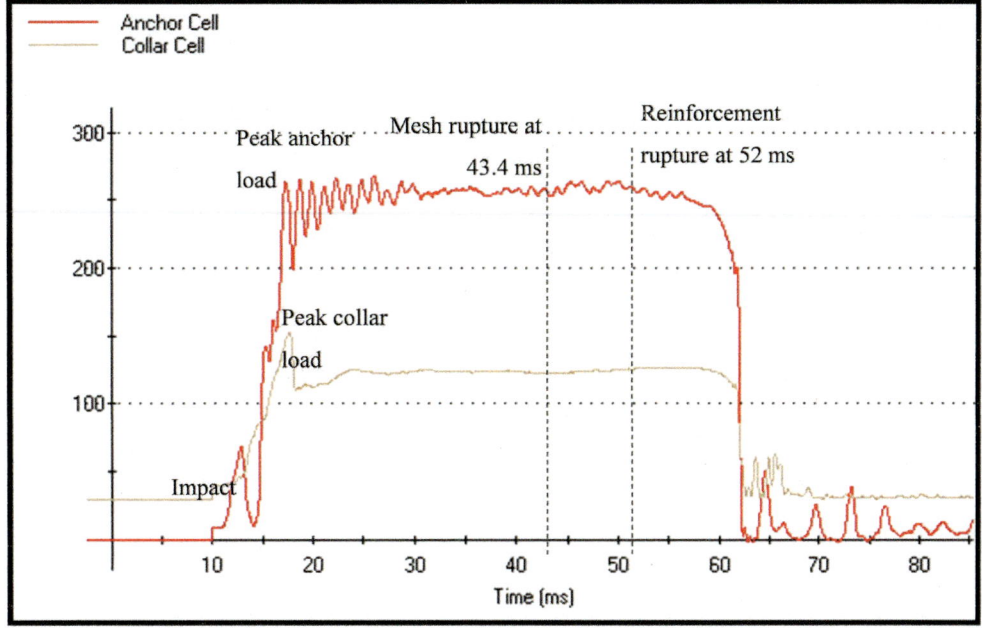

FIGURE 11.12   Anchor and collar load cell responses for fully cement encapsulated threaded bar and G80/4 mesh—scheme #196.

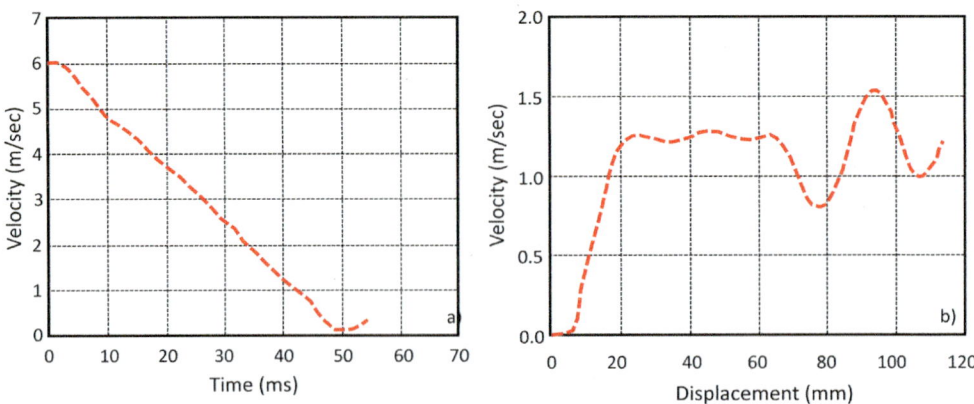

FIGURE 11.13 Details of violent failure of ground support scheme following dynamic impact testing.

FIGURE 11.14 Velocity of (a) loading mass and (b) reinforcement loading across simulated discontinuity (fully cement encapsulated unstable threaded bar and G80/4 mesh—scheme #196).

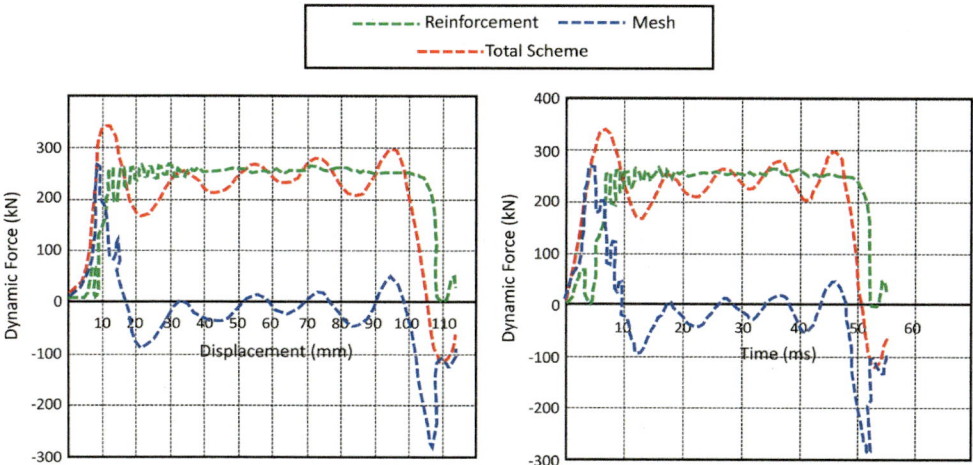

FIGURE 11.15 Reinforcement, mesh and ground support scheme dynamic response for sample #196 (unstable scheme—fully cement encapsulated threaded bar and G80/4 mesh).

for the ground support scheme. The ground support scheme was unable to slow down and stabilize the loading mass. The energy balance is shown in Figure 11.16, with most of the response provided by the reinforcement system and the scheme failing at 23.3 kJ.

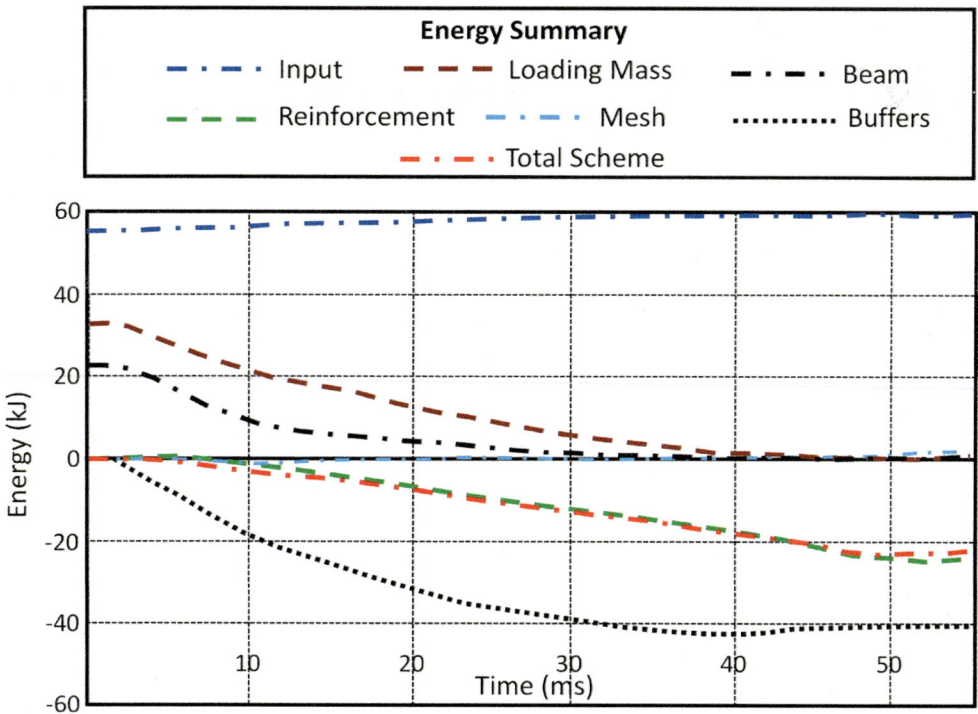

FIGURE 11.16 Energy balance for an unstable ground support scheme (fully cement encapsulated threaded bar and G80/4 mesh—scheme #196).

11.4.3.2 Cement-Encapsulated Rebar and Welded Wire Mesh

As indicated in Table 11.1, galvanized welded wire mesh (5.6 mm diameter) was used in conjunction with a fully cement encapsulated 20-mm diameter threaded bar. Figure 11.17 shows a typical

FIGURE 11.17 Details of pre-test set-up for ground support scheme #197 comprising one fully encapsulated threaded bar and a 5.6-mm diameter galvanized welded wire mesh.

set-up prior to testing, while Figure 11.18 shows the anchor and collar load cells' response to the dynamic loading for scheme #197. The beam impacted the buffers at 0.011 seconds, and the anchor load reached 252 kN and remained nearly constant until the end of the test. The bar yielded at a maximum of 59 mm, and the combined system was considered stable (even with four ruptured wires) (Figure 11.19). Figure 11.20 shows the loading mass velocity and the reinforcement loading velocity across the simulated discontinuity, while Figure 11.21 shows the force versus time and displacement response for the ground support scheme. The energy balance is shown in Figure 11.22 and shows that most of the response was provided by the reinforcement system, with the scheme stable at 16.0 kJ.

Figure 11.23 shows the set-up of sample #198 prior to testing, while Figure 11.24 shows the anchor and collar load cell responses to the dynamic loading. The beam impacted the buffers at 0.010 seconds, and the anchor load rapidly reached 260 kN and remained nearly constant until the end of the test. The reinforcement reached a maximum displacement of 85.4 mm at the simulated discontinuity, and the mesh ruptured at 39.4 ms with a total of ten damaged wires (Figure 11.25). The scheme was considered to have reached stability. Figure 11.26 shows the loading mass velocity and the reinforcement loading velocity across the simulated discontinuity. Figure 11.27 shows the force versus time and displacement response. The energy balance is shown in Figure 11.28 and indicates that most of the work was provided by the reinforcement system, with the complete scheme dissipating 23.3 kJ.

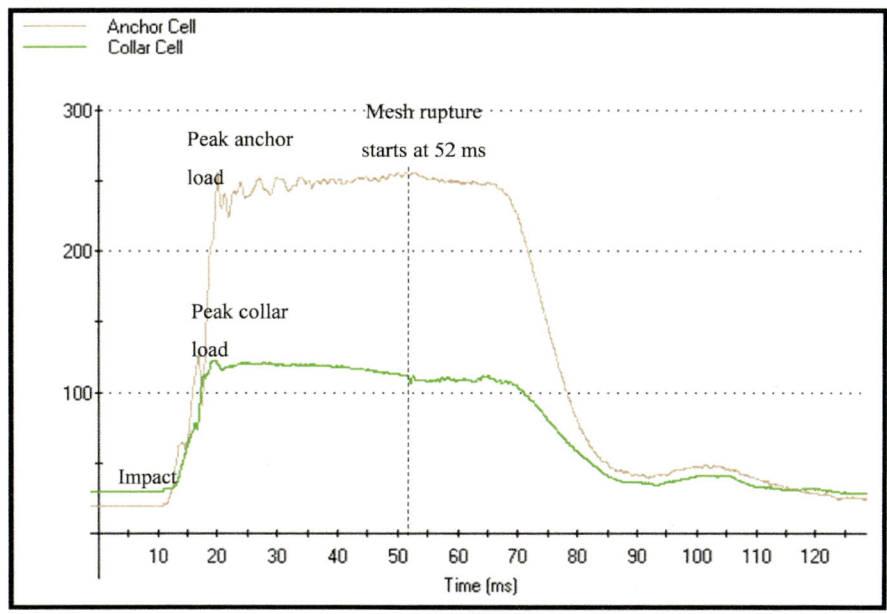

FIGURE 11.18 Anchor and collar load cell response for a fully cement encapsulated threaded bar and 5.6 mm welded wire mesh (scheme #197).

FIGURE 11.19 Details of post-test conditions for ground support scheme #197 showing yielded reinforcement element, with the loading mass stabilized and limited mesh damage.

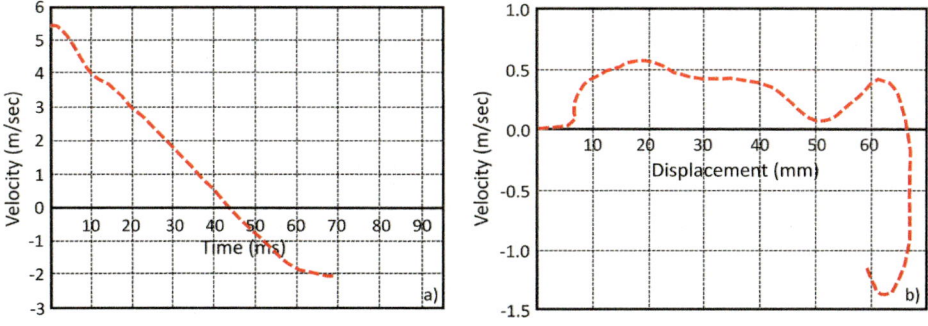

FIGURE 11.20 Velocity of (a) loading mass and (b) reinforcement loading across the simulated discontinuity (fully cement encapsulated threaded bar and welded wire mesh—scheme #197, which reached stability).

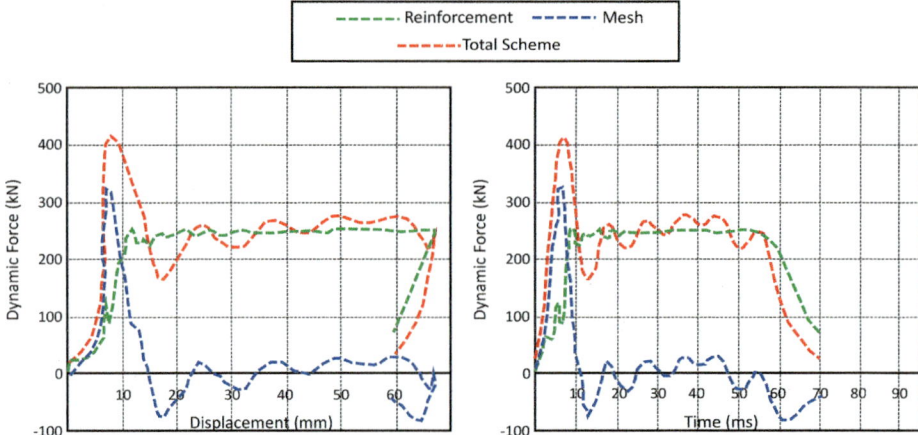

FIGURE 11.21 Reinforcement, mesh and ground support scheme dynamic response, sample #197 (fully cement encapsulated threaded bar and welded wire mesh, which reached stability).

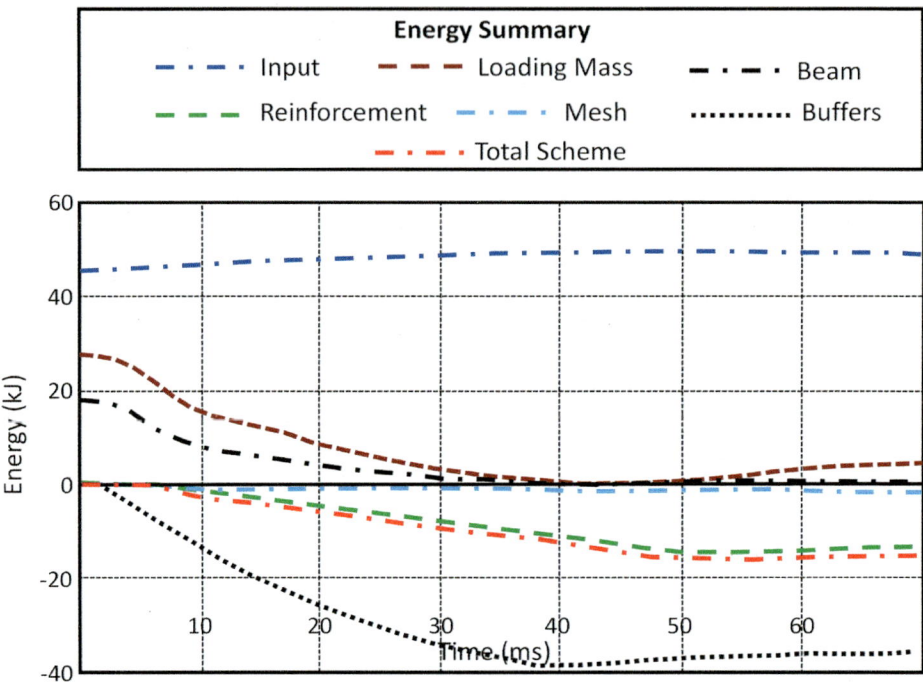

FIGURE 11.22 Energy balance for stable ground support scheme #197 (fully cement encapsulated threaded bar and welded wire mesh).

11.4.3.3 Decoupled Posimix and Welded Wire Mesh

As indicated in Table 11.1, several 20-mm diameter, decoupled threaded bars encapsulated in cement grout were used in conjunction with welded wire mesh. As stated in Chapter 7, cement encapsulation allows reinforcement displacement by releasing the element at the grout-steel interface. Therefore, the results presented within this section cannot be directly compared with jumbo-installed, resin-encapsulated Posimix. Figure 11.29 shows the anchor and collar load cells' response to the dynamic loading of scheme #199.

FIGURE 11.23 Details of pre-test set-up for ground support scheme #198 comprising of 1 fully encapsulated threaded bar and 5.6-mm diameter galvanized welded wire mesh.

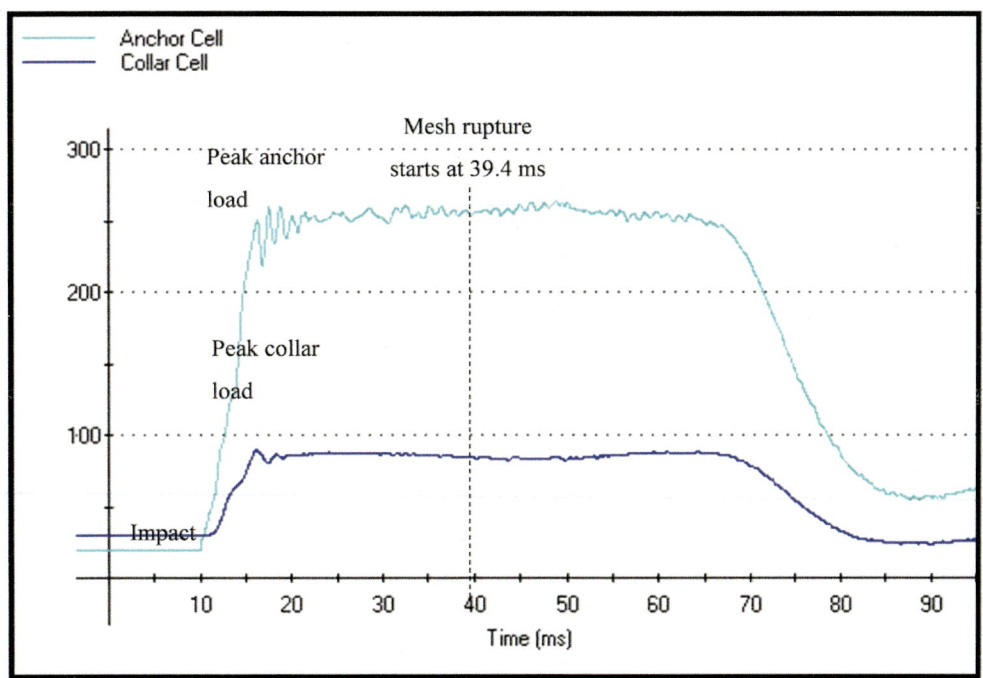

FIGURE 11.24 Anchor and collar load cell responses for a fully cement encapsulated threaded bar with 5.6-mm welded wire mesh (scheme #198).

The beam impacted the buffers at 0.008 seconds, with the anchor load reaching 220 kN and then gradually increasing to 250 kN, when the reinforcement element failed by completely pulling out of the toe anchor region (Figure 11.30). The mesh totally ruptured at 35.7 ms, followed by reinforcement pull-out at 55.6 ms. The toe anchor region length was 490 mm, but in cement grout, this was insufficient, and the ground support scheme failed catastrophically. Failures of similar nature have been observed at some mine sites where welded wire mesh has simply failed at the edges, being unable to transfer load (Figure 11.31).

FIGURE 11.25 Details of post-test conditions for ground support scheme #198 showing a yielded reinforcement element and moderate mesh damage with the loading mass stabilized.

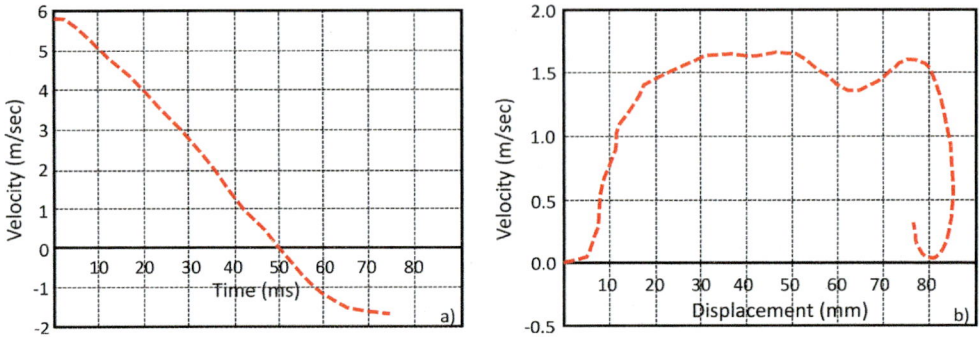

FIGURE 11.26 Velocity of (a) loading mass and (b) reinforcement loading across the simulated discontinuity (fully cement encapsulated threaded bar and welded wire mesh—scheme #198).

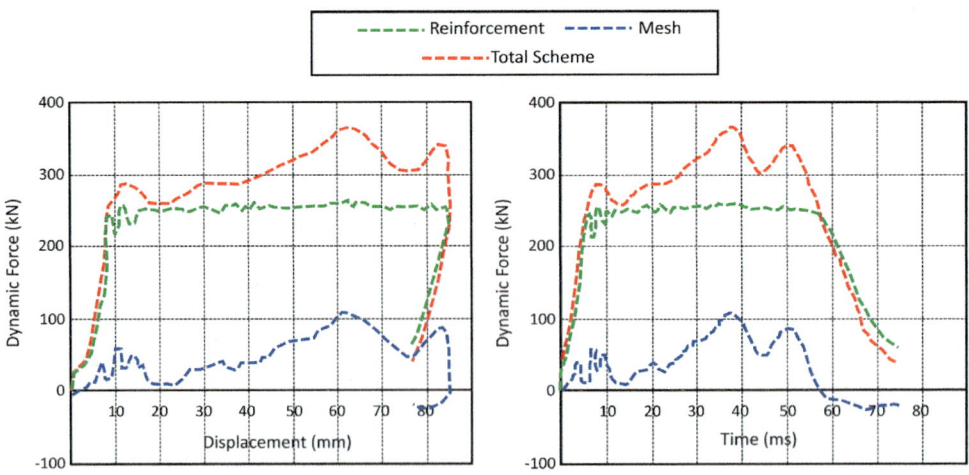

FIGURE 11.27 Reinforcement, mesh and ground support scheme dynamic response, sample #198 (fully cement encapsulated threaded bar and welded wire mesh which attained stability).

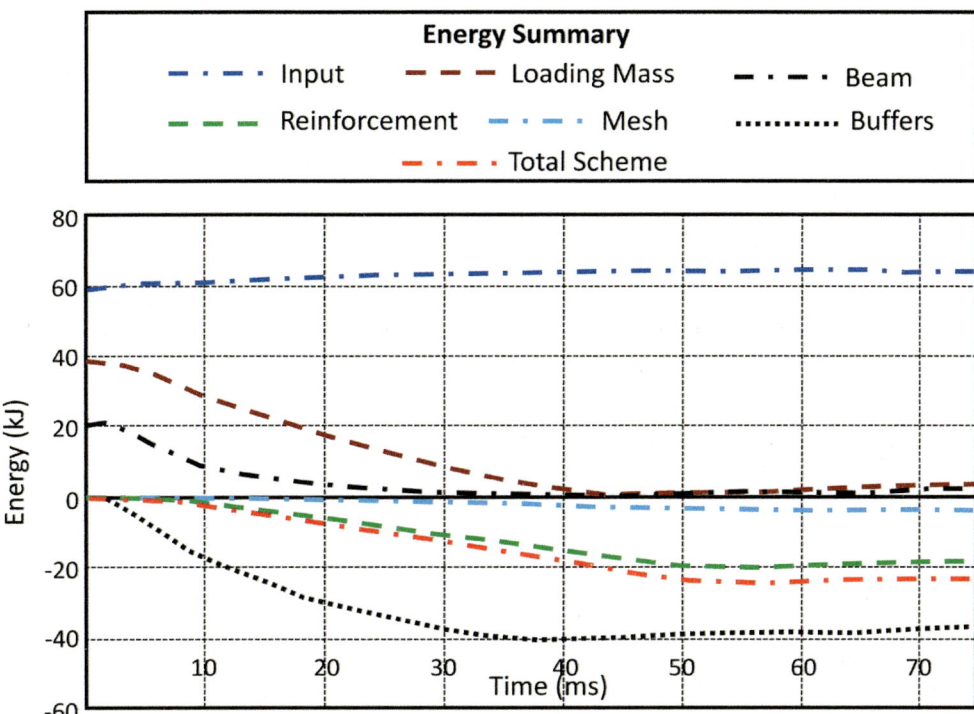

FIGURE 11.28 Energy balance for stable ground support scheme #198 (fully cement encapsulated threaded bar and welded wire mesh).

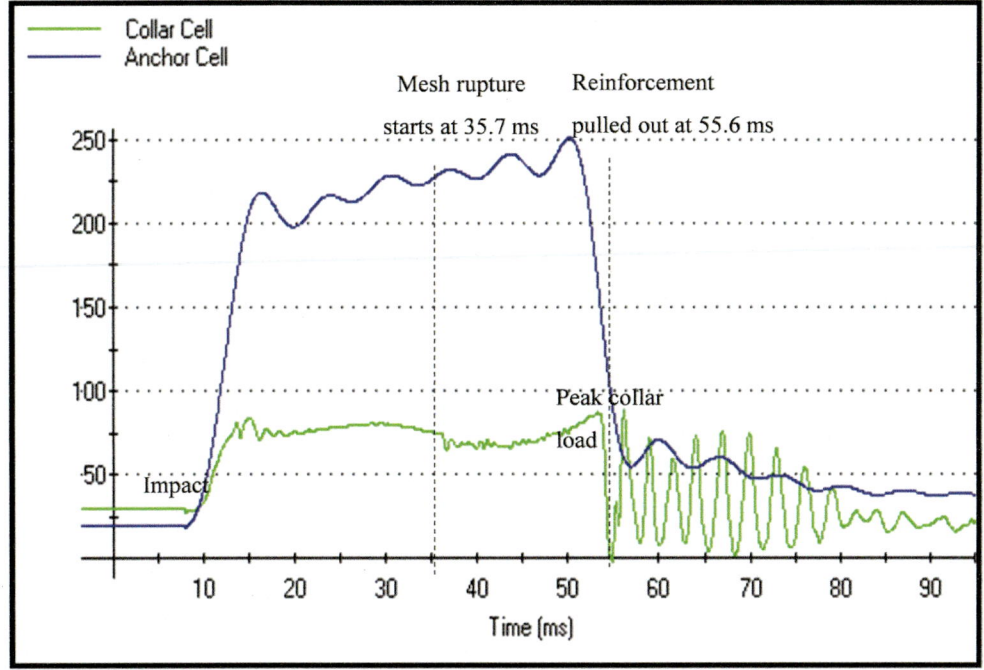

FIGURE 11.29 Anchor and collar load cell responses for a fully cement encapsulated decoupled threaded bar and 5.6-mm welded wire mesh (scheme #199).

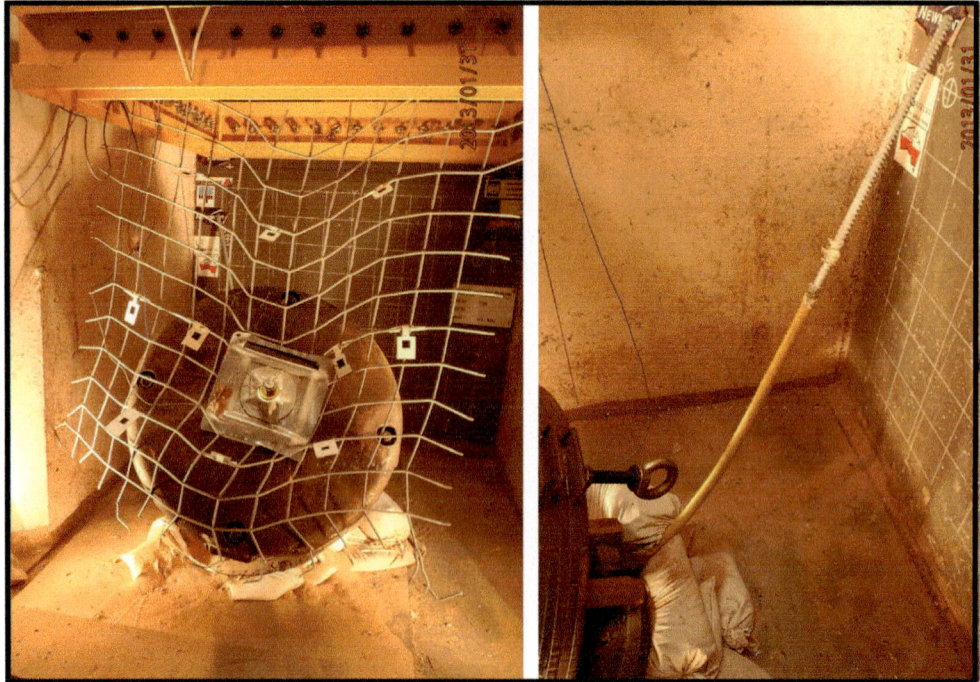

FIGURE 11.30 Details of post-test conditions for ground support scheme #199 showing completely broken mesh and reinforcement pulled out from the toe anchor region.

FIGURE 11.31 Example of catastrophic failure of resin-encapsulated Garford bolts and weld mesh.

Figure 11.32 shows the loading mass velocity and the reinforcement loading velocity across the simulated discontinuity, while Figure 11.33 shows the force versus time and displacement response for the ground support scheme.

The energy balance is shown in Figure 11.34. The data show that the reinforcement system dissipated nearly 30 kJ prior to pulling out, with the total scheme dissipating nearly 40 kJ.

Figure 11.35 shows the anchor and collar load cell responses of ground support scheme #202. The beam impacted the buffers at 0.010 seconds, with the anchor load rapidly reaching 251 kN after impact. The reinforcement was stable after 121 mm of displacement at the simulated discontinuity, located within the decoupled region (Figure 11.36). A total of six wires ruptured on the welded wire mesh; however, the combined scheme was stable.

Figure 11.37 shows the loading mass velocity and the reinforcement loading velocity across the simulated discontinuity, while Figure 11.38 shows the force versus time and displacement responses for the ground support scheme. The energy balance is shown in Figure 11.39. The data indicate that the ground support scheme dissipated nearly 32 kJ. Most of the energy was dissipated by the decoupled reinforcement element.

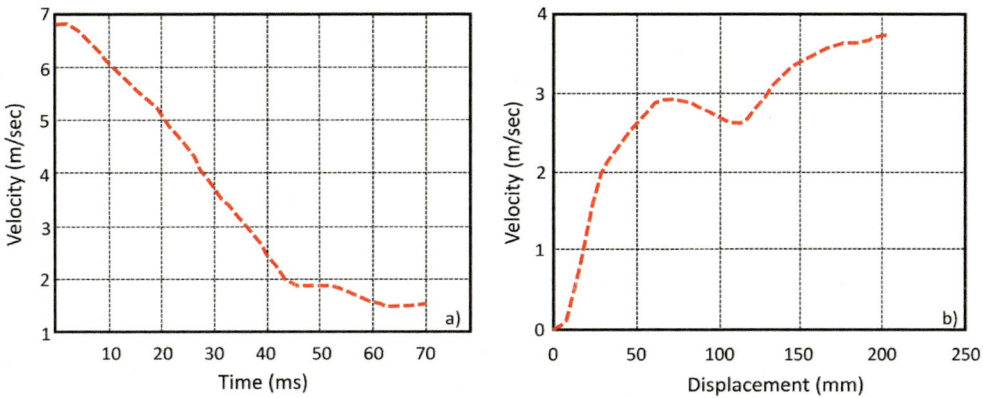

FIGURE 11.32 Velocity of (a) loading mass and (b) reinforcement loading across simulated discontinuity (cement-encapsulated decoupled threaded bar and welded wire mesh—scheme #199, which failed).

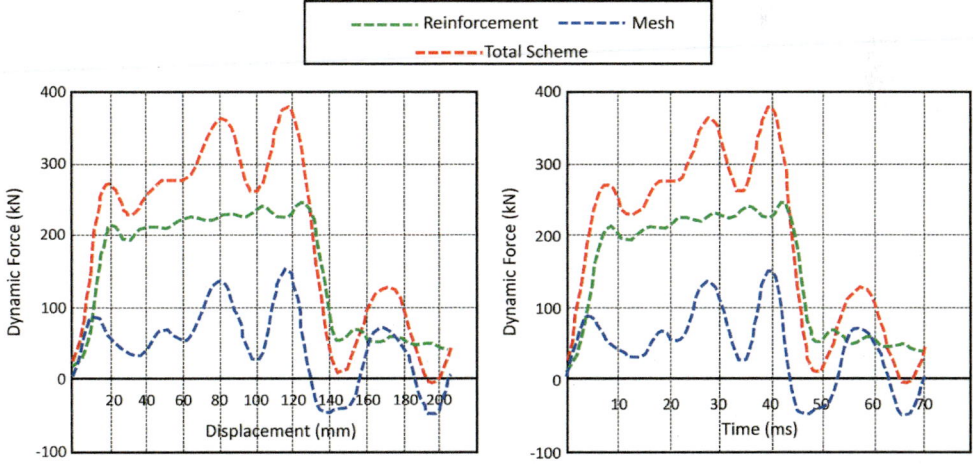

FIGURE 11.33 Reinforcement, mesh and ground support scheme dynamic response, sample #199 (cement-encapsulated decoupled threaded bar and welded wire mesh, which failed).

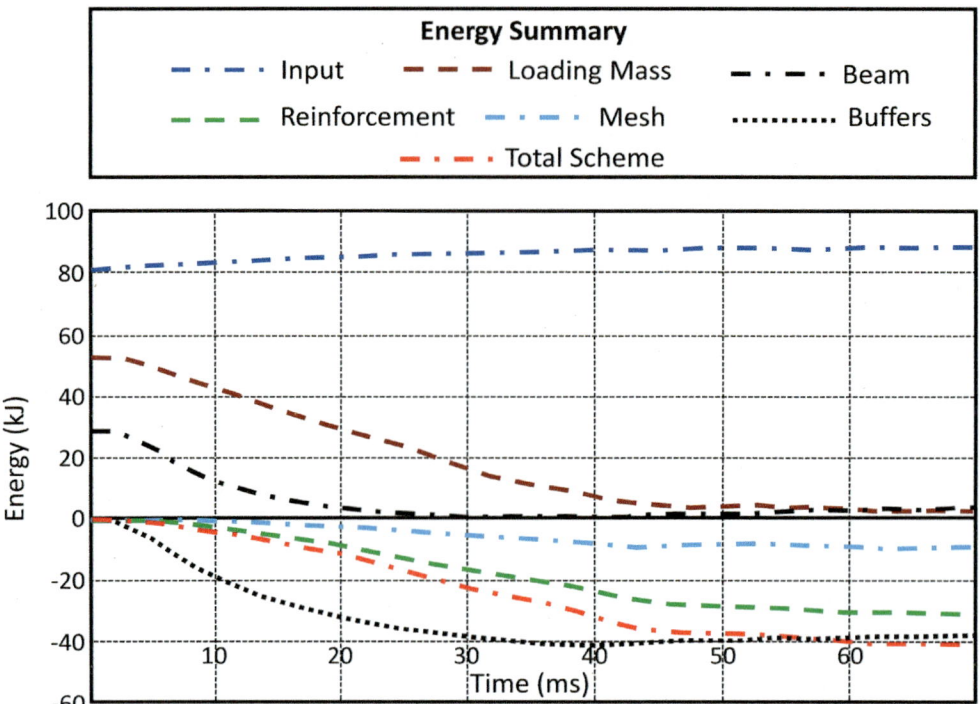

FIGURE 11.34 Energy balance for unstable ground support scheme #199 (cement-encapsulated decoupled threaded bar and welded wire mesh).

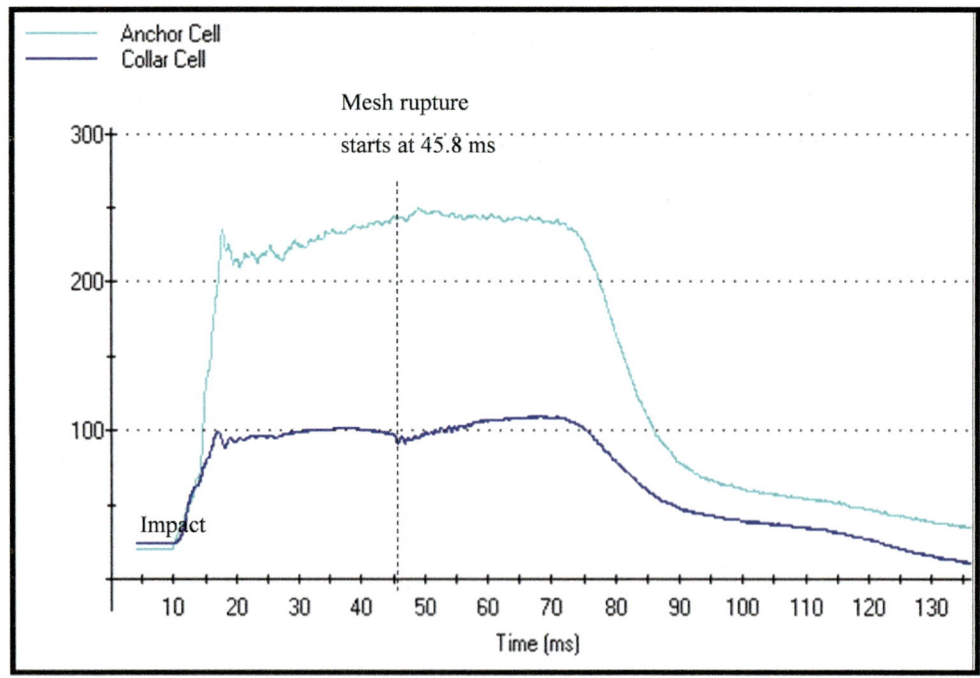

FIGURE 11.35 Anchor and collar load cell responses for a fully cement encapsulated decoupled threaded bar with 5.6 mm welded wire mesh (scheme #202).

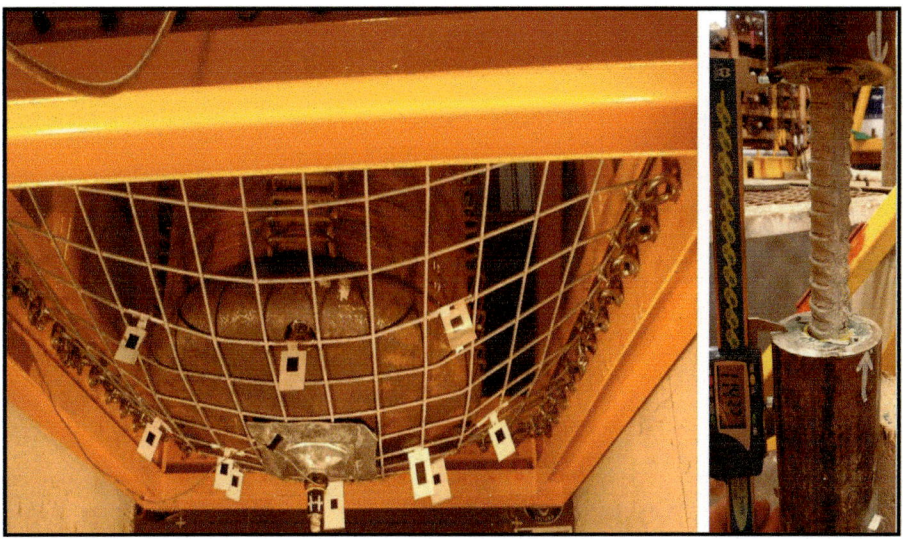

FIGURE 11.36 Details of post-test conditions for stable ground support scheme #202, with the reinforcement yielding within the decoupled region.

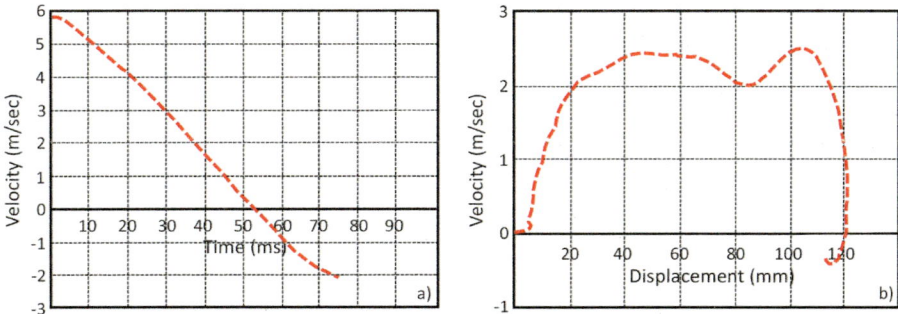

FIGURE 11.37 Velocity of (a) loading mass and (b) reinforcement loading across simulated discontinuity (cement-encapsulated decoupled threaded bar and welded wire mesh—scheme #202, which attained stability).

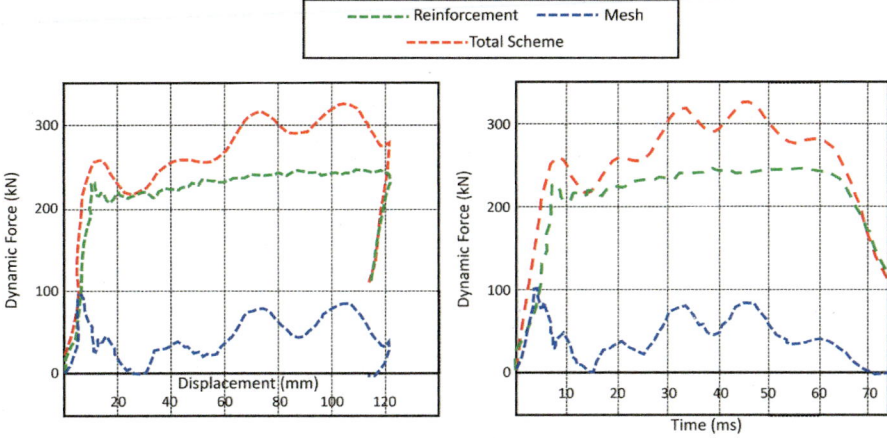

FIGURE 11.38 Reinforcement, mesh and ground support scheme dynamic response for sample #202 (cement-encapsulated decoupled threaded bar and welded wire mesh, which attained stability).

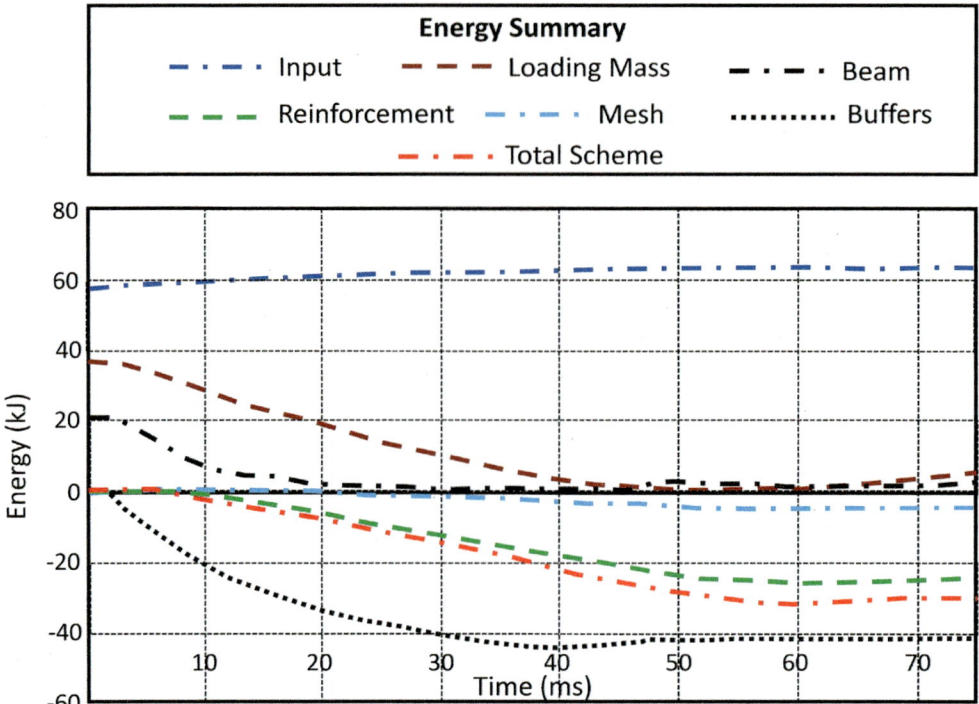

FIGURE 11.39 Energy balance for stable ground support scheme #202 (cement-encapsulated decoupled threaded bar and welded wire mesh).

11.4.3.4 Decoupled Posimix and Woven Mesh

The decoupled threaded bar was also tested in conjunction with G80/4 woven mesh (Figure 11.40). Figure 11.41 shows the anchor and collar load cell responses of scheme #200. The beam impacted the buffers at 0.009 seconds, and the anchor load rapidly reached 262 kN during the impact. The bar started to pull out at 50.8 ms. A total of 15 wires failed in the mesh. The combined system reached stability at 174.5 mm displacement (Figure 11.42).

Figure 11.43 shows the loading mass velocity and the reinforcement loading velocity across the simulated discontinuity, while Figure 11.44 shows the force versus time and displacement response for the ground support scheme. The energy balance is shown in Figure 11.45. The data indicates that the ground support scheme dissipated 23 kJ prior to reinforcement element rupture at 50.8 seconds.

Figure 11.46 shows the anchor and collar load cell responses for scheme #201. The beam impacted the buffers at 0.009 seconds, and the anchor load rapidly reached 251 kN after the impact. The decoupled reinforcement survived after a maximum displacement of 128 mm at the simulated discontinuity (Figure 11.47). A total of five wires ruptured in the mesh, but overall, the ground support scheme was stable.

Figure 11.48 shows the loading mass velocity and the reinforcement loading velocity across the simulated discontinuity, while Figure 11.49 shows the force versus displacement response for the ground support scheme. The energy balance is shown in Figure 11.50. The data indicate that the ground support scheme dissipated 30 kJ, achieved mostly through the decoupled reinforcement element.

As indicated in Table 11.2, several 3 m long decoupled Posimix elements were tested in conjunction with woven mesh. The elements had a substantial decoupled length of 1.4 m, which is in the upper range for rock bolt elements. Two types of mesh were used: A mild steel mesh provided by CODELCO and the high tensile G80/4 mesh provided by Geobrugg. Figure 11.51 presents the test

FIGURE 11.40 Details of pre-test set-up for ground support scheme #200 comprising one cement-encapsulated decoupled threaded bar and G80/4 woven mesh.

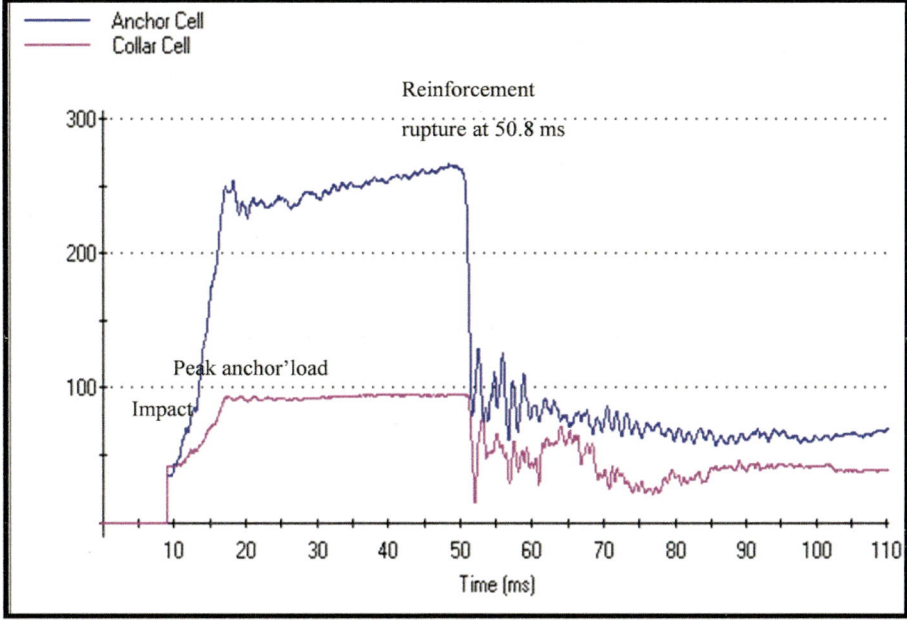

FIGURE 11.41 Anchor and collar load cell responses for a cement-encapsulated decoupled threaded bar and G80/4 woven mesh (scheme #200).

FIGURE 11.42 Details of post-test conditions for ground support scheme #200, which attained stability with localized mesh damage.

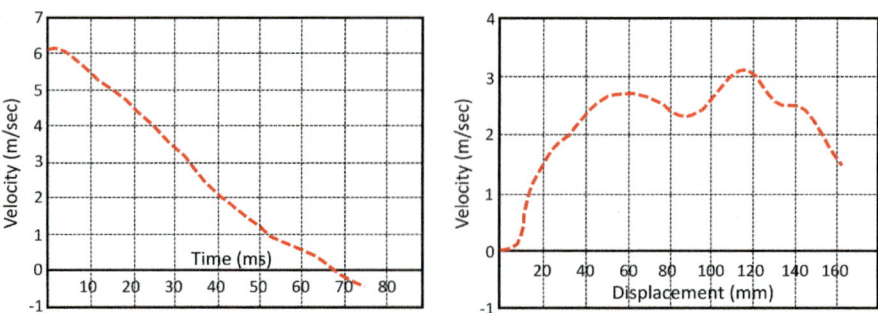

FIGURE 11.43 Velocity of (a) loading mass and (b) reinforcement loading across simulated discontinuity (cement-encapsulated decoupled threaded bar and G80/4 woven mesh—scheme #200, which attained stability).

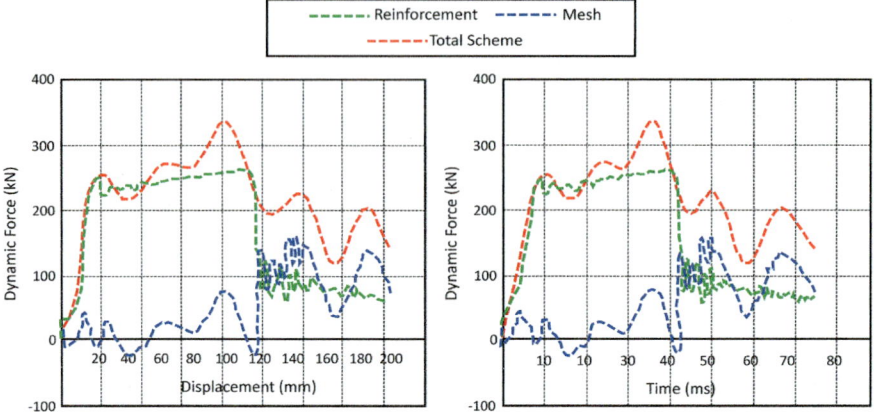

FIGURE 11.44 Reinforcement, mesh and ground support scheme dynamic response for scheme #200 (cement-encapsulated decoupled threaded bar and G80/4 woven mesh).

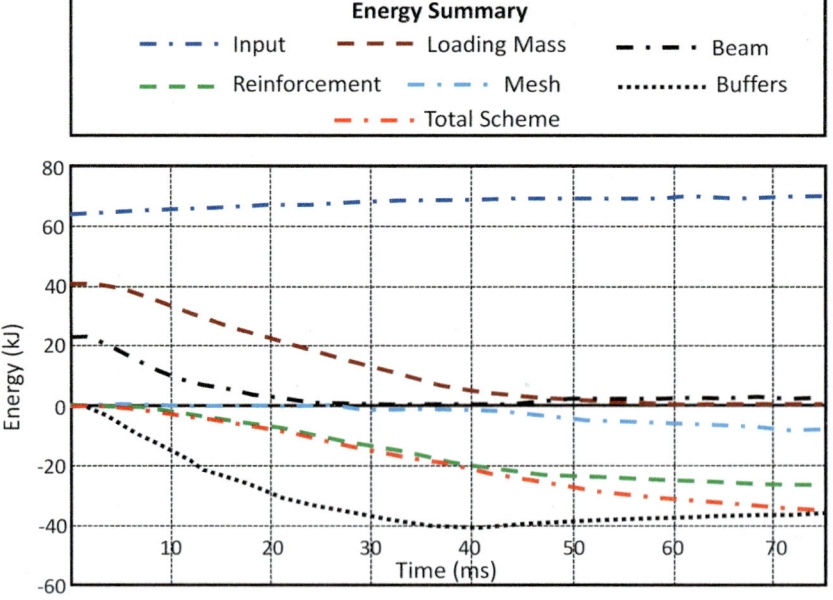

FIGURE 11.45 Energy balance for stable ground support scheme #200 (cement-encapsulated decoupled threaded bar and G80/4 woven mesh).

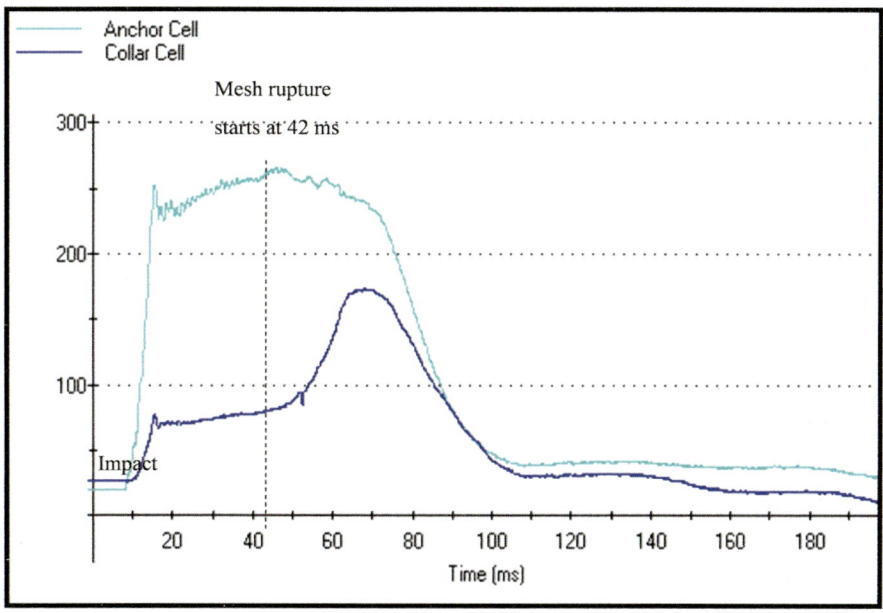

FIGURE 11.46   Anchor and collar load cell responses for a cement-encapsulated decoupled threaded bar and G80/4 woven mesh (scheme #201).

FIGURE 11.47   Details of post-test conditions for ground support scheme #201, which attained stability with localized mesh damage and reinforcement yielding within the decoupled region.

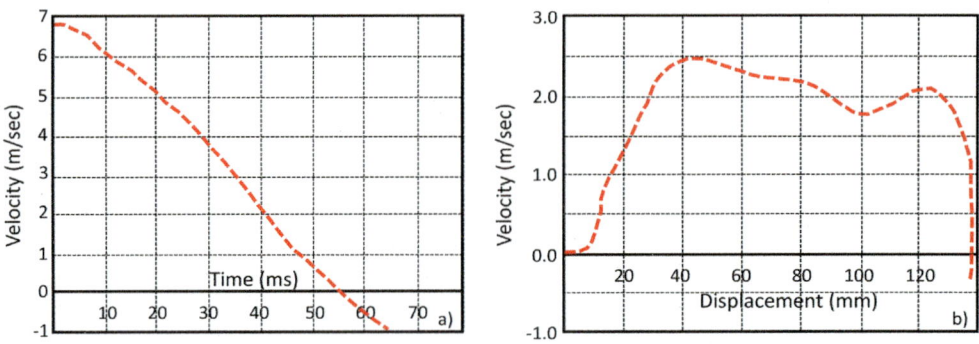

FIGURE 11.48 Velocity of (a) loading mass and (b) reinforcement loading across simulated discontinuity (cement-encapsulated decoupled threaded bar and G80/4 woven mesh—scheme #201, which attained stability).

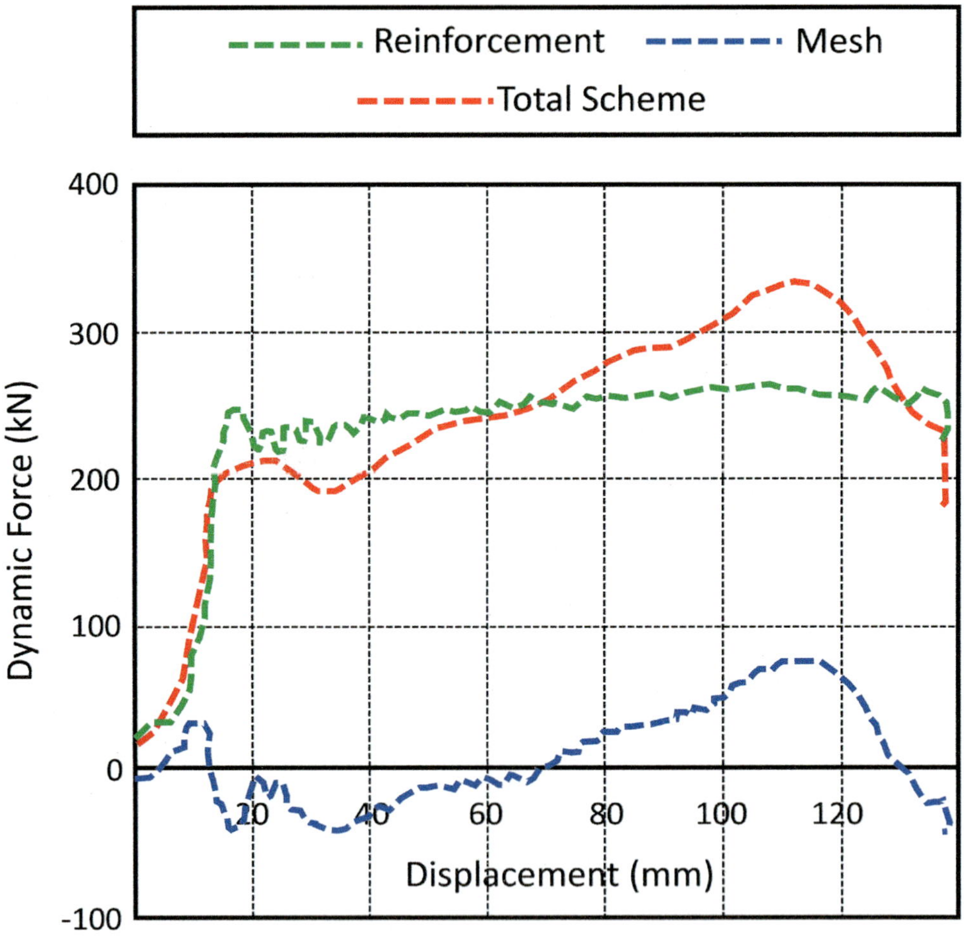

FIGURE 11.49 Reinforcement, mesh and ground support scheme dynamic response for scheme #201 (cement-encapsulated decoupled threaded bar and G80/4 woven mesh, which attained stability).

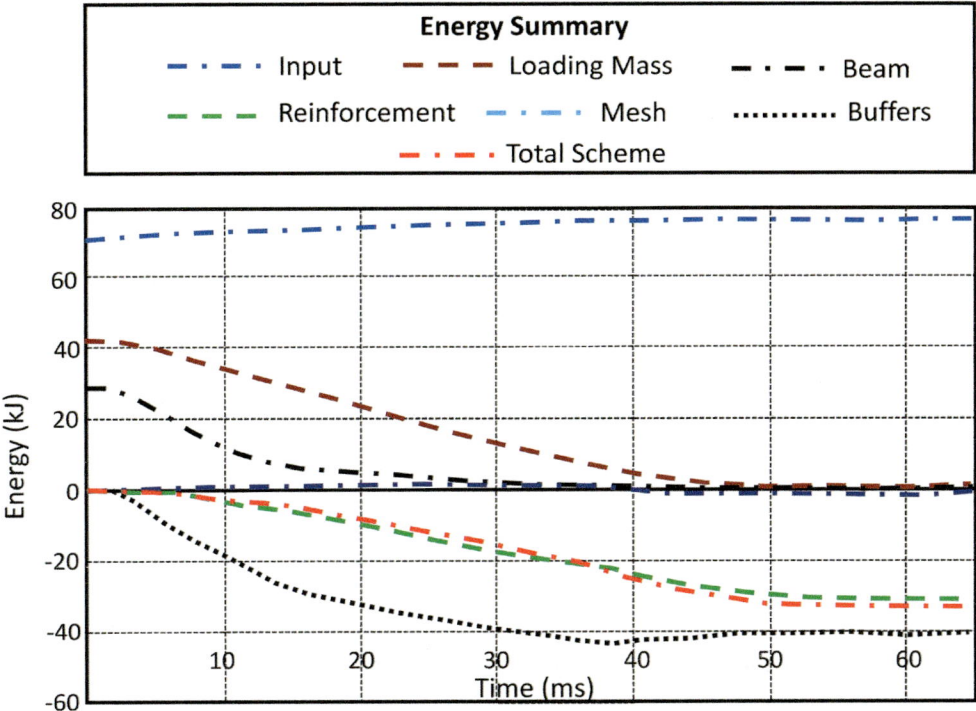

FIGURE 11.50 Energy balance for stable ground support scheme #201 (cement-encapsulated decoupled threaded bar and G80/4 woven mesh).

FIGURE 11.51 Details of ground support scheme #231 showing (a) pre-test set-up and (b) stable scheme with localized mesh damage and the reinforcement yielding within the decoupled region.

details for scheme #231, while Figure 11.52 shows the anchor and collar load cell responses. The beam impacted the buffers at 0.03 seconds, and the anchor load rapidly reached 301 kN after the impact. The reinforcement was stable after 142 mm of displacement at the simulated discontinuity. A single wire was ruptured in the mesh, and the scheme was deemed to be stable.

Figure 11.53 shows the force versus displacement response for ground support scheme #231. The energy balance is shown in Figure 11.54. The data indicate that the ground support scheme dissipated 29 kJ, mostly through the decoupled reinforcement element. The division of support and reinforcement performance is depicted in Table 11.4.

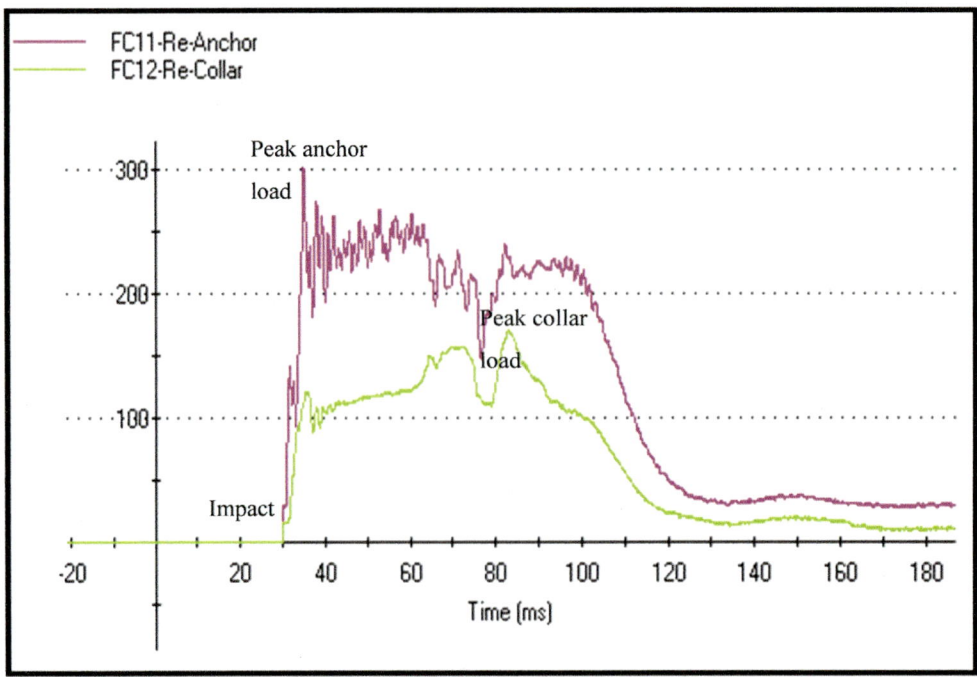

FIGURE 11.52 Anchor and collar load cell responses for a cement-encapsulated decoupled threaded bar and mild steel woven mesh (scheme #231).

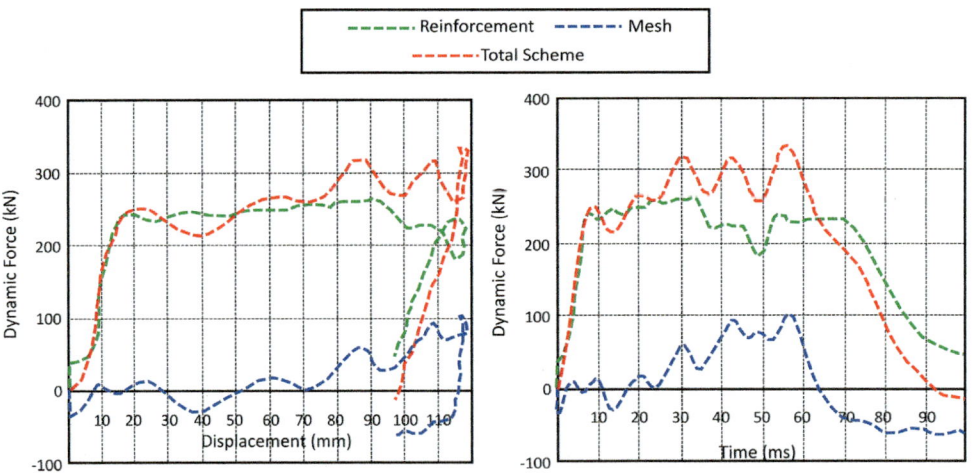

FIGURE 11.53 Reinforcement, mesh and ground support scheme dynamic response, sample #231 (cement-encapsulated decoupled threaded bar and mild steel woven mesh, which attained stability).

Figure 11.55 presents the test details for scheme #234, while Figure 11.56 shows the anchor and collar load cell responses. The beam impacted the buffers at 0.030 seconds, while the anchor and collar loads reached 350 kN and 155 kN, respectively. The reinforcement element was stable after 155 mm of displacement at the simulated discontinuity. A single wire ruptured in the mesh, and the combined scheme was stable.

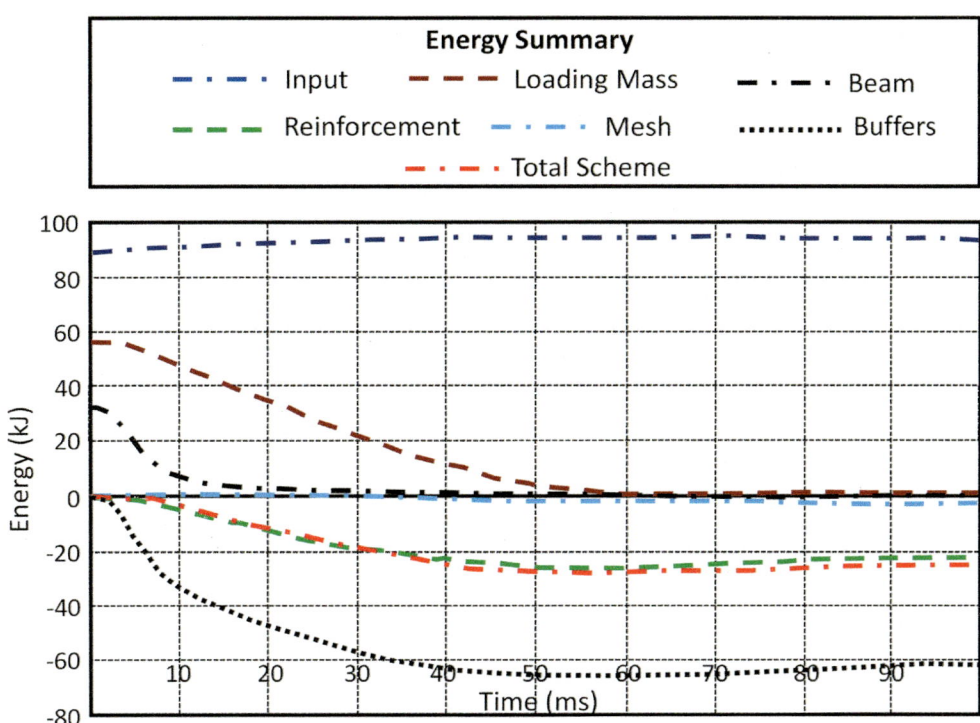

FIGURE 11.54 Energy balance for stable ground support scheme #231 (cement-encapsulated decoupled threaded bar and mild steel woven mesh).

TABLE 11.4

Quantification of Performance Parameters—Scheme #231.

Reinforcement System	
Maximum displacement at the simulated discontinuity	118 mm
Peak dynamic force on reinforcement element simulated discontinuity	262 kN
Maximum loading mass deceleration	14.3 g
Peak reinforced mass ejection velocity	3.7 m/s
Energy dissipated by reinforcement system	26.4 kJ
Reinforcement stable @	66.64 ms

Mesh System	
Maximum mesh displacement	100 mm
Peak dynamic force on mesh	102 kN
Maximum mesh deceleration	13.8 g
Peak reinforced mass ejection velocity	3.7 m/s
Energy dissipated by mesh system	2.8 kJ
Mesh ruptured @	36.8 ms

FIGURE 11.55 Details of ground support scheme #234 showing (a) pre-test set-up and (b) stable scheme with localized mesh damage and reinforcement yielding within the decoupled region.

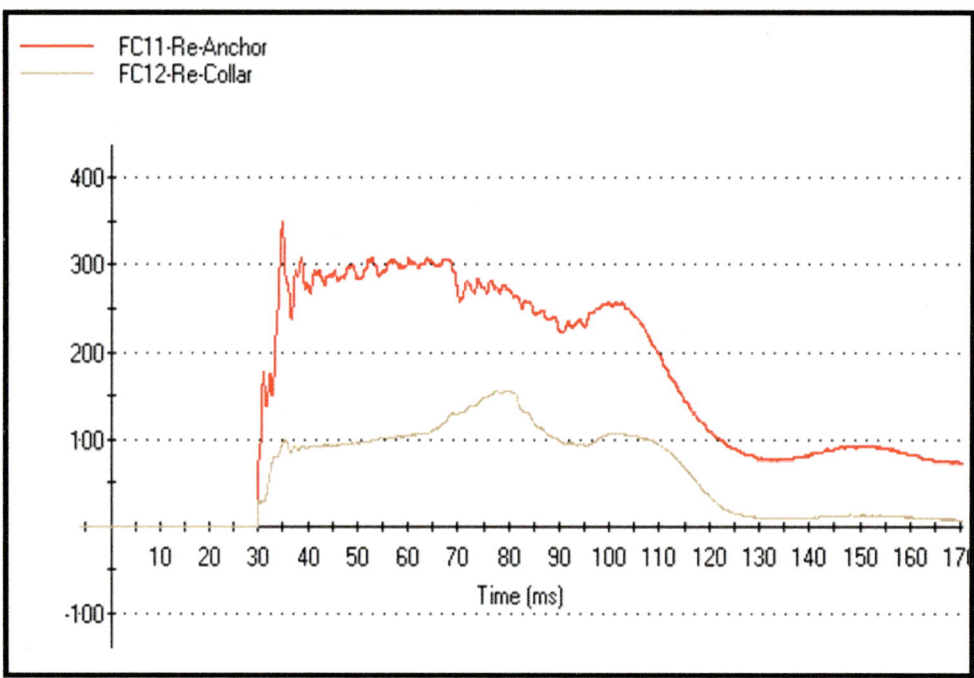

FIGURE 11.56 Anchor and collar load cell responses for a cement-encapsulated decoupled threaded bar and mild steel woven mesh (scheme #234).

Figure 11.57 shows the force versus displacement response for ground support scheme #234. The energy balance is shown in Figure 11.58. The data indicate that the ground support scheme dissipated 27.3 kJ, mostly through the decoupled reinforcement element. The breakdown of the support and reinforcement performance is depicted in Table 11.5.

Figure 11.59 presents the test details for scheme #235, while Figure 11.60 shows the anchor and collar load cell responses. The beam impacted the buffers at 0.030 seconds, while the anchor and collar loads reached 310 kN and 230 kN, respectively. The reinforcement element was stable after

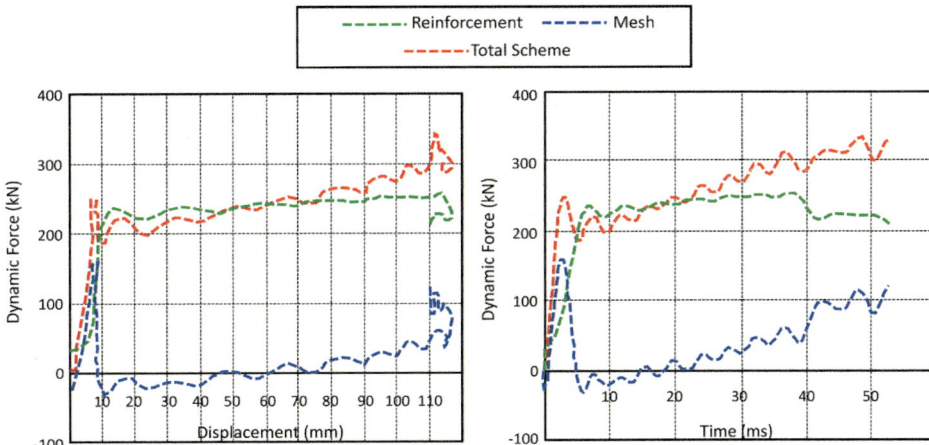

FIGURE 11.57 Reinforcement, mesh and ground support scheme dynamic response for scheme #234 (cement-encapsulated decoupled threaded bar and mild steel woven mesh, which attained stability).

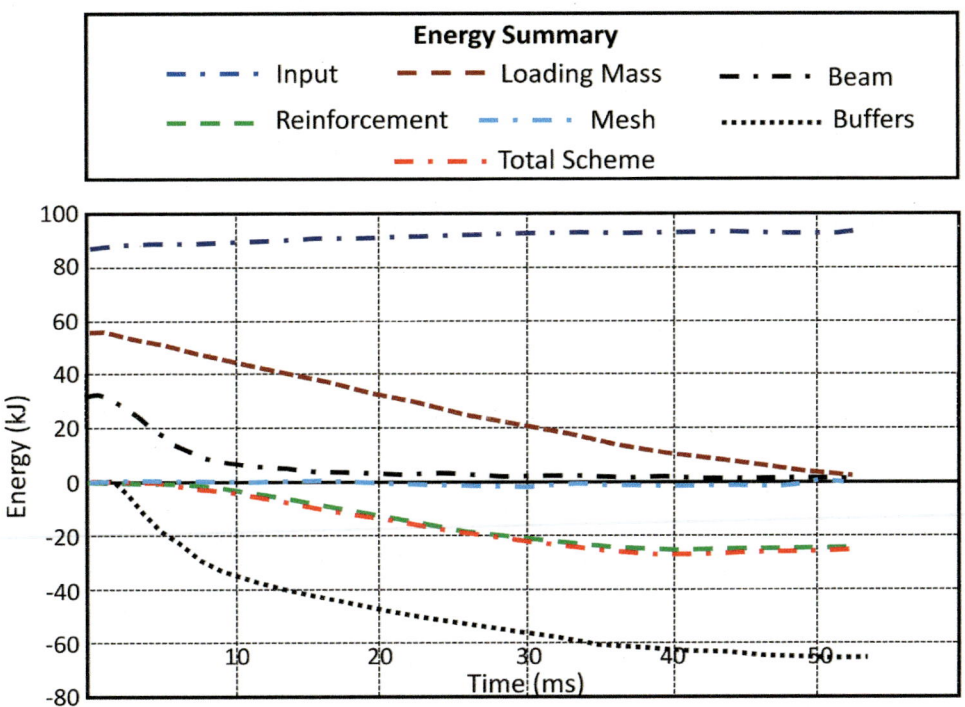

FIGURE 11.58 Energy balance for stable ground support scheme #234 (cement-encapsulated decoupled threaded bar and mild steel woven mesh).

149 mm of displacement at the simulated discontinuity. A total of nine wires ruptured in the mesh, and the combined scheme was stable.

Figure 11.61 shows the force versus displacement response for ground support scheme #235. The energy balance is shown in Figure 11.62. The data indicate that the ground support scheme dissipated 30.9 kJ, mostly through the decoupled reinforcement element. The division of support and reinforcement performance is depicted in Table 11.6.

TABLE 11.5

Quantification of Performance Parameters—Scheme #234.

Reinforcement System	
Maximum displacement at the simulated discontinuity	111 mm
Peak dynamic force on reinforcement element simulated discontinuity	255 kN
Maximum loading mass deceleration	14.7 g
Peakreinforced mass ejection velocity	3.5 m/s
Energy dissipated by reinforcement system	26.3 kJ
Reinforcement stable @	68.08 ms

Mesh System	
Maximum mesh displacement	111 mm
Peak dynamic force on mesh	158 kN
Maximum mesh deceleration	11.8 g
Peak reinforced mass ejection velocity	3.5 m/s
Energy dissipated by mesh system	1.0 kJ
Mesh ruptured @	55.3 ms

FIGURE 11.59 Details of ground support scheme #235 showing (a) pre-test set-up and (b) stable scheme with localized mesh damage and reinforcement yielding within the decoupled region.

Figure 11.63 presents the test details for scheme #236, while Figure 11.64 shows the anchor and collar load cell responses. The beam impacted the buffers at 0.031 seconds, while the anchor and collar loads reached 300 kN and 227 kN, respectively. The reinforcement element was stable after 149 mm of displacement at the simulated discontinuity. A total of four wires ruptured in the mesh, and the combined scheme was stable.

Figure 11.65 shows the force versus displacement response for ground support scheme #236. The energy balance is shown in Figure 11.66. The data indicate that the ground support scheme dissipated 37.6 kJ, mostly through the decoupled reinforcement element. The division of support and reinforcement performance is depicted in Table 11.7.

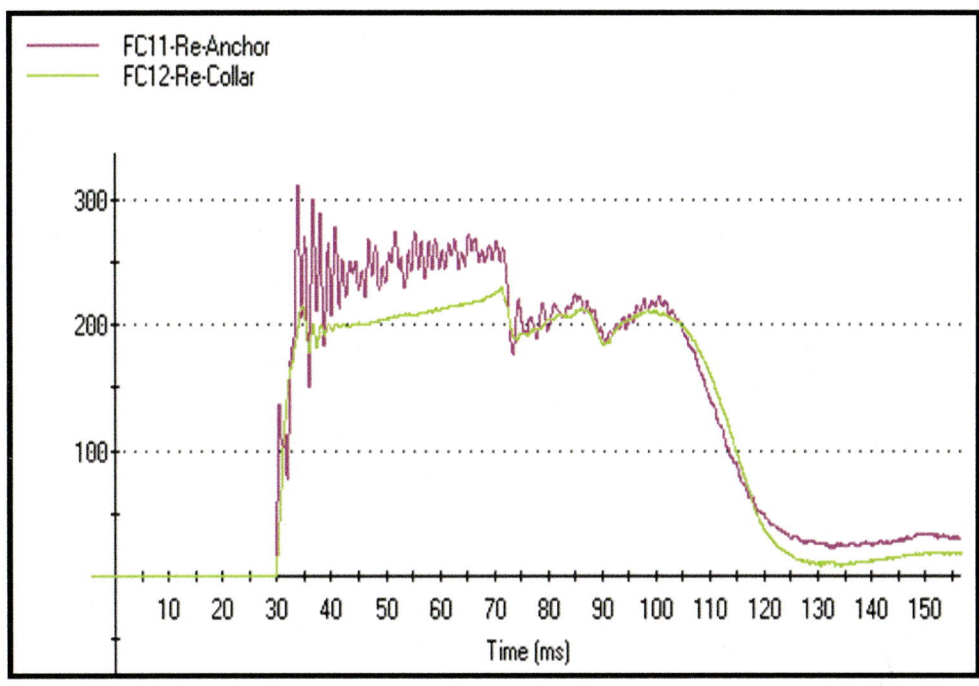

FIGURE 11.60 Anchor and collar load cell responses for a cement-encapsulated decoupled threaded bar and G80/4 woven mesh (scheme #235).

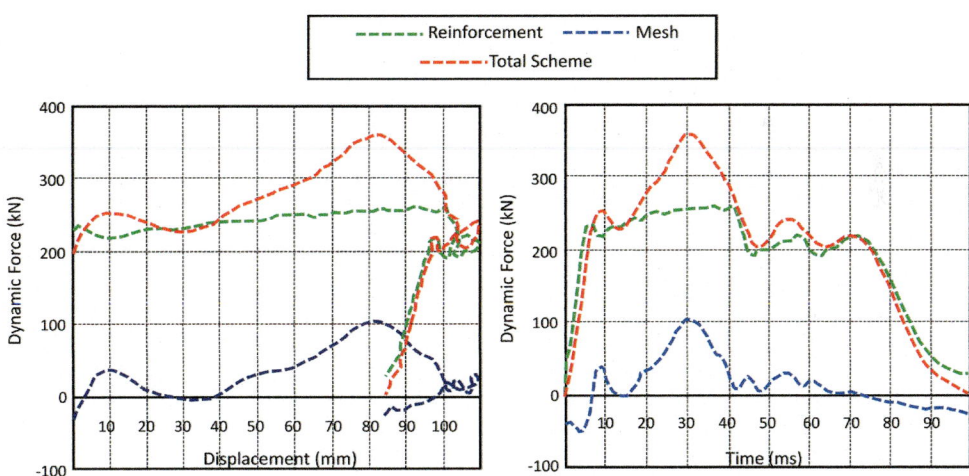

FIGURE 11.61 Reinforcement, mesh and ground support scheme dynamic response for scheme #235 (cement-encapsulated decoupled threaded bar and G80/4 woven mesh, which attained stability).

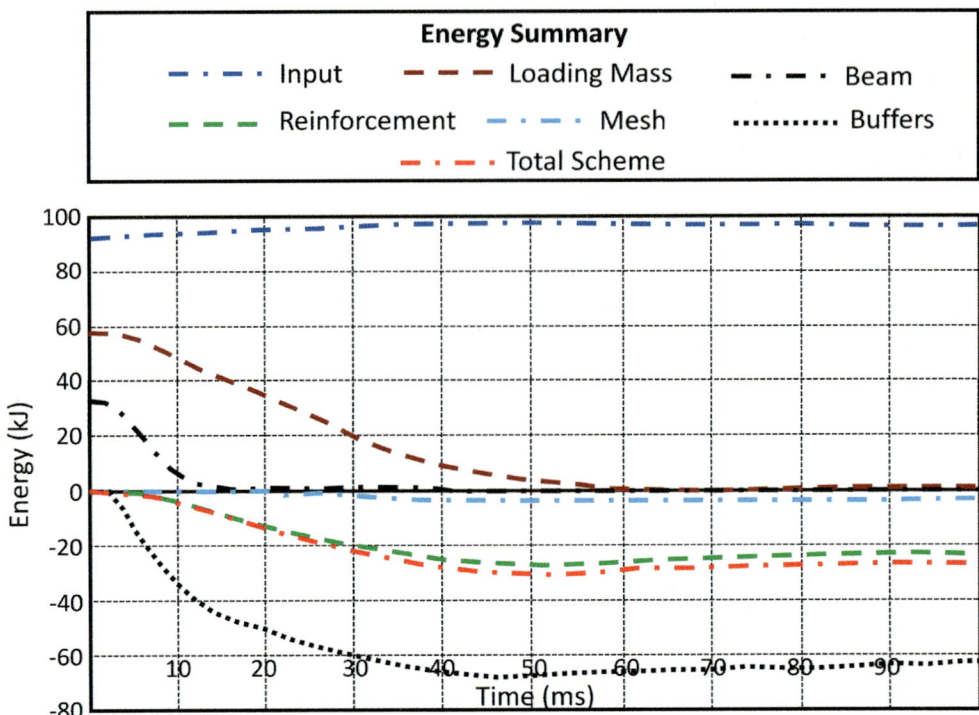

FIGURE 11.62 Energy balance for stable ground support scheme #235 (cement-encapsulated decoupled threaded bar and G80/4 woven mesh).

TABLE 11.6

Quantification of Performance Parameters—Scheme #235.

Reinforcement System

Maximum displacement at the simulated discontinuity	110 mm
Peak dynamic force on reinforcement element simulated discontinuity	260 kN
Maximum loading mass deceleration	15.5 g
Peakreinforced mass ejection velocity	4.8 m/s
Energy dissipated by reinforcement system	27.4 kJ
Reinforcement stable @	73.84 ms

Mesh System

Maximum mesh displacement	86 mm
Peak dynamic force on mesh	102 kN
Maximum mesh deceleration	24.4 g
Peak reinforced mass ejection velocity	4.8 m/s
Energy dissipated by mesh system	3.5 kJ
Mesh ruptured @	33.0 ms

FIGURE 11.63 Details of ground support scheme #236 showing (a) pre-test set-up and (b) stable scheme with localized mesh damage and reinforcement yielding within the decoupled region.

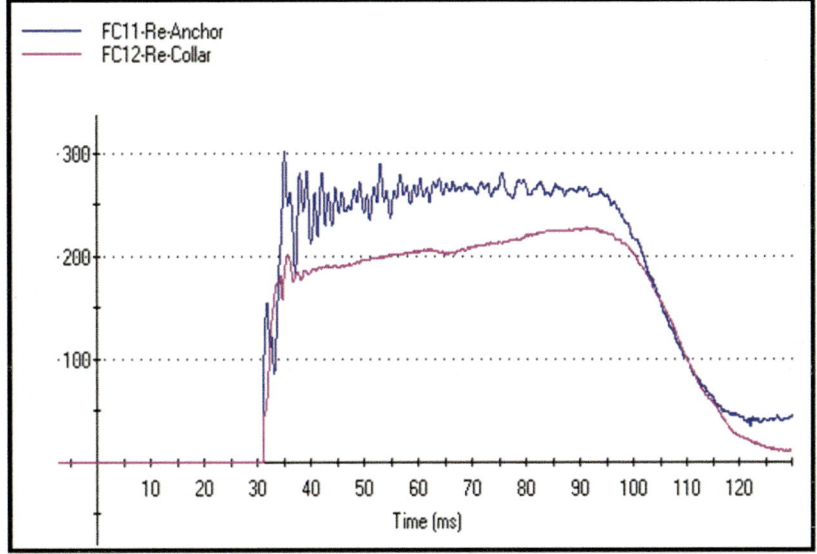

FIGURE 11.64 Anchor and collar load cell responses for a cement-encapsulated decoupled threaded bar and G80/4 woven mesh (scheme #236).

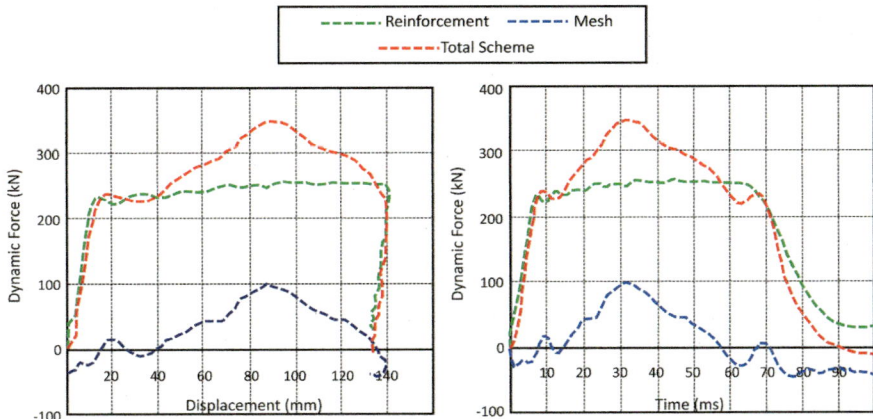

FIGURE 11.65 Reinforcement, mesh and ground support scheme dynamic response for scheme #236 (cement-encapsulated decoupled threaded bar and G80/4 woven mesh, which attained stability).

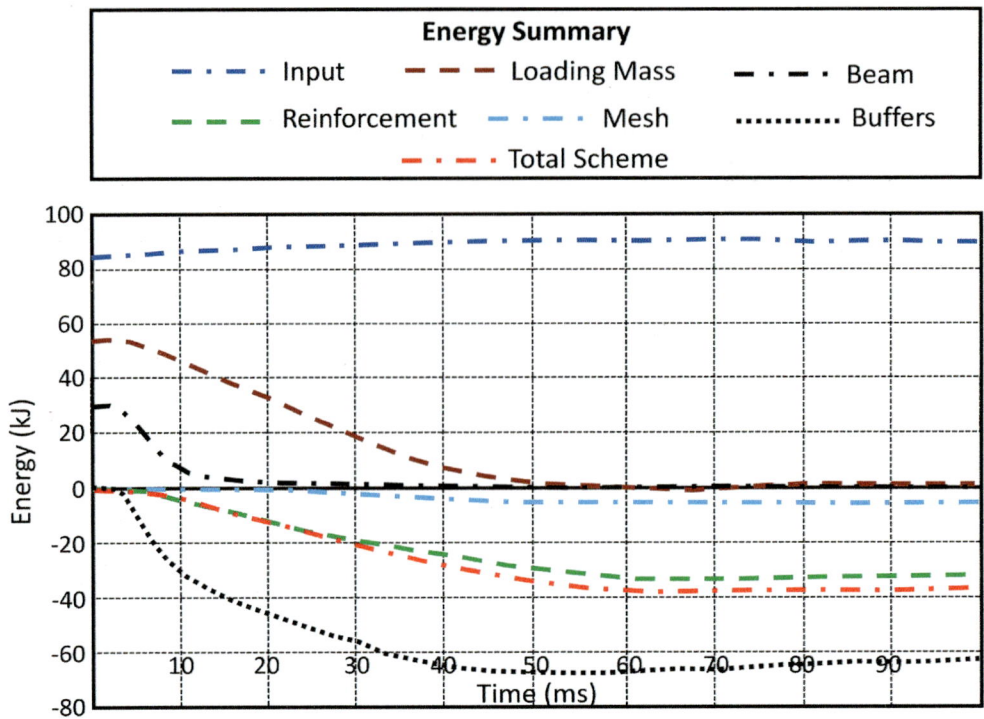

FIGURE 11.66 Energy balance for stable ground support scheme #236 (cement-encapsulated decoupled threaded bar and G80/4 woven mesh).

TABLE 11.7

Quantification of Performance Parameters—Scheme #236.

Reinforcement System	
Maximum displacement at the simulated discontinuity	141 mm
Peak dynamic force on reinforcement element simulated discontinuity	256 kN
Maximum loading mass deceleration	15.0 g
Peakreinforced mass ejection velocity	3.6 m/s
Energy dissipated by reinforcement system	32.9 kJ
Reinforcement stable @	65.52 ms

Mesh System	
Maximum mesh displacement	90 mm
Peak dynamic force on mesh	97 kN
Maximum mesh deceleration	13.2 g
Peak reinforced mass ejection velocity	3.6 m/s
Energy dissipated by mesh system	4.7 kJ
Mesh ruptured @	33.0 ms

11.4.4 Summary of Energy Dissipation

A comparison of the dynamic force-displacement responses for the ground support schemes assessed in Test Programs 1 and 2 are given in Figures 11.67–11.69, respectively. The tests with ruptured or pulled-out threaded bars are marked with end stars. The schemes using decoupled cement-encapsulated threaded bars clearly elongated more than those using fully coupled threaded bars. As expected, the displacement was controlled by the length of the decoupled regions. The peak dynamic forces for all schemes were in the range of 300–410 kN.

The energy dissipated in relation to the deformation at failure for all the ground support schemes is illustrated in Figure 11.70 (Villaescusa et al., 2016). The graph is divided into regions described as low, medium, high and very high energy dissipation according to the typical rock mass demand for ground support scheme design described earlier in Chapter 6.

The results presented here clearly indicate that the scheme displacement has been largely controlled by the performance and response of the reinforcement elements. The damage to the mesh has been clearly affected by the boundary conditions, which include very tight mesh installation and very small loading areas. However, it is expected that a much larger displacement would be experienced by *in situ* ground support schemes (Figure 11.71). Observations indicate that this is due to the sequential failure of reinforcement elements in which the load transfer from ruptured/damaged elements to other undamaged elements nearby is achieved by the continuity of a high-capacity mesh.

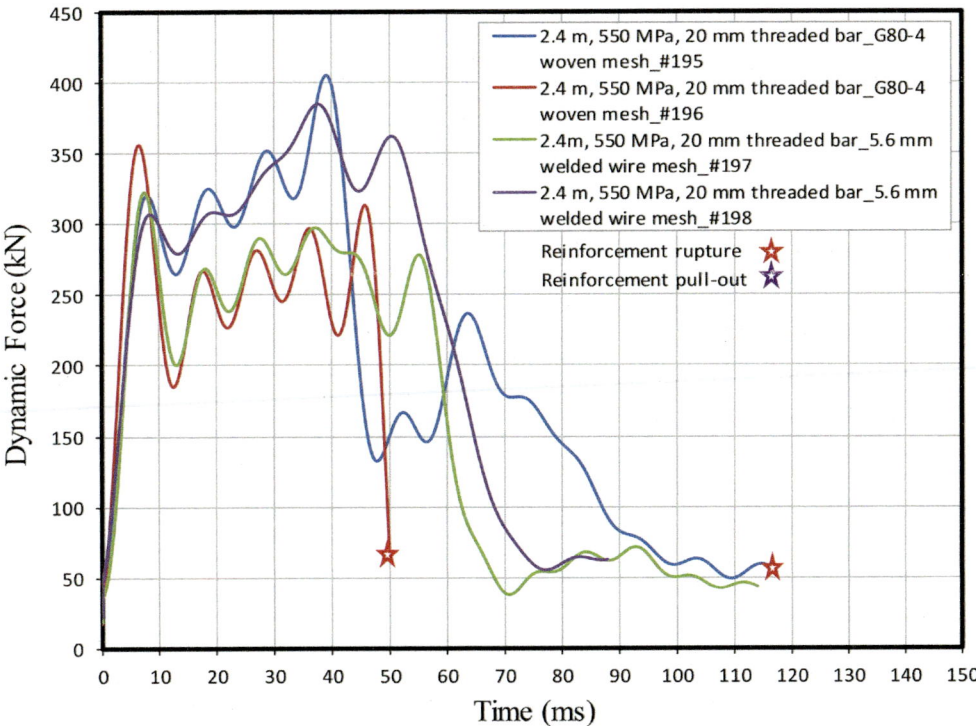

FIGURE 11.67 A comparison of dynamic force-time responses for ground support schemes consisting of fully coupled, cement-encapsulated threaded bars in conjunction with a weld mesh and G80/4 woven mesh.

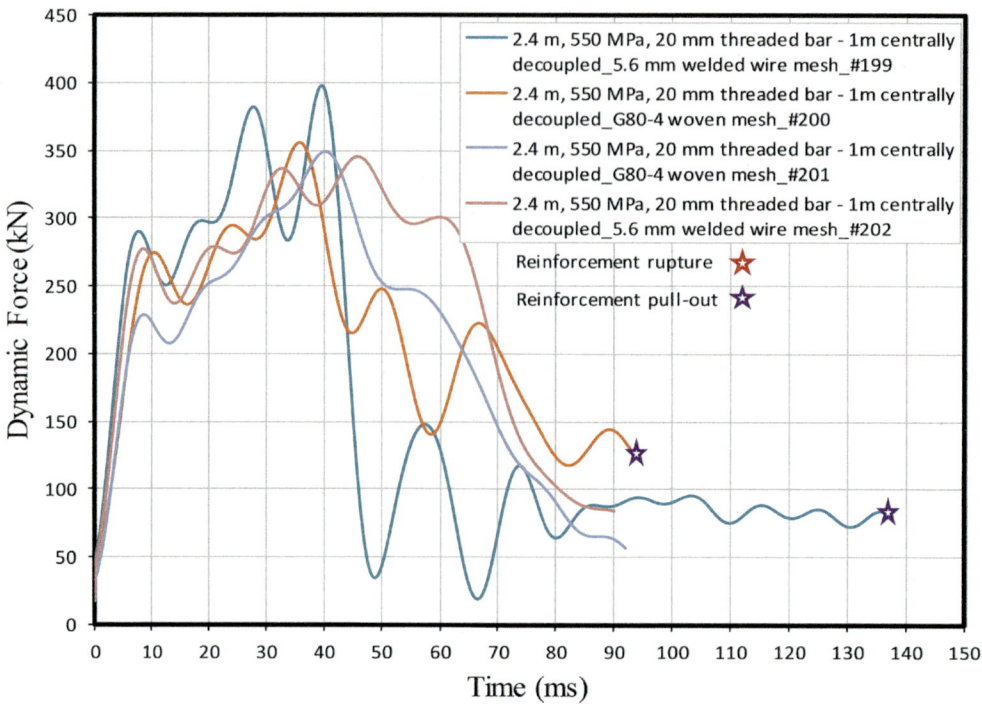

FIGURE 11.68 A comparison of dynamic force-time responses for ground support schemes consisting of decoupled (1-m region), cement-encapsulated threaded bars in conjunction with a weld mesh and G80/4 woven mesh.

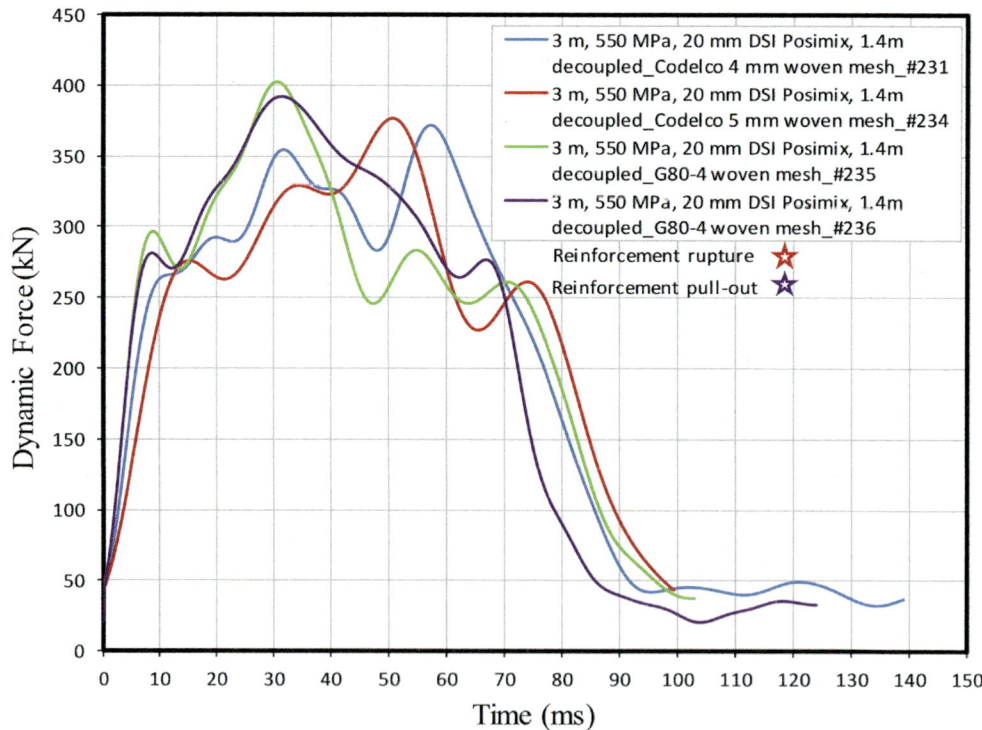

FIGURE 11.69 A comparison of dynamic force-time responses for ground support schemes consisting of decoupled (1.4-m region) cement-encapsulated threaded bars in conjunction with a woven mesh.

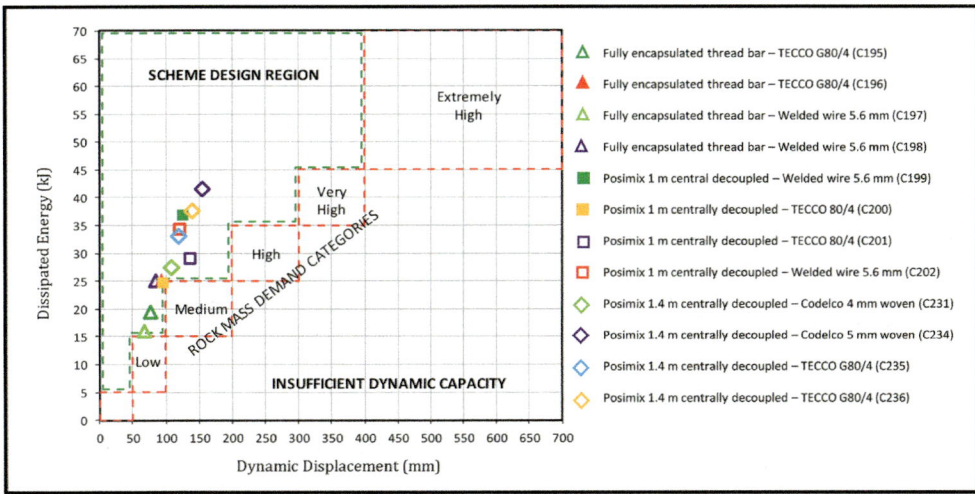

FIGURE 11.70 Energy dissipation of ground support schemes comprising cement-encapsulated threaded bar reinforcement and mesh support.

FIGURE 11.71 Example of large deformation due to sequential failure of reinforcement elements connected by a high-capacity mesh.

11.5 LARGE-SCALE TESTING OF FULL-SCALE SCHEMES

On-going research at the WA School of Mines has led to the development of a large-scale dynamic test facility capable of input energies of up to 600 kJ (Roth, 2023). The set-up involves simulation of the rock mass, as well as the installation of a pattern (typically spaced at 1m x 1m) of reinforcement consisting of 9 full-scale elements, installed in conjunction with mesh reinforced shotcrete as shown in Figure 11.72. The reinforcement elements are installed within split-pipes that simulate unstable collar and stable toe anchor regions. The dynamic impact load is distributed over an area

FIGURE 11.72 View (grid 1m x 1m) from under the dynamic test facility showing the surface plates of 9 decoupled threaded bars installed in conjunction with G80/4.5 mesh and 100 mm thick layer of shotcrete (Roth, 2023).

of more than 6 m2 with the main area of impact being the centre element followed by four elements immediately adjacent to it (Dice-5) as well as the four corner elements.

The facility is fully instrumented with load cells, accelerometers, high speed cameras and a laser scanner which allows an analysis of force-time response of the reinforcement elements and the surface support. The dynamic displacements from the high-speed camera footage and the force-displacement plots can be used to calculate the dissipated energies (Roth, 2023).

Figure 11.73 shows the result of a 193-kJ dynamic impact upon 9 cement encapsulated decoupled threaded bars (2.4 m long, 20 mm diameter) having a 1 m decoupled region. The scheme was stable with localized damage near the centre element, which was broken. Shotcrete ejection was measured at velocities more than 7 m/s and load transfer from the broken centre bolt to the adjacent bolts was successfully achieved via the surface support (Figure 11.74). The loading area was defined by a damage radius where more than 50 mm of deformation were experienced.

FIGURE 11.73 Views under the facility showing a stable ground support scheme which dissipated 193 kJ (Roth, 2023).

Figure 11.75 shows a plan view of the reinforcement pattern where the ruptured centre bolt is indicated with a red dot. Similarly, the adjacent bolts (Dice-5) are shown in blue dots with the

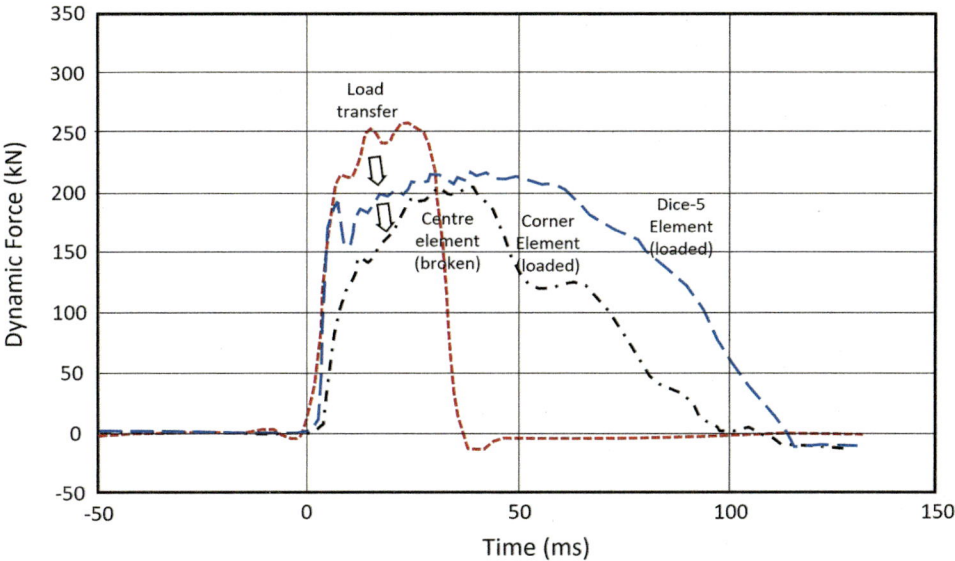

FIGURE 11.74 Experimental results showing a sequential load transfer from a ruptured centre element to elements immediately adjacent (Dice-5) and then to the corner elements further away from the central loading point (Roth, 2023).

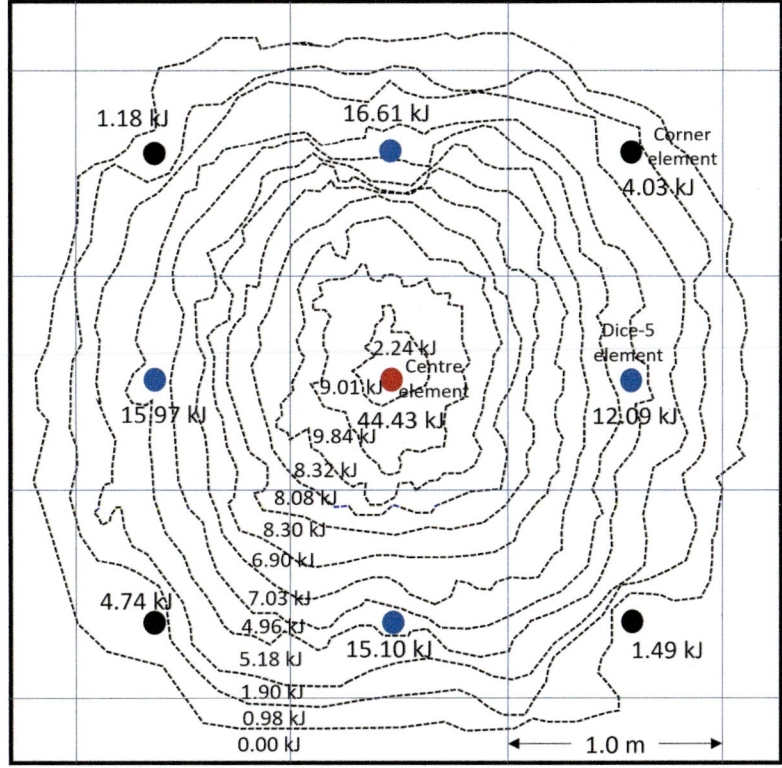

FIGURE 11.75 Experimental results showing a sequential load transfer from a ruptured centre element to elements immediately adjacent (Dice-5) and then to the corner elements further away from the central loading point (Roth, 2023).

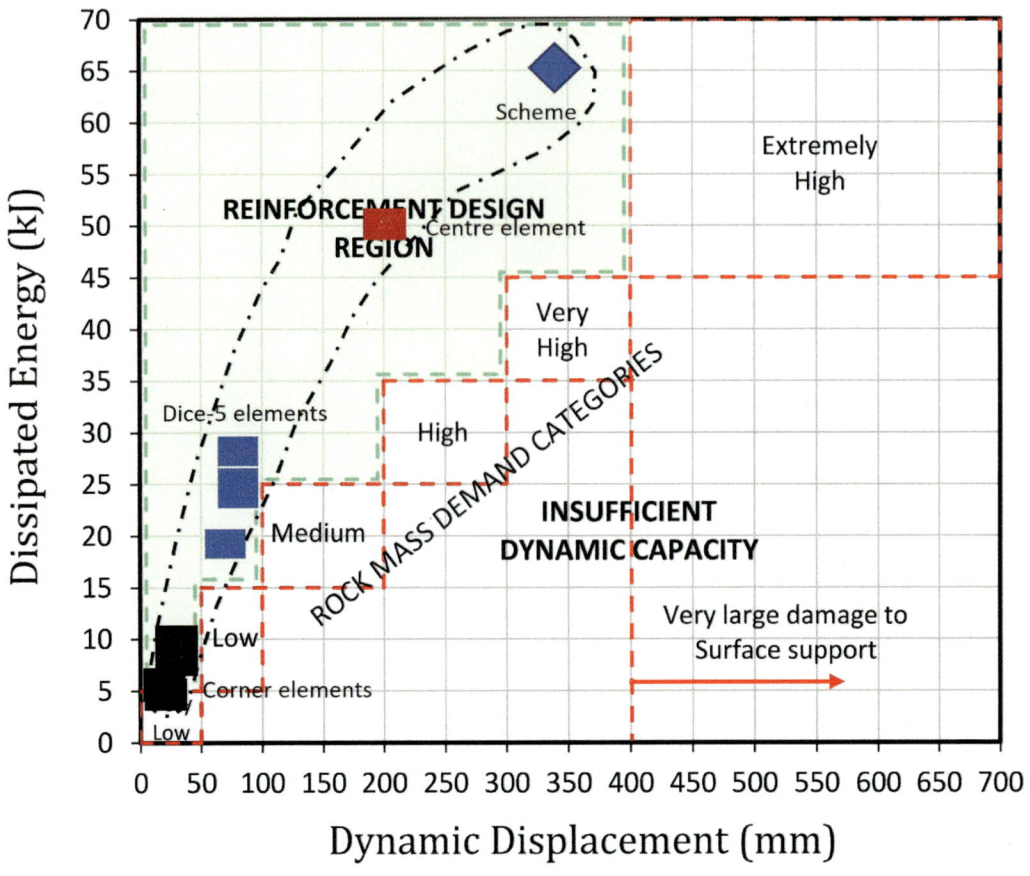

FIGURE 11.76 Experimental results showing a sequential load transfer from a ruptured centre element to elements immediately adjacent (Dice-5) and then to the corner elements further away from the central loading point (Roth, 2023).

corner holes shown in black dots. The dissipated energy at each element location is shown with a number and indicates symmetrical loading with respect to the central impact point. A total of 115.6 kJ was dissipated by the decoupled reinforcement elements. Additionally, and based on the results from dynamic testing of shotcrete reinforced with woven mesh shown previously in Chapter 10, the response of the surface support the energy the energy dissipation of the surface support was distributed over the full area of deformation and shown by contour lines. The assumption is that mesh reinforced shotcrete behaves linear elastic up to 50 mm of deformation and then ideally plastic on the same load level.

Roth (2023) combined the dissipated energy values of the reinforcement elements and the support layers using a 1.0 m x 1.0 m grid. The ground support scheme dissipated 66.1 kJ/m2 at a dynamic displacement of 340 mm and remained stable. The surrounding square meters (Dice-5 elements) dissipated 19.2–27.7 kJ/m2 at displacements of 70-80 mm and the corner areas dissipated 4.6–8.8 kJ/m2 at displacements of 20-30 mm (Figure 11.76).

12 Reinforced Block Analysis

12.1 INTRODUCTION

Previous chapters of this book have explained that the shapes and sizes of potential instabilities are usually defined by combinations of discontinuities within the rock mass and the free faces of an excavation. The discontinuities may exist prior to the creation of the excavation or may initiate and propagate due to stress redistribution after the construction of an excavation (Figure 12.1). The forces causing detachment of potential instabilities may be classified as being imposed statically or dynamically. Consequently, the responses from reinforcement systems are required to resist both types of imposed forces. Also, the methods of analysis for reinforced blocks need to account for both statically and dynamically applied loadings.

12.2 DESCRIPTION OF THE PROBLEM

The general problem considered here is a reinforced block, as shown in Figure 12.2. While the tetrahedral block shown is one of the simplest block shapes that can be analysed, it is important to note that although not common, a block may have many more than just four faces. A feature of the problem often overlooked is that the reinforcement systems may involve different devices, and their orientations may vary widely. Both these possibilities have important implications for analysis, where the reinforcement responses need to relate force to displacement and, more specifically, when energy dissipation is required if the destabilizing forces are imposed dynamically. It is also important to note that the block has six degrees of freedom of movement: three translations and three rotations. In most analyses, this is also ignored.

The analysis of a reinforced block requires recognition that the reinforcement system response is usually generated at the block face adjacent to a discontinuity between the unstable block and the assumed stable zone of the rock mass, as shown in Figure 12.2. The reinforcement system described in the following sections will be assumed to be a steel element encapsulated within a cement grout annulus within a borehole. Readers are referred to earlier chapters for details of different reinforcement systems and their generic classifications.

12.3 REINFORCEMENT RESPONSE AT A BLOCK FACE

The reinforcement response at the block face is influenced by the different reinforcement configurations on each side of the discontinuity. For example, Figure 12.3 shows that the response within the block is influenced by the presence of a plate and fixture at the collar of the reinforcement system, whereas the response in the anchor zone relies only on the load transfer between the reinforcement element and borehole wall provided by the annulus of grout.

Thompson et al. (2011) have described how the separate collar and anchor responses can be derived from Figure 12.4 and Figure 12.5, respectively. This can then be combined to form the force-displacement response at the block face, as shown in Figure 12.6. This particular example represents an anchor length of 1 m combined with a collar length of 2 m and demonstrates that the anchor length limits the force capacity to be less than the element strength.

This section implies that a database of reinforcement anchor and collar responses is required to enable formal analyses of reinforced blocks of rock where each reinforcement system may have different combinations of anchor and collar lengths, as shown previously in Figure 12.2. In fact,

DOI: 10.1201/9781003357711-12

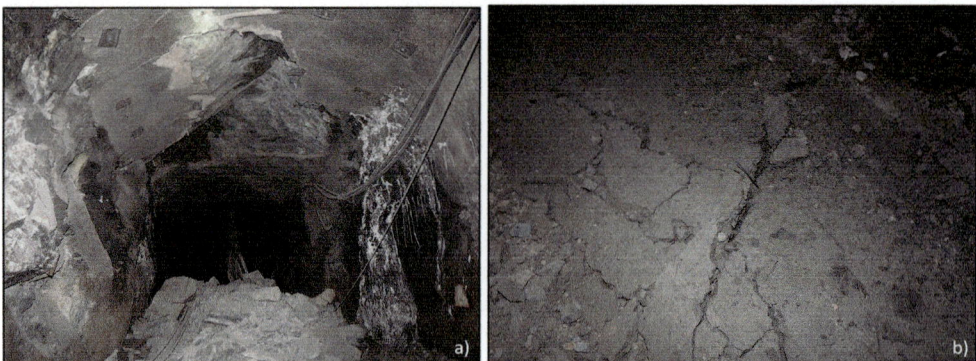

FIGURE 12.1   Examples of instability occurring several years after the creation of the excavation: (a) wedge failure and (b) floor rupture.

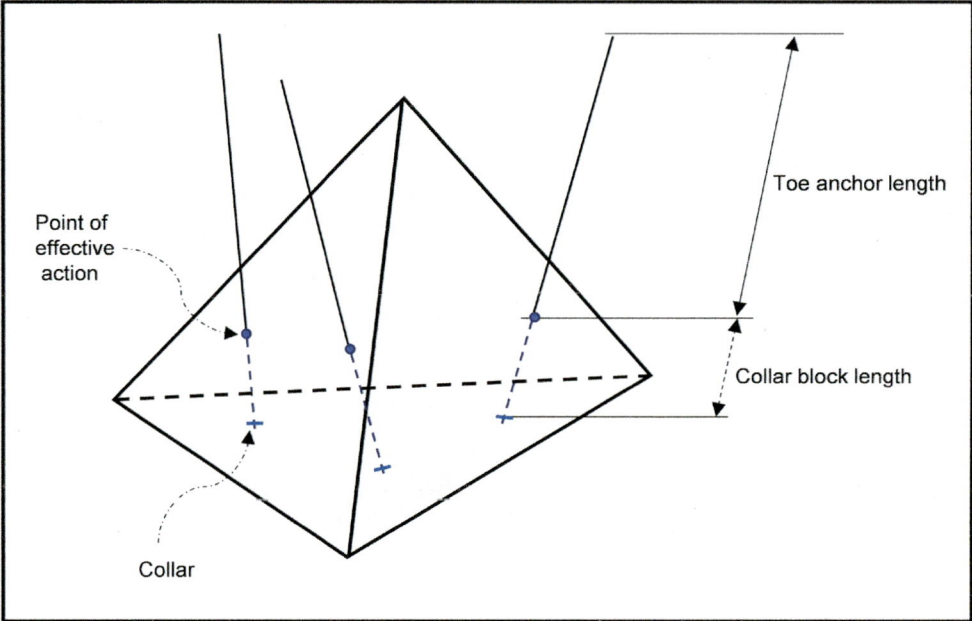

FIGURE 12.2   Arbitrarily-shaped block with intersecting reinforcement systems (after Windsor and Thompson, 1997).

two databases are required: one applicable to responses of reinforcement systems subjected to static loading and the other applicable to responses of reinforcement systems subjected to dynamic loading.

12.4   REINFORCEMENT DATABASES

Unfortunately, for many reinforcement systems, the yield and ultimate force capacities of the 'element' may be the only reliable parameters available from the manufacturer. Note that the term 'element' was developed by Windsor and Thompson (1993) to describe the axial bar, rod, bolt, strand or tube component of a rock reinforcement system. Consequently, it is then necessary for a

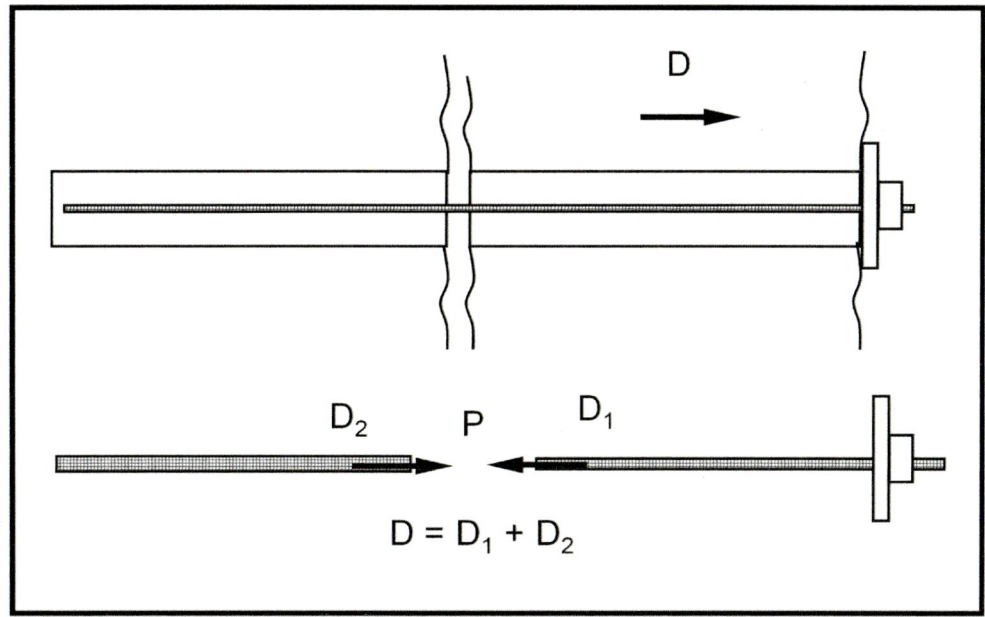

FIGURE 12.3   Reinforcement response at the point of effective reinforcement action.

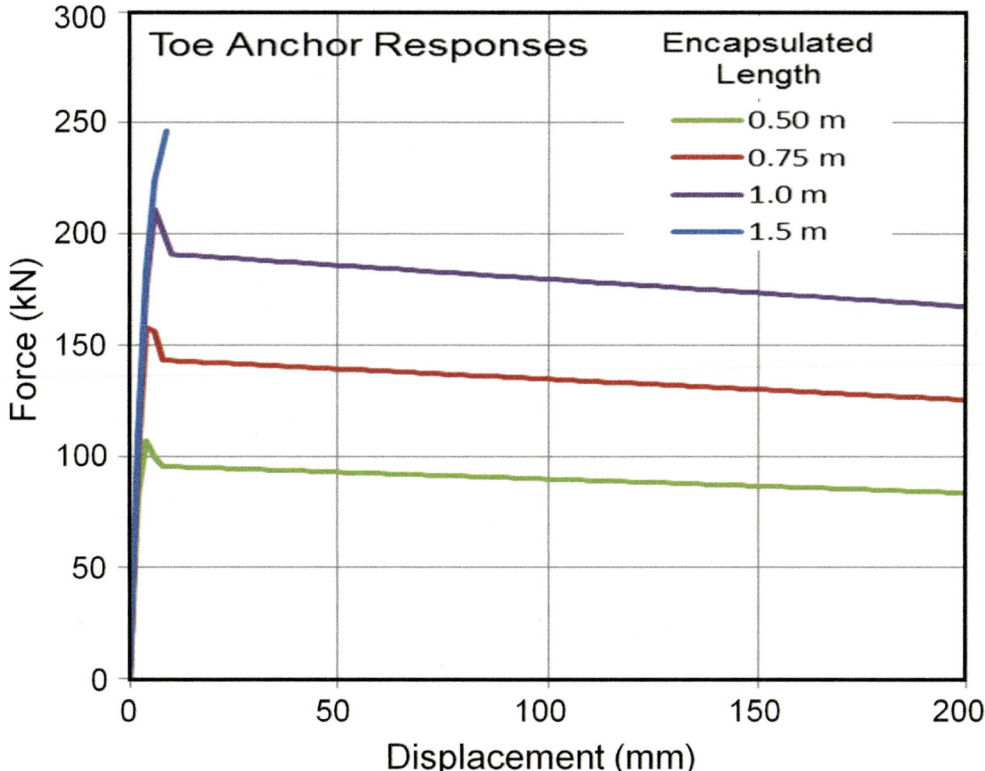

FIGURE 12.4   Reinforcement toe anchor force-displacement responses at a block face for different embedment lengths.

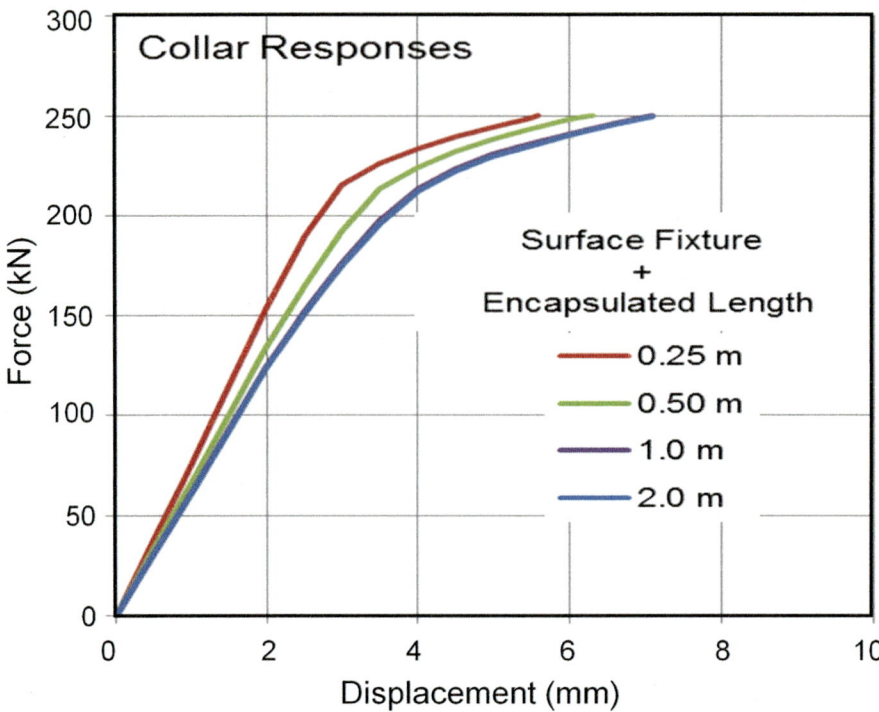

FIGURE 12.5 Reinforcement collar force-displacement responses at a block face for different embedment lengths.

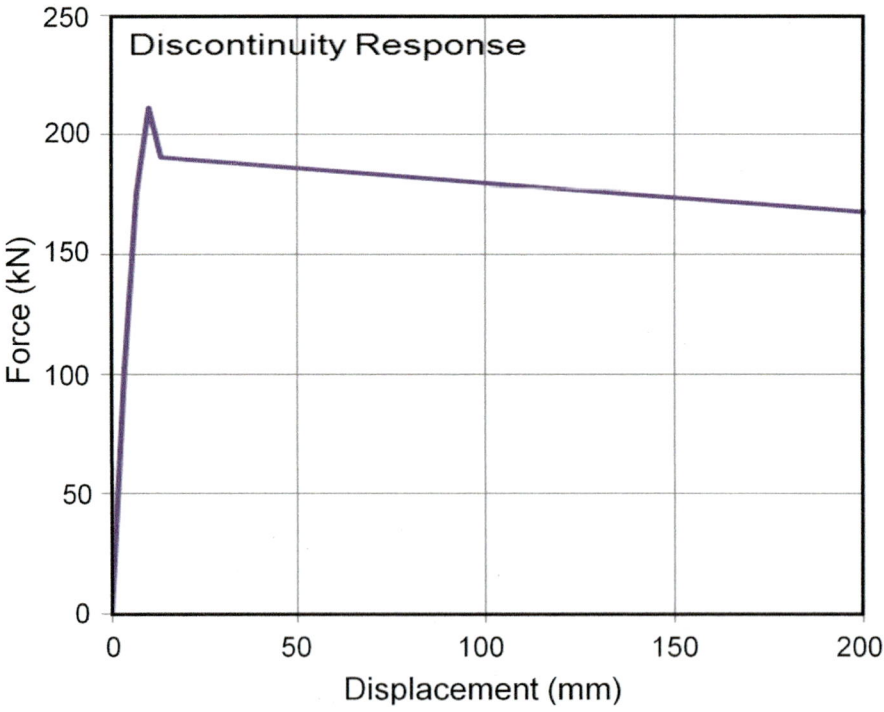

FIGURE 12.6 Combined toe anchor, collar and element force-displacement response of reinforcement at a block face.

potential client to perform specific testing for each configuration in local ground conditions and at a particular grout strength (see Chapter 8). This is very difficult for clients without the necessary understanding of what tests are required and how the results should be analysed and interpreted for the purposes of design.

Consequently, in part of the work presented here, additional databases had to be developed to enable the reinforcement design process for a wide range of reinforcement system configurations and loading conditions. Firstly, the databases are based on a generic, geometric configuration that may be used to describe all reinforcement systems. Secondly, the databases are based on results from actual reinforcement testing programs. It was also necessary to supplement these databases with computer simulations in order to use the testing data in design for both static and dynamic imposed loadings.

The possible responses of reinforcement systems to rock movements are shown schematically in Figure 12.7. It is apparent from this figure that reinforcement systems are required to respond to a wide variety of loading mechanisms, in most cases, unrelated to any tests that may have been performed to quantify their performance.

Previous discussion and Figure 12.2 indicate that the response of a reinforcement system at a discontinuity between unstable and stable rock is defined by the individual responses to loading either side of the discontinuity, and these responses are likely to be quite different. Firstly, the most obvious difference is the presence of the plate and fixture at the surface of the excavation. Secondly, the length between the toe of the element and the discontinuity and that between the discontinuity and the collar are unlikely to be equal.

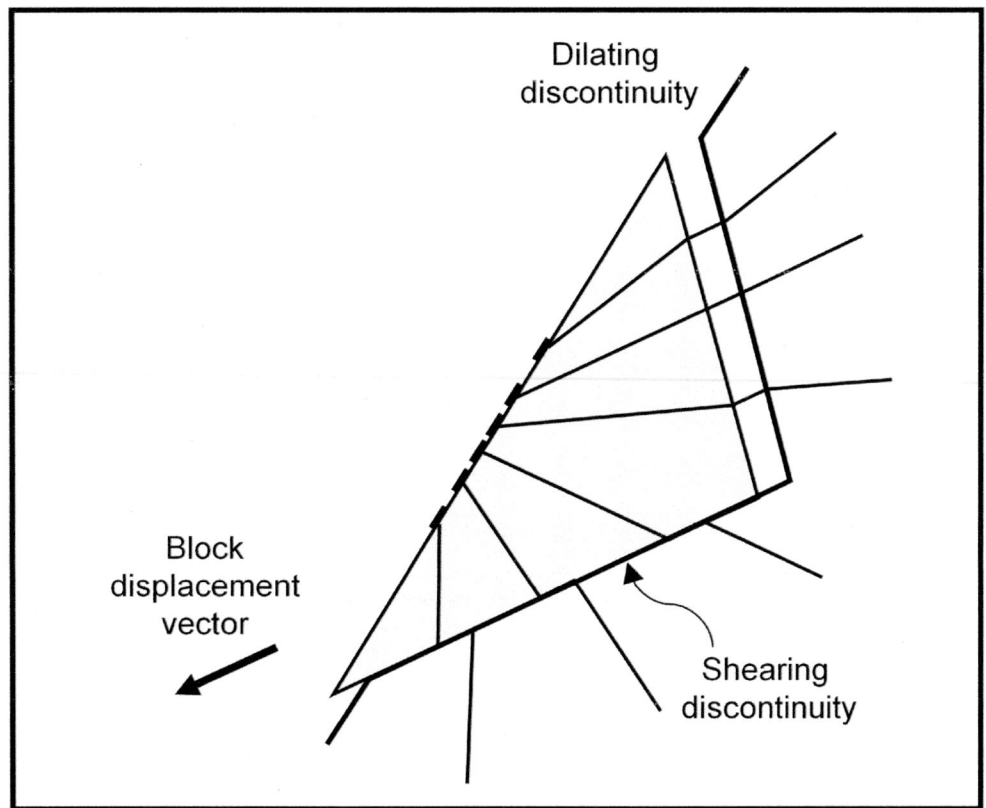

FIGURE 12.7 Modes of reinforcement deformations in response to block movements (Windsor and Thompson, 1997).

Consequently, any reinforcement response database needs to be structured to separately characterize the performance of the toe anchor and collar regions of a reinforcement system. Secondly, a computational method is required to characterize the collar region's performance. Finally, an additional computational method is required to take into account the variable embedment lengths on either side of a potentially unstable discontinuity.

12.4.1   A Generic Reinforcement System

A description of a reinforcement system classification based on the load transfer concept between the element and the rock at the borehole wall was provided in Chapter 2 (Thompson and Windsor, 1992). A complementary definition is possible using the generic reinforcement element configuration, as shown in Figure 12.8.

This generic configuration of the reinforcement element enables the definition of any reinforcement system in terms of its components, geometry and physical properties. This is illustrated in Figure 12.9, where the element configuration is complemented with a borehole, the presence or absence of encapsulation (e.g., cement or resin grout) and the presence or absence of internal fixtures (e.g., expansion shell anchor) and/or external fixtures (e.g., domed plate, washer and nut, flat plate or barrel and wedge anchor).

In an analysis, it is necessary to consider the effects of a discontinuity intersecting the axis of the reinforcement system, as shown in Figure 12.10. The different respective lengths will each influence the force-displacement response at the discontinuity.

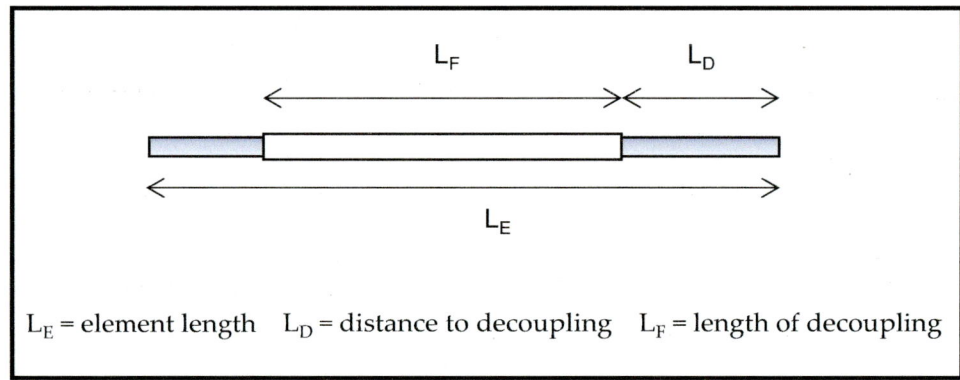

L_E = element length L_D = distance to decoupling L_F = length of decoupling

FIGURE 12.8   Generic reinforcement element configuration.

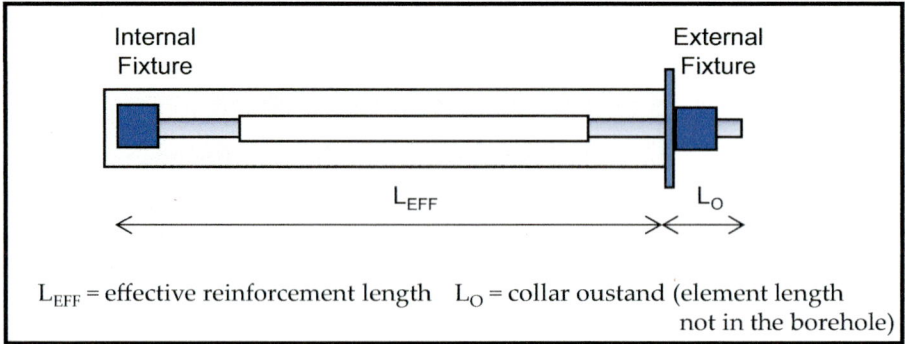

L_{EFF} = effective reinforcement length L_O = collar oustand (element length not in the borehole)

FIGURE 12.9   Generic reinforcement system configuration.

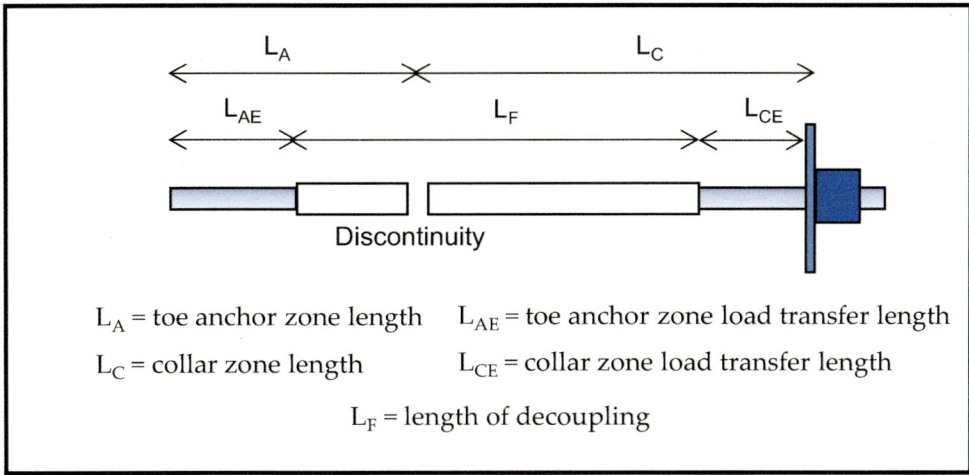

FIGURE 12.10   Schematic showing the configuration of a reinforcement system at a discontinuity.

12.4.2   MEASUREMENT OF REINFORCEMENT SYSTEM RESPONSES

Thompson et al. (2013) described the requirements for planning and documenting reinforcement system test programs. A well-planned testing program needs to consider testing for the mechanical properties of all components and the respective interactions between them (Chapter 6). The authors have been involved in many systematic testing programs. These programs have involved static and dynamic imposed loadings, individual component tests, small encapsulation length tests, single and double embedment length tests in the laboratory and field pull-out tests. A critical outcome from any testing program must be full, detailed documentation of the tests performed and, for pull-out tests, must include all the dimensions included in Figures 12.9 and 12.10.

12.4.3   REINFORCEMENT SYSTEM SIMULATIONS

Thompson (2011) described a method for predicting the responses of reinforcement systems when subjected to complex loadings involving both axial and shear deformations at discontinuities. The model developed for simulating reinforcement system responses is shown in Figure 12.11.

The software associated with this model can simulate all geometric configurations that can be arranged in the generic model of Figure 12.10 combined with the mechanical responses of the components and the interactions between them.

12.4.4   REINFORCEMENT SYSTEM RESPONSES

The database of reinforcement system responses is based on extensive testing programs performed by the authors over many years (e.g., Villaescusa et al., 1992), with some examples of case studies reported by Thompson and Villaescusa (2013, 2014). In this section, the response of cable bolt systems based on a 15.2-mm diameter, 7-wire steel strand will be used to demonstrate the constituents of the database. In order to provide consistency and comparison between different configurations for the same reinforcement elements, the system responses in the database are derived from computer simulations.

12.4.4.1   Component Properties

The force-strain response for a plain, 15.2-mm, 7-wire steel strand is shown in Figure 12.12. This response is typically provided as a certificate by the strand manufacturer and is very important, particularly if any length of a reinforcement system is decoupled.

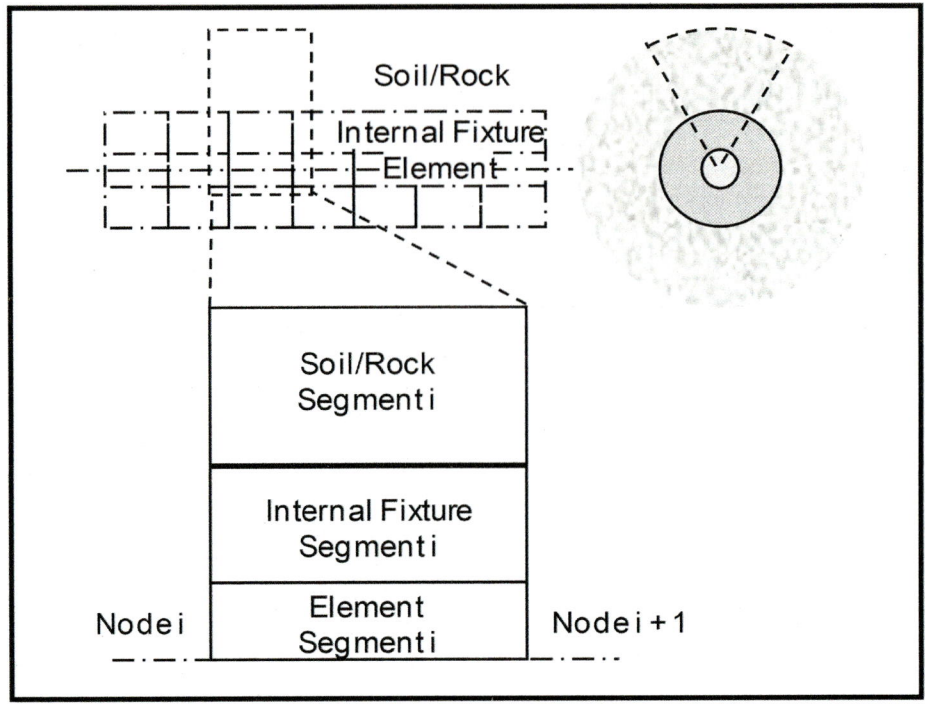

FIGURE 12.11 The basic elements used in the simulation of reinforcement force-displacement responses.

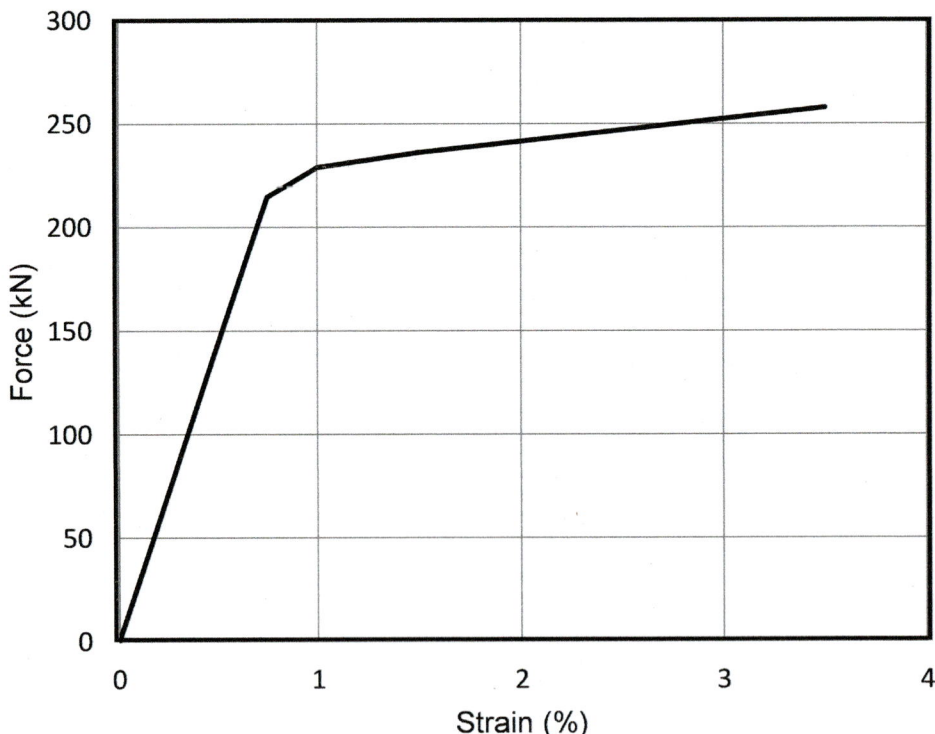

FIGURE 12.12 Force-strain response for a plain 15.2-mm diameter, 7-wire steel strand.

12.4.4.2   Interface Properties

It is well-established that the force resisting pull out of reinforcement is related to both the strength of the grout between the borehole and element and the roughness of both borehole rock/grout and grout/element interfaces. Figure 12.13 shows the stress-displacement response assumed for the interface between the strand and grout with the WCR of 0.40. This response is based on the results from small encapsulation length tests.

12.4.4.3   Variations of Axial Force-Displacement Responses with Encapsulation Length

Systematic testing over many years has demonstrated that the total resistance to pull out of elements encapsulated within cement grout-filled boreholes is related to embedment length. This phenomenon has been simulated using the SAFEX software, and the results are given in Figure 12.14. This shows that the response which results in rupture of the element (i.e., for the longest embedment length of 1.5 m) forms an envelope that bounds all responses at shorter embedment lengths.

12.4.4.4   Axial Force-Displacement Responses with External Fixture and Encapsulation Lengths

Without a fixture at the collar of a strand reinforcement system, the force-displacement responses would be the same as that given in Figure 12.14. The force-displacement responses when a barrel and wedge anchor is fitted at the collar are quite different, as shown in Figure 12.15.

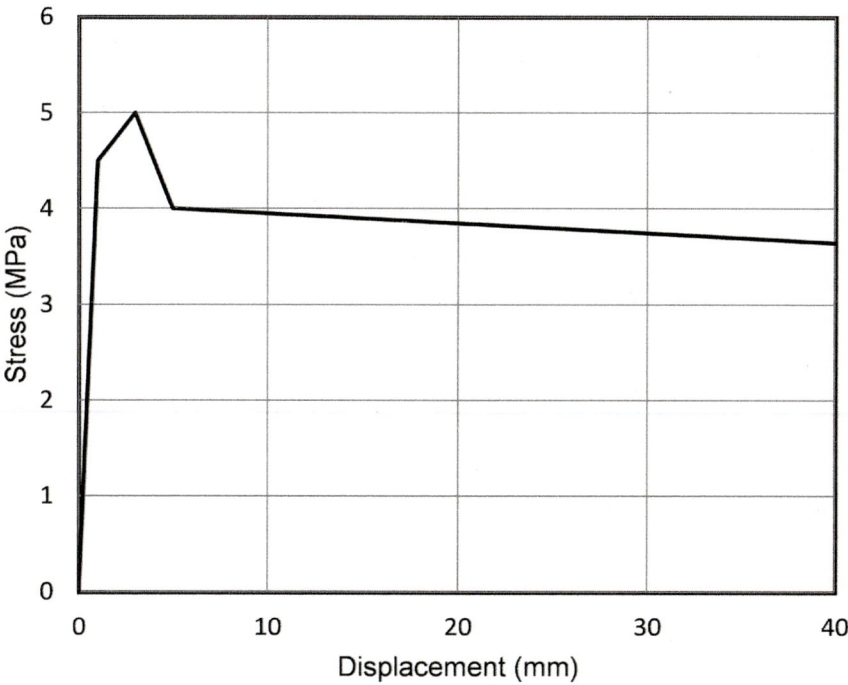

FIGURE 12.13   The stress-displacement response at the interface between the strand and grout with a water/cement ratio of 0.40.

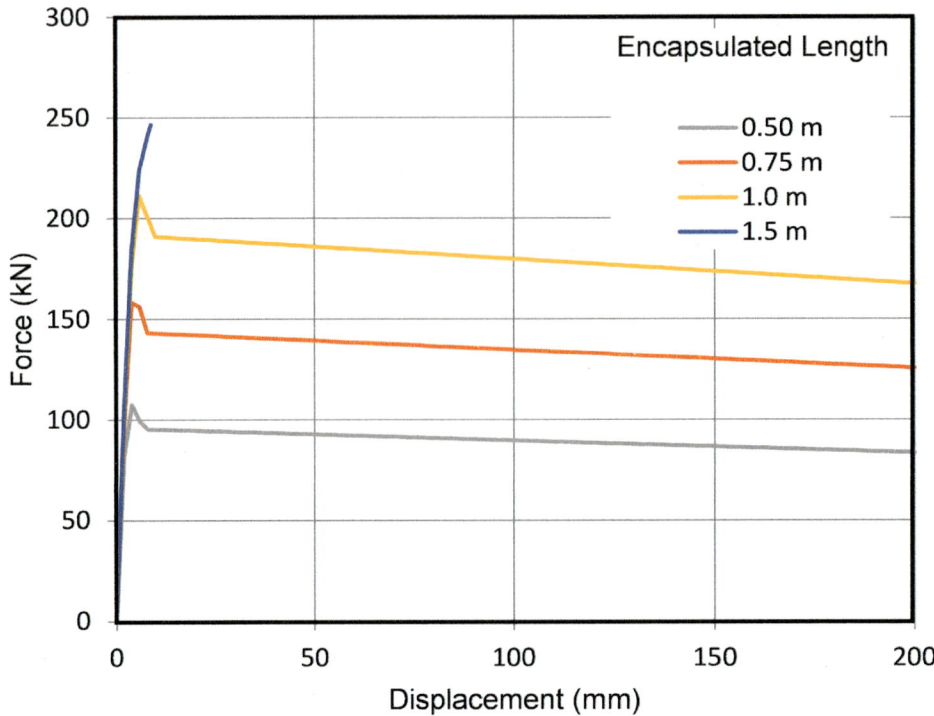

FIGURE 12.14 Force-displacement responses predicted for several encapsulation lengths.

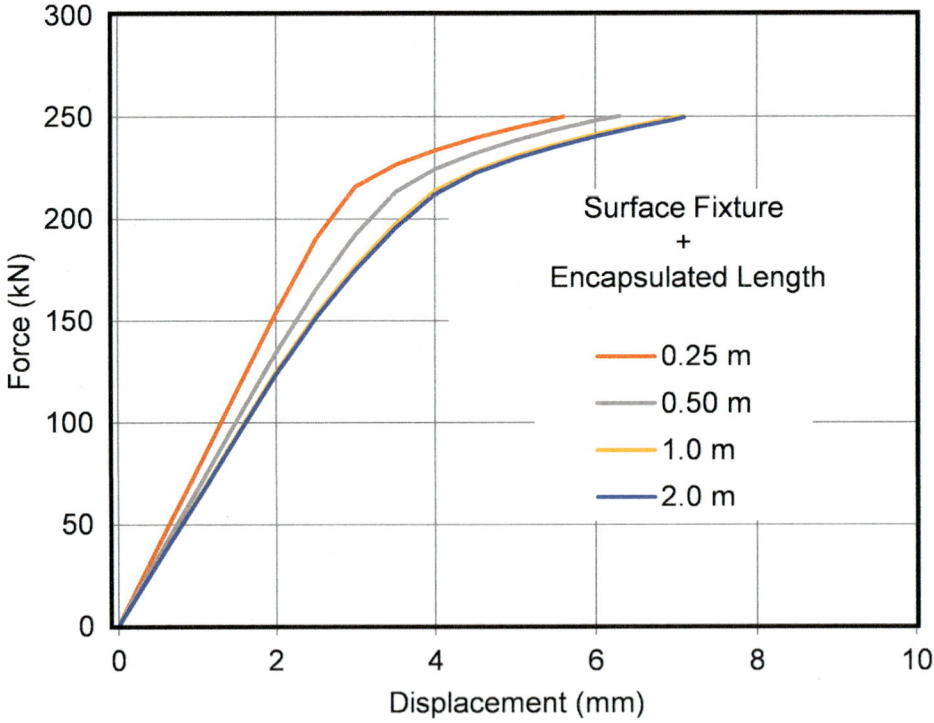

FIGURE 12.15 Force-displacement responses predicted for several encapsulation lengths when a barrel and wedge anchor is fitted at the collar.

12.4.5   APPLICATION OF THE REINFORCEMENT DATABASES IN DESIGN FOR A REINFORCED BLOCK

Thompson and Villaescusa (2022) have described a methodology for the analysis of reinforced blocks subjected to dynamic loading. A similar methodology for the static equilibrium of a reinforced block was described in greater detail by Thompson (1989). The software based on that theory will be used here to demonstrate how the database is used in practice.

12.4.5.1   Description of the Analysis

A single reinforcement system from a reinforced block analysis is selected. The geometric details for the reinforcement are (see Figures 12.8 and 12.9 for notation):

- Element length (L_E) = 4.0 m
- Depth to decoupling (L_D) = 1 m
- Length of decoupling (L_F) = 2.1 m
- Collar outstand (L_O) = 0.3 m
- Effective reinforcement length (L_{EFF}) = 3.7 m

In the analysis, the block face (discontinuity) is located 1.37 m from the collar; leading to the following geometric details (see Figure 12.10 for notations):

- Collar zone length (L_C) = 1.37 m
- Collar zone load transfer length (L_{CE}) = 0.7 m
- Toe anchor zone length (L_A) = 2.33 m
- Toe anchor load transfer length (L_{AE}) = 0.9 m

12.4.5.2   Estimation of Force-Displacement Responses

The anchor force-displacement responses established for the toe anchor and collar load transfer regions are shown in Figure 12.16 and Figure 12.17, respectively. These responses are interpolated

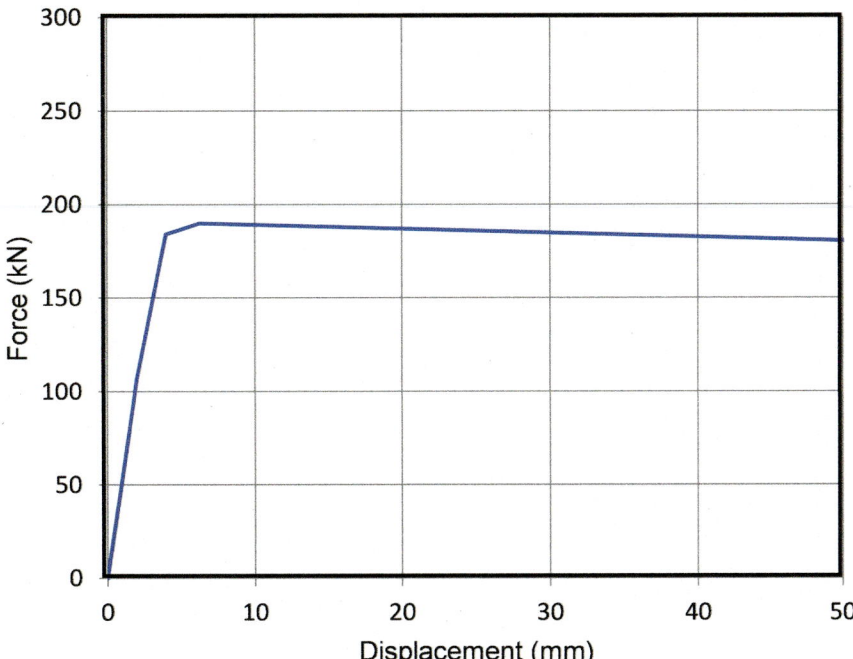

FIGURE 12.16   Example of the toe anchor load transfer force-displacement response.

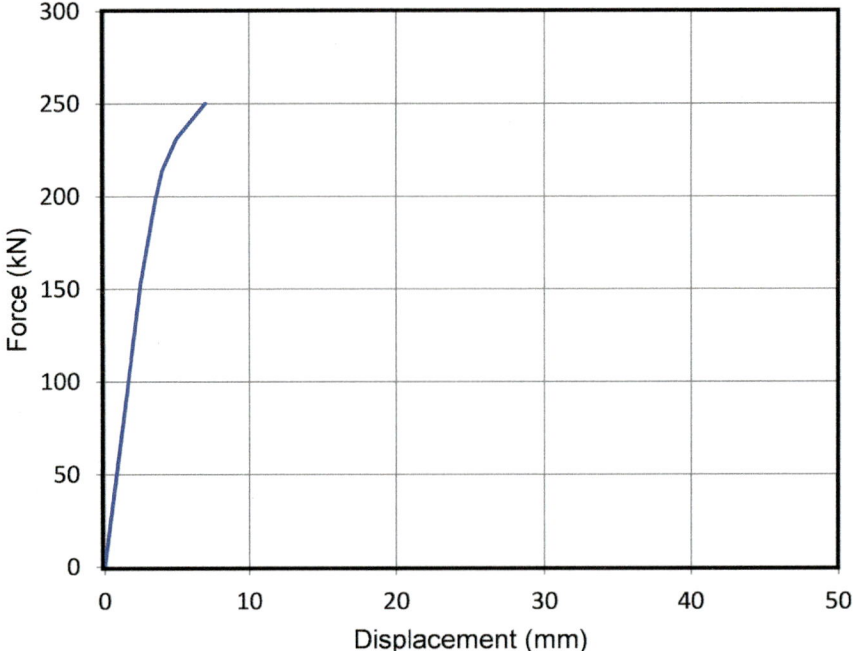

FIGURE 12.17   Example of the collar load transfer force-displacement response.

from the database responses given previously in Figure 12.4 and Figure 12.5. It is noteworthy that the toe load transfer response shown in Figure 12.16 is truncated at 50 mm. For the purposes of analysis, the maximum displacement would be in excess of 200 mm, as indicated in Figure 12.4. For static analysis, the maximum displacement of 50 mm would be appropriate. On the other hand, for dynamic analysis, it is important to consider the complete force- displacement responses required to calculate dissipated energy capacities.

The force-displacement response of the decoupled element length shown in Figure 12.18 is derived from the force-strain response given previously in Figure 12.12. The force-displacement response at the discontinuity is then estimated by combining the anchor and free length responses, as shown in Figure 12.19. Finally, this response is combined with the collar response to obtain the overall force-displacement response, as shown in Figure 12.20. This response is then used in the reinforced block analysis described later in Section 12.8.

12.5   STATIC ANALYSIS OF A REINFORCED ARBITRARILY-SHAPED BLOCK

A three-dimensional analysis technique for the design of reinforcement of arbitrarily-shaped rock blocks with a combination of free faces and faces that interact with the adjacent rock mass has been described by Thompson (1989). The rock blocks can have three translational and three rotational modes of displacement. This differs from the previously published work of a number of workers (e.g., Warburton, 1993; Goodman and Shi, 1985) in which translations alone were considered. Reinforcement elements are modelled with non-linear load-displacement characteristics, as described earlier. The solution technique for block translations and rotations is unique because joint interactions are modelled as constraints on block displacements (Thompson, 1989).

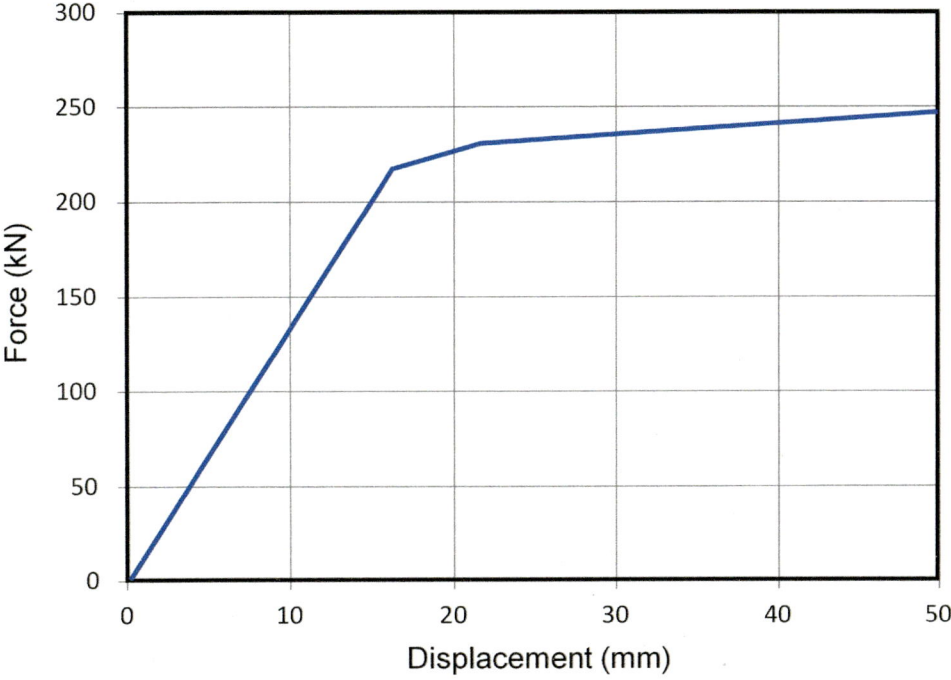

FIGURE 12.18 Example of the decoupled element free length force-displacement response.

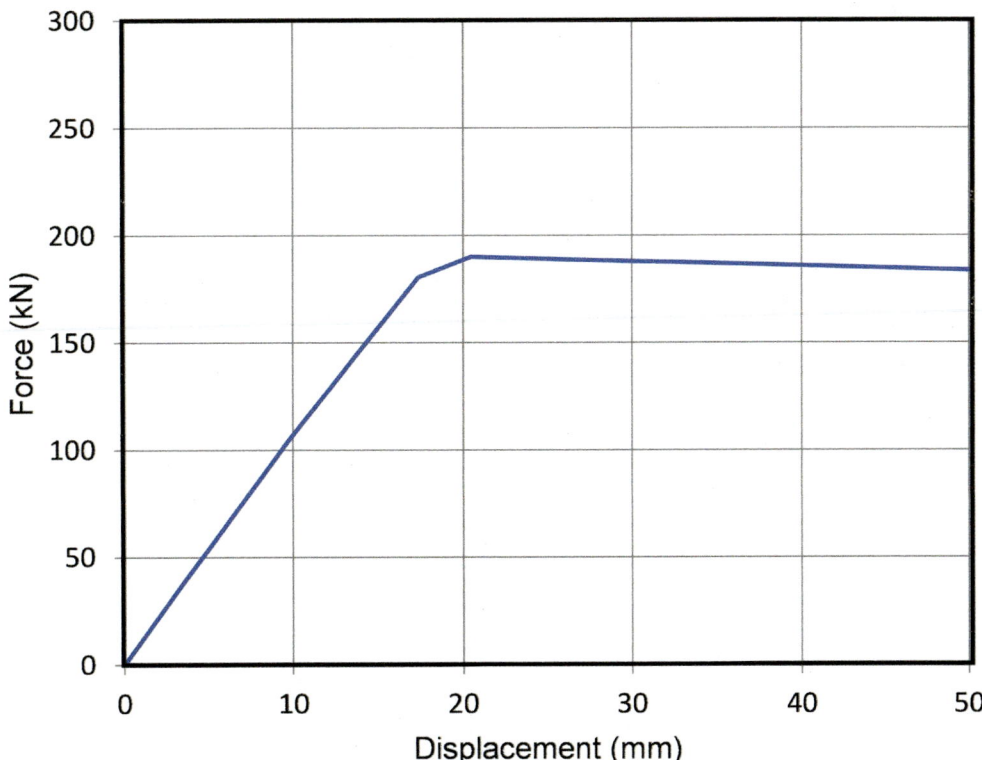

FIGURE 12.19 Example of the combined anchor and free length force-displacement responses.

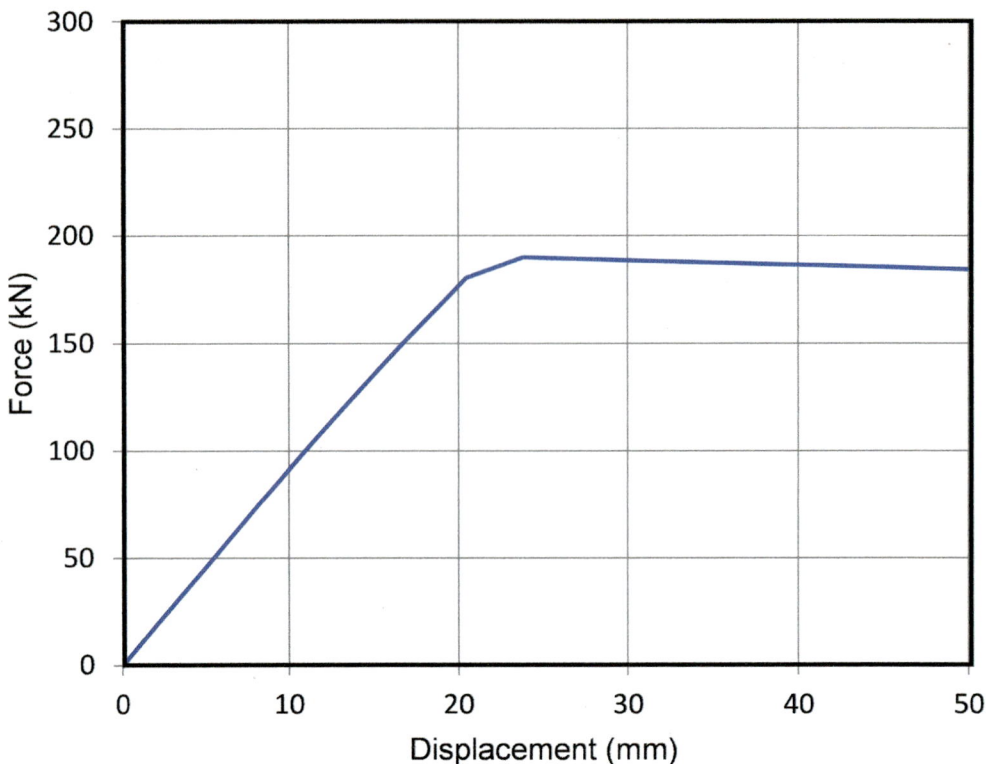

FIGURE 12.20 Example of overall force-displacement response at a block face.

12.5.1 THE DESIGN PROBLEM

The type of problem that can be analysed was given previously in Figure 12.2. The blocks can have an arbitrary shape, and the reinforcement elements can be located anywhere on the block with any orientation. The major assumptions made in developing the theory (Thompson, 1989) are as follows:

1. The analysis is applicable to a single rigid block.
2. Block displacements are small.
3. The driving force on the block is due solely to gravity.
4. The passive forces acting on the block are due to the reinforcement and interactions across the joints forming the block.

12.5.2 DESCRIPTION OF THE ANALYSIS METHOD

The solutions for reinforcement loads and joint reactions in problems such as those shown in Figure 12.2 are generally statically indeterminate. However, with the assumptions detailed previously, it is possible to obtain a unique solution by simultaneously satisfying force equilibrium and displacement compatibility. Also, because of the non-linear nature of the problem, an iterative procedure using manipulation of matrix equations needs to be used to solve for the increments in displacements of the block, reinforcement forces and joint reactions. The analysis was described in detail by Thompson (1989). This analysis involved six degrees of freedom for block movement (i.e., three translations and three rotations) with force-displacement responses specific to each reinforcement system.

Simplistically, the known applied forces [F] are related to the unknown displacements [d] through a 'stiffness' matrix [K] related to the force-displacement responses of the reinforcement systems, that is (in matrix form):

$$[F] = [K][d] \tag{12.1}$$

Then, the unknown displacements are determined using the following equation:

$$[d] = [K]^{-1}[F] \tag{12.2}$$

In practice, as described in detail by Thompson (1989), the solution is much more complex than that implied by Equation 12.2. These complexities relate to the non-linear, force-displacement responses of the reinforcement systems and the potential for block translations and rotations to cause interference between the block faces and the surrounding rock mass. However, the static analysis technique can be used as the basis of the novel dynamic analysis methodology described in the following sections.

12.6 DYNAMIC BLOCK LOADING

Three alternative approaches may be used to analyse a moving block retarded by reinforcement forces. A moving mass retarded by a constant force, as illustrated in Figure 12.21 will be used to show that these three approaches may be used to achieve identical outcomes.

The destabilizing force will be characterized by an ejection velocity, that is, speed and a direction. The following symbols will be used for the analyses described in the following sections:

- A = Acceleration
- D = Distance travelled before mass achieves zero velocity
- F = Constant retarding force
- m = Mass
- t = Time to arrest mass movement
- v_0 = Initial velocity
- v_a = Average velocity

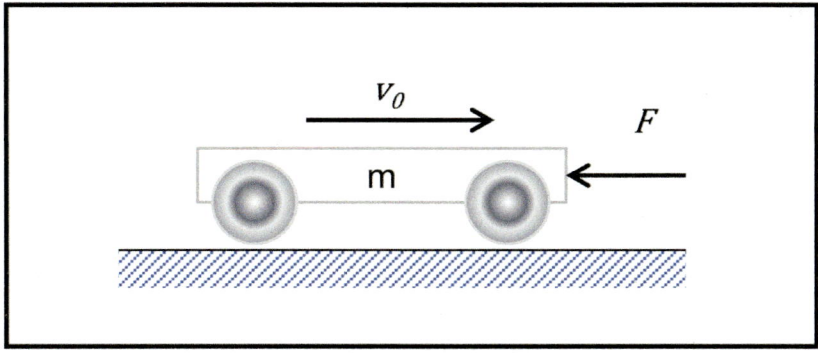

FIGURE 12.21 Moving mass retarded by a constant force.

12.6.1 NEWTONIAN MECHANICS-BASED ANALYSIS

The distance (d) travelled by the mass before being brought to rest is given by:

$$d = v_0 t + \frac{1}{2} a t^2 \qquad 12.3$$

where

$$a = -\frac{F}{m} \qquad 12.4$$

and

$$t = -\frac{v_0}{a} \qquad 12.5$$

Then,

$$t = \frac{v_0 m}{F} \qquad 12.6$$

Finally,

$$d = \frac{m v_0^2}{2 F} \qquad 12.7$$

12.6.2 MOMENTUM-BASED ANALYSIS

To bring the mass to rest, the product of F and t is set equal to the change in momentum from the initial mass velocity to the mass at rest, that is:

$$Ft = m v_0 \qquad 12.8$$

where

$$t = \frac{m v_0}{F} \qquad 12.9$$

and

$$v_a = \frac{1}{2} v_0 \qquad 12.10$$

Then,

$$d = v_a t \qquad 12.11$$

Finally,

$$d = \frac{m\,v_0^2}{2\,F}$$

12.12

12.6.3 ENERGY-BASED ANALYSIS

The energy-based equation is based on the principle that the energy dissipated by a constant force is equal to the original kinetic energy. That is:

$$F\,d = \frac{1}{2}m\,v_0^2$$

12.13

Thus,

$$d = \frac{m\,v_0^2}{2\,F}$$

12.14

12.6.4 SUMMARY OF DYNAMIC ANALYSIS METHODS

In summary, Equations 12.7, 12.12 and 12.14 are identical for the three alternative methods of analysing the distance travelled by a mass being retarded by a constant force. Because of this finding, a dynamic analysis methodology involving the energy dissipation-based analysis technique was chosen as the most suitable for adaptation from the static analysis technique developed by Thompson (1989), which will be described in the following section.

12.7 DISPLACEMENT CONTROLLED DYNAMIC ANALYSIS METHODOLOGY

The starting point for the dynamic analysis is equivalent to Equation 12.1, where a stiffness matrix [K] is established for each reinforcement system to relate axial and shear forces to block displacements [d]. The block is assumed to have an initial ejection velocity and direction. The direction is defined using the plunge and azimuth angles. A total displacement increment (d_i) is defined from which the component displacements are applied to the block in the directions of the three-dimensional coordinate system. A basic assumption that enables simplification of the solution is that block rotations are assumed to be negligibly small and are set to zero.

Complex relationships are established between the block translations and the individual reinforcement system local displacements and forces. These forces are then resolved into coordinate directions. At this stage, the energy dissipated by each reinforcement system during the block displacement increment may be determined and aggregated. It is also noteworthy at this stage that the block may displace vertically downward, and the loss of 'potential' energy may contribute to an increase in kinetic energy. Thus, at the end of the displacement increment (i):

$$KE_i = KE_0 + PE_i - DE_i$$

12.15

where

- KE_i = Block kinetic energy
- KE_0 = Initial block kinetic energy

- PE_i = Potential energy loss
- ED_i = Energy dissipated by the reinforcement

Then, the velocity after the displacement increment is given by:

$$dv_i = \sqrt{\frac{2\,K\,E_i}{m}}$$

12.16

The average velocity during the displacement increment is given by:

$$v_a = \frac{1}{2}\left(v_i + v_{i-1}\right)$$

12.17

And the time after the displacement increment is:

$$t_i = t_{i-1} + \frac{d_i}{v_a}$$

12.18

The block displacement is incremented by d_i in the direction of the velocity vector until:

1. The dissipated energy DE_i exceeds $KE_0 + PE_i$, or,
2. The reinforcement systems are predicted to fail.

In the first case, the calculations are interrupted at the previous total displacement, and the displacement increment is reduced until the block velocity is predicted to be zero.

In the latter case, the calculations are stopped, and failure is reported due to insufficient energy dissipation capacity of the chosen reinforcement. The potential solutions are to change the reinforcement systems or to increase the number of systems involved in reinforcing the block.

12.8   EXAMPLE OF IMPLEMENTED THEORY

The following sections provide a summary of the implementation of the theory within the SAFEX software package developed by the writers over many years. It was first described by Windsor and Thompson (1992c).

12.8.1   DESCRIPTION OF THE ANALYSIS

The reinforced block analysed is shown in Figure 12.22. The tetrahedral-shaped block has a mass of 40.9 tons, and the reinforcement systems are single 15.2-mm diameter, 7-wire plain steel strands with a total element length (L_E) of 4.0 m, a decoupled length (L_F) of 2.1 m starting at a depth (L_D) of 1.0 m from collar end of the element, as shown earlier in Figure 12.8. When installed, there is a collar outstand (L_O) of 0.3 m and an effective internal length (L_{EFF}) of 3.7 m with a barrel and wedge anchor, as shown in Figure 12.9. The reinforcement is assumed to be untensioned. For the analysis, there are a total of seven active reinforcement systems.

Figure 12.23 shows the software user interface implemented to define the reinforcement systems used in the analysis. The reinforcement systems are assumed to be installed in rings spaced at 2 m, and the pattern is generated and the intersections with the block-free face and internal faces are determined automatically. The reinforcement systems that are involved in the analysis are detailed in Figure 12.24.

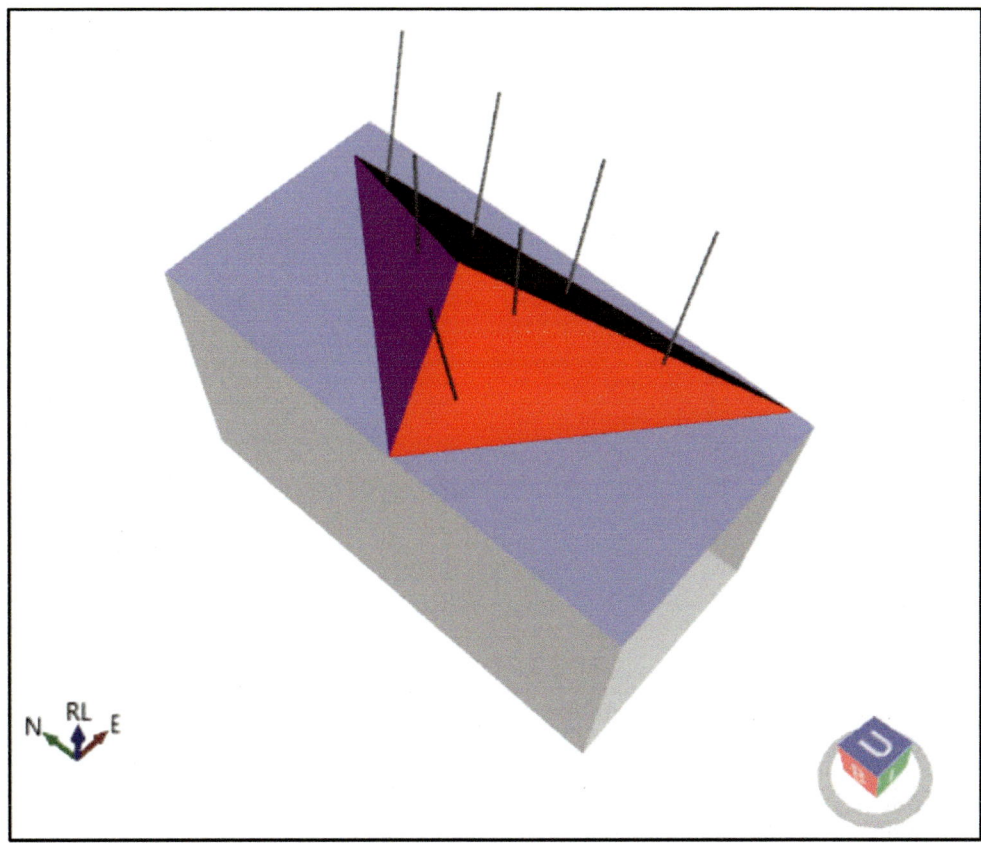

FIGURE 12.22 Example of reinforced block configuration.

	Element	Element Length (m)	Collar oustand (m)	Decoupled	Decoupling Length (m)	Distance to Decoupling (m)	Encapsulation	Internal Fixture	External Fixture	Internal Length (m)	Toe Transfer Length (m)	Collar Transfer Length (m)	Inclination (deg)	Azimuth (deg)	Distance from left (m)	Distance from floor (m)	Ring Spacing (m)	Start Position (m)
1	Strand 15.2mm	4.00	0.3	yes	2.1	1	0.4 WCR	None	Plate and B-W	3.7	0.9	0.7	80	270	1.5	5.0	2.0	0.3
2	Strand 15.2mm	4.00	0.3	yes	2.1	1	0.4 WCR	None	Plate and B-W	3.7	0.9	0.7	90	000	2.5	5.0	2.0	0.3
3	Strand 15.2mm	4.00	0.3	yes	2.1	1	0.4 WCR	None	Plate and B-W	3.7	0.9	0.7	80	090	3.5	5.0	2.0	0.3

FIGURE 12.23 Specification of pattern for the chosen reinforcement system.

12.8.2 IMPOSED LOADING AND RESULTS

The example dynamic loading for the reinforced block is a velocity of 2 m/s directed vertically downwards, as shown in Figure 12.25. The kinetic energy of ejection is 81.9 kJ. The results of the analysis are summarized in Figure 12.26 and Figure 12.27. The vertical displacement to the time of zero block velocity is 109.4 mm.

Additional results of the analysis are detailed in Figure 12.28, which shows the variations with time of the analysed reinforced block's displacement, velocity, deceleration and net energy. Figure 12.29 shows the variations with block displacement of the deceleration force, dissipated energy and block net energy.

It is noteworthy that the block initially accelerates while the reinforcement develops force in response to the block displacement. This means the block will displace further before being arrested

| Reinforcement Analysis Bolt Results | | | | | | | |

Excavation Surface Number	Excavation Surface Area (m²)	Block Number	Block Shape	Block Face Face Area (m²)	Total Number of Bolts	Number of Intersecting Bolts	Number of Reinforcement Points
1	50.000	1	Tetrahedra	21.301	15	7	7

Bolts:

Reinforcement Point	Point 1	Point 2	Point 3	Point 4	Point 5	Point 6	Point 7
Reinforcement Number	6	7	8	9	11	12	15
Row Number	2	3	3	3	4	4	5
Number in Row	3	1	2	3	2	3	3
Element Type	Strand 15.2	Strand 15.2	Strand 15.2	Strand 15.2	Strand 15.2	Strand 15.2	Strand 15.2
Encapsulation	0.40 WRC	0.40 WRC	0.40 WRC	0.40 WRC	0.40 WRC	0.40 WRC	0.40 WRC
Internal Fixture	None	None	None	None	None	None	None
External Fixture	Plate and B-W	Plate and B-W	Plate and B-W	Plate and B-W	Plate and B-W	Plate and B-W	Plate and B-W
Element Length - L_t (m)	4.00	4.00	4.00	4.00	4.00	4.00	4.00
Collar Outstand - L_O (m)	0.30	0.30	0.30	0.30	0.30	0.30	0.30
Effective Length - L_{EFF} (m)	3.70	3.70	3.70	3.70	3.70	3.70	3.70
Decoupling Length - L_r (m)	2.10	2.10	2.10	2.10	2.10	2.10	2.10
Distance to Decoupling - LD (m)	1.00	1.00	1.00	1.00	1.00	1.00	1.00
Toe Transfer Length – L_{TE} (m)	0.90	0.90	0.90	0.90	0.90	0.90	0.90
Collar Transfer Length – L_{CE} (m)	0.70	0.70	0.70	0.70	0.70	0.70	0.70
Azimuth (deg)	090	270	000	090	000	090	090
Inclination (deg)	80	80	90	80	90	80	80
Toe Zone Length – L_T (m)	2.86	3.04	2.06	2.85	2.28	3.01	3.16
Toe Transfer Capacity (kN)	189.84	189.84	189.84	189.84	189.84	189.84	189.84
Element Capacity (kN)	257.65	257.65	257.65	257.65	257.65	257.65	257.65
Collar Zone Length – L_C (m)	0.84	0.66	1.64	0.85	1.42	0.69	0.54
Collar Transfer Capacity (kN)	250.00	250.00	250.00	250.00	250.00	250.00	250.00
Effective Capacity (kN)	189.84	189.84	189.84	189.84	189.84	189.84	189.84
Dissipation Capacity (kJ)	19.0	19.0	19.0	19.0	19.0	19.0	19.0

FIGURE 12.24 Details for the reinforcement systems involved in the reinforced block analysis.

than when the reinforcement systems are tensioned during installation. It should also be noted that vertical displacement downward after the block movement is initiated adds to the energy demand. For the analysis used, the initial block kinetic energy of ~82 kJ is supplemented by an additional ~44 kJ, which results in the reinforcement requiring to dissipate ~126 kJ.

It must also be noted that the analysis of a reinforced block subjected to dynamic loading is the final step in an integrated set of modules involving the analysis of structural geological data and the probabilistic prediction of block shapes of various sizes forming at the various faces of underground excavations. For excavations formed within highly stressed rock masses at large depths, a procedure described by Zuo et al. (2005) is used to predict rock mass demand in terms of energy (kJ) per square metre for the estimated depths of failure. The estimate of energy demand depends on the physical and mechanical properties of the rock from which an estimate of ejection velocity may be made (see Chapter 5).

Finally, whilst the analysis method assumes a single block, field observations have shown that dynamic failures due to localized overstressing involve many smaller blocks. However, the extent of the failure is usually associated with pre-existing discontinuities (Figure 12.30). The analysis of a single block is, therefore, considered to be valid only if the installed support provides restraint between the reinforcement collar points. This requires that the mesh can provide an effective contact point at the collars of the reinforcement systems. The field observations have also demonstrated that welded wire mesh and fibre-reinforced shotcrete are ineffective in response to dynamic loading (Chapter 4). However, high-strength steel wire woven mesh slightly embedded within shotcrete (Figure 12.31) is effective in providing broken rock mass restraint between reinforcement systems that have high energy dissipation capacities (Villaescusa et al., 2016; Arcaro et al., 2021).

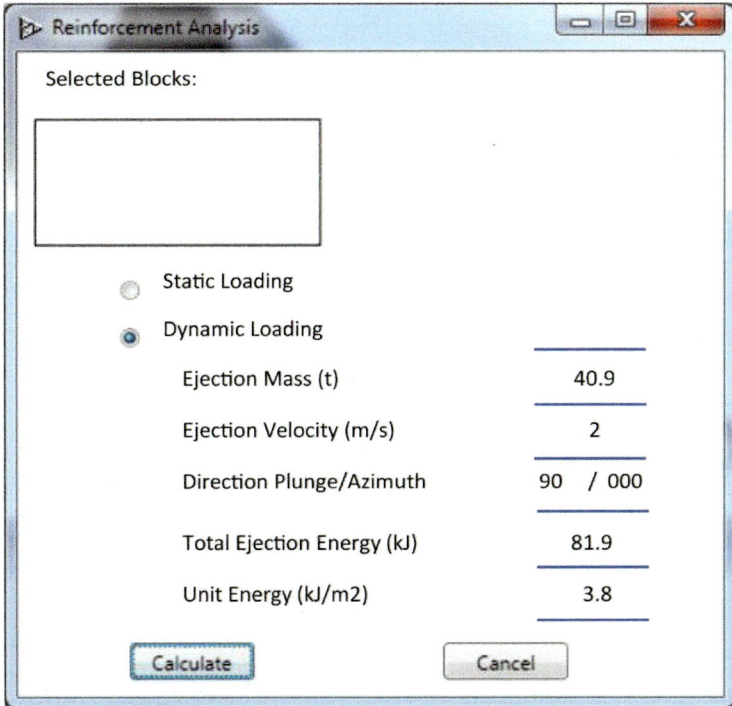

FIGURE 12.25   Details of imposed dynamic loading.

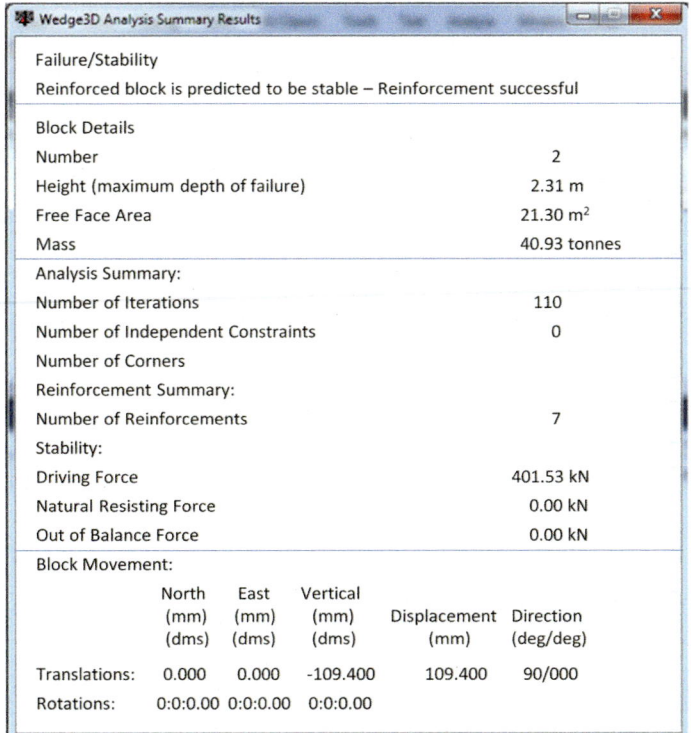

FIGURE 12.26   Summary of results after dynamic loading of the reinforced block.

| Reinforcement Analysis Bolt Results | | | | | | | |

Excavation Surface Number	Excavation Surface Area (m²)	Block Number	Block Shape	Block Face Face Area (m²)	Total Number of Bolts	Number of Intersecting Bolts	Number of Reinforcement Points
1	50.000	1	Tetrahedra	21.301	15	7	7

Bolts:

Reinforcement Point	Point 1	Point 2	Point 3	Point 4	Point 5	Point 6	Point 7
Reinforcement Number	6	7	8	9	11	12	15
Row Number	2	3	3	3	4	4	5
Number in Row	3	1	2	3	2	3	3
Element Type	Strand 15.2	Strand 15.2	Strand 15.2	Strand 15.2	Strand 15.2	Strand 15.2	Strand 15.2
Encapsulation	0.40 WRC	0.40 WRC	0.40 WRC	0.40 WRC	0.40 WRC	0.40 WRC	0.40 WRC
Internal Fixture	None	None	None	None	None	None	None
External Fixture	Plate and B-W	Plate and B-W	Plate and B-W	Plate and B-W	Plate and B-W	Plate and B-W	Plate and B-W
Element Length - L_E (m)	4.00	4.00	4.00	4.00	4.00	4.00	4.00
Collar Outstand - L_O (m)	0.30	0.30	0.30	0.30	0.30	0.30	0.30
Effective Length - L_{EFF} (m)	3.70	3.70	3.70	3.70	3.70	3.70	3.70
Decoupling Length – L_f (m)	2.10	2.10	2.10	2.10	2.10	2.10	2.10
Distance to Decoupling – LD (m)	1.00	1.00	1.00	1.00	1.00	1.00	1.00
Toe Transfer Length – L_{TE} (m)	0.90	0.90	0.90	0.90	0.90	0.90	0.90
Collar Transfer Length – L_{CE} (m)	0.70	0.70	0.70	0.70	0.70	0.70	0.70
Azimuth (deg)	090	270	000	090	000	090	090
Inclination (deg)	80	80	90	80	90	80	80
Toe Zone Length – L_T (m)	2.86	3.04	2.06	2.85	2.28	3.01	3.16
Toe Transfer Capacity (kN)	189.84	189.84	189.84	189.84	189.84	189.84	189.84
Element Capacity (kN)	257.65	257.65	257.65	257.65	257.65	257.65	257.65
Collar Zone Length – L_C (m)	0.84	0.66	1.64	0.85	1.42	0.69	0.54
Collar Transfer Capacity (kN)	250.00	250.00	250.00	250.00	250.00	250.00	250.00
Effective Capacity (kN)	189.84	189.84	189.84	189.84	189.84	189.84	189.84
Dissipation Capacity (kJ)	19.0	19.0	19.0	19.0	19.0	19.0	19.0

FIGURE 12.27 Summary of reinforcement system properties and analysis results.

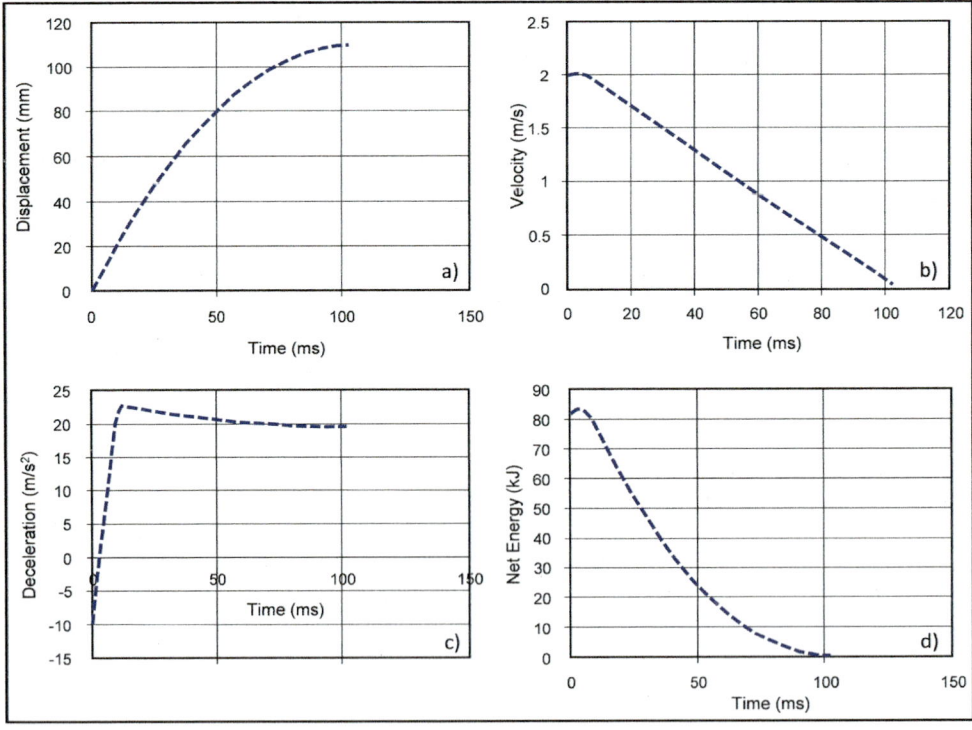

FIGURE 12.28 Calculated (a) block displacement-time, (b) block velocity-time, (c) block deceleration-time and (c) block net energy-time.

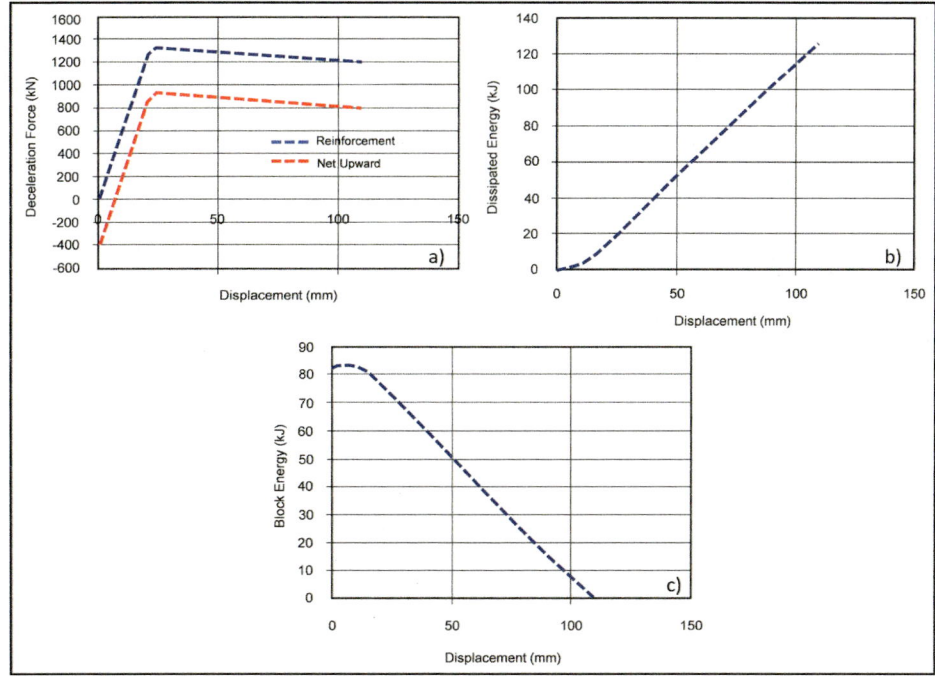

FIGURE 12.29   Calculated block displacement and (a) deceleration force, (b) dissipated energy and (c) block energy.

FIGURE 12.30   Example of failed rock blocks (with joint-created faces) following a violent, seismic event.

FIGURE 12.31   Example of a stabilized bench stop draw point conditions following a 2.4 magnitude seismic event.

13 Construction and Monitoring

13.1 INTRODUCTION

The worldwide progressive depletion of near-surface mineral resources is resulting in an increasing trend to access deeply buried orebodies. As such, the mining operations require development access into highly stressed rock masses where structurally-controlled, dynamic shear failure can occur either at the early development phase or later during the production stages or both. This tendency to mine deep orebodies inevitably results in increased rock mass demand, which, in some cases, can exceed the installed ground support capacity at the boundary of the excavations. The problem is exacerbated when the construction practices employed at great depth simply duplicate those employed in low or moderate *in situ* stress geotechnical environments. As a result, stress-driven failures of a dynamic nature are often experienced and linked to practices involving conventional ground support design methods, low energy dissipation ground support devices and materials used to secure rectangular excavation shapes that are not harmonic with the *in situ* stress field.

13.2 INDUCED STRESS

The induced stresses are particularly important in assessing the resulting stability of the different excavation shapes and orientations at great depth. The presence of both low and high stresses at particular locations around an excavation is important. Low compressive and less common tensile stresses may enable gravity-driven loosening and detachment of blocks from the backs and walls of excavations. On the other hand, very high stresses may exceed the strength of the rock mass and result in progressive failure at any location around the excavation boundary. Often a progressive failure terminates when the stress-driven, tangential cracking reaches a major geological discontinuity (Figure 13.1).

Figure 13.2 shows how the major (σ_1) and minor (σ_3) principal stresses are redistributed around a development excavation. The major principal stress and the deviatoric stress (σ_1–σ_3) are the central factors influencing rock mass behaviour around excavations formed in the rock.

It is worth remembering that deviatoric stresses are greatest near the excavation boundaries and that shear stress results from the deviatoric stress. Shear stress in excess of shear strength results in the propagation of pre-existing geological features and sliding on any unstable discontinuities. Additionally, intact rock failure may be initiated by cracking due to tensile stresses close to the excavation boundary with eventual coalescence in the fracture network.

At some distance into the rock mass from the excavation surface, the rock mass is expected to be more stable and less likely to fail. To minimize the depth of potential failure, ground support that provides immediate reactions to rock mass movement is required. Its task is to decrease the depth of damage (or failure) by providing confining stress (i.e., increase in minor principal stress) and limiting rock mass displacements at the excavation boundary. It is well known that the maximum shear stress occurs in the plane of the major and minor principal normal stresses. When stresses change due to the creation of the excavation, the orientations of the major, minor and intermediate principal stresses also change, and the orientation of the maximum shear stress changes in accordance. These stress changes, combined with the dynamic effects of blasting, may result in an increase in fracture frequency within the rock mass. Failure does not necessarily occur at the same time as the formation of the excavation. The stress changes may induce a gradual increase in micro fracturing and eventually failure at an unknown time later. It is, therefore, important to consider the complete

FIGURE 13.1   Example of stress-driven failure propagating to a large geological discontinuity.

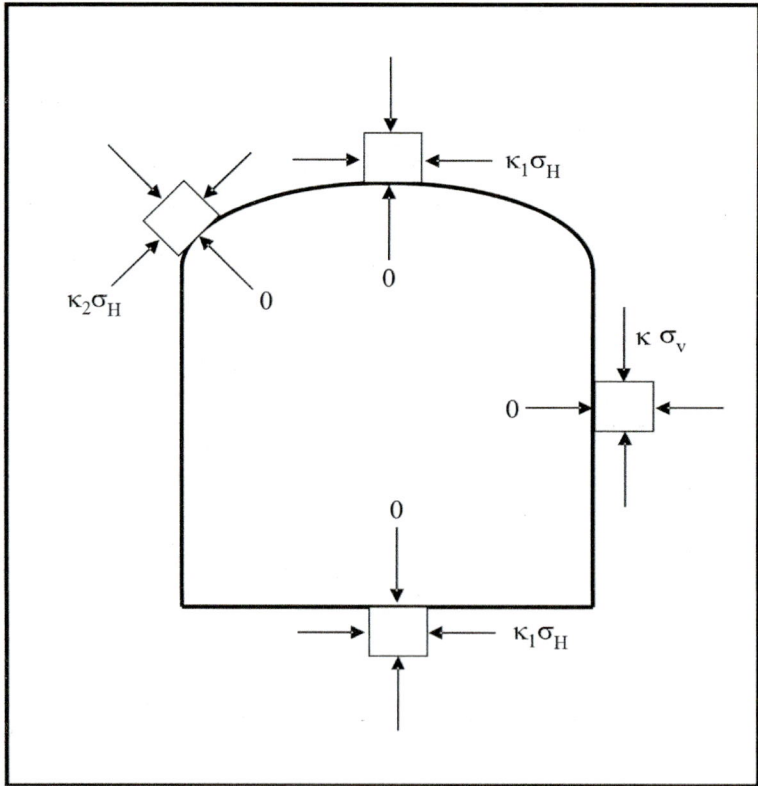

FIGURE 13.2   Schematic of stress redistribution around an underground excavation.

'stress path' that results from the initial creation of the excavation together with subsequent mining activities.

It is very important to realize that it may be possible to minimize or eliminate rock failure by modifying the excavation from conventional regular shapes (e.g., square or rectangular) with flat surfaces to excavations incorporating concave surfaces. Analyses have shown that the stresses in the rock in the backs, shoulders, walls and floors of excavations with flat surfaces often fail at lower stresses than when the excavations are created with concave surfaces. Thus, excavation shape becomes an important design parameter when mining at great depth.

A secondary consideration concerns the local 'curvature' and 'roughness' of the excavation surface, which can either assist or compromise the installation and effectiveness of both support and reinforcement.

In the context of sudden, stress-driven instability of deep mine tunnels experiencing an associated violent release of strain energy at the tunnel boundary, four generic mechanisms of excavation failure have been identified (Figure 13.3). These mechanisms range from shallow spalling to ejection of large structurally controlled blocks, rupture and displacement on significant geological features (i.e., faults) intersecting the excavations and shear/crushing failure of pillars formed between adjacent excavations. Scaled-down laboratory experiments by Kusui (2015) have shown that spalling of excavation surfaces subjected to very high compressive stress occurs prior to the initiation of crushing or shear failure in any adjoining pillars. Furthermore, observational evidence of excavation failures in many underground mines indicates that ejection of large structurally controlled blocks occurs as an intermediate consequence, that is, after spalling, but before pillar failure.

Empirical evidence indicates that when development mining experiences very high induced stresses, the instability of large, structurally defined blocks takes time to develop, often occurring once the development face has advanced by 20 m or more beyond the affected area. By contrast, fault rupture may potentially occur at any time during the development cycle, as this mechanism occurs over a large scale, often controlled by factors such as global changes in mine extraction geometry and the associated stress field adjustments (Drover and Villaescusa, 2015a). The conceptual progression of stress-driven excavation damage mechanisms is illustrated in Figure 13.4 as a load-displacement graph that is based on the experimental results obtained by Kusui (2015) as well as recent observational evidence from underground mines (Drover and Villaescusa, 2022).

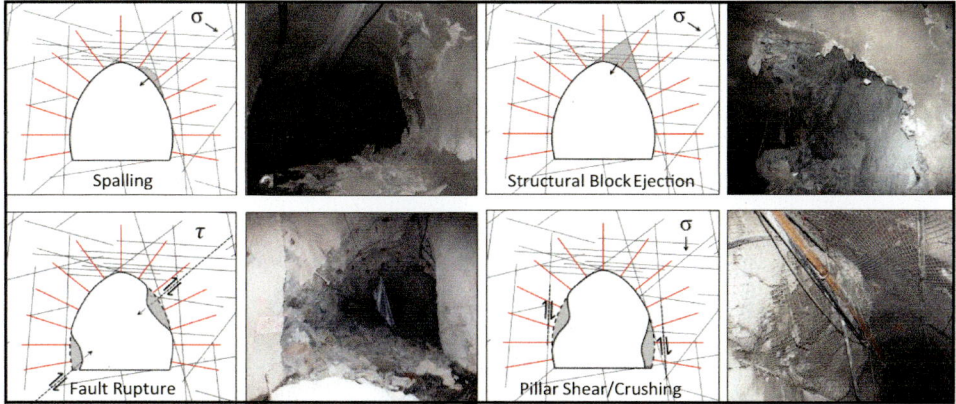

FIGURE 13.3 Modes of tunnel failure involving violent strain energy release at the excavation boundary (Drover and Villaescusa, 2022).

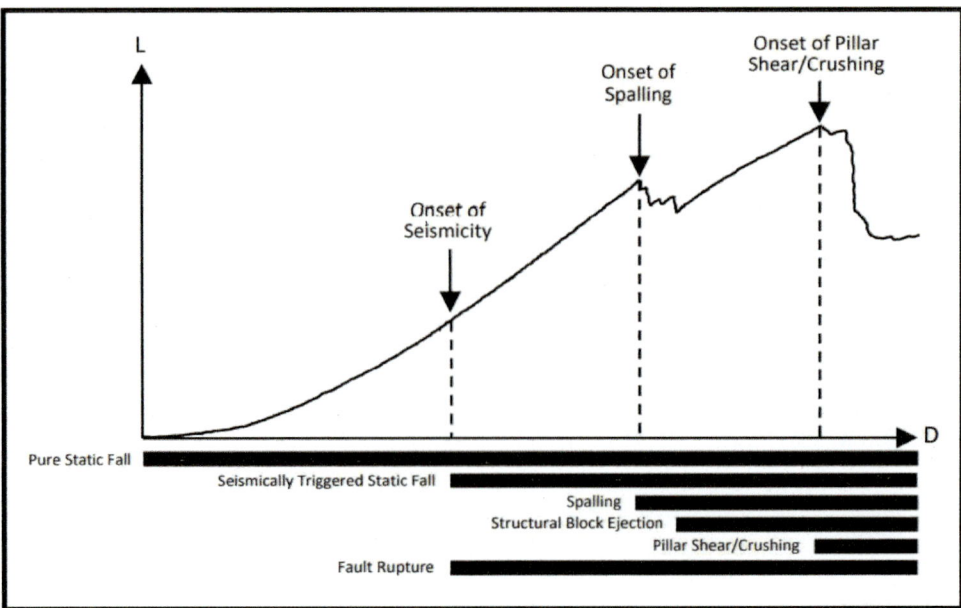

FIGURE 13.4 Suggested progression of rock mass failure modes at the excavation boundary presented in load (L)-displacement (D) space (Drover and Villaescusa, 2022).

13.3   CONSTRUCTION

A key construction objective of development blasting is to maximize advance rate while consistently achieving at least 95% utilization of the drilled metres. The advancing rate must be achieved while the extent of the disturbed zones, as well as the visible damage around the excavation, such as overbreak, is minimized. The variables influencing the development rate are numerous and include, among other things, a sudden change in rock mass conditions, water in holes, hole collapse leading to non-uniform charging with hole depth, difficulty in drilling due to floors not being adequately cleaned, poor cut performance, development jumbos not fit for purpose, poor explosive performance, as well as the type of ground support being implemented.

The blast holes are drilled into the rock faces, generally parallel to the development axis and to depths typically ranging from 3.0 to 4.1 m. Three critical groups of holes can be defined, including the burn cut, stripping holes and perimeter holes (Figure 13.5). The actual advances are largely influenced by the performance of the cut, as poorly drilled and charged cuts correlate well with poor advances. The uncharged relief holes at the cut provide the initial void to minimize 'freezing'. For an average hard rock mass, typically mined at great depth, an approximate 20% initial void is required. Experience indicates that wall stability and ground support requirements are, in part, controlled by stripping and perimeter holes (i.e., charge concentration, burden and spacing). Stripping holes also control fragmentation (an optimum charge of about 2/3 of the borehole length is often employed) with the lifter holes, drilled at a reduced burden and fired last to control both floor conditions and muck pile shape.

For deep mining tunnels with widths up to 12 m, a complete round of blast holes is typically detonated in a pre-established order so that the entire face is broken out to the drilled depth. The perimeter holes need to be drilled accurately and charged using heavily decoupled explosives. In general, Short Period delays result in a lower, longer and flatter profile, but air-blast damage may be experienced. Therefore, the use of Long Period non-electronic delays (ranging from 200 to 9500 ms) is recommended unless a sulphide dust explosion is thought likely. In general, the number of holes per

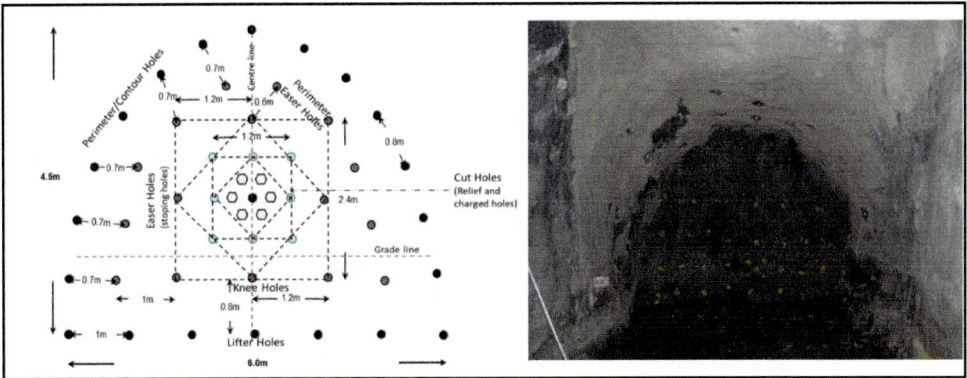

FIGURE 13.5 Example of development drill pattern and the resulting shape for a mine main decline development to produce a semi-elliptical excavation shape.

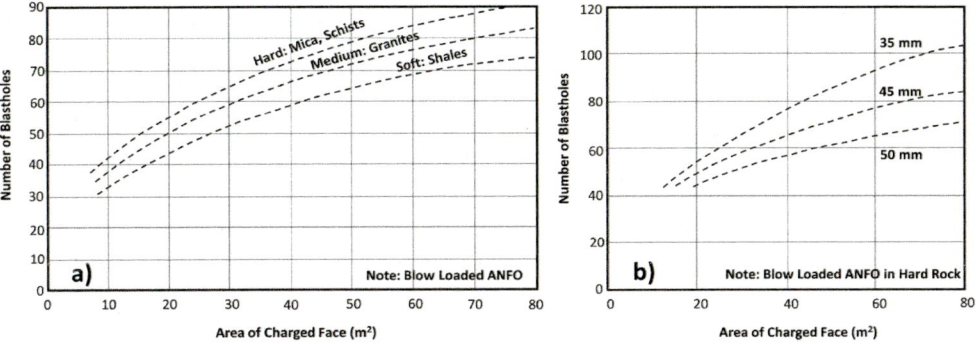

FIGURE 13.6 Relations between the number of blast holes required and charged face area for (a) 45-mm diameter holes and three rock types and (b) range of hole diameters in hard rock conditions (Elith, 1990).

face depends upon the rock type, the hole diameter and the excavation face area (Figure 13.6). These findings are supported mainly by rules of thumb developed from vast experience in competent rock masses (Elith, 1990). However, complex geologies require continuous site-specific improvement strategies supported by field monitoring and local observation.

13.3.1 EXCAVATION SHAPE

The historical development excavation profiles employed by the mining industry have been largely rectangular-shaped. The perceived advantages are the minimization of waste rock removal and the ease of subsequent production blast hole drilling. Figure 13.7 shows typical rectangular-shaped excavations where ground support practices ranging from spot bolting alone to the systematic arrangement of friction bolts and weld mesh have been implemented. The ground support is preferentially installed on the roof of the excavations. Variations of this excavation shape and ground support strategy can be effective at shallow depths below the surface. However, as the depth of mining increases to moderate depths, say 500–800 m below the surface, the tunnel shape is often changed to rectangular with an arched roof, as shown in Figure 13.8.

Figures 13.9–13.11 show a comparison of resulting shear stress at the boundaries of deep excavations with different excavation shapes. The non-linear modelling results have been calculated

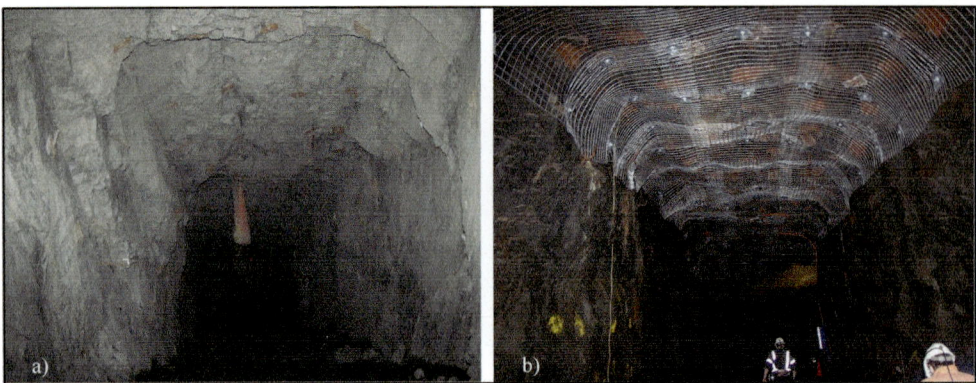

FIGURE 13.7 Examples of rectangular-shaped excavations with the roof stabilized using (a) spot bolting and (b) a pattern of friction bolts and weld mesh installed in the roof.

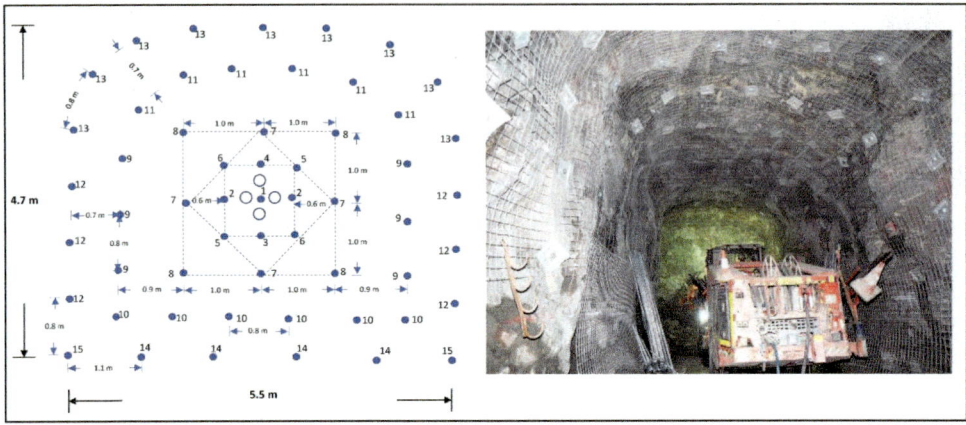

FIGURE 13.8 Drill design and detonation sequence for development blasting in hard rock to form a rectangular excavation with an arched roof.

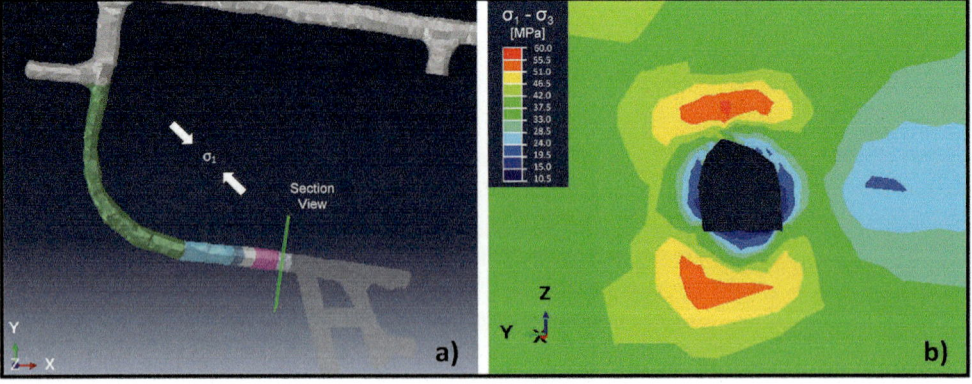

FIGURE 13.9 Plan view of a deep tunnel showing (a) the location of the cross-section views where the induced stresses were calculated for several excavation shapes and (b) resulting stress concentration on the actual constructed shape.

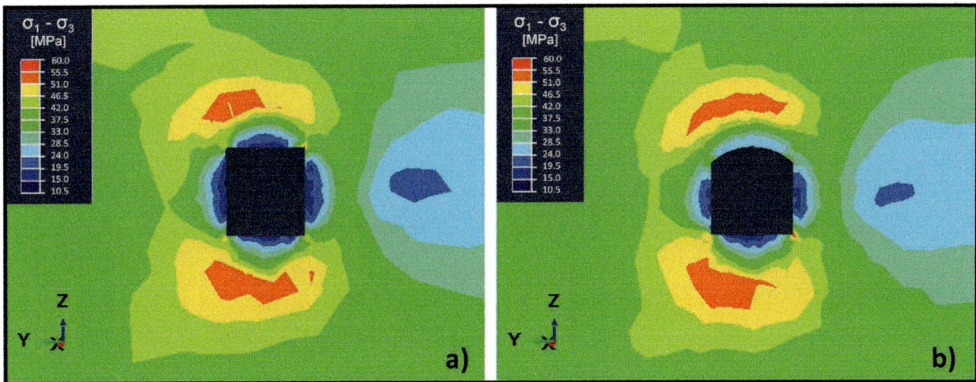

FIGURE 13.10 Cross-sectional view showing the calculated induced stress on (a) a simulated rectangular shape and (b) stress concentration on a simulated excavation with an arched back.

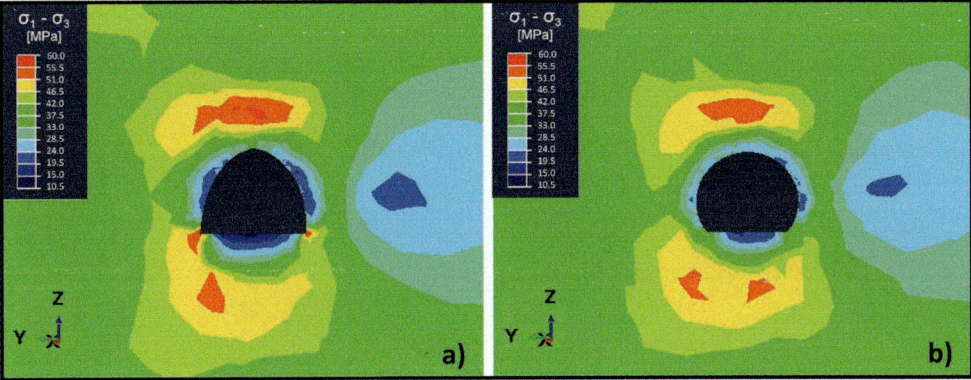

FIGURE 13.11 Cross-sectional view showing the calculated induced stress on (a) a simulated ovaloid shape and (b) stress concentration on a simulated semi-circular shape.

using the Abaqus software. The results show that semi-circular excavations are more stable (i.e., experience less shear stress). The results also suggest that additional stability could be achieved by mining a tunnel invert, which can be subsequently filled in. Therefore, when mining at great depth, the stress concentrations at the excavation corners can be minimized by modifying the excavation shapes from having an arched roof profile with sub-vertical walls to a semi-circular excavation shape (13.12). These 'rounded' shape excavations perform well when located at a great depth where stress-driven failures are common. Many of the results shown in this book relate to semi-circular excavations, which are recommended for mining at great depth, as they tend to be self-supporting and reduce subsequent spalling while preferentially responding in compression when suddenly loaded by a large dynamic event.

13.3.2 De-Stress Blasting

Destress blasting is a tunnel construction technique implemented to fracture the rock so that strain energy is dissipated away from the excavation to a deeper heading within the rock mass while also controlling the resulting deformation (Drover et al., 2018). The technique is implemented to reduce the intensity or frequency of strain bursting or spalling events in high-stress mining environments (Figure 13.13). It relies on small explosive charges used to propagate or mobilize geological

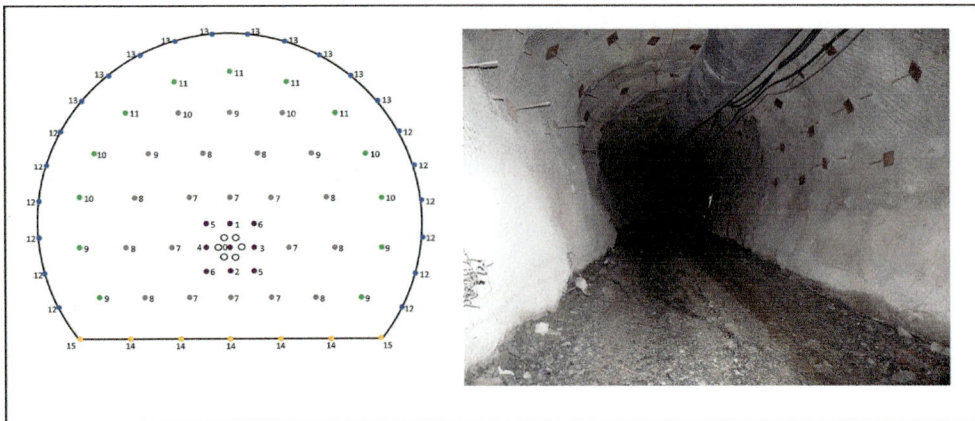

FIGURE 13.12 Example of drill design, detonation sequence and the resulting shape for a tunnel driven in hard rock using a semi-circular excavation.

FIGURE 13.13 Example of violent, sudden ejection of rock fragment spalling from an unsupported, highly stressed excavation face.

structures present within the rock mass environment in the immediate vicinity of the development headings.

The success of a destress blasting program depends on many factors. These include the location of the spalling events, the orientation of the geological discontinuities with respect to the prevailing stress magnitudes and orientations as well as the number of de-stressing holes and the ability of the charges to mutually interact. Several items of operational, geotechnical and safety concerns should be considered prior to the implementation of a development de-stress blasting program. These include:

- Effect on cycle time and advance rate in high-priority development excavations,
- Immediate influence on overbreak and subsequent application of increased surface support,
- The requirement of additional deep reinforcement if excessive de-stressing damage occurs,
- Additional exposure of personnel to the charging face and
- The requirement to monitor the long-term stability of pre-conditioned drives.

The operational impacts are well-documented increases in the development cycle time due to additional drilling, charging and stemming of de-stressing holes and clearing of sockets or butts upon re-entry.

13.3.2.1   Mechanics

The methodology involves detonating short charge lengths at relatively large blast hole spacings to create small 'pockets' of disturbed material confined within the rock mass either ahead of the new face or immediately around it. For de-stressing to be effective, the continuity of the small-fractured pockets must be ensured before any re-compaction of the fractures occurs that may limit the effect on the overall behaviour. It must be noted that de-stressing does not create any stress shadows; instead, the broken zones sufficiently pre-condition the rock mass to induce immediate damage from localized stress concentrations that mobilize geological structures.

Thus, the ideal outcome of the de-stress blasting is that small-scale damage allows for the induced stresses and accumulated strain energy at the face to be relieved through a slip of discontinuities or fracture propagation near the excavation boundary immediately following de-stress blasting. It is desirable that rapid stress redistribution and small-scale rock mass failure occur prior to personnel returning to the face, thus reducing the risk to exposed workers. It was the experience and returned the desirable measure of success for a development de-stress blasting program conducted at the Inco Creighton Mine in Canada (O'Donnell, 1992), where approximately 85% of the strain bursting events were experienced immediately following a drift blast with de-stress holes.

Approximation of the extent and orientation of the confined blast-induced fractures is difficult to predict accurately due to the interaction of complex rock mass conditions with excavation orientation, *in situ* stresses and the dynamic effect of explosive detonation. The University of Queensland's Hybrid Stress and Blasting Model, HSBM (Furtney et al., 2009), can be used to provide theoretical validation of design parameters, such as confining stresses, borehole diameter, spacing and explosive type prior to any field trials of de-stress blasting (Drover et al., 2018). The creation of adequate fractured zones around the points of de-stress blasting is a key ingredient in ensuring rock mass destabilization and induced seismicity at the time of blasting. The fracture zones within the confined rock mass must be large enough to destabilize the rock fabric but not intersect the development boundary, which may result in immediate overbreak (Figure 13.14).

Many previous studies have confirmed that the magnitudes and orientations of the principal stresses strongly influence the explosive-induced fracturing of rock materials (e.g., Kutter and Fairhurst, 1971). The existing data indicate a preferential fracture direction parallel with the orientation of the main principal stress. However, due to often highly confined fracture tips, a limited extent of explosive-induced fresh fracturing should be expected in a high-stress environment. Thus, gas penetration into freshly formed or pre-existing fractures would be limited, thereby reducing the extension of radial fractures as would be expected to occur in unconfined conditions.

Figure 13.15 presents HSBM modelling results, indicating that the extent of fracturing around detonating charges is dependent upon the induced stress. When the applied stresses are set to $\sigma_1 = 54$ MPa and $\sigma_3 = 27$ MPa (typical of the transition to great mining depths), the results indicate that the zones of micro-fracturing damage do not interact between widely spaced de-stressing charges. The limited degree of fresh fracturing due to confinement would indicate that the likely mechanism

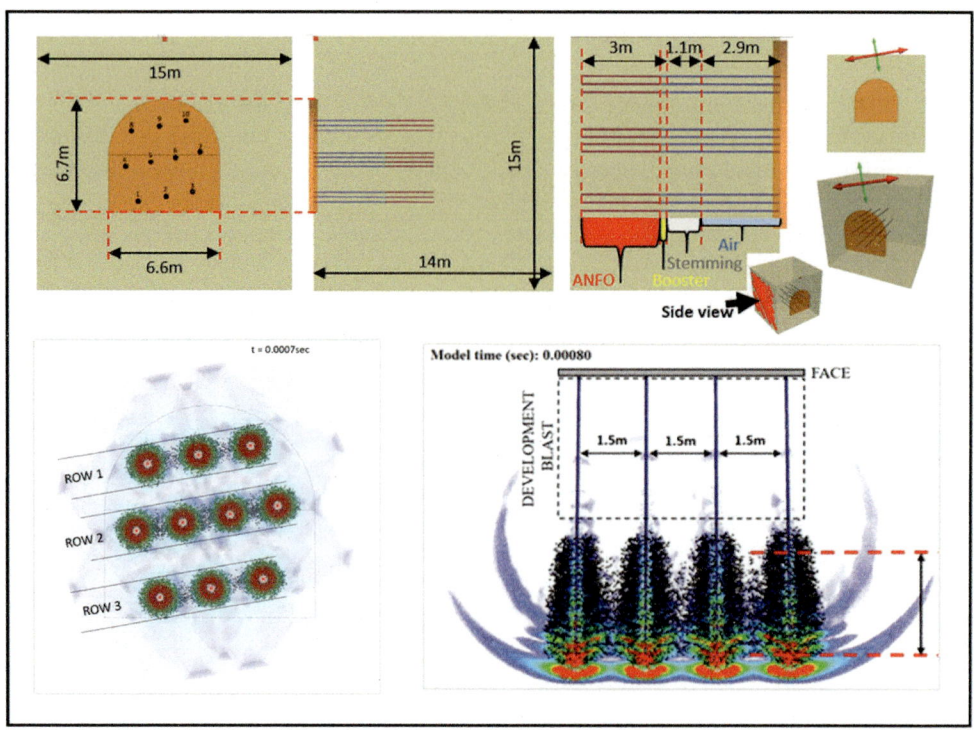

FIGURE 13.14 HSBM modelling results indicate fracture interaction limited to within rows of de-stressing holes, with no crack formation across the burden (Drover et al., 2018).

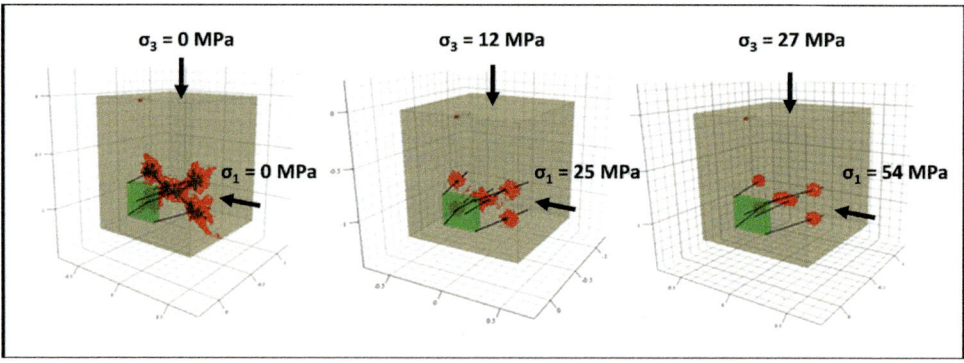

FIGURE 13.15 HSBM model of damage interaction under induced stress level that should be expected as the excavations progress to greater mining depths.

of de-stress blasting would be degradation or dilation of existing low-strength or preferentially-oriented geological discontinuities. This suggestion is supported by the observation that de-stress blasting has been more effective in highly jointed rock masses compared with massive rock masses (O'Donnell, 1999). More intensive de-stress blasting would, therefore, be required for more competent rock masses to achieve comparable results.

13.3.2.2 De-stress Blasting Patterns

A review of de-stress blasting patterns for development mining indicates that at least four corner holes and two face holes are required (Carr et al., 1999). The corner and wall holes are typically drilled across the length of the advancing round and angled into the walls and backs by 30°–45°. These holes are typically toe-charged with ANFO or a cartridge emulsion for less than 25% of the blast hole length. The face holes typically extend twice the cut depth to achieve a conditioned zone ahead of the face. These holes usually have a larger diameter, given that drill rods often need to be coupled to reach the area of interest ahead of a conventional cut length. The holes are typically charged over 1.5–2.5 m with ANFO (Figure 13.16).

Drover et al. (2018) optimized the geometry and sequence for a 12-hole de-stress blasting pattern in a hard rock mass structured with tightly healed discontinuities (Figure 13.17). The UCS of

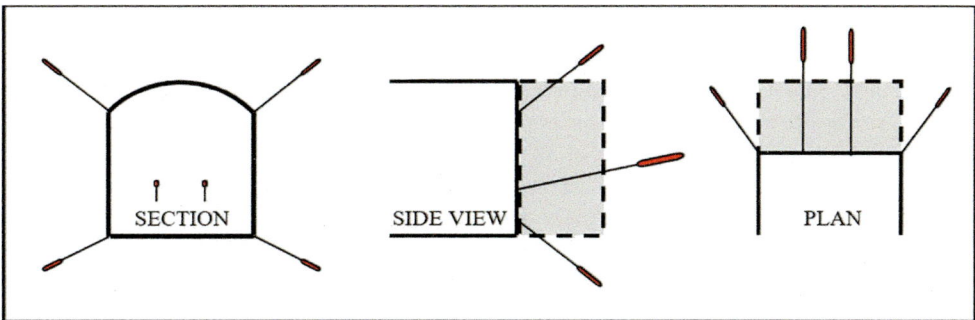

FIGURE 13.16 Conventional de-stress blast pattern for mining development incorporating four corner holes and two face holes (Carr et al., 1999).

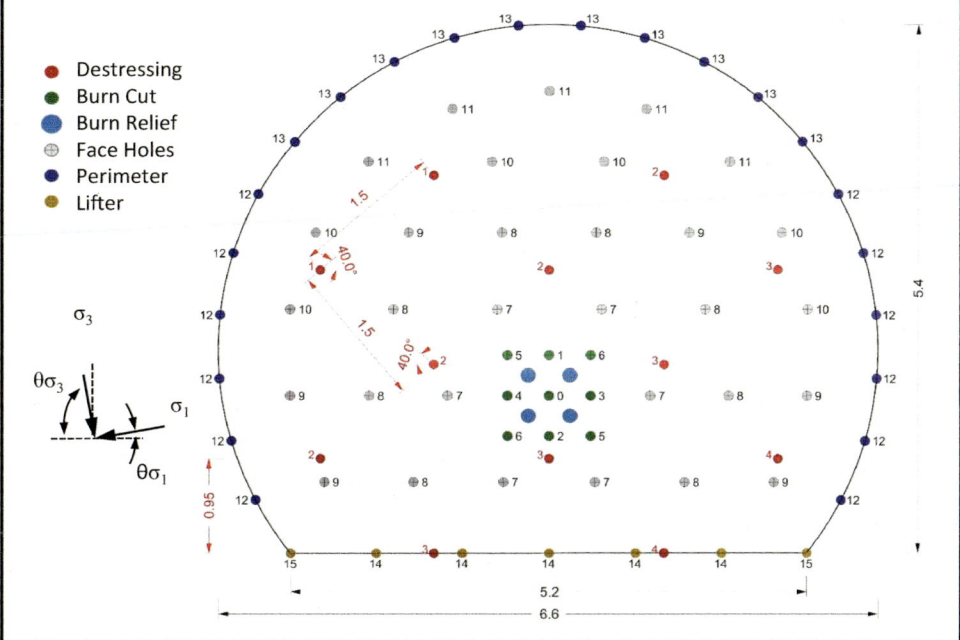

FIGURE 13.17 De-stress blast pattern and detonation sequence, optimized for case-specific rock structure, strength and stress conditions at a deep mine (Drover et al., 2018).

the intact rock was 150 MPa, and the major principal stress was 60 MPa, oriented sub-horizontally and perpendicular to the tunnel axis. The pattern design was based upon the orientation of the joint sets at the excavation face, the plunge of the major principal stress ($\theta\sigma_1$), the inclination of the rows of de-stressing holes (θr), as well as the optimal charge spacing (s) and burden (b) from HSBM modelling. To promote localized shear deformation, the de-stressing charges were aligned in rows with a 30° inclination with respect to the plunge of the main principal stress. As mentioned earlier, it would be ideal if the de-stressing holes also aligned with respect to at least one set of geological discontinuities for the gas penetration to be more effective in crack propagation.

For a circular excavation of a width of 6.4 m and a height of 5.2 m, a 1.5-m burden requires four inclined rows of de-stressing charges to be positioned within the limits of the development blast. This would generate four continuous parallel de-stressing fractures (Figure 13.18). Fracture interaction is unlikely to occur sub-perpendicular to the major principal stress (Jung et al., 2001), and therefore, the spacing is the most important design parameter when assessing crack interaction. The spacing also dictates the amount of explosive work required for in-row fracture propagation and coalescence between the de-stressing charges. The maximum length of radial fracturing can also be estimated using HSBM modelling or field trials. The magnitude of the major principal stress controls fracture tip confinement, whereas its plunge indicates the optimal inclination angle (θ_r) for the rows of de-stressing charges (Drover et al., 2018).

It is emphasized that, other than for the face and floor, the de-stress blast design methodology suggested by Drover et al. (2018) deliberately avoids the placement of explosive charges beyond the planned excavation perimeter. This is to minimize interaction with subsequent rock reinforcement drilling and installation and to minimize mobilization of potentially unstable block geometries. The recommended practice here is to offset the de-stressing charges from the desired excavation perimeter by a minimum distance equal to that of the burden between the contouring charges and the outer row of stripping holes.

13.3.2.3 Explosive Energy
The recommended length for the de-stressing charges may range anywhere from 65% to 80% of the length of the holes in the main development round (Drover et al., 2018). The actual length may be controlled by the capabilities of available drilling equipment and consumables. It is critical to ensure that an uncharged buffer zone, approximately 0.3–0.5 m thick, exists between the toe of the development round and the collar of the de-stressing charges. This explosive-free 'buffer' zone

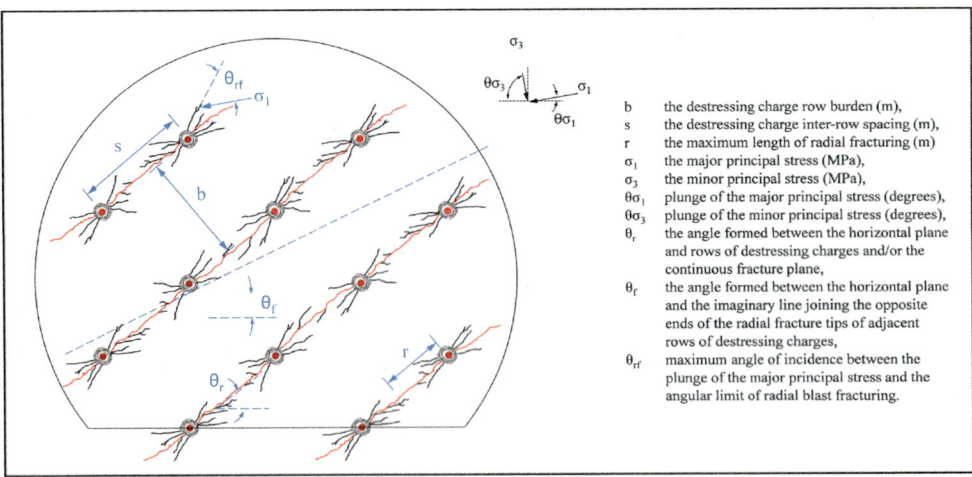

FIGURE 13.18 Development face de-stressing concept and design parameters for a suitable de-stress blast hole pattern (Drover et al., 2018).

is required to prevent sympathetic and undesirable detonations of the toe-primed main development charges initiated after the de-stressing holes. That is, if a 3.5 m long development round, with a 0.5-m buffer zone, is considered, a typical de-stressing charge should be 2.5 m in length (Figure 13.19). The required borehole length would be 6.5 m, and rod couplers may be needed to allow standard drilling equipment to achieve this hole length. This may necessarily increase the de-stressing borehole diameter to 63 mm.

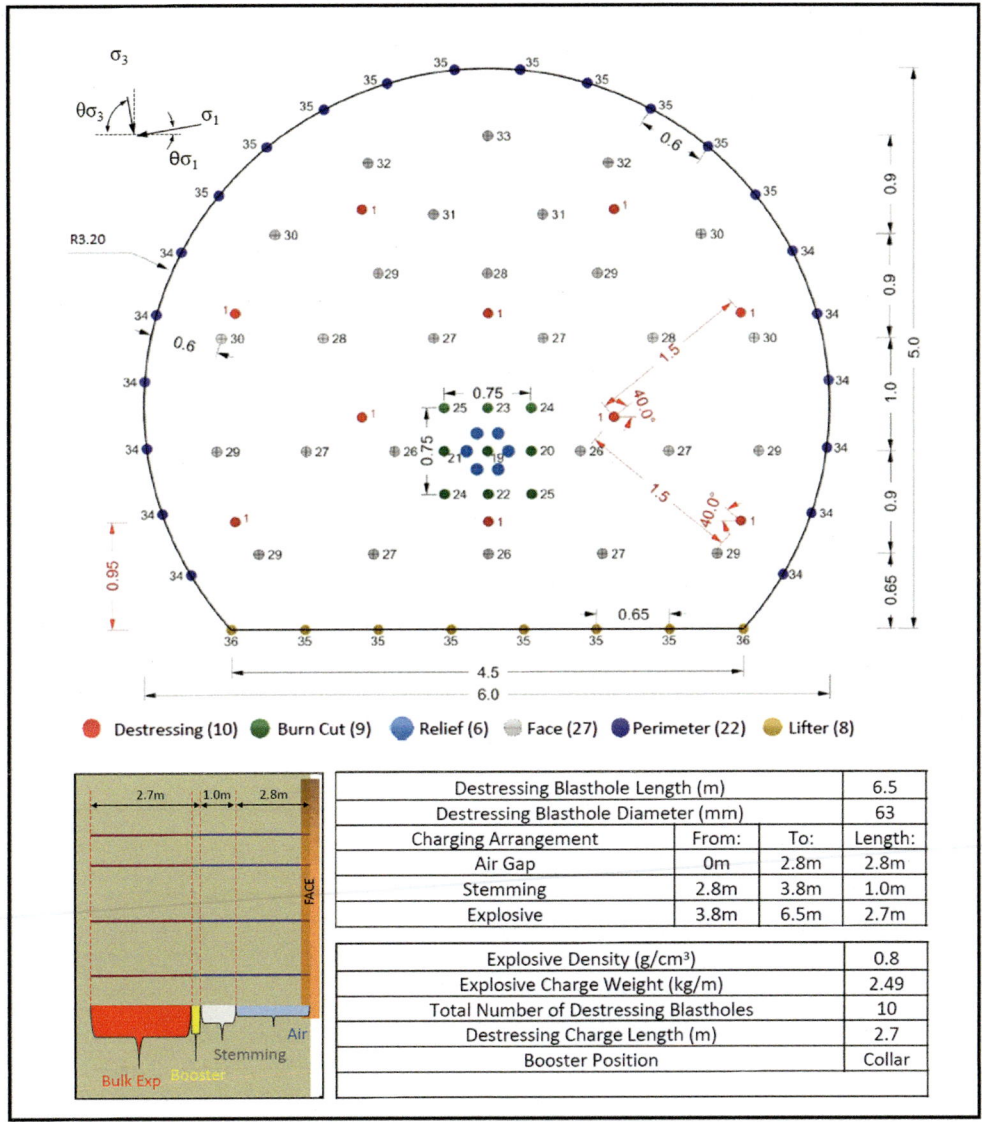

Destressing Blasthole Length (m)			6.5
Destressing Blasthole Diameter (mm)			63
Charging Arrangement	From:	To:	Length:
Air Gap	0m	2.8m	2.8m
Stemming	2.8m	3.8m	1.0m
Explosive	3.8m	6.5m	2.7m

Explosive Density (g/cm³)	0.8
Explosive Charge Weight (kg/m)	2.49
Total Number of Destressing Blastholes	10
Destressing Charge Length (m)	2.7
Booster Position	Collar

FIGURE 13.19 Example of a face de-stressing charge arrangement for a 4.5 m wide × 5.0 m high excavation with a semi-circular profile.

The explosive charges significantly affect the outcome of a de-stress blast. Considering that the main objective is to create new fractures and to penetrate the network of pre-existing geological discontinuities, explosives generating high gas volumes, such as ANFO products, are recommended. In addition, HSBM modelling indicates that collar priming of the de-stressing charges ensures a detonation front is developed, which radiates the energy towards the area of interest ahead of the final face of the conventional development round (Drover et al., 2018).

The choice of stemming material and its length is also important to ensure adequate confinement of the explosive gases, as high borehole pressures maximize the likelihood of fracture connectivity between adjacent charges. Fieldwork reported by Drover et al. (2018) suggests that a 1.5-m column of tamped wet clay packs can be used to seal the gases for a typical de-stress blast. Alternatively, a more expensive cement grout plug could also be used to contain the explosive energy of the gas. Operationally, it is easier to leave the holes unstemmed; however, this would result in a significant loss of borehole pressure leading to a reduced length of radial fracturing.

13.3.2.4 Micro Seismic Activity

A face de-stress blasting program implemented during the construction of two experimental tunnels at great depth has been reported by Drover and Villaescusa (2019). One tunnel was constructed using de-stress blasting, and another was developed nearby, in the same geotechnical environment, using conventional blasting to act as the experiment control. Both tunnels were developed parallel to each other and perpendicular to a high sub-horizontal stress field. High-resolution seismic monitoring instrumentation was used to record the data used to compare the seismic response generated by each excavation. Analysis of the seismic data from the control tunnel showed a developed seismogenic zone extending up to 3.6 m ahead of the face. De-stress blasting within the corresponding zone of the adjacent tunnel had the effect of reducing the rock mass stiffness, primarily due to the weakening of geological discontinuities. This was inferred from the spatial broadening of the seismogenic zone, as shown in Figure 13.20.

Figure 13.21 shows a quantitative comparison of the seismic event count between the conventional and the face de-stressing development excavations. The figure shows a longitudinal view of the average number of seismic events per development cycle, occurring at a set distance from both tunnel faces. The characteristic seismic response to conventional development blasting results in a relatively small number of seismic events adjacent to the excavation surfaces where permanent ground support had been installed in previous cuts. The number of seismic events then increased exponentially adjacent to the unsupported ground, including on the floor of the excavation. The number of events in the seismogenic zone peaked at approximately 1.5 m ahead of the face and

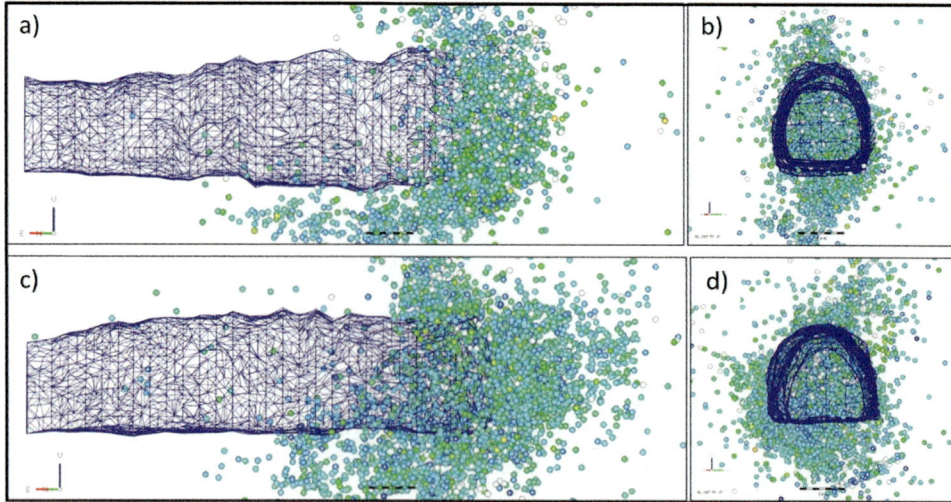

FIGURE 13.20 Spatial characteristics of the seismogenic zone for response to conventional tunnel blasting (a and b); and response to de-stress tunnel blasting (c and d) (Drover and Villaescusa, 2019).

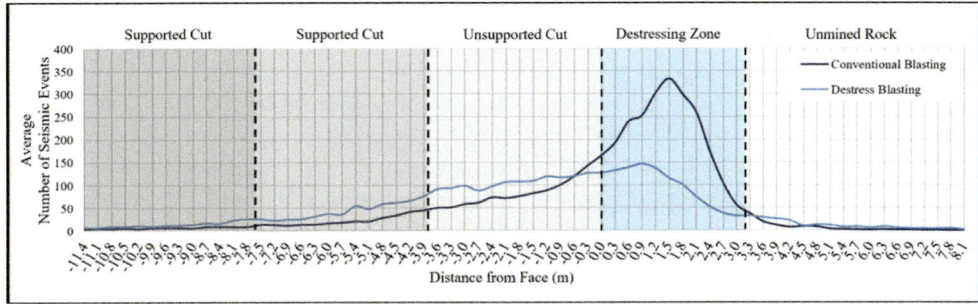

FIGURE 13.21 Examples of the average number of seismic events per development cycle for both conventional and face de-stressing blasting techniques (Drover and Villaescusa, 2019).

then decayed rapidly over an additional distance of +1.5 m to +3.6 m ahead of the face. Very little seismicity was recorded beyond this distance, thus defining the length of the zone that required de-stressing to about half the excavation width. The rate of change in the spatial frequency of events on either side of the event count peak was more gradual in the de-stressing tunnel, as indicated by the lower slope in the data plot. This indicates that the strain gradients adjacent to the de-stressed face were relatively low compared with the steep increase and decay gradients in the event count curve for the conventional development.

This is a very significant difference, as a higher potential for violent strain energy release is a clear possibility from the rock mass around the conventionally blasted tunnel.

A reduction in rock mass stiffness following de-stress blasting was also evident in the wider range of seismic source mechanisms recorded adjacent to the de-stressed tunnel (Figure 13.22). The source mechanisms data showed that de-stress blasting induced instability on all pre-existing geological discontinuity sets (Drover and Villaescusa, 2019). Weakening or mobilization of geological discontinuities facilitates a greater diversity of failure modes, effectively reducing the rock mass stiffness. This minimizes the likelihood of high strain accumulation, which can result in a single large block instability, usually involving a face-forming discontinuity set or other adversely oriented sets.

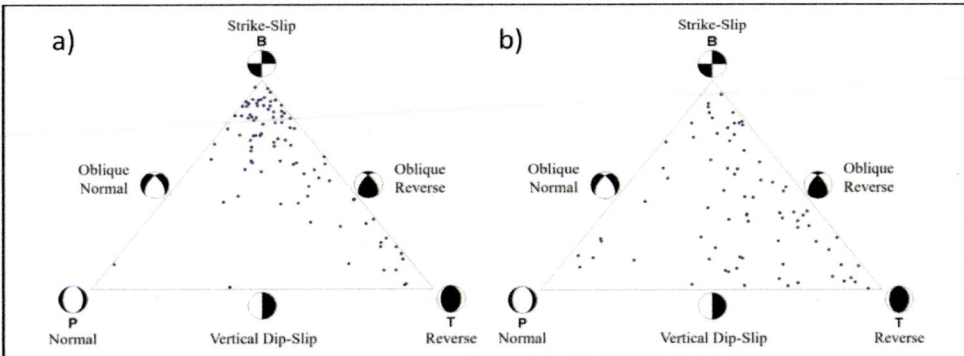

FIGURE 13.22 Ternary diagrams of source mechanisms following (a) conventional and (b) face de-stress blasting (Drover and Villaescusa, 2019).

13.3.2.5 Geotechnical Concerns

Several operational, blast performance and safety risks can be identified for a development de-stress blasting program. These risks include drilling into undetonated explosives, misfire or sympathetic detonation of the main blast round and an additional overpressure in the drive from an instantaneous firing of multiple over-confined blast holes among other risks. Detailed procedures for assessing and mitigating these risks are mandatory prior to the implementation of a de-stress blasting program.

In addition to operational and safety concerns related to de-stress blasting programs, the short-term and long-term geotechnical behaviour of the resulting excavation must be considered. The requirement for additional ground support to ensure long-term stability of the damaged rock mass should be considered, along with the time-dependent response of the excavation to changing stress conditions during subsequent mining. Stress relaxation leading to a loss of confinement around the excavation could propagate wedge formation or trigger unravelling if the damage zone extends to the drive surface.

13.3.2.6 Case Study—Kanowna Belle Mine, WA, Australia

The Kanowna Belle deposit is located 20 km northeast of Kalgoorlie, in Western Australia. The ore body is hosted in the Fitzroy structural zone, which delineates a hangingwall and footwall sequence. The hangingwall sequence comprises felsic, volcaniclastic siltstones and pebble conglomerates intruded by basaltic, andesitic and porphyritic dykes. The footwall sequence comprises volcanogenic conglomerates separated by thin lenses of arkosic grit, sandstone and felsic units. The Fitzroy structural zone hosts the ore body and consists of the Fitzroy Mylonite, the Fitzroy shear zone and the Fitzroy fault. It is interspersed with gold-hosting porphyritic intrusions (Ross et al., 2004). The mine has been divided into stoping blocks based on depth, lettered from A to E. Varden and Esterhuizen (2012) have outlined the history of seismicity at Kanowna Belle, suggesting that seismicity is structurally controlled and that its magnitude and intensity have both increased as the mine was deepened. The results presented here relate to the E block, which is located at a depth of 1,135 m below the surface (Figure 13.23).

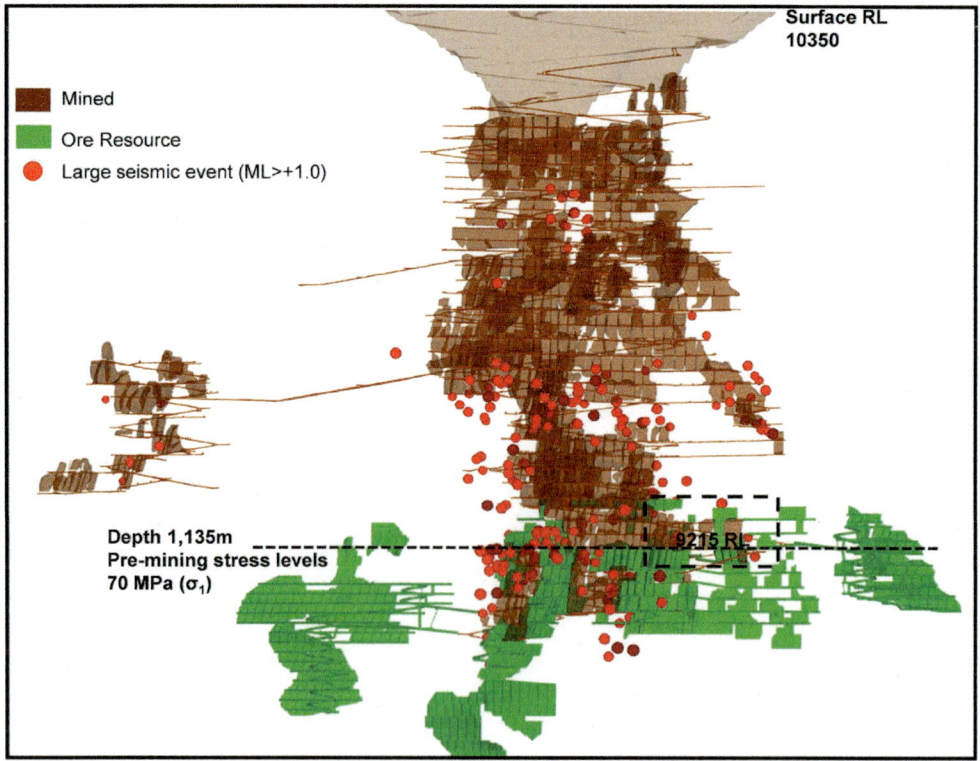

FIGURE 13.23 A long section view of the Kanowna Belle Mine showing stoping blocks with large seismic events (E block is situated at RL 9215 marked by the dashed line).

RL 9215 is a development and stoping level in the Kanowna Belle Mine E-Block and marks the first mining activity to be undertaken in that part of the mine for several years following a period of intense seismic response to mining during 2014/2015 (Figure 13.24). RL 9215 forms a trial area for the implementation of High Energy Dissipation Ground Support techniques (Arcaro et al., 2021) and the development of de-stress blasting, conducted as part of the 'Mine Development at Great Depth' Research Project conducted by WASM and Northern Star Resources.

The modelled (magnitude and vectors orientation) of the mining-induced major principal stress (σ_1) adjacent to the 9215 FW (Footwall) access development take-off point is illustrated in Figure 13.25. The model indicates that the early decline development is sub-parallel to the induced σ_1, passing through a zone of high abutment stress of greater than 90 mPa below the pre-existing RL 9245 stope panels. At this point, the access development grade flattens parallel to the orebody. By that stage, the development is oriented sub-perpendicular to the sub-horizontal σ_1. This is reduced relative to that for the early access but still consistently above 70 mPa.

A high seismic potential was forecast throughout the RL 9215 precinct during both development and subsequent production. Figure 13.26 shows a plan view of modelled Rate of Energy Release (RER) from the rock mass, as defined by Levkovitch et al. (2010). The modelling results indicate the zones of High (>200,000 J/m³/s), Medium (>100,000 J/m³/s) and Low (>25,250 J/m³/s) energy release for the bulk rock mass (as distinct from RER due to fault slip). It must be noted that the modelled RER reflects all forms of rock mass energy release, of which seismic activity is only one component (Levkovitch et al., 2010). The results indicate similar levels of seismicity are generated along the entire length of the development, starting at the access stockpile. Due to the orientation of the major principal stress, development parallel to the orebody is likely to experience a higher risk of violent face spalling requiring de-stress blasting.

FIGURE 13.24 Examples of excavation and ground support damage following a large seismic event.

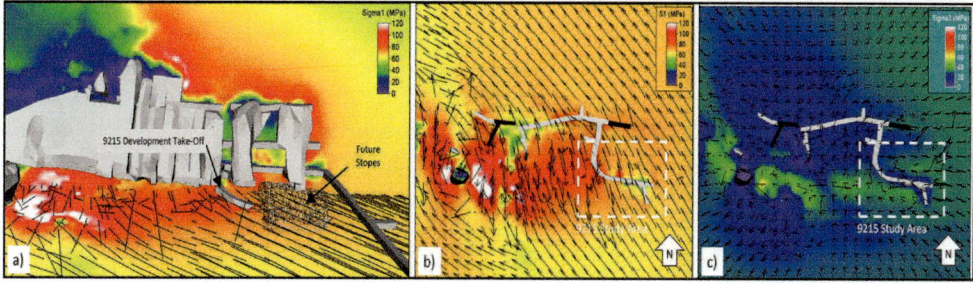

FIGURE 13.25 Mining induced stress conditions for initial 9215 planned development computed using the program Abaqus providing (a) 3D view, Major principal stress, (b) Plan View, Major principal stress and (c) Plan View, Minor principal stress.

FIGURE 13.26   Modelled seismic potential (Beck, 2008) during the planned sequential development of the 9215 Footwall Access drift and the first stope access cross-cut.

The predicted zone of high seismic potential envelops both the stope access cross-cuts as well as the footwall access drift, extending around one stope diameter from the immediate stope boundary into the footwall. A high potential for Fault-Slip RER occurs on an oblique shear structure bisecting the first two scheduled stopes at 9215. The incremental strains and their associated displacement vectors for the initial production stage were computed as shown in Figure 13.27. The areas that experience the largest strains are intersected by the oblique shear structure and are, arguably, the locations of the greatest damage potential.

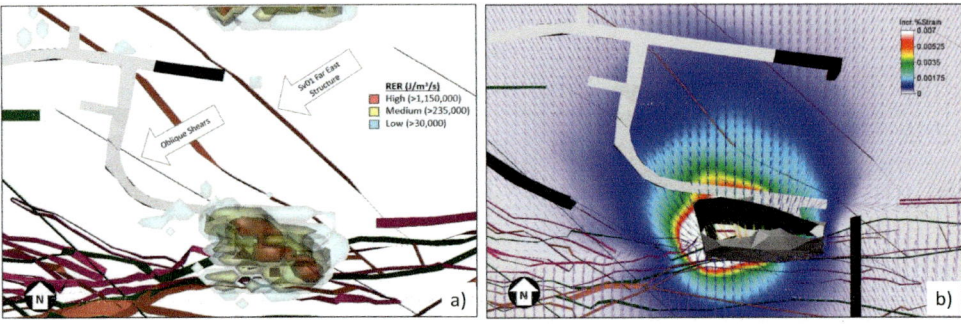

FIGURE 13.27   Model of planned stope production showing (a) fault-slip seismic potential and (b) incremental strain (Beck, 2008).

The experimental research in the 9215 FWD (E Block) is ongoing, with continued systematic seismic data collection and analysis. Figure 13.28 shows the location of the array of triaxial accelerometers used to monitor seismicity around the 9215 FW access drift, access cross-cut and the retreating stoping front.

As the semi-circular FW access drift was being developed (150 m long at completion), a large number of development blasts were undertaken. Some blasts were conventional, and some included de-stressing patterns and explosive charging sequences similar to those shown earlier in Figures 13.17 and 13.19. The research results summarized here refer to the last section of the drift (Zone C), which was driven at a high angle of incidence with respect to the orientation of the *in situ* stress (Figure 13.29).

13.3.2.6.1   Spatial Characteristics of the Seismogenic Zone

When measured along the tunnel chainage, the seismogenic zone typically enveloped the last supported cut, the subsequent unsupported cut and then continued up to 3 m ahead of the face (Figure 13.30). Occasionally, seismic events of relatively high energy occurred ahead of the

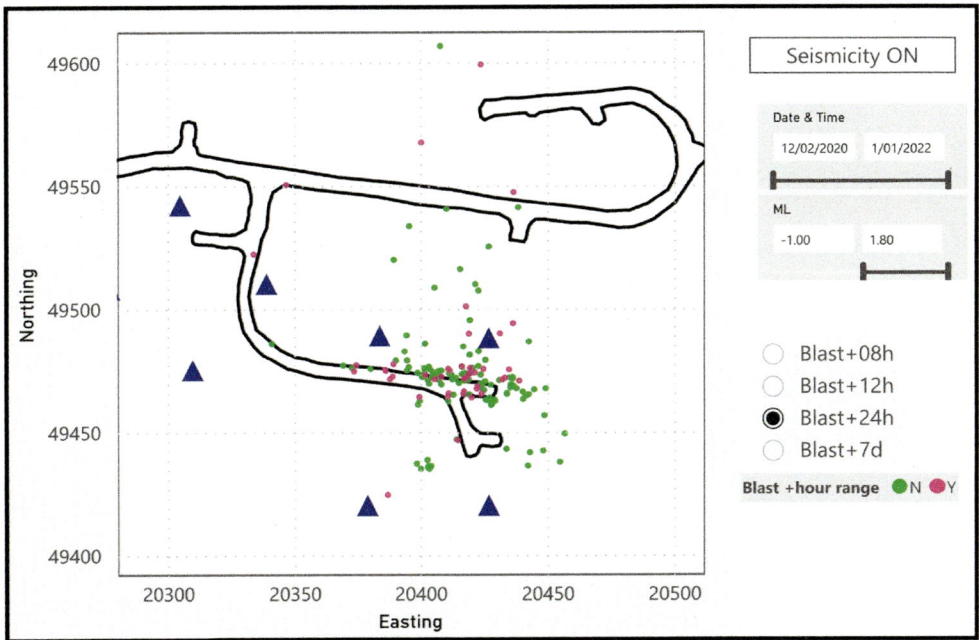

FIGURE 13.28 A plan and 3D view of 9215 tunnel access drift and the first extracted stope showing the location of the seismic sensors (marked by triangles).

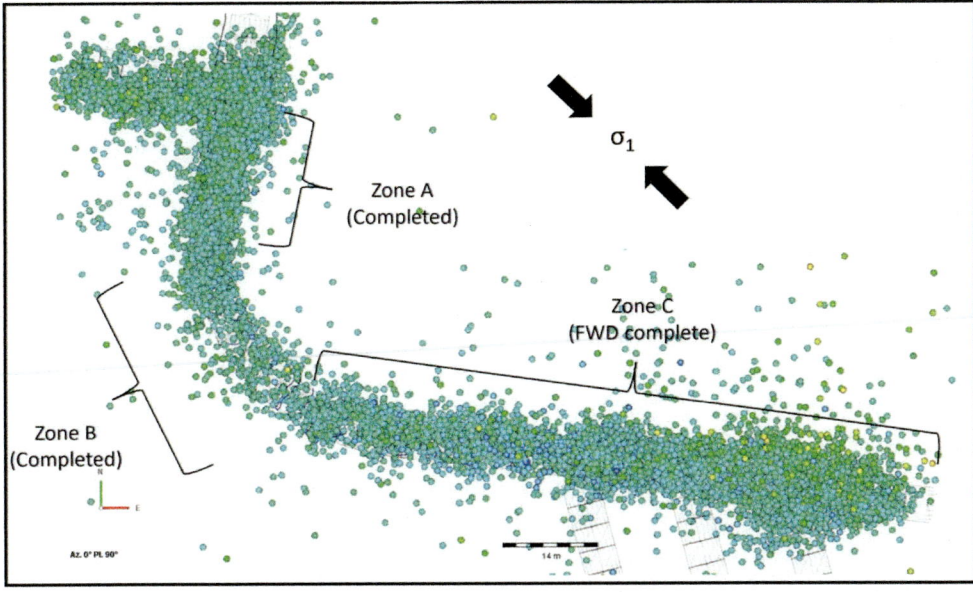

FIGURE 13.29 A plan view of 9215 Level development showing seismicity and the selected zones where detailed analyses were undertaken.

de-stressed zone, probably linked to geological structures. Intense seismicity usually extends laterally into the walls by only 1–2 m. On some occasions, the event clustering extended 3–4 m into the wall, especially on the north side. This was due to shearing on structures by the NW-SE oriented major principal stress component. Vertically, the seismogenic zone typically extends above the tunnel roof and below the floor by a distance equivalent to approximately one tunnel diameter (~5 m).

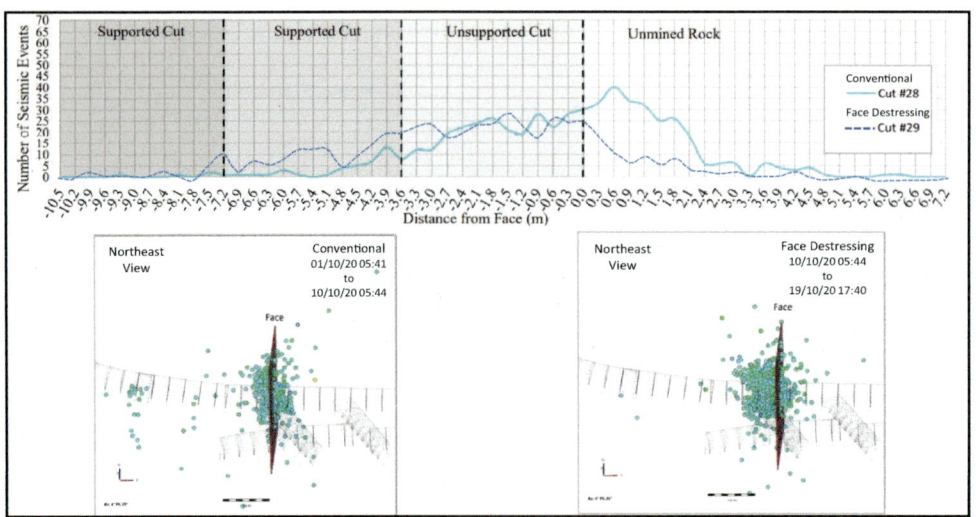

FIGURE 13.30 A long section of view showing seismic event count for conventional versus de-stressing development blasting for the 9215 FW access tunnel.

In around half of the cuts, the seismogenic zone had a slight peak in event count just ahead of the face (0.6 m). Otherwise, the number of events ahead of the face was similar to that for the adjacent walls/roof/floor (per 0.3 m increment chainage). On average, for the entire tunnel, the shape of the curve trend was similar for both development methodologies. The event count increased through the last supported cut, and the seismic activity peaked at less than 1 m ahead of the face, with the seismogenic zone largely decaying about 3 m from the tunnel face. In general, the results were similar to those reported by Drover et al. (2018).

Figure 13.31 provides a comparison of the radiated seismic energy for both conventional (Cut #28) and face de-stress blasting (Cut #29) for the same period as that considered in Figure 13.30. Less radiated energy was detected within the de-stressed region. However, as reported previously by Drover et al. (2018), an increase over the unsupported and previously supported cuts was observed. Also, the average data from seven conventional and 11 de-stress blasts across different locations indicated low levels of radiated energy, in the range of 10–1000 J per 0.3 m increment of chainage, across the development in general. Many of the larger energy peaks in the de-stressing cuts occurred well ahead of the de-stressing zone, likely indicating an advance in the stress concentration zone ahead of the face (Figure 13.32).

13.3.2.6.2 De-stressing Effect

The seismic data from sequential cuts often displayed large variability (Figure 13.33). Direct comparison between conventional and de-stress blasting across alternating cuts did not reveal a consistent effect on event count, energy release or potency due to the two blasting techniques. In general, the seismic activity did display an increasing upward trend as the tunnel advanced eastward through the high-stress stoping abutment. Also, due to the high stress and the presence of oblique geological features, this area had seven consecutive de-stressing cuts (Cuts #29d to #35d), and hence the rock mass response to conventional blasting could not be quantified. Overall, it appears that variable rock mass conditions (especially the presence of a medium-scale geological structure) rendered the seismic development response highly variable for the duration of construction. The highest seismic energy was released when the tunnel intersected a medium-scale oblique feature in Cut #35d. Although the effect of face de-stressing was unclear, importantly, no violent face spalling events took place, and the FW access drive was completed safely and efficiently according to schedule.

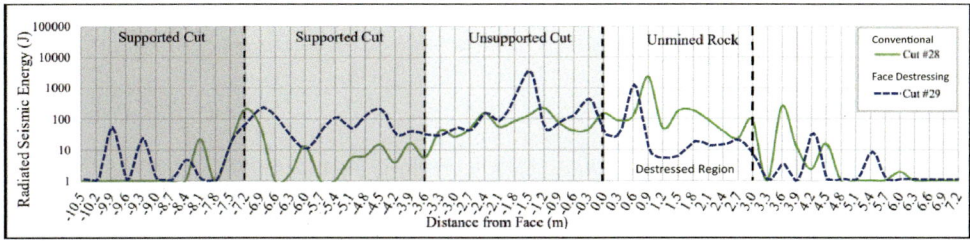

FIGURE 13.31 A long section view showing the radiated seismic energy for conventional versus de-stressing development blasting in the 9215 FW access tunnel.

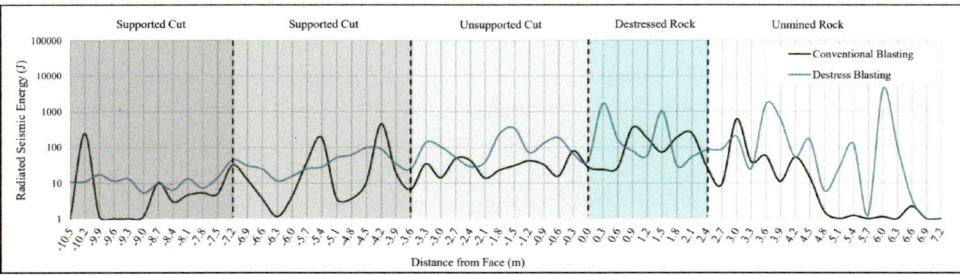

FIGURE 13.32 A long section view showing the average radiated seismic energy for conventional versus de-stressing development blasting in the 9215 FW access tunnel.

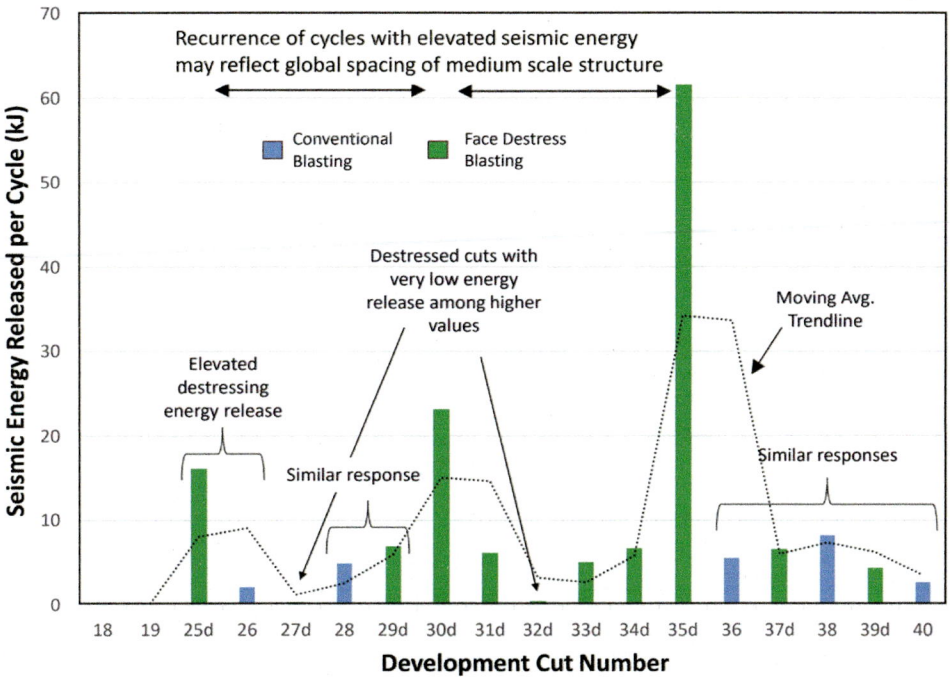

FIGURE 13.33 Seismic energy released along the tunnel axis for 18 consecutive cuts reflecting a significant effect from the response of the pre-existing geological discontinuities.

13.3.2.6.3 Exclusion Period Response

The seismic data was analysed to determine whether de-stressing triggered more seismicity during a six-hour exclusion period that was implemented following each development blast. Ideally, it would be of considerable advantage if most of the seismicity occurred during the exclusion period. However, the data was inconclusive, suggesting that the effect of face de-stressing alternated between reduced and increased seismic responses within the exclusion periods. Figure 13.34 shows the data expressed as a percentage of the total seismic response for several development cycles. Some conventional blasts had a larger percentage of the seismicity occurring during the exclusion period; however, sometimes, the reverse occurred. The data also suggests that a longer exclusion period was applicable towards the end of the tunnel (Cuts #32 to #40), where seismicity was influenced by the response of the geological structures (as displayed previously in Figure 13.27).

13.3.2.6.4 Event Decay to Background Seismicity

The key factor controlling the variability in seismic event decay rate from cut to cut relates to differences in geological structure. Within the 9215 FW access, both de-stressed and conventionally blasted cuts displayed rapid and gradual seismic event rate decays before the activity returned to background levels (Figure 13.35). The experimental data do not indicate that de-stressing always affects the event rate decay or promotes a return to the background seismicity levels sooner. Nevertheless, most blasts returned to the background seismicity levels after 12 hours, with few exceptions. Some data (i.e., Cut numbers 29, 30 and 31) suggests that de-stressing triggers an immediate seismic response and more events in the first hour after blasting.

13.3.2.6.5 Seismicity during Construction

Figure 13.36 shows that significant spikes in energy can occur many days after blasting, likely triggered by construction activities. The sudden onset of event spikes occurred during both conventional

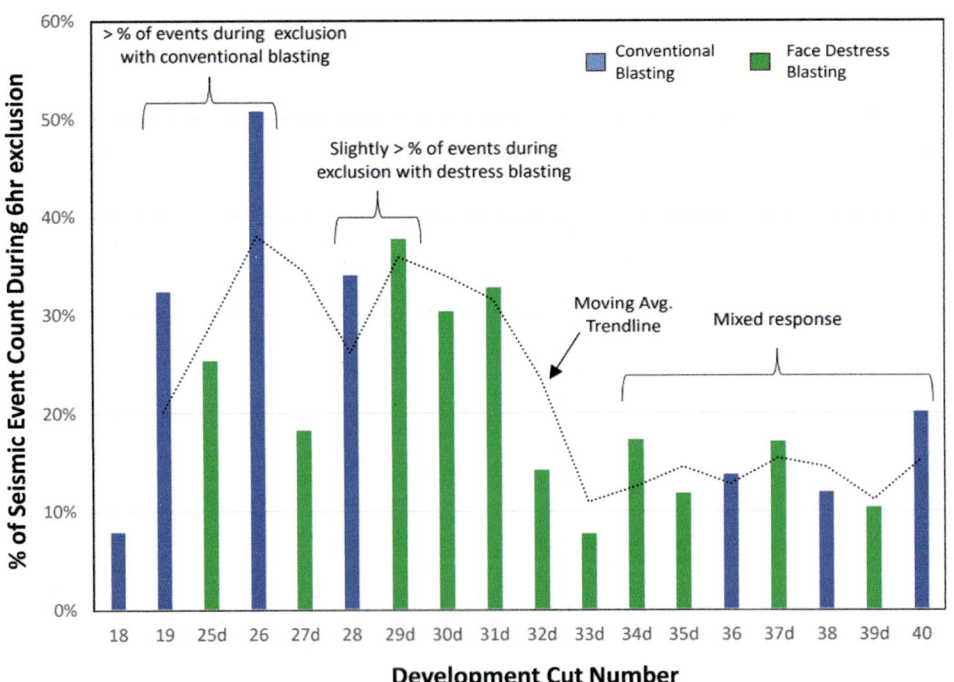

FIGURE 13.34 An example of seismic response within a six-hour exclusion period following conventional and de-stress blasting.

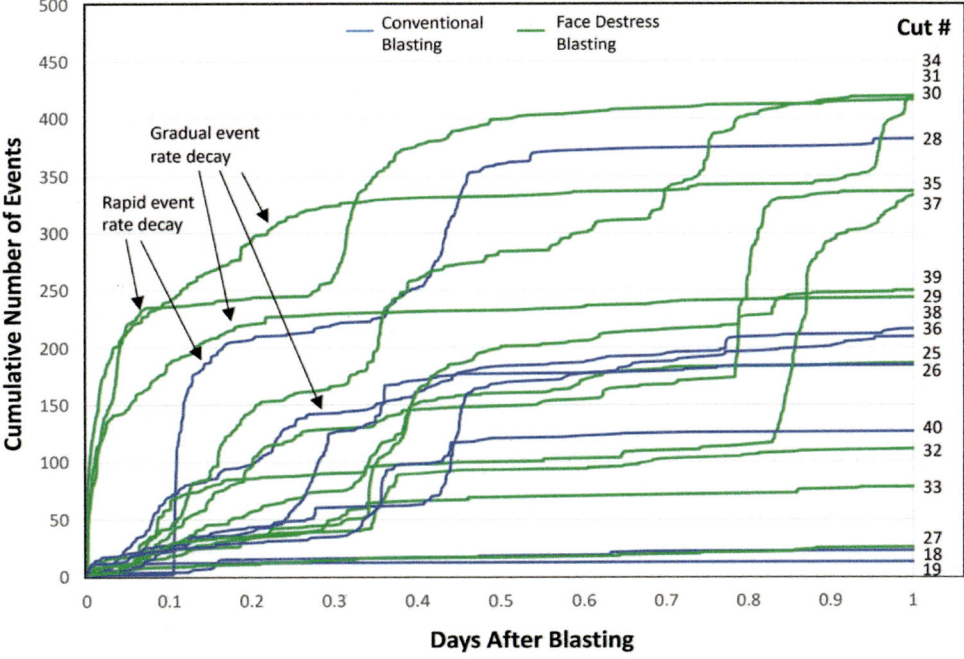

FIGURE 13.35 The event count time history for conventional and de-stressing development blasting.

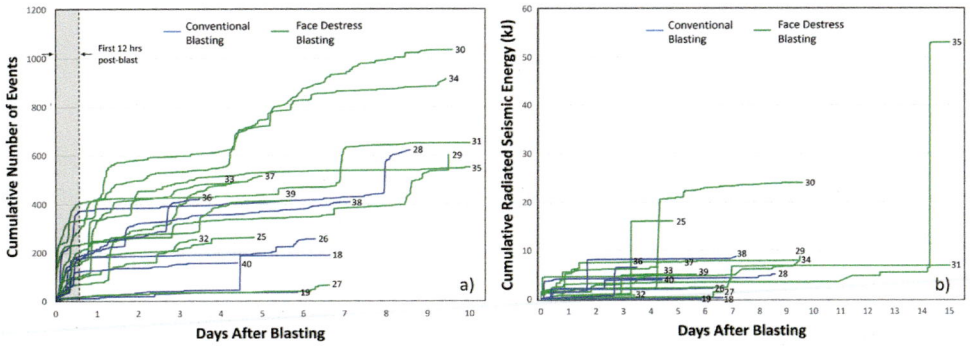

FIGURE 13.36 Evidence of (a) spikes in seismic event count and (b) increases in radiated seismic energy several days after blasting.

and face de-stressing development cycles. The energy spikes were associated with events that were located by the seismic system to occur mostly above the roof of the tunnel or else adjacent to the north wall. That is, they were adjacent to the supported excavation surfaces and not directly ahead of the unsupported face. The event counts during activity spikes are often similar in size. For example, view the results for Cuts 28 (conventional) and 31 (de-stressed) 7–8 days post-blast. The spikes occur well after the blasting exclusion period. Additionally, sudden energy spikes of varying sizes (Cut #25, #30 and #35) suggest tunnel interactions with local geological structures during development. De-stressing does not aim to reduce the likelihood of instability in the geological structure located outside the volume preconditioned by de-stressing charges. The intent of de-stressing is to reduce the potential for violent failure directly ahead of the face, such that the risk of face ejection is reduced.

13.3.2.6.6 Source Mechanisms

A few of the seismic events close to the 9215 FW access drive provided sufficient information to enable a source mechanism solution, as shown in Figure 13.37. Due to the orientation of the major principal stress, almost all the events occurred in the north wall. Importantly, the seismic source mechanism solutions agreed with the *in situ* stress measurements determined using the oriented core (Villaescusa et al., 2002) and the predictions from Abaqus modelling. Indication of stress re-orientation and plunge rotation associated with the effect of the tunnel excavation on the local stress field was also observed (Figure 13.38). Nodal plane orientations indicated that a moderately-steep, southeast dipping joint set may have been repeatedly activated in the north wall of the 9215 FWD. This set could be intermediate-scale structures with a spacing equivalent to around five cuts (≈17 to 18 m). Structures at this orientation have previously been observed in the face, causing overbreak.

   The seismic source data indicates that both discontinuity slip and overstressing mechanisms of failure occurred during the 9215 FW access development (Figure 13.39). Several events are consistent with pure structural shear/fault slip (deviatoric component dominant). The remainder are characteristics of compressive overstressing (implosive isotropic component dominant).

13.3.2.6.7 Tunnel Wall Seismic Instability

The seismic activity generated adjacent to the 9215 FWD excavation surfaces was analysed using spatial filters. Over the entire tunnel construction (for both conventional and de-stress blasting), the face and roof of the 9215 FWD experienced a similar number of seismic events. However, the total energy radiated above the tunnel roofline was 8.3 times greater than the energy radiated directly ahead of the face (Figure 13.40). Considering that most of the cuts were de-stressed, this is a positive observation that validates the effectiveness of the de-stressing approach. The floor exhibited approximately twice the number of events at the north wall, but the energy released was slightly

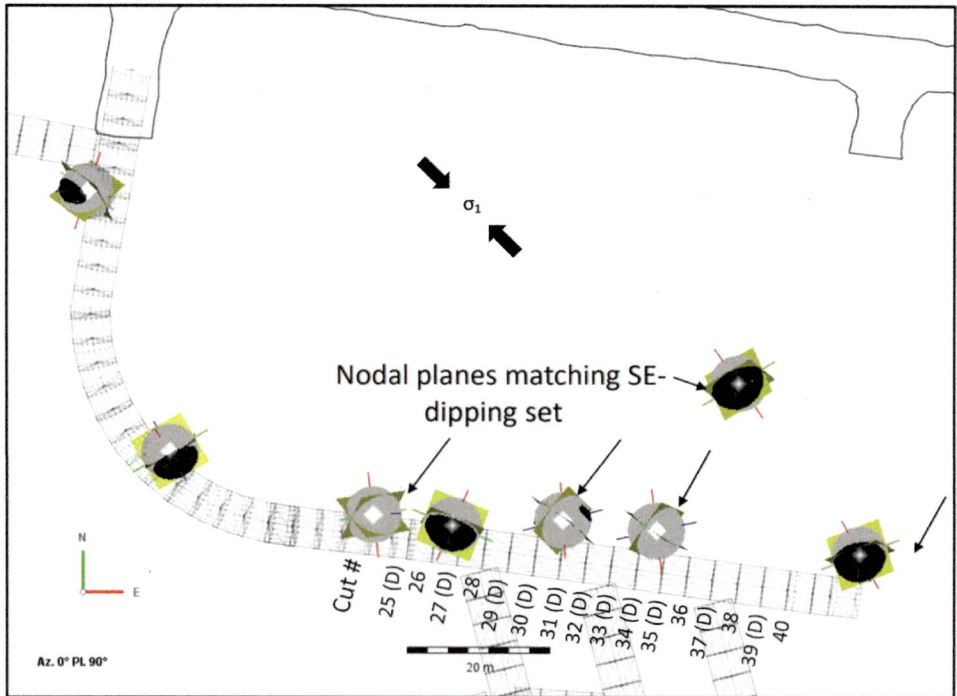

FIGURE 13.37 The principal strain axes, nodal planes and mechanism beachballs for a number of seismic events with respect to the tunnel axis.

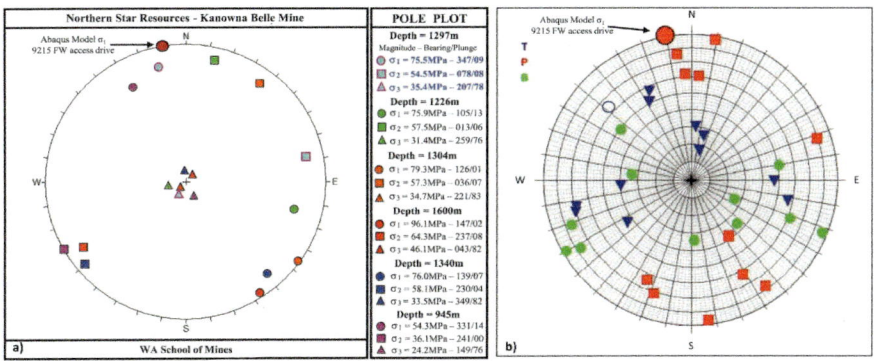

FIGURE 13.38 Kanowna Belle Mine (a) measured (WASM AE) and modelled stress and (b) P-axis directions from seismic source mechanism solutions.

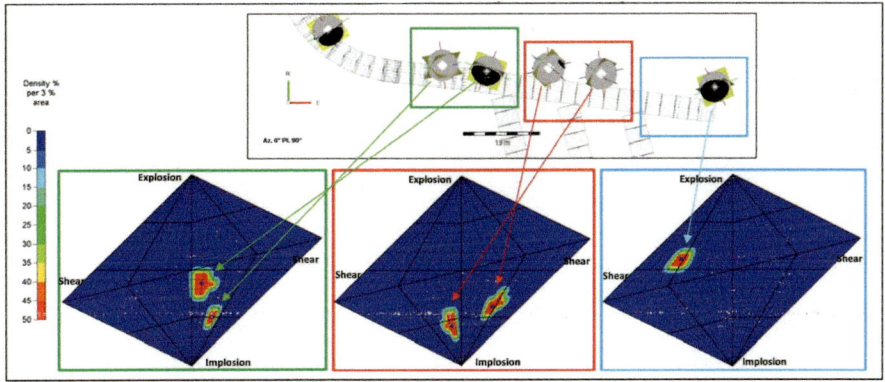

FIGURE 13.39 Examples of source mechanisms for five seismic events around 9215 FW Access Drive.

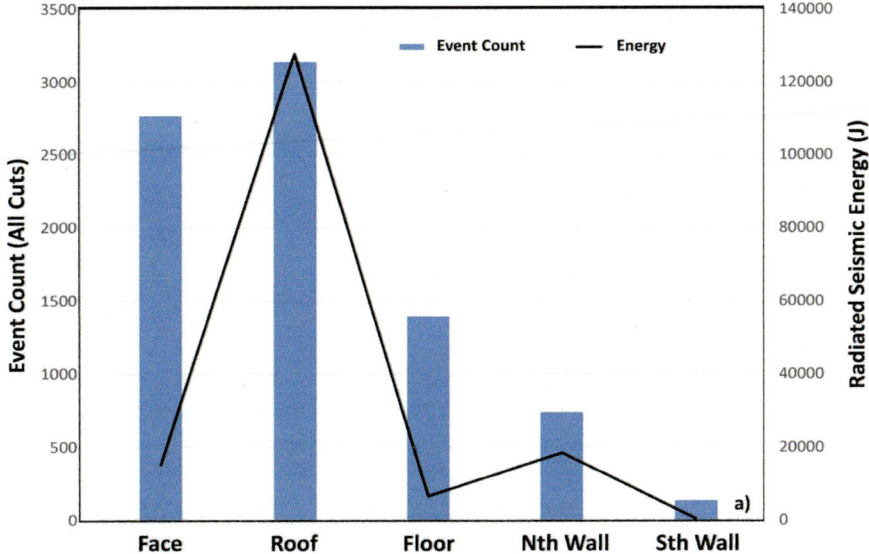

FIGURE 13.40 Event counts and spatially filtered radiated seismic energy from tunnel walls for all cuts in the tunnel.

higher on the wall. In comparison, the south wall was very stable, with far fewer events and very small energy releases.

Also, data recorded within Cuts 26–36 (mostly created using the de-stressing method) indicate more events and greater energy at each excavation surface when de-stressing was applied (Figure 13.41). The raw statistics suggest that de-stressing caused far greater energy radiation from the roof and that this was likely due to the interaction of the tunnel with medium-scale geological structures.

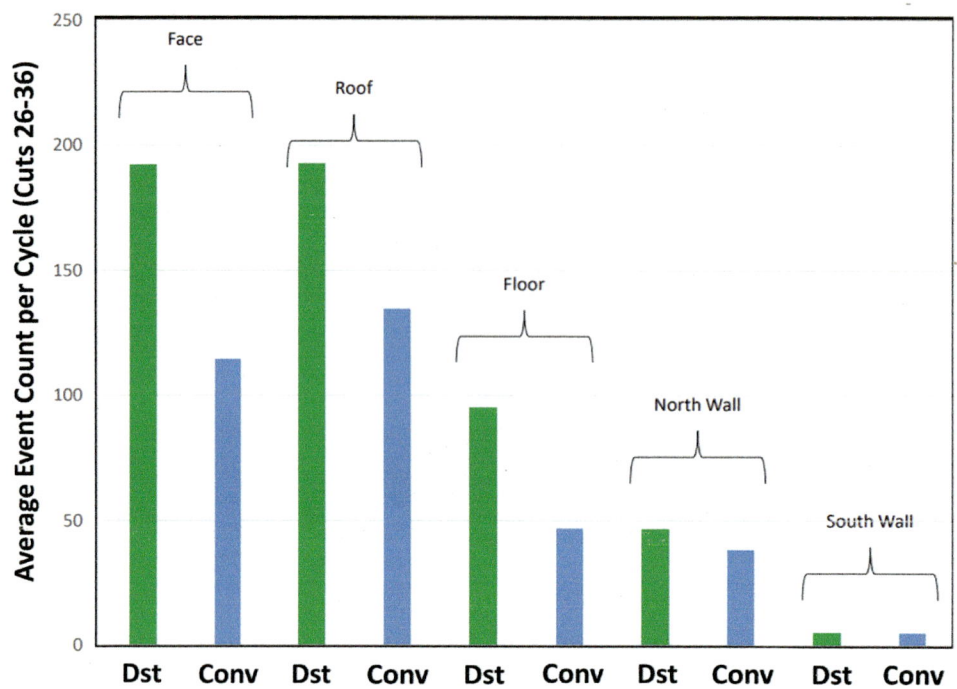

FIGURE 13.41   Event counts per wall for tunnel regions constructed using de-stress blasting techniques.

13.3.2.6.8   Seismicity during Sublevel Open Stoping

In sublevel open stoping at great depth, the requirement for any access infrastructure is to remain stable not only during its construction but also during the subsequent production blasting operations. Many stopes are usually blasted nearby, and ideally, the access development must undergo minimum to no ground support rehabilitation. Figure 13.42 shows the location of the seismicity resulting from the typical activities required during sublevel open stoping. Most of the seismicity (and potential for damage) occurs due to the stress change resulting from the creation of the sublevel open stoping geometry (Figure 13.43).

Additionally, the ground support capacity must account for large events occurring several days after the blasting events. Although re-entry time controls that are usually implemented immediately after blasting do help seismic activity along geological structures can occur several days later, as shown in Figure 13.44. The amount of energy radiated during open stoping is clearly larger than during the development access, and the ground support must remain stable without rehabilitation to optimize the production of the stoping front. Figure 13.45 shows the location and magnitude of the large events with respect to the access development and sublevel open stoping geometries.

The need for a robust (tough and resilient) ground support performance can also be indicated by analysing the amount of seismicity it has experienced at a particular location within development

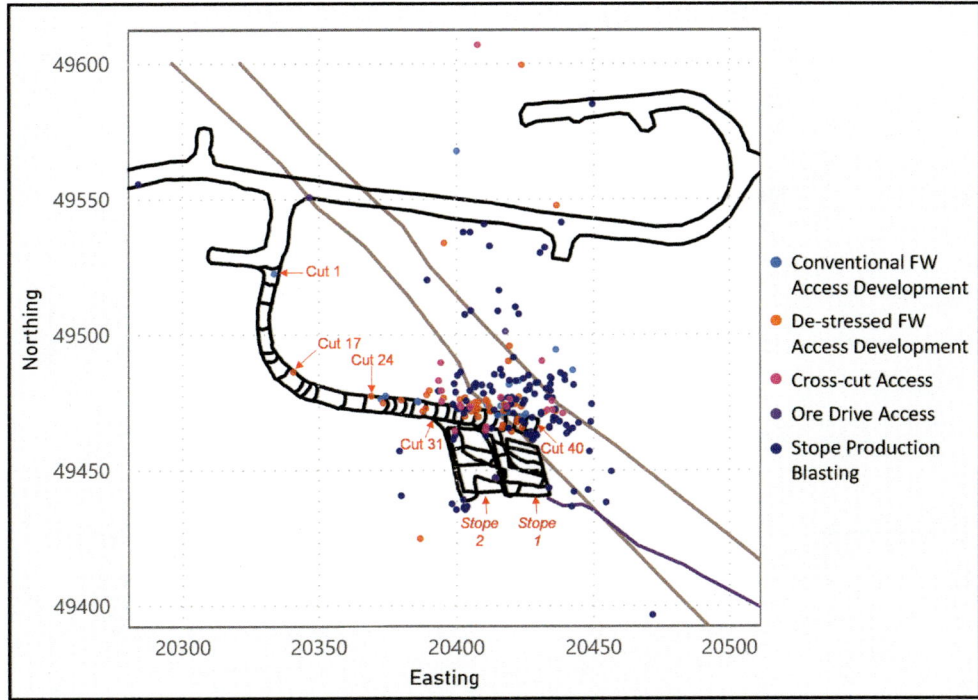

FIGURE 13.42   Plan view of event location coloured by type of blasting.

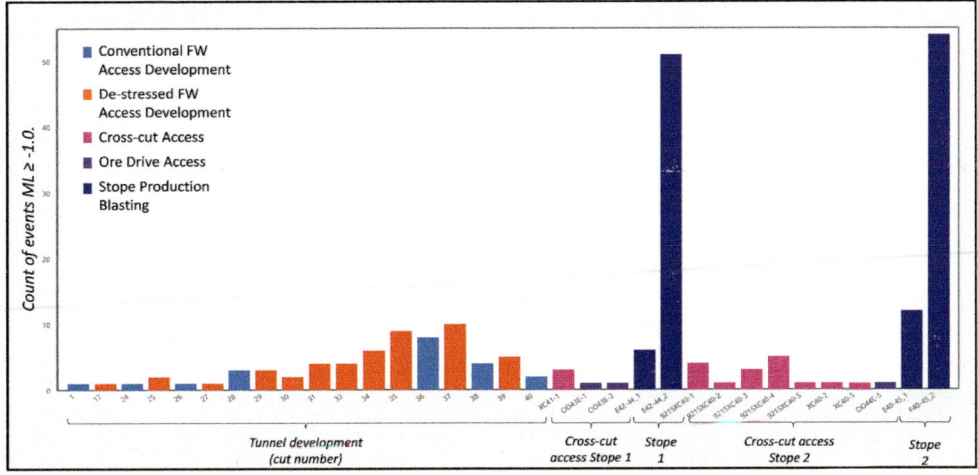

FIGURE 13.43   Number of events with ML ≥ −1.0.

access. Figure 13.46 shows the history of seismicity and related energy experienced at the exact position of Cut #36. The data indicates this position to be experiencing seismicity when the tunnel face was located at Cut #30, which was positioned 20 m away. The data also indicates that most of the seismicity at this particular location occurred when the tunnel was developed (i.e., during and immediately after Cut #36). Nevertheless, a significant amount of seismicity also occurred at the same location when the nearby sublevel open-stoping geometries were created. This included two large seismic events within the 20 m radius from the development Cut #36, indicating that the

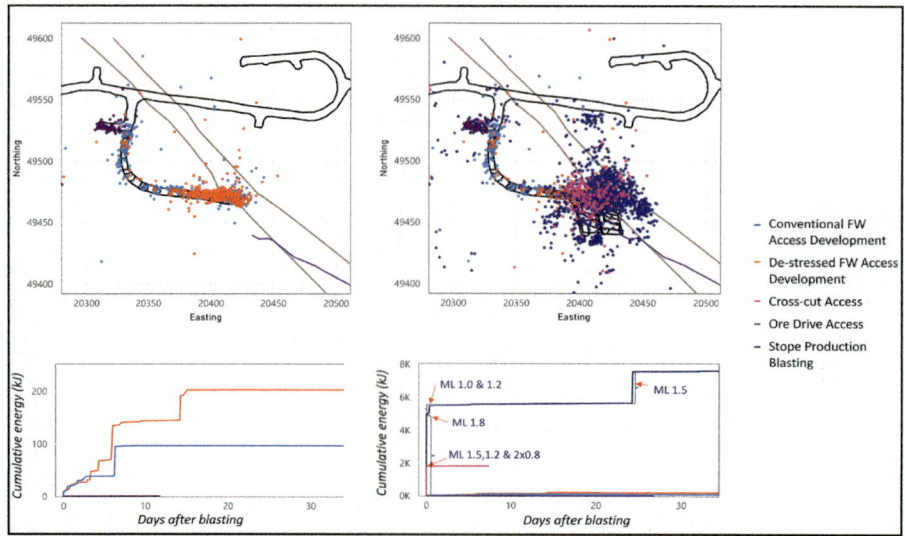

FIGURE 13.44   Location of seismicity, cumulative radiated energy and large events with time.

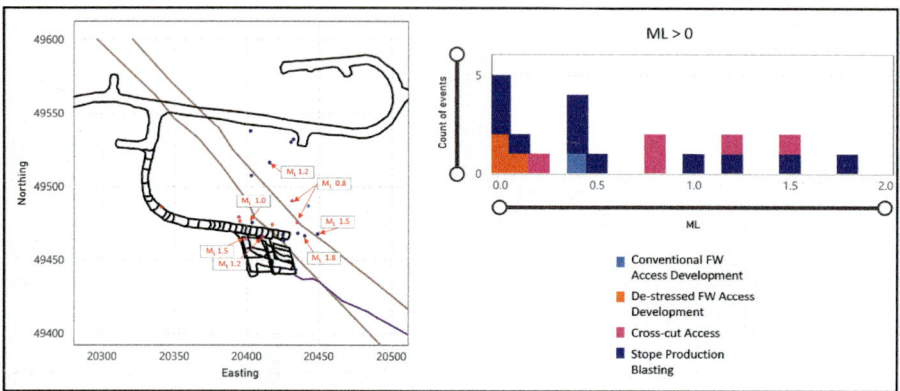

FIGURE 13.45   Location of large seismic events with respect to development and stoping geometries.

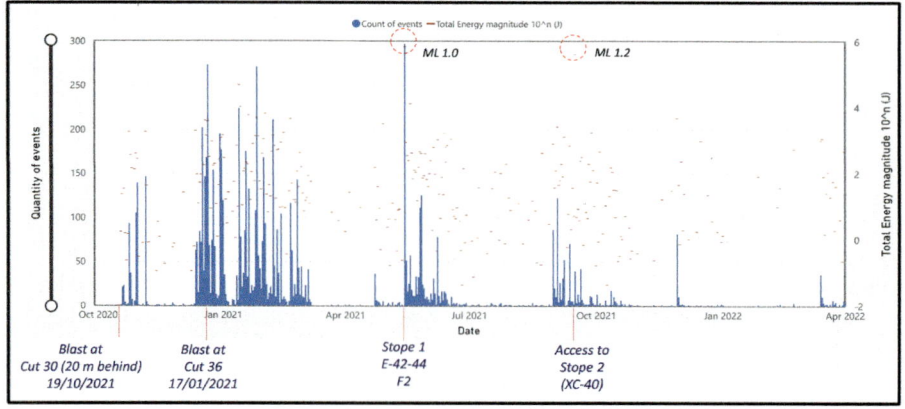

FIGURE 13.46   Frequency of events and energy magnitude calculated using a sphere with a 20-m radius centred at Cut #36.

ground support capacity needs to be sufficient during both the development and the subsequent sublevel open-stoping activities.

Figure 13.47 shows the type of damage experienced around Cut #36 after the completion of the initial stope blast in the area. Extraction of the first two open stopes was completed without any ground support rehabilitation being necessary, even though damage from blasting was experienced because some production holes were drilled from the 9215 FW access drive itself (Figure 13.48).

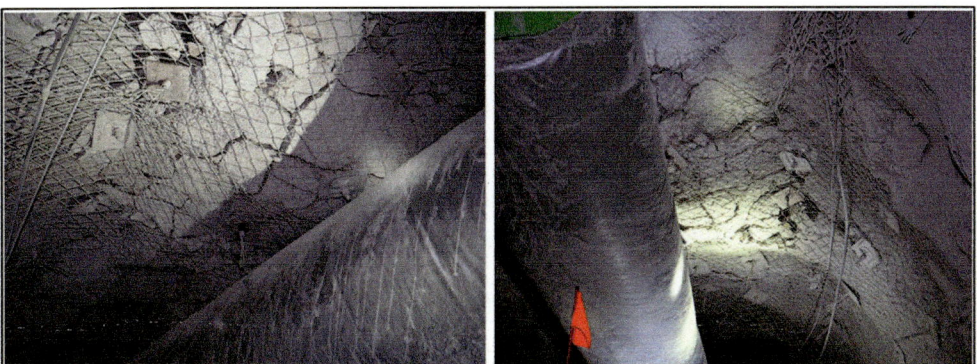

FIGURE 13.47   An example of the worse excavation damage experienced within the 9215 FW access drive following production blasting of a nearby stope.

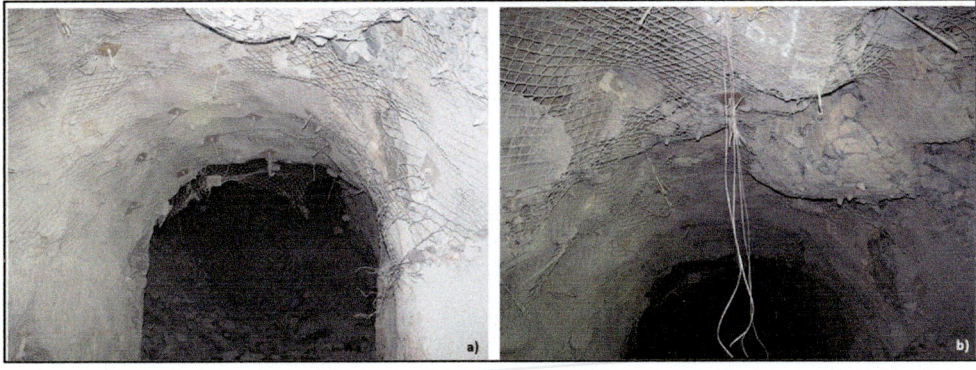

FIGURE 13.48   Details of initial stope extraction at the 9215 Level showing (a) draw point and (b) damage in the cross-cut back where production blasting took place immediately above the cross-cut.

13.3.3   CONSTRUCTION OF A HIGH ENERGY DISSIPATION GROUND SUPPORT SCHEME

It is well established that ground support schemes cannot markedly change induced stresses or prevent rock mass failure. However, even at great depth, the failures experienced are often shallow, with rupture and ejection in most cases with a depth of failure restricted to less than 2 m. This means that the retention capacity of the ground support at the boundary of the excavations is critical to maintaining the serviceability of excavations after rock mass failure. Hence, the dynamic performance of the surface connections of the ground support scheme is critical to the overall energy dissipation at a particular location within the rock mass. This includes strategies to minimize the violent ejection of any shotcrete at the surface of the excavations.

The following sections discuss several experimental steps that can be used as a step-wise guide to construct and implement a high-energy dissipation ground support strategy (Arcaro et al., 2021).

The support is provided in sequential layers to limit the combined rock reinforcement and support displacement at the boundaries of excavations. The sequence of construction activities depends upon the methodology of ground support installation, which may range from manual to highly mechanized.

13.3.3.1 Clearing of Temporary Face Support

The ground support installation described here was implemented during a design and construction arrangement that was part of a WASM industrial research and development project (Drover, 2018). The ground support was installed using a highly mechanized method compared to the conventional installation methods used in Australia and elsewhere.

The requirement for face mesh installation prior to drilling and blasting operations can result in a significant amount of damaged temporary ground support hanging from the backs and adjacent sidewalls following blasting (Figure 13.49). This ground support is deliberately sacrificial in order to provide temporary protection from face ejections for the drilling and blast crews during their activities. The damaged ground support must be carefully removed following extractions of the blasted muck pile to provide a clear work area for mechanical scaling. Mesh sheets and protruding rock bolt tails need to be cut flush to the rock surface to minimize interference with the subsequent installation of permanent ground support.

13.3.3.2 Mechanical Scaling

Mechanical scaling of the roof and sidewalls of the excavation is performed to remove loose and broken rock disturbed by the development blasting. Scaling serves several important purposes. Firstly, scaling provides a more uniform and stable surface for the application of the initial shotcrete layer. Secondly, removing broken rock on the excavation perimeter reduces the unnecessary static load on the ground support. Thirdly, removing this broken rock ensures that the ground support scheme is installed as close as possible to solid intact rock. This limits the amount of radial deformation that

FIGURE 13.49   Example of damaged temporary face support that must be removed following blasting.

could occur prior to loading of the ground support scheme. Importantly, the removal of this broken layer allows the ground support to provide an immediate containment effect to the rock surface.

Mechanical scaling can be performed using purpose-designed mobile equipment (Figure 13.50). Observations during construction indicate that scaling usually generates a significant seismic response, often with spalling occurring from the walls of the excavation. This inherently hazardous phase of construction is made safe using equipment that removes the operator from the danger zone. In this case, the operator is stationed within the machine, 4–5 m behind the last row of permanent reinforcement within a cabin that provides protection from chance ejections and simple rock falls.

FIGURE 13.50   Example of dedicated machinery for mechanical scaling of unsupported excavation walls.

13.3.3.3   Structural Geological Mapping with Photogrammetry

The freshly exposed rock faces can be readily mapped using digital photogrammetry. The objective is to collect sufficient structural information to conduct a reinforced block analysis. Photogrammetry involves washing down the face with water to remove dust and expose a clear rock mass surface. A high-resolution digital camera can then be stationed on one side of the tunnel. This camera is set well back from the face to limit the exposure from any face instability. Additional lighting is required to highlight the joint planes (Figure 13.51). A series of photographs are taken from one side of the excavation, the camera is then moved to the opposite side and the process is repeated. It must be noted that the compatibility of photogrammetry techniques with SAFEX data collection requirements is the subject of ongoing investigations.

13.3.3.4   Shotcrete Application

A primary layer of sprayed shotcrete (typically 50 mm thick) is applied using a diesel-hydraulic-powered mobile spraying machine equipped with an air-jet spray nozzle. The wet-mix method of shotcrete installation is conventional in the mining industry worldwide. However, it may expose the operator to the hazard of face instability. Therefore, the operator is located close to the edge of unsupported ground when controlling the spray jet nozzle using a handheld control unit (Figure 13.52). Potentially the operator is exposed to scats ejected or simply falling from the face, sidewall and especially the roof. Although the period of exposure is often relatively brief, perhaps no more than 45 minutes, this activity usually occurs within 60 minutes of mechanical scaling. As such, caution is required when deciding to commence this activity following a field assessment of the seismic activity in the heading. Shotcreting must be delayed whenever instability, such as rock noise or spalling, is detected near the face.

Following shotcrete application, a pause is required to allow for curing of the shotcrete. This necessary step in the development cycle affects construction efficiency. The ground support installation cannot commence while the shotcrete develops sufficient strength. Historically, in many mines, early-age shotcrete with compressive strength of 1 MPa has been used as the threshold for re-entry. However, this criterion does not realistically consider the fact that shotcrete early age failure is

FIGURE 13.51   Typical digital photogrammetry station for face structural mapping.

FIGURE 13.52   Example of the application of a primary shotcrete layer.

typically due to failure in shear (Chapter 10). Saw et al. (2017) recommended that an alternative strength threshold be applied. Specifically, the shear strength of shotcrete should be sufficient to stabilize an unstable free-falling tetrahedral block of rock with 1 m edge lengths. For a 50 mm thick layer of typical shotcrete composition, this equates to a required shear strength of approximately 20 kPa, typically achieved within an hour.

13.3.3.5 Primary Reinforcement Mark-Up

The optimal load transfer characteristics of a ground support scheme rely heavily on the spacing regularity and symmetry in the reinforcement pattern and mesh sheet overlaps. As such, practical implementation of the reinforcement and mesh placement is an important area of focus during ground support installation. As such, reinforcement pattern compliance can be achieved in one of two ways. The first method involves marking the pattern on the shotcrete surface using 'paint poles' such that the operators are several metres from the edge of the unsupported ground. Figure 13.53 illustrates a typical paint markup for a 1 m × 1 m staggered reinforcement pattern used to guide the bolting jumbo operators. This shows the primary ground support layer from a previous development round (to the right in the figure), fresh shotcrete and the reinforcement pattern paint markup of the most recent round (to the left). For a 3.5-m development round, approximately 20 minutes is required to accurately define and mark the complete reinforcement pattern in this manner.

A second method of controlling the position of holes for the reinforcement involves unfurling each mesh roll on a flat surface and marking its position using paint. The mesh can then be rolled back up and placed on the jumbo mesh handler arm for installation. The visual markers on the mesh can guide the jumbo operator to position the drill bit between mesh wires when collaring the reinforcement holes.

Both methods of establishing the reinforcement pattern are relatively simple but necessary to ensure compliance with the ground support scheme design. Jumbo technology is also available,

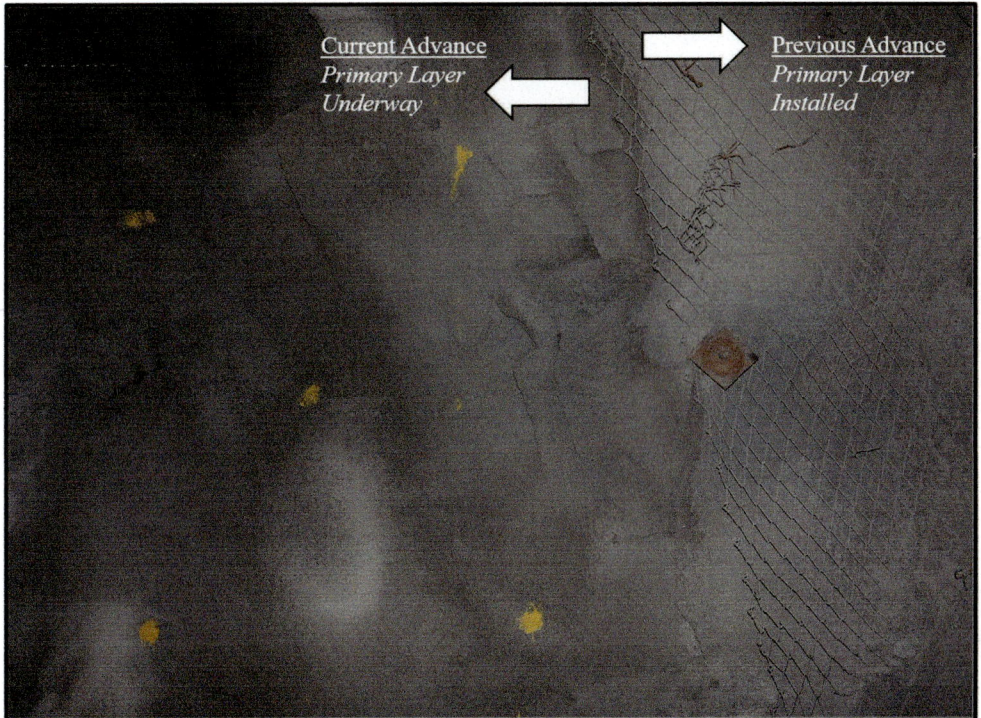

FIGURE 13.53 Example of hole collar marking (yellow dots) for rock reinforcement installation.

which assists guiding of the jumbo boom into the correct collaring position for reinforcement installation. An example of this technology is the Sandvik Bolting Instrumentation (SBI) system (Sandvik, 2018), which may be installed on the various Sandvik DS model, bolting jumbos. This system uses a fan laser or theodolite to optically locate the correct collaring position for the reinforcement. It is expected that, in the future, similar types of sophisticated methods of reinforcement pattern guidance will become standard.

13.3.3.6 Installation of Primary Reinforcement and Mesh

The installation of reinforcement elements and mesh as a scheme can be completed using fully mechanized ground support jumbos. Example equipment includes the Atlas Copco BOLTEC (Figure 13.54) and the Sandvik model DS411-C jumbo (Figure 13.55). Both jumbos are diesel

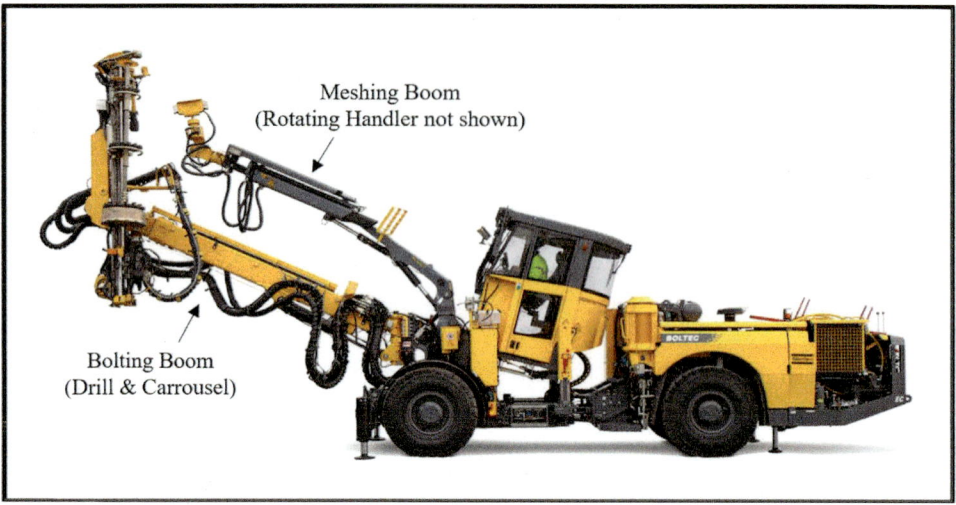

FIGURE 13.54   The mechanized Boltec Jumbo used for ground support installation (Atlas Copco, 2017).

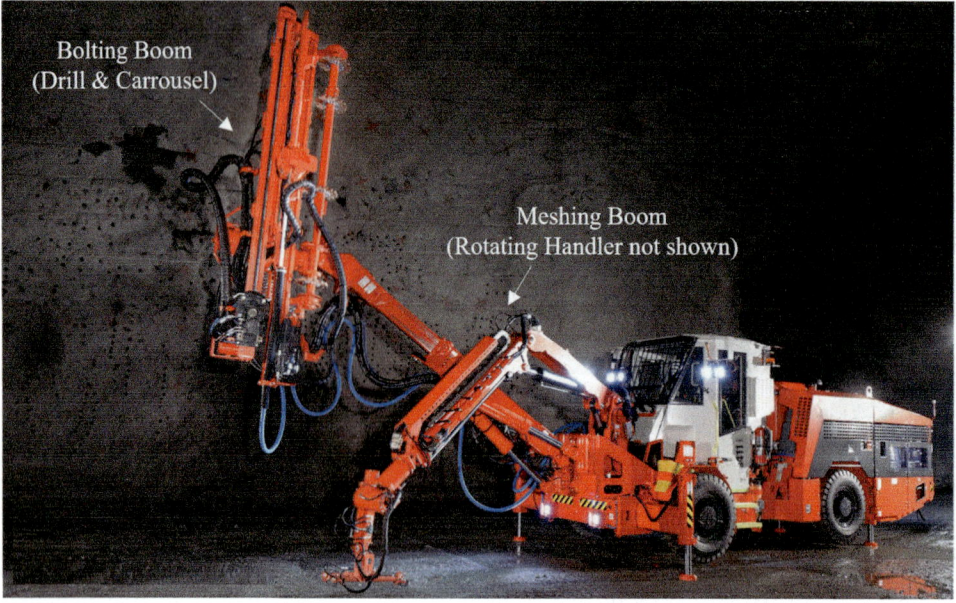

FIGURE 13.55   The mechanized DS411 bolting jumbo used for ground support installation (Sandvik, 2018).

engine-powered and connect to the mine electricity supply. They are based on similar design principles and involve a hydraulically operated twin-boom arrangement. One of the booms on each jumbo is dedicated to the task of installing the reinforcement. The other is dedicated to mesh handling (Figure 13.56).

The bolting boom (Figure 13.57) comprises a hydraulic rock drill, a rotating carrousel for storing and installing the reinforcement elements and a cement grout insertion tube connected to a grout mixing bowl (Figure 13.58). The grout insertion tube functions in the same fashion as that of a regular cable bolting machine. The grout mixing bowl is in-built on the jumbo (Figure 13.59). Grout is prepared by the machine and pumped into the borehole via an insertion tube. The bolt carrousel can store between 8 and 16 reinforcement elements, depending on the plate size. Bolts are loaded onto the carrousel prior to the jumbo approaching the face and commencing installations.

The mesh handling arm is typically an optional attachment. The fitted arm is designed to manoeuvre up to 20 m long × 4 m wide, woven mesh sheets of G80/4 to G80/4.6 specification (see Chapter 9). The arm has independently controllable rotation such that the mesh sheet can be unfurled to cover the excavation surface as the reinforcement pattern is progressively installed. The arm is fully manoeuvrable from floor-to-floor to permit full mesh coverage of the excavation perimeter. A close-up view of the mesh handling arm of the Sandvik machine is shown in Figure 13.60. Figure 13.61 shows the Atlas Copco jumbo in operation, installing reinforcement and mesh.

Figure 13.62 shows a Sandvik jumbo designed to install chain link mesh and resin-encapsulated reinforcement systems. The jumbo uses compressed air to push the cartridges into the drilled holes until they are full, prior to the insertion of the reinforcement elements.

One of the main objectives of a ground support scheme is to control spalling (and another progressive unravelling) failure as quickly as possible. As such, the mesh installation and reinforcement should be initially targeted toward any visible areas of stress-driven instability in the tunnel profile.

FIGURE 13.56   Detail of the reinforcement and mesh installation booms.

FIGURE 13.57   Details of a mechanized jumbo aim for drilling, grouting and reinforcement installation.

FIGURE 13.58   Details of drilling, cement grouting and bolting attachments fitted to a mechanized jumbo.

FIGURE 13.59 Details of an integrated grout mixing bowl on the mechanized ground support jumbo.

FIGURE 13.60 The mesh handling arm for mechanized ground support installation (mesh roll not shown).

FIGURE 13.61   Mechanized ground support jumbo in operation (Sandvik DS411-C).

FIGURE 13.62   Example of mechanized installation of chain link mesh and resin-encapsulated, fully coupled Posimix bolts.

Evidence of notching in the profile due to spalling (Figure 13.63) can be used as a guide for the initial installation of reinforcement and mesh.

The process of installing cement-encapsulated reinforcement elements follows a series of distinct steps. After drilling the borehole, the grout tube is inserted into the borehole and the grout is pumped into hole starting at the toe and gradually progressing towards the collar by slowly retracting the grout tube. Retraction is at constant machine speed, ensuring that no gaps are left along

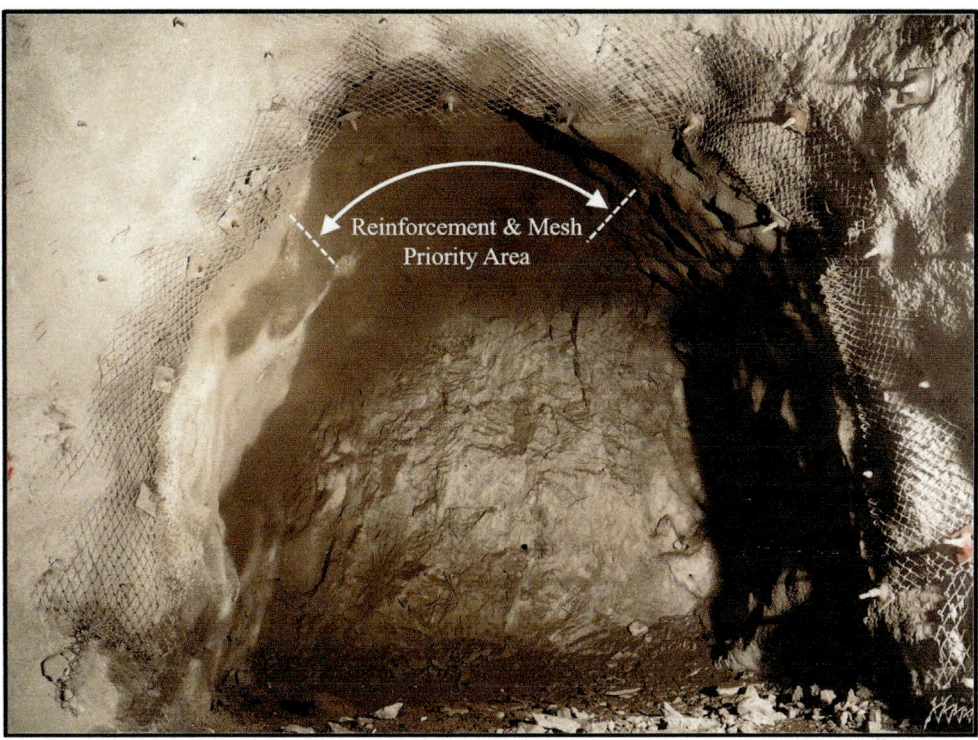

FIGURE 13.63 The ground support installation should be prioritized in zones of stress-driven overbreak.

the borehole axis. Once a borehole is fully grouted, the threaded bars are slowly inserted into the borehole. Closed circuit video cameras on the jumbo boom assist the operator in aligning each reinforcement element with the borehole for insertion. The operator is not required to leave the protective cabin for these tasks.

On-board camera technology removes the need for a jumbo 'off-sider' to perform the function of a 'spotter', again eliminating human exposure to potentially hazardous conditions on the face. An expansion shell point anchor (see Chapter 6, Figure 6.11) can be used to secure the threaded bar in place during the curing of the cement grout. A surface fixture is installed mechanically by the jumbo using a torqued nut at the collar. The threaded tails of all the reinforcement elements can be lightly greased prior to installation to prevent shotcrete material from adhering to the tails. Once the areas identified to be at risk of spalling instability are reinforced and meshed, the jumbo is backed away from the face and reloaded with a full carrousel of bolts. The jumbo can then return to the face and continue to install ground support in the adjacent walls and roof to complete the floor-to-floor pattern.

Figure 13.64 shows a typical view of the primary reinforcement and surface support system (in this case, installed in the lower wall) and the advance of the primary ground support layer for consecutive development cycles. Continuity of load transfer between the ground support installed in consecutive development rounds is ensured by securing the mesh sheet overlaps with reinforcement. Due to different mine site safety policies, it is sometimes required to install permanent reinforcement, up to and including the last row of bolts, in each development round. This may result in double the number of bolts at the overlaps. A recommended approach is to stagger the elements from one development round to another, as shown in Figure 13.65.

Figure 13.66 shows both a close-up view of the primary reinforcement, shotcrete and mesh layer prior to application of the shotcrete, as well as a view looking towards the face.

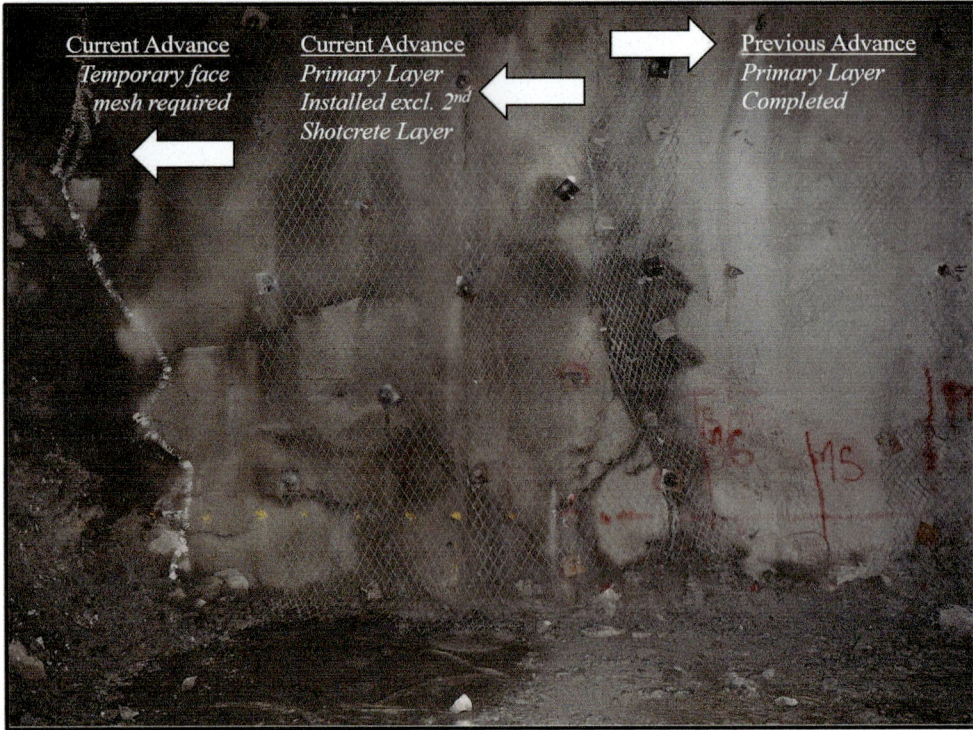

FIGURE 13.64   Example installation of primary reinforcement in a consistent pattern using a high-capacity woven mesh.

FIGURE 13.65   Example of reinforcement installation within the mesh overlap area showing (a) staggered pattern and (b) interacting pattern from cut to cut.

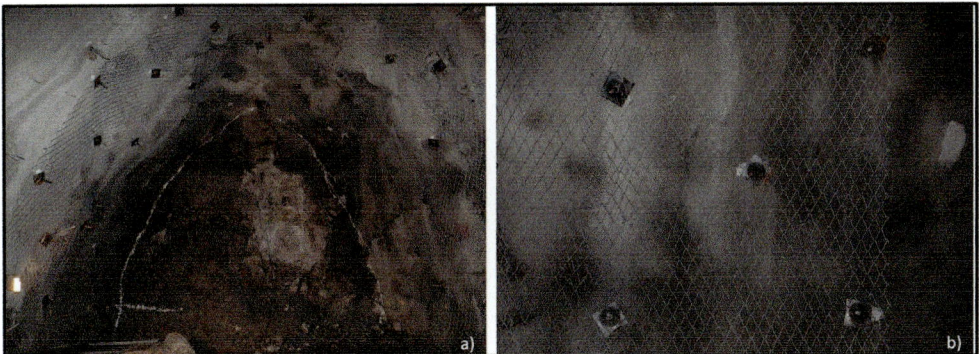

FIGURE 13.66 Example of primary reinforcement prior to shotcrete overspray: (a) general view and (b) detailed view.

The primary ground support scheme is completed by applying a 25-mm thick second application of shotcrete over the existing component (i.e., 50 mm thick shotcrete layer and G80/4 or G80/4.6 high tensile mesh as discussed previously in Chapter 4). This final layer of shotcrete (fill-in) is designed to fully encapsulate the mesh. As demonstrated in Chapter 4, mesh-reinforced shotcrete has significantly greater energy dissipation capacity than shotcrete with an exposed (external) mesh layer. Mesh-reinforced shotcrete also has superior load transfer characteristics. Since both surface support components are rigidly connected, they deform at the same time and rate. It allows the ultimate capacity of both components to be achieved over the range of displacement compatibility.

In situations where freshwater is available, the primary shotcrete layer and exposed mesh can be hosed down prior to the application of the secondary shotcrete spray. This removes all dust from the primary shotcrete layer, improving adhesion between the two layers. The addition of water to the surface between the two shotcrete layers also aids in the hydration of the second shotcrete layer during curing. Where hyper-saline water is used to wash the first layer of shotcrete and exposed mesh, this may be detrimental due to corrosion of the mesh and the presence of mineral salts affecting the chemical reactions of bonding between layers. Both of these may compromise the final scheme capacity. In all cases, it is important to control the thickness of the second layer of shotcrete to minimize the potential for shotcrete ejection at the excavation boundary.

It must be noted that the ground support installation process should not assume that development face de-stress blasting will totally eliminate the potential for stress-driven spalling at the face. Therefore, temporary ground support needs to be installed to control potential instabilities and provide a barrier between the face and the drilling and blasting crews. Typical face support includes either mild steel woven or welded wire mesh from roof-to-floor level, secured with 2.4 m long, SS46 friction stabilizers on a 1 m × 1 m spacing. Furthermore, the development of face de-stressing has been observed to cause some remnant damage to the face, typically with the dislodgement of a limited number of small, loose blocks. This poses a risk, particularly to the hands and arms of the charge crew operators, due to scats falling down the face. Therefore, when securing the mesh to the face using split sets, the mesh is not initially installed flush to the face. A gap of 100–150 mm between the rock and mesh is required during blast hole drilling. This allows any loose scats generated by drilling vibrations to fall behind the mesh to floor level and also prevents later interference when loading the explosives (Figure 13.67). Prior to charging the face, the jumbo returns to install the split sets to full embedment, thus securing the mesh flush against the rock face.

13.3.3.7 Primary to Secondary Support Installation Sequence

One of the main objectives of ground support construction is to install a primary ground support layer as soon as possible after development blasting. Subsequent ground support layers consisting of

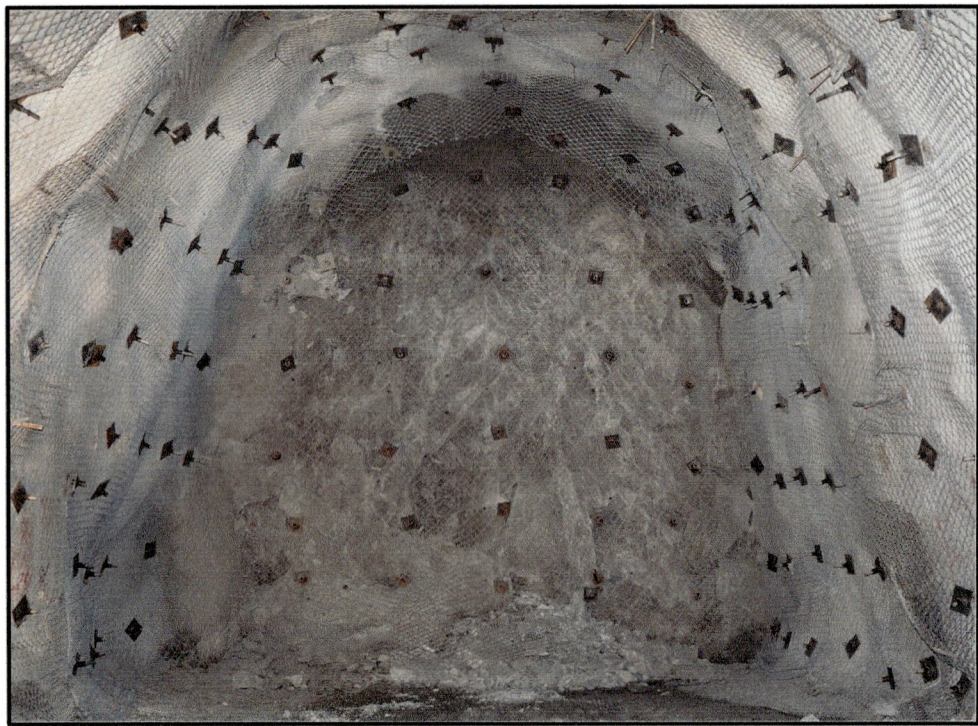

FIGURE 13.67 An example of temporary face support to contain any small face instabilities during drilling and blasting activities.

additional mesh and cable bolt reinforcement can be installed in later campaigns at certain distances from the advancing tunnel face (Figure 13.68). The second layer of mesh is optional and depends upon the rock mass demand (Villaescusa et al., 2016) and the capacity of the initial mesh installed during the primary stage. Ideally, the secondary ground support layer needs to be installed prior to the possible onset of structurally controlled instability around the excavation. The potential timing of structural instability can be determined by a detailed study of the onset of seismicity at that site and general observations of structural failures from other areas and other mines.

Field observations indicate that the instantaneous ejection of large structurally controlled blocks that do not fail with the development blast occurs at a later stage after the excavation surface has been exposed. Typically, structural failures occur after the development face has advanced by some distance beyond the zone of instability. The reasons for this are not well understood. However, it is speculated that stress redistributions on the tunnel boundary may involve time-dependent behaviour of the block-forming joint surfaces and that the associated damage develops progressively as the excavation is advanced. Nevertheless, it is very important that the cable bolt reinforcement installation is not delayed beyond the onset of structural instability. In practice, however, the secondary ground support installations depend upon the availability of faces and cable bolt installation equipment. This phase is often scheduled to be completed in short, repetitive campaigns. Figure 13.69 shows a ground support scheme in which the primary support includes a layer of shotcrete with bolts and mesh followed by a secondary layer of cable bolt reinforcement and shotcrete fill-in (usually the walls only) installed very close (two cuts or approximately 8 m) from the advancing face.

Another example is given in Figures 13.70 and 13.71, which illustrate a sequence of ground support installation, where an 11.4 m development was stabilized with a primary ground support scheme consisting of an initial layer of shotcrete plus mesh and bolts. Subsequently, a secondary

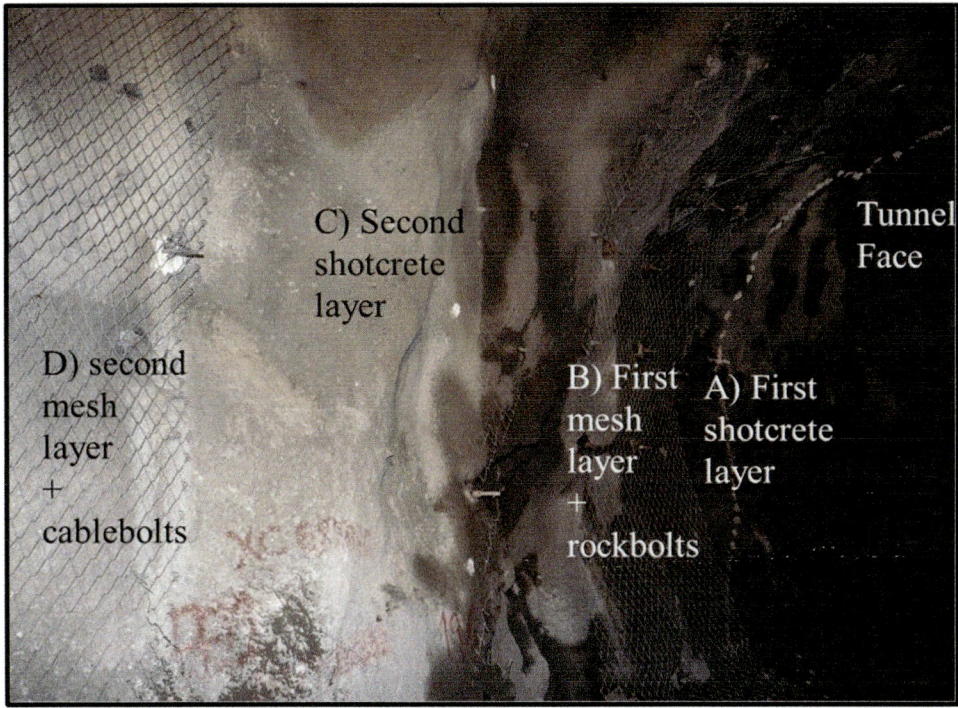

FIGURE 13.68 Example of a ground support scheme with two integrated layers of woven mesh.

FIGURE 13.69 Example of sequential installation of layers of ground support that include cable bolt reinforcement at a distance very close to the advancing face.

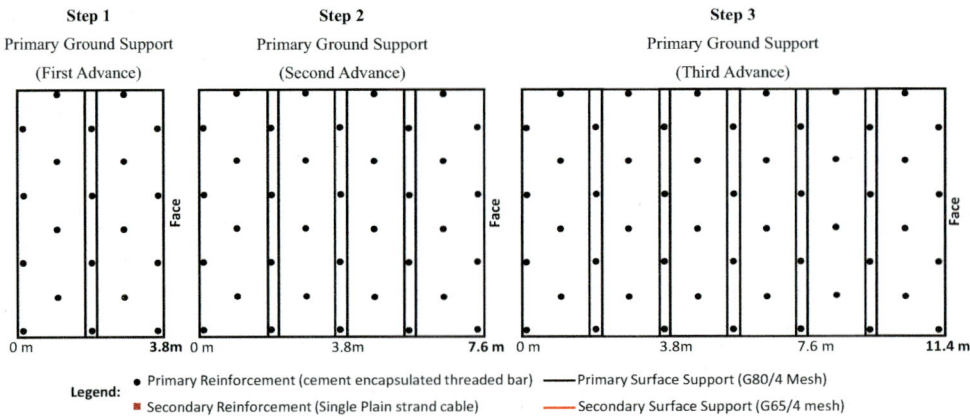

FIGURE 13.70   Long section views of an example of the primary ground support installation sequence.

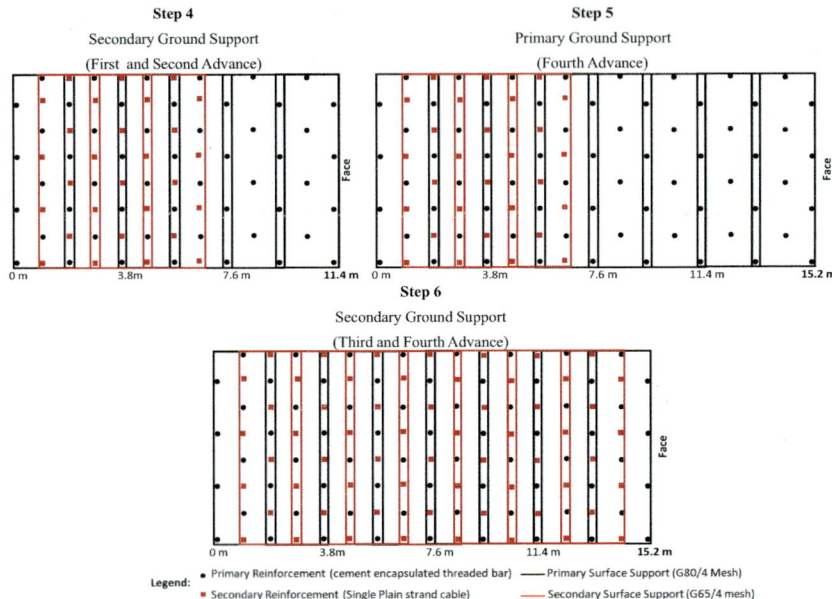

FIGURE 13.71   Long section views of an example of the secondary ground support installation sequence.

layer of cable bolts and the additional (external) mesh was installed for the first two of the three cuts. The strategy was to maintain a minimum offset distance of one cut, or approximately 3.8 m, between the end of the secondary ground support pattern and the advancing face. The purpose of maintaining this offset was to reduce the potential for the blasted flyrock to damage the exposed second mesh layer.

The choice of sequential ground support installation depends upon the rock mass demand and the expected mode of instability, with the layers of shotcrete applied in immediate contact with the rock surface or as shotcrete fill-in within a secondary layer. Figure 13.72 shows an example in

FIGURE 13.72 Example of sequential ground support installation in moderate stress conditions with no immediate instability.

which the rock bolts and mesh installation are delayed by one cut, with boltless shotcrete used as the primary support.

Figure 13.73 shows an innovative set-up that allows the installation of rolls of a woven mesh using a conventional twin boom jumbo. The methodology involves the introduction of an 'L Pin' device that allows mesh installation without losing the functionality of one of the booms (Arcaro

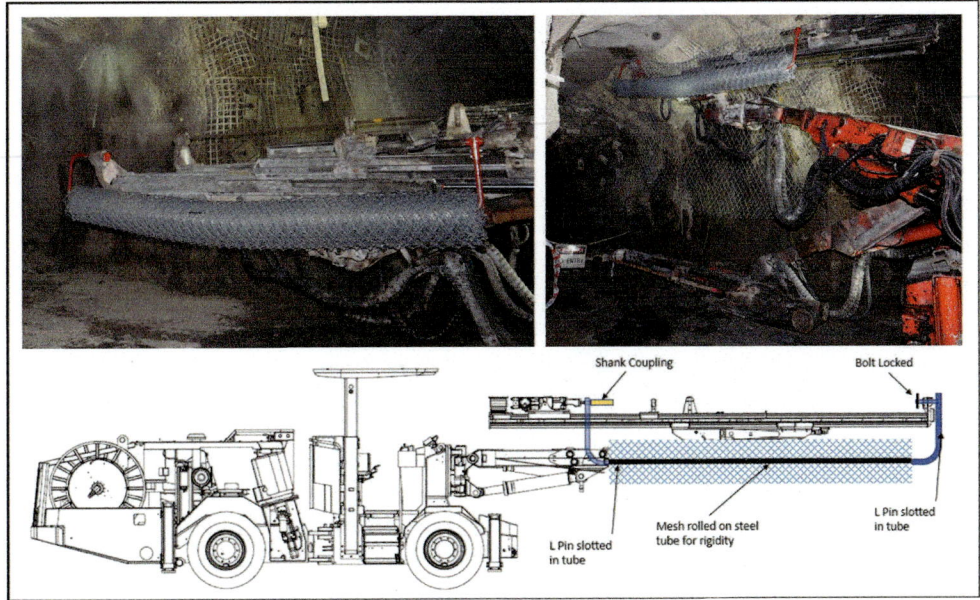

FIGURE 13.73 the 'L Pin' Jumbo arrangement for installation of woven mesh (Arcaro et al., 2021).

et al., 2021). While one boom is fitted with drilling and bolting equipment, the mesh can be installed using the other boom, which can then return to normal functionality after pinning the mesh. Note that the majority of high-energy dissipation ground support schemes constructed in Australia from 2019 to 2022 have been achieved using this methodology. This includes most of the photographs that have been presented in this book.

13.3.3.8   Final Ground Support Scheme Arrangement

In summary, a completed installation of a mechanized ground support scheme for high energy dissipation would include the following components (see a typical example illustrated in Figure 13.74).

13.3.3.8.1   *Primary Surface Support*

A 75 mm thick plain shotcrete layer (50 mm initially + 25 mm fill-in), internally reinforced with high tensile G80/4.0 or G80/4.6 woven or articulated mesh (floor to floor).

13.3.3.8.2   *Primary Reinforcement*

A mild steel threaded bar, 20–25 mm in diameter, 2.4–3 m long on a 1.2 m × 1.2 m square pattern, continuously mechanically coupled, fully cement encapsulated or fully resin encapsulated (with a decoupled region exceeding 1 m) and plated using 200 m × 200 m × 6–8 mm thick domed plates.

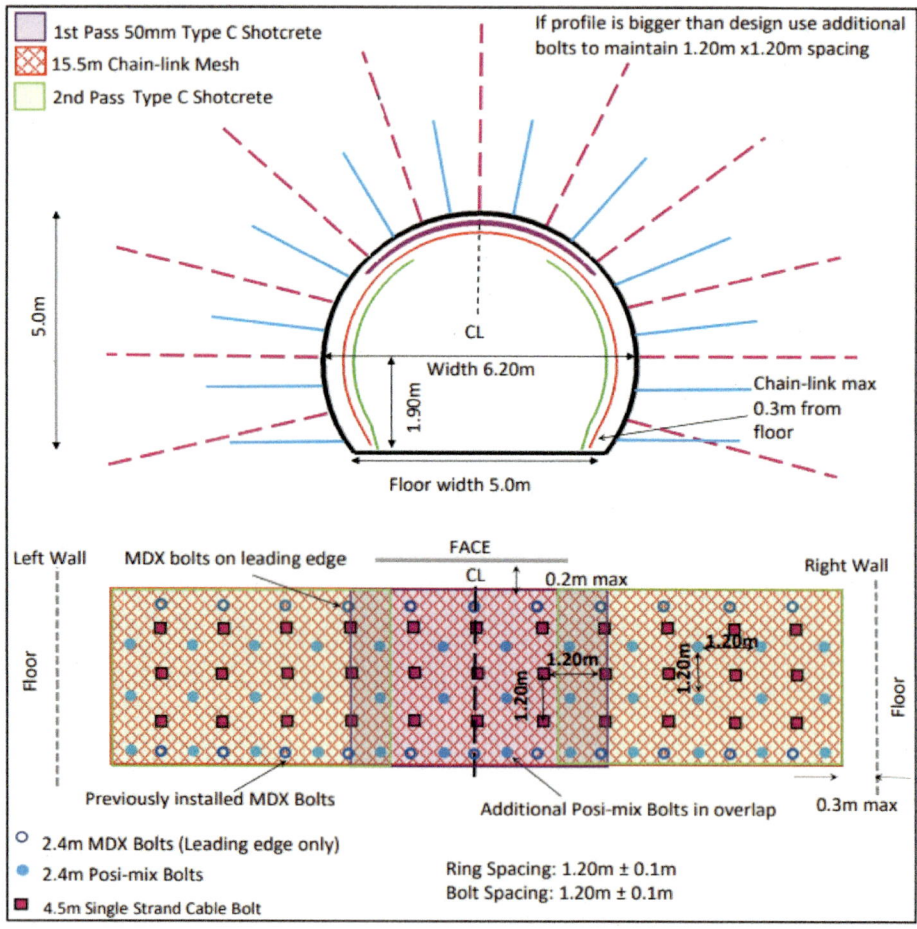

FIGURE 13.74   Example of a high energy dissipation ground support scheme standard implemented in a fresh hard rock development (Arcaro et al., 2021).

13.3.3.8.3 Secondary Surface Support

In cases where rock mass demand requires a second layer of mesh, either G65/4 or G80/4.6 high tensile woven mesh would be installed (floor to floor).

13.3.3.8.4 Secondary Reinforcement

High tensile, plain, single-strand cable bolts, 15.2–17.8 mm in diameter, 4.5–5.5 m long on a 1.2 m × 1.2 m square pattern spacing staggered with respect to the rock bolt collars, continuously mechanically coupled, fully cement encapsulated, with 300 mm × 300 mm × 10 mm face plates, and pre-tensioned to 5 tons. Note that proper pre-tensioning requires the application of a grease film at the barrel-wedge interface.

13.4 MONITORING

In the mining context, development performance can be measured using two methodologies. The first is from an economic perspective, where the main interest is in how much development is being undertaken and its cost (Figure 13.75). This is usually measured in terms of the rate of advance (distance per unit time frame) and the overall cost per unit metre of excavation. The second is to construct a tunnel of the required size and shape that is serviceable for the complete duration of the mining activity for which it is designed. It includes controlling damage from either time-dependent deformations or from repetitive dynamic loading events.

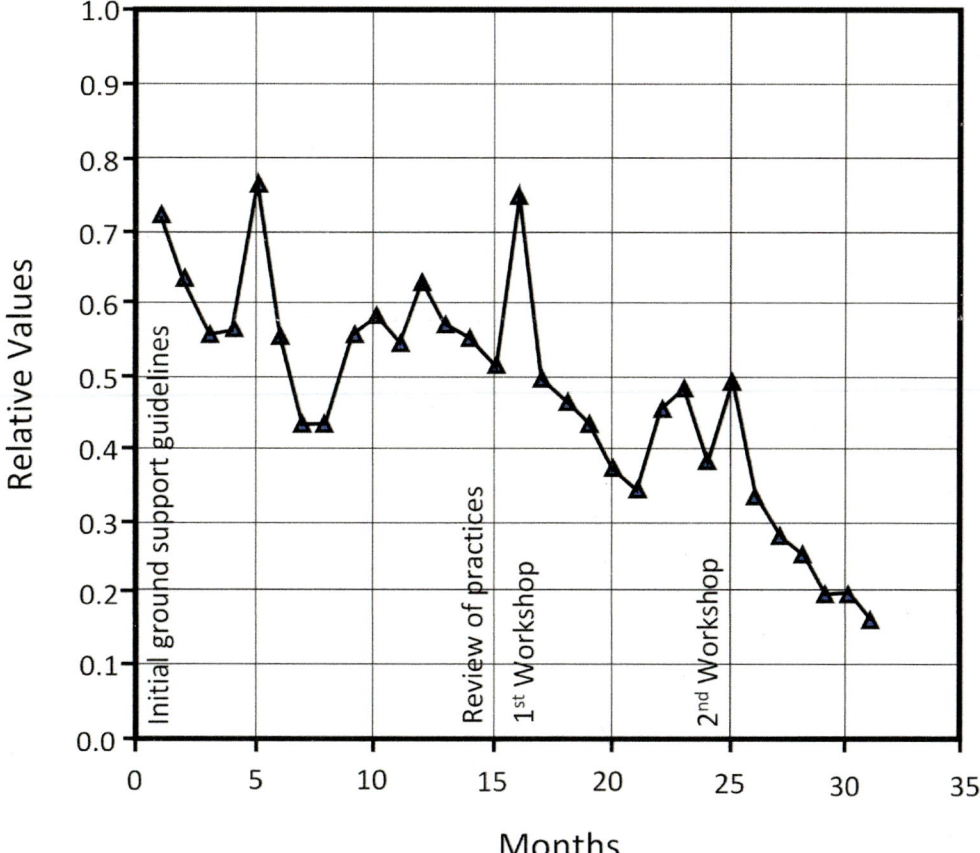

FIGURE 13.75 Ground support costs calculated over a three-year period of sequential design improvements (Villaescusa et al., 1994).

From an overall engineering perspective, both methods must be considered when assessing the overall excavation performance. Furthermore, measuring the performance of mining development practices is only the first step, as it is essential to understand the parameters affecting performance and how these parameters can be managed and controlled. One of the most important parameters is the blasting performance and, in particular, the interaction of the blasting energy with the geological discontinuities present within the rock mass. Other parameters include the orientation, shape and size of the development excavation as well as the induced stresses. The interaction of all these variables will control both the fragmentation of rock within the excavation boundary and damage to the walls, which ultimately control the performance of the whole excavation.

Other important parameters include organization and training of the mining crews, construction and ground support equipment types and availability, as well as consumable materials and costs. These are managerial issues and will not be discussed further in this book. Nevertheless, when looking at the overall performance, these management issues do need to be addressed in conjunction with the technical issues presented here to achieve optimal performance.

13.4.1 DRILLING AND BLASTING

One of the typical problems experienced is the over-excavation (or overbreak) of the development headings (Figure 13.76). It causes particular concern since larger openings are inherently more unstable, thus requiring increased ground support. Also, the development takes longer to achieve if more rock needs to excavated from an opening that becomes larger than its design dimensions. It would result in an overall decrease in the development rate, which, as explained earlier, is one of the main measures of overall performance.

Other problems identified include excavation alignment, particularly in the vertical plane, resulting in uneven floors and backs. Uneven floors on tramming routes result in a rougher ride for the operators, slowing mucking cycle times, and an increased need for road and equipment maintenance. Uneven floors produced by poor development blasting practices may also result in problems with respect to the alignment of holes during subsequent production drilling, caused by difficulties in setting up the drilling rigs correctly. Ponding water is another problem that results from uneven floors (Figure 13.77). It adversely affects the whole mining environment, including increasing the atmospheric corrosion of the exposed rock support and reinforcement elements (Hassell, 2007).

Systematic data collection is required to monitor practices and to assess the results being experienced at a particular mining location. Emphasis is required on the drilling and explosive charging practices to identify what impact changes are having on the final excavation profiles. The main blasting parameters to be monitored include:

FIGURE 13.76 Examples of excessive development overbreak showing (a) structurally controlled failure and (b) stress-driven instability.

FIGURE 13.77   Example of uneven floors leading to water ponding.

- Grade taking and face mark up,
- Cut geometry,
- Drilling performance and alignment of holes,
- Number and location of holes throughout the entire face,
- Type and amount of explosives used,
- Detonation sequence used,
- Presence of butts and/or misfires,
- Post blasting excavation profile and
- Muckpile fragmentation and profile.

Grade taking and face mark-up can contribute to over-excavation and misalignment. The failure of some operators to pay adequate attention to the initial marking up of the face means that in many cases, even well-drilled faces may not have the correct alignment or the correct size. Any poor drilling and blasting practices further exacerbate these problems by increasing the amount of over-excavation. Also, uneven faces become more unstable when high induced stresses are present. Drilling of a development round usually starts at the bottom of the face, ensuring that the lifter holes are positioned as close to the floor as possible. Drilling then proceeds up the face using both jumbo booms to achieve maximum productivity (Figure 13.78).

The cut geometry needs particular attention given to the alignment of the holes, which is facilitated by using a guide rod usually placed in the first cut hole drilled. It provides a simple reference for lining up the remainder of the holes in the cut, including the relief holes. For a typical blast (38–54-mm diameter charged holes), the cut usually consists of four to six relief holes (89–127 mm diameter). The relief holes are placed in conjunction with eight charged holes arranged in square patterns around them, as shown in Figure 13.79. The location of the cut on the face usually alternates from side to side (or up to down) to avoid having drill holes near the butts from the previous cut. It must be noted that the proximity of a cut in relation to the boundaries of an excavation has the potential to cause damage to the walls, thus impacting the overall stability of the excavation.

Poor positioning and alignment of the perimeter holes often cause damage to the excavation walls, resulting in an uneven excavation profile and excessive shotcrete requirements (Figure 13.80). Of particular concern is the occurrence of poorly aligned lifters in the lower corners of the development drives (Figure 13.81). It creates wall undercutting that may result in instability of the upper portion of a wall, which may then affect the roof.

Most of the holes in a face are typically charged with ANFO, apart from the perimeter holes, which must be charged with decoupled explosives to reduce damage and roughness of the excavation walls. Usually, the firing sequence is very much operator-dependent, with several slightly

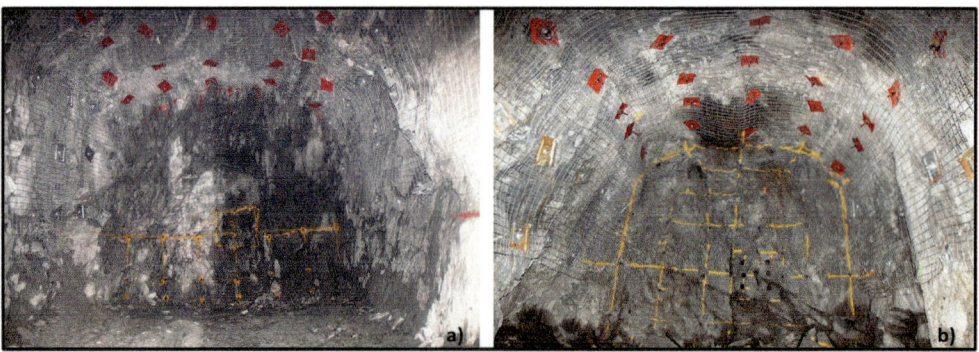

FIGURE 13.78 Examples of development face drilling showing (a) progress drilling up to lifter, knee and cut holes and (b) completed blast pattern showing the cut location on the right-hand side of the face.

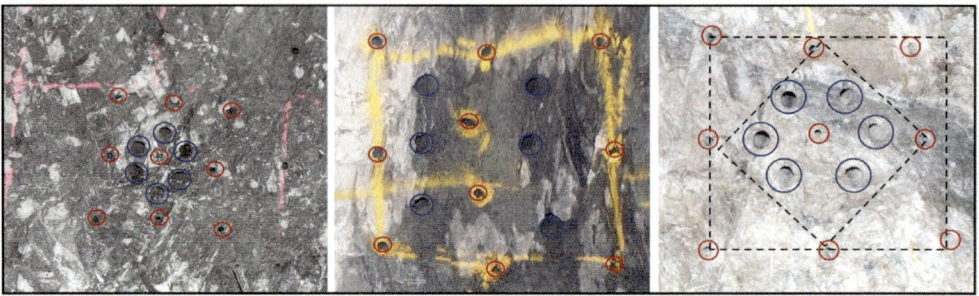

FIGURE 13.79 Examples of drilled cut geometries showing relief holes (blue) and charged holes (red).

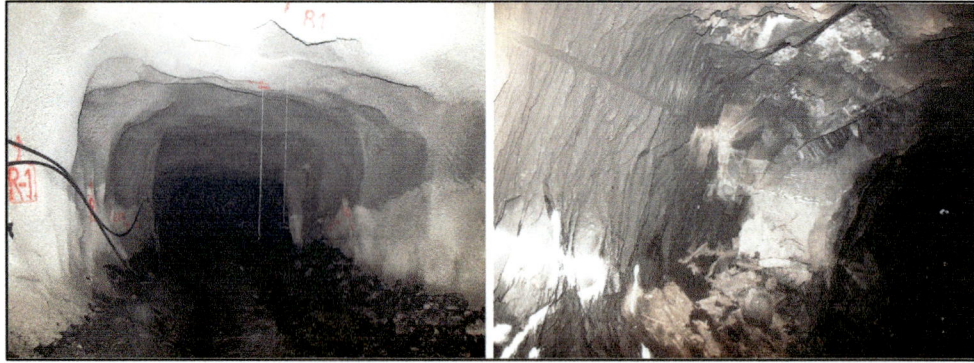

FIGURE 13.80 Examples of excessive hole misalignment at the perimeter of a development excavation resulting in excessive shotcrete requirements.

different designs being typical of any given mine site precinct. Unfavourable mechanics relating to sequential void creation is apparent in some detonation sequences with the potential for very confined holes when the delay scatter is considered.

The implementation of current blast technology and knowledge with respect to the factors affecting development blasting performance has the potential to significantly improve excavation quality. In particular, the importance of correct hole positioning and alignment must be impressed on the development crews, along with the safety and efficiency advantages that such attention to detail

FIGURE 13.81 Examples of undercutting the lower walls due to hole deviation and the subsequent blast damage.

provides to them. Such changes do not require any advances in our current knowledge but rather improved implementation of our current technology. Figure 13.82 shows results from semi-autonomous blast performance analysis. This technology (www.e-fidbak, 2021) can detect potential detonation issues in development blasting, such as missing holes, holes firing out of sequence or having a significant variation in the energy released. The technology relies on wireless accelerometers that can be readily coupled to existing rock bolt plates using magnets (Figure 13.83). The technology includes algorithms for fast signal processing and analysis to evaluate blast performance. This should be used by mine sites as part of their standard operating procedures, which also reduces the need for blasting specialists.

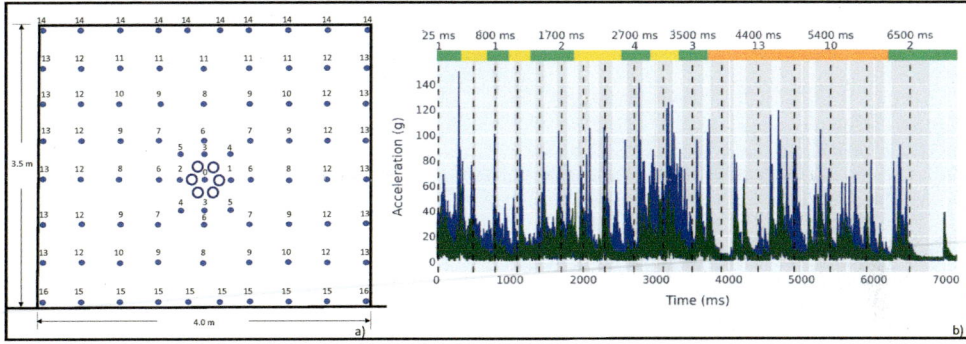

FIGURE 13.82 A cross-sectional view of square development showing (a) drill holes and detonation sequence and (b) monitored near field acceleration.

Figure 13.84 shows the results from blast monitoring for a trial de-stress blasting configured with a four de-stress hole geometry. The de-stressing holes were charged with a reduced amount of explosives compared with conventional charges, which is reflected in the vibration traces. As indicated earlier, isolated holes are unlikely to achieve any crack propagation along the rows of the de-stressing holes (Figure 13.85), as documented by Drover (2018).

13.4.2 Deformation

The geometry of the mining development shapes is generated by surveying a limited number of precisely located points around the excavation periphery. The process is repeated on a cut-by-cut basis along the axis of a drive, with the results discretized into a three-dimensional mesh (Jones et al.,

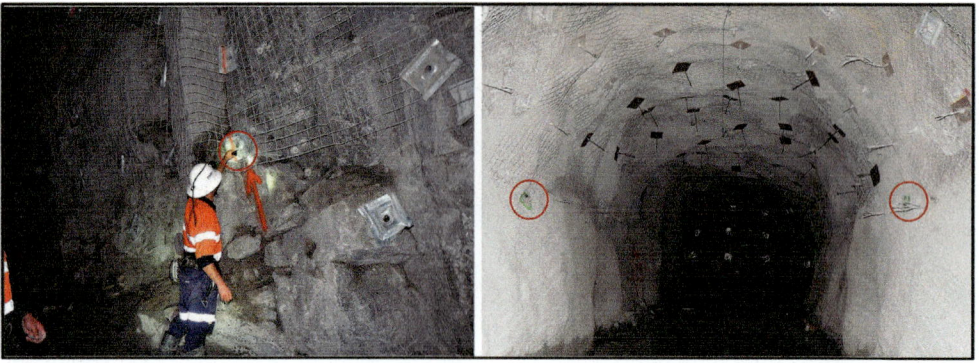

FIGURE 13.83   Placement of portable, wireless blast monitoring sensors with respect to a detonating face.

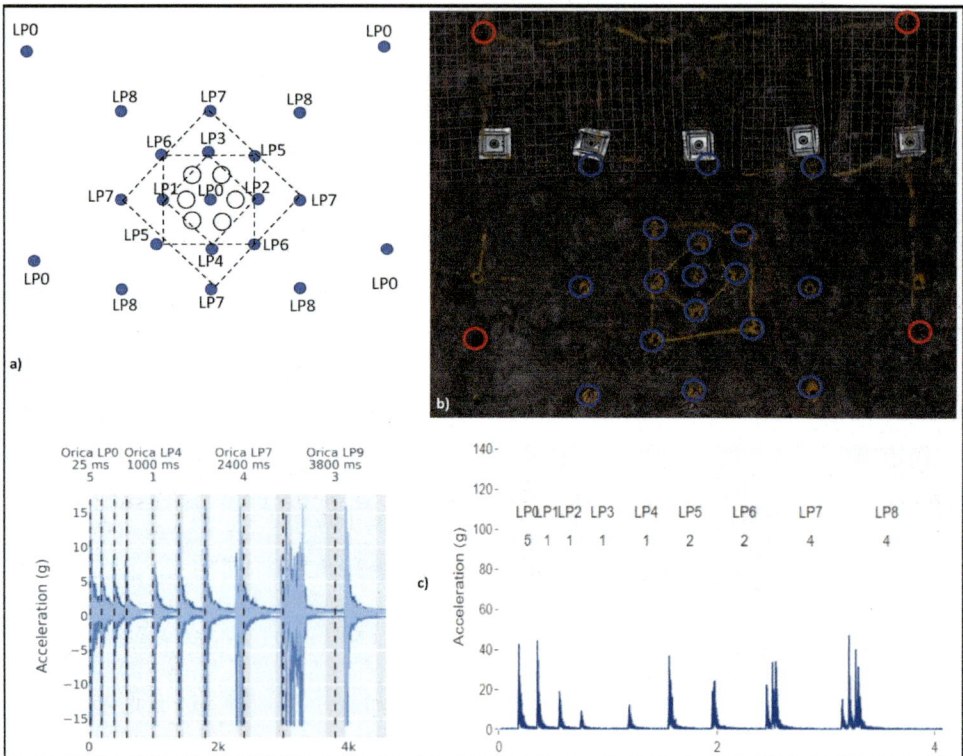

FIGURE 13.84   Details of face de-stress blasting showing (a) design, (b) implementation and (c) monitored near field accelerations.

2018). This process is data-limited but sufficient to determine the position of an advancing face. However, it may not be able to capture the entire geometry, which is important when a rectangular tunnel is modified to an oval or semi-circular shape. Additionally, when assessing ground support effectiveness, including any depth of failure determinations, the use of point cloud surveying technology is required. Jones et al. (2018) describe the use of mobile laser scanners that allow for a rapid generation of a point cloud survey of tunnel excavation. The technology described by Jones et al. (2018) utilizes a combination of Simultaneous Location and Mapping (SLAM) algorithms and an

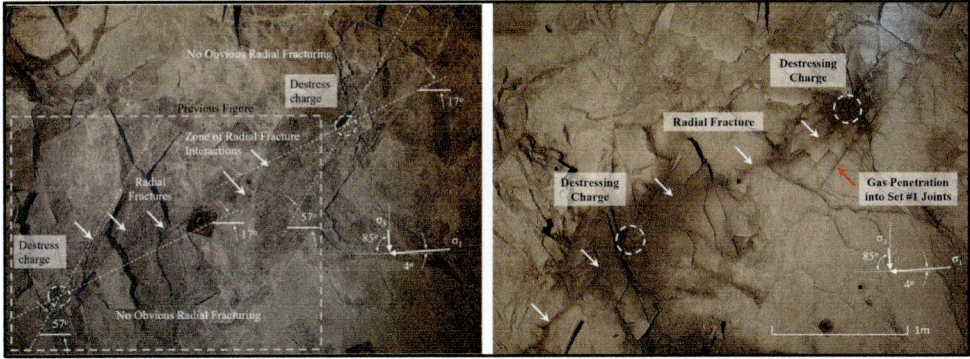

FIGURE 13.85 Details of radial fracture interaction due to de-stress blasting under very high stress conditions (Drover, 2018).

Inertial Measurement Unit (IMU) to generate a 3D point cloud while moving through an environment such as walking along an access tunnel. The methodology requires a baseline scan and its comparison with results from subsequent scans. Figure 13.86 shows the results from a monitoring campaign undertaken after a large seismic event that resulted in large induced compressive stresses resulting in wall damage and floor heave.

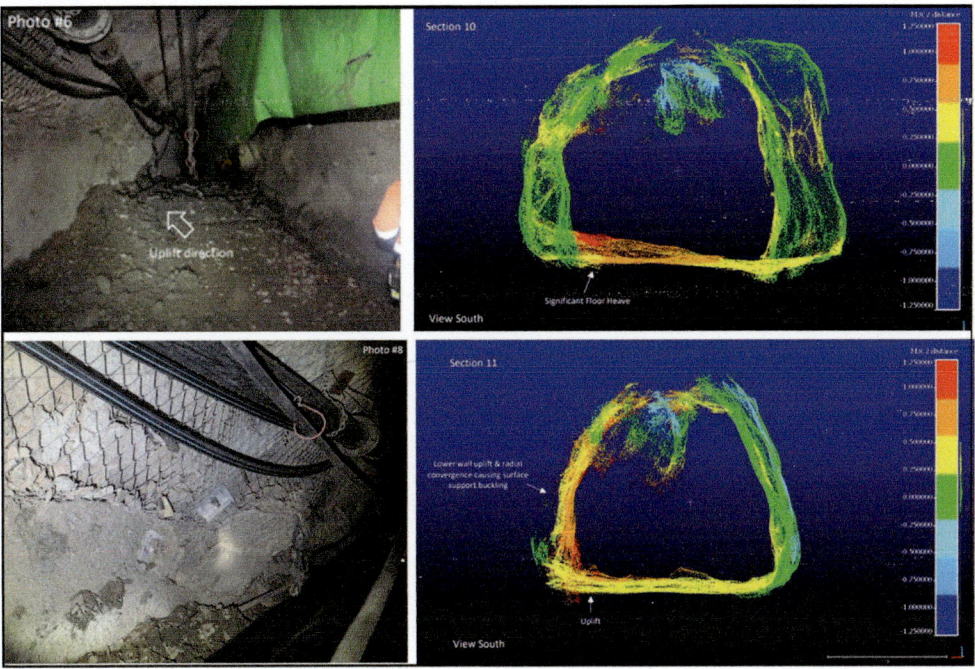

FIGURE 13.86 Example of point cloud deformation monitoring showing a reduction in tunnel cross-sectional area due to floor heave and wall damage.

References

Abdun-Nur, E. 1961. Fly ash in concrete—An evaluation (Bulletin No. 284, Highway Research Board). Washington, DC, 138.

ACI 212.3R-10, 2010. Report on chemical admixtures for concrete. Farmington Hills, MI: American Concrete Institute (ACI) Committee 212.

ACI 232.2R-03, 2003. Use of fly ash in concrete (ACI committee 233). Farmington Hills, MI: American Concrete Institute.

ACI 234R-06, 2006. Guide for use of silica fume in concrete (ACI committee 234). Farmington Hills, MI: American Concrete Institute.

ACI 506.4R-04, 2004. Guide for the evaluation of Shotcrete. Farmington Hills, MI: American Concrete Institute (ACI) Committee 506.

ACI 506.5R-09, 2009. Guide for specifying underground shotcrete (ACI committee 506). Farmington Hills, MI: American Concrete Institute.

ACI 544.1R-96, 1996. State-of-the-Art report on fibre reinforced concrete. Farmington Hills, MI: American Concrete Institute.

Andreas, B., and Schubert, W. 2017. Calculating the spacing of discontinuities from 3D point clouds. Procedia Engineering, 191:270–278.

Arcaro, C., Villaescusa, E., Hassell, R., Talebi, R., and Kusui, A. 2021. Implementation of a High Energy Dissipation ground support scheme (Procc. Underground Operators Conference). Perth: The AusIMM.

Armelin, H. S., and Banthia, N. 1998. On the rebound—Minimizing shotcrete rebound with fibres. Concrete International, 20(9):74–79.

AS 1289.3.6.1, 2009. Methods of testing soils for engineering purposes. Method 3.6.1: Soil classification tests - Determination of the particle size distribution of a soil—Standard method of analysis by sieving. Australian Standards.

Ashby, M. F., and Jones, D. R. H. 1986. Engineering materials 2. London: Pergamon, 369.

ASTM 2002, Standard Test Method for Flexural Toughness of Fiber-Reinforced Concrete (Using Centrally Load Round Panel., West Conshohocken: ASTM, 6.

ASTM A820/A820M-04, 2004. Standard specification for steel fibers for fiber-reinforced concrete. West Conshohocken, PA: ASTM International.

ASTM C 1550 – 05, 2005. Standard test method for flexural toughness of fibre reinforced concrete (using centrally loaded round panel). West Conshohocken, PA: ASTM International.

ASTM C1141/C1141M-15, 2015. Standard Specification for Admixtures for Shotcrete. West Conshohocken, PA: ASTM International.

ASTM C1240-10a, 2010. Standard specification for silica fume used in cementitious mixtures. West Conshohocken, PA: ASTM International.

ASTM C125-10a, 2010. Standard terminology relating to concrete and concrete aggregates. West Conshohocken, PA: ASTM International.

ASTM C1583-04, 2004. Standard test method for tensile strength of concrete surfaces and the bond strength or tensile strength of concrete repair and overlay materials by direct tension (Pull-off method). West Conshohocken, PA: ASTM International.

ASTM C1602/C1602M-06, 2006. Standard Specification for mixing water used in the production of hydraulic cement concrete. West Conshohocken, PA: ASTM International.

ASTM C219-07a, 2007. Standard terminology relating to hydraulic cement. West Conshohocken, PA: ASTM International.

ASTM C39/C39M-09a, 2009. Standard test method for compressive strength of cylindrical concrete specimens. West Conshohocken, PA: ASTM International.

ASTM C494/C494M-08, 2008. Standard specification for chemical admixtures for concrete. West Conshohocken, PA: ASTM International.

ASTM C618-08a, 2008. Standard specification for coal fly ash and raw or calcined natural pozzolan for use in concrete. West Conshohocken, PA: ASTM International.

ASTM C989-10, 2010. Standard specification for slag cement for use in concrete and mortars. West Conshohocken, PA: ASTM International.

ASTM, 1995. Standard specification for roof and rock bolts and accessories. F432: ASTM.

ASTM, 2010. F432-10 Standard Specification for Roof and Rock Bolts and Accessories, 18, West Conshohocken: ASTM.

Atlas Copco, 2017. Boltec E—Technical specification. Australia: Atlas Copco.

Barla, G. 2014. TBM tunnelling in deep underground excavation in hard rock with spalling behaviour. Proceedings of the 43rd Geomechanics Colloquium, 25–39.

Barley, A. D., and Windsor, C. R. 2000. Recent advances in ground anchor and ground improvement technology with reference to the development of the art. GeoEng 2000, Proc Int Conf on Geotechnical and Geological Engineering, Melbourne, IEAust, Melbourne, 1048–1094.

Barrett, S., and McCreath, D. 1995. Shotcrete support design in blocky ground: Towards a deterministic approach. Tunnelling and Underground Space Technology, 10(1):79–89.

Barton, N., Lien, R., and Lunde, J. 1974. Engineering classification of rock masses for design of tunnel support. Rock Mechanics, 6(4):189–236.

Bawden, W. F., A. J. Hyett and Lausch, P. 1992. An experimental procedure for the in situ testing of cable bolts. International Journal of Rock Mechanics and Mining Science & Geomechanics Abstracts, 29(1):525–533.

Beaupré, D., 1994. Rheology of high performance shotcrete. PhD thesis, Department of Civil Engineering, University of British Columbia, Vancouver, BC, Canada, 250.

Beck, D. A. 2008. Multi-scale, non-linear numerical analysis of mining induced deformation. Proceedings of the 42nd US Rock Mechanics Symposium and 2nd U.S.-Canada Rock Mechanics Symposium, held in San Francisco, June 29–July 2, 2008. ARMA, American Rock Mechanics Association.

Bernard, E. S. 2003. Release of new ASTM Round Panel Test, Shotcrete, Spring, 20–23, Farmington Hills, MI: American Shotcrete Association.

Bernard, E. S. 2008. Early-age load resistance of fibre reinforced shotcrete linings. Tunneling and Underground Space Technology, 23:451–460.

Bernard, E. S., Jerlin, O., and Lönnesjö, K. 2006. High dosages of set accelerator in fibre reinforced shotcrete. In: Morgan, D.R., and Parker, H. W. (eds) *Proceeding of 10th International Conference on Shotcrete for Underground Support*, Whistler, British Columbia, Canada, American Society of Civil Engineers, 110–119.

Brady, B. H. G., and Brown, E. T. 2004. Rock mechanics for underground mining, 3rd edition, Dordrecht: Kluwer, 628.

Brennan, E. 2005. Testing shotcrete for bond. American Shotcrete Association, Shotcrete Magazine, Winter, 18–19.

British Standards, 2007. BS 7861-1:2007 Strata reinforcement support system components used in coal mines – Part 1: Specification for rock bolting. London: BSi, 44.

British Standards, 2009. BS 7861-2:2009 Strata reinforcement support systems components used in coal mines – Part 2: Specification for flexible systems for roof reinforcement, London: BSi, 48.

Broch, E., and Sørheim, S. 1984. Experiences from the planning, construction and supporting of a road tunnel subjected to heavy rockbursting. Rock Mechanics and Rock Engineering, 17(1):15–35. doi: 10.1007/BF01088368.

Brown, E. T. 1999a. The evolution of support and reinforcement philosophy and practice for underground mining excavations. In: Villaescusa, W. T. (eds) Rock support and reinforcement practice in mining, Balkema, Rotterdam, 3–17.

Brown, E. T. 1999b. Rock mechanics and the Snowy Mountains Scheme In: The Spirit of the Snowy Fifty Years On, Proc 1999 Invitation Symposium, Cooma NSW, Academy of Technological Sciences and Engineering, Melbourne.

Brown, E. T. 2004. The dynamic environment of ground support and reinforcement. In: Villaescusa, E., and Potvin, Y. (eds) Ground support in mining and underground construction. Balkema, London, pp 3–16.

Brzovic A. 2010. Characterization of primary copper ore for caving at the El Teniente Mine, Chile. Curtin University of Technology, PhD Dissertation, 273.

Bullard, J. W., H. M. Jennings, R. A. Livingston, A. Nonat, G. W. Scherer, J. S. Schweitzer, K. L. Scrivener, and Thomas, J. J. 2011. Mechanisms of cement hydration. Cement and Concrete Research, 41(12):1208–1223.

Bürge, T. A. 2001. Mode of action of alkali-free sprayed shotcrete accelerators. In: Bernard, E.S. (ed) Proceeding of International Conference on Engineering Developments in Shotcrete. Hobart, Tasmania, Balkema Publishers, The Netherlands, 79–85.

Bustos, N. 2022. An integrated approach for design and construction of drawbells in block cave mines. PhD Thesis: Western Australian School of Mines, Curtin University, Perth, Western Australia, Australia, 293.

Bywater, S., and Fuller, P. G. 1983. Cable support of lead open stope hangingwalls at Mount Isa Mines Limited. Proc Int Symp on Rock Bolting Abisko Sweden Balkema, Rotterdam, 539–555.

Call R. D. 1972. Analysis of geologic structure for open pit slope design. Tucson, University of Arizona, PhD Dissertation, 245.

Call R. D., J. P. Savely and D. E. Nicholas, 1976. Estimation of joint set characteristics from surface mapping data. In: Hustrulid, W. A. (ed) Monograph on rock mechanics, applications in mining, New York: AIME, 65–73.

Carlton, R., B. Darlington B., and P. A. Mikula 2013. In situ dynamic drop testing of the MD bolt at Mt. Charlotte Gold Mine. In Potvin, Y., and Brady, B. (eds) Procc Ground Support 2013. Australian Centre for Geomechanics, Perth, ISBN 978-0-9806154-7-0.

Carr, C. J., D. R. Rankin and J. J. Fuykschot 1999. Development of advanced blasting practices at Forrestania Nickel Mines, in Proceedings of EXPLO '99, AusIMM, pp.239–246.

Chacos, G.P. 1993. Wedge forces on post-tensioning strand anchors. PTI Technical Note 2, Phoenix: Post-Tensioning Institute, 4.

Christiansson, R., Hakala, M., Kemppainen, K., Siren, T., and Martin, C. D. 2012. Findings from large scale in-situ experiments to establish the initiation of spalling. Paper presented at the ISRM International Symposium-EUROCK 2012, Stockholm, Sweden.

Clements, M. 2004. Comparison of methods for early age strength testing of shotcrete. In: Bernard, E. S. (ed) Proceedings of the 2nd International Conference on Engineering Developments in Shotcrete, Cairns, Queensland, Australia, Taylor and Francis Group, London, 81–87.

Clifford, R. L. 1974. Long rockbolt support at New Broken Hill Consolidated Limited. Proc AusIMM, AusIMM, Melbourne, 25(1):21–26.

Collepardi, S., L. Coppola, R. Troli, and Collepardi, M. 1999. Mechanism of actions of different superplasticisers for high performance concrete. Proceedings of the 2nd CANMET/ACI Conference, Gramado, Brazil, SP 186, 503–524.

Darlington, B. M. Rataj, and Roach, W. 2019. A new method to evaluate dynamic bolts and the development of a new dynamic rock bolt. Procc Deep Mining Conference, Muldersdrift, The Southern African Institute of Mining and Metallurgy, 205–216.

De Zoysa, A. U. 2015. Dynamic testing of mesh and rock bolt support systems, MPhil Thesis, WA School of Mines, Curtin University: Kalgoorlie, Western Australia, 136.

De Zoysa, A. U. 2022. Dynamic testing of cablebolt reinforcement, PhD Thesis, WA School of Mines, Curtin University: Kalgoorlie, Western Australia, 213, In Preparation.

Deere, D. U., and R. P. Miller, 1966. Engineering classification and index properties of rock. Technical Report No. AFNL-TR-65-116, Albuquerque, NM: Air Force Weapons Laboratory.

Deere, U., R. Peck, J. Monsees and B. Schmidt 1969. Design of tunnel liners and support systems. Final report for office of high speed ground transportation. Washington, DC: Department of Transportation.

Diederichs, M.S. 2007. The 2003 Canadian Geotechnical Colloquium: Mechanistic interpretation and practical application of damage and spalling prediction criteria for deep tunnelling. Canadian Geotechnical Journal, 44(9):1082–1116.

Dolch, W. L. 1984. Air entraining admixtures. In Ramachandran, V. S. (ed) Concrete admixtures handbook: Properties, science, and technology. Park Ridge, NJ: Noyes Publications, 269–302

Dolinar, D. 2006. Load capacity and stiffness characteristics of screen materials used for surface control in underground coal mines. 25th International conference on ground control in mining. Morgantown, August.

Drover, C., and Villaescusa, E., 2015a. Estimation of dynamic load demand on a ground support scheme due to a large structurally controlled violent failure – a case study. Journal of Mining Technology, 125(4).

Drover, C., and Villaescusa, E., 2015b. Performance of shotcrete surface support following dynamic loading of mining excavations. Proceedings of Shotcrete for Underground Support XII, Singapore.

Drover, C., and Villaescusa, E., 2022. A design and construction methodology for deep mine development. Proceedings of AusRock 2022, AusIMM, Melbourne, Australia.

Drover, C. 2018. Design, construction and monitoring of hard rock tunnels at great depth. PhD thesis, WA School of Mines, Curtin University, Kalgoorlie, Australia, 391.

Drover, C., and E. Villaescusa 2019. A comparison of seismic response to conventional and face destress blasting during deep tunnel development. Journal of Rock Mechanics and Geotechnical Engineering: https://doi.org/10.1016/j.jrmge.2019.07.002.

Drover, C., E. Villaescusa and E. Onederra 2018. Face de-stressing blast design for hard rock tunnelling at great depth. Journal of Tunnelling and Underground Space Technology, 80:257–268.

Edmeades, R. M., and Hewlett, P. C., 1988. Cement admixtures. In: Hewlett, P.C. (ed) Lea's Chemistry of Cement and Concrete, 4th. Edition, Oxford: Elsevier Ltd, 841–905.

EFNARC, 1996. European Specification for Sprayed Concrete, 30.

Elith, N. 1990. Safe and efficient blasting in metalliferous mining. Chapter 11 Tunnelling and Mine Development. ICI Australia Operations Pty Limited.

Fernandez-Delgado, G, Mahar, J and Parker, H 1976. Structural behaviour of thin shotcrete liners obtained from large scale tests. Shotcrete for ground support. Proceedings of the engineering foundation conference. October 4–8, 399–442. ACI.

Fukuhara, M., S. Goto, K. Asaga, M. Diamon and R. Kondo, 1981. Mechanisms and kinetics of C4AF hydration with gypsum. Cement and Concrete Research, 11:407–414.

Fuller, P. G. 1983. Cable bolt support in mining. Proc. Int Symp on Rock Bolting, Abisko Sweden, Balkema, Rotterdam, 511–522.

Fuller, P. G., and R. H. T. Cox 1975. Mechanics of load transfer from steel tendons to cement based grout. In Proc. 5th Aust. Conf. on the Mechanics of Structures and Materials, 189–203, Melbourne.

Furtney, J., Cundall, P., and G. Chitombo (2009). Developments in numerical modelling of blast induced rock fragmentation: updates from the HSBM project. International Symposium on Rock Fragmentation by Blasting, Granada, Spain, Taylor & Francis Group, pp.335–342.

Gardner, F. J. 1971. History of rock bolting. Symp on Rock Bolting, AusIMM, Illawarra Branch, Paper No. 2, 11.

Gaudreau, D., M. Aubertin and R. Simon 2004. Performance assessment of tendon support systems submitted to dynamic loading. In E. Villaescusa and Y. Potvin (eds) Ground Support in Mining & Underground Construction, Proc 5th Int Symp on Ground Support, Perth, 299–312. Leiden: Balkema.

Gere, J. M., and S. P. Timoshenko 1990. Mechanics of materials. Third edition, PWS-Kent Publishing Company, Boston, 809.

Goodman R. E., and G-h. Shi, 1985. Block Theory and Its Application to Rock Engineering. London: Prentice-Hall, 365.

Goris, J. M. 1990. Laboratory evaluation of cable bolt supports (in two parts) 1. Evaluations of supports using conventional cables. USBM RI 9308.

Goris, J. M. 1990. Laboratory evaluation of cable bolt supports (in two parts) 2. Evaluations of supports using conventional cables with steel buttons, birdcage cables and epoxy-coated cables. USBM RI 9342.

Greenelsh, R. 1985. The N663 stope experiment at Mount Isa Mine. International Journal of Mining and Mineral Engineering, 3:183–194.

Hahn, T., Holmgren, J. 1979. Adhesion of shotcrete to various types of rock surfaces. 4th ISRM Congress, International Society for Rock Mechanics.

Hassell, R. C. 2007. Corrosion of rock reinforcement in underground excavations. PhD Thesis, Western Australian School of Mines, Curtin University of Technology, Perth, Western Australia, Australia, 277.

Hassell, R. C., and Villaescusa, E. 2005. Overcoring techniques to assess in situ corrosion of galvanized friction bolts. In: Peng, S. S., et al. (eds) Proceedings, 24th international conference on ground control in mining, Balkema, Morgantown, 349–356.

Hassell, R. C., Villaescusa, E., and Thompson, A. G. 2006. Testing and evaluation of corrosion on cable bolt anchors. In 41st US Rock Mechanics Symposium, Golden, June 17–21, Paper 06–996, 11, Washington, ARMA.

Hassell, R. C., Villaescusa, E., Ravikumar, A., and Cordova, M. 2010. Development of a corrosivity classification for weld mesh support at the Cannington Mine. In: Hagan, P., and Saydam, S. (eds) Proc 2nd Australasian Conf on Ground Control in Mining, AusIMM, Sydney, Melbourne.

Hauck, C. J., and Kristiansen, M. G. 2010. The influence of air content on sprayed concrete quality and sprayability in a civil tunnel. Proceedings of the 3rd International Conference on Engineering Developments in Shotcrete, 119–124.

Heal, D. 2010. Observations and analysis of incidents of rockburst damage in underground mines, PhD Thesis, The University of Western Australia, 357.

Higginson, E. C., Wallace, G. B., and Ore, E. L. 1963. Effect of maximum size of aggregate upon compressive strength of mass concrete. Symposium On Mass Concrete, ACI SP-6, Detroit, Michigan, 219–56.

Hoek, E. 1999. Support for very weak rock associated with faults and shear zones. In: Villaescusa, E., Windsor, C. R., and Thompson, A. G. (eds) Rock support & reinforcement practice in mining, Balkema, Rotterdam, 19–32.

Hoek, E., and Brown, E. 1980. Underground excavations in rock. London: IMM.

Hoek, E., Kaiser, P. K., and Bawden, W. F. 1995. Support of underground excavations in hard rock. Rotterdam: Balkema.

Hogan, F. J., and Meusel, J. W. 1981. Evaluation for durability and strength development of a ground granulated blast furnace slag. Cement, Concrete and Aggregates, 3(1):40–52.

Holmgren, J. 1976. Thin shotcrete layers subject to punch loads. Shotcrete for ground support. Proceedings of the engineering foundation conference, ACI, October, 443–459.

Holmgren, J. 2001. Shotcrete linings in hard rock. Underground mining methods. In: Hustrulid, W., and Bullock, R. (ed) Engineering fundamentals and international case studies. Colorado: Colorado Society for Mining, Metallurgy and Exploration, 569–577.

Hudson, J. A., and Priest, S. D. 1979. Discontinuities and rock mass geometry. International Journal of Rock Mechanics and Mining Sciences & Geomechanics Abstracts, 16(6):339–362.

Hutchins, W. R. Bywater, S., Thompson, A. G., and Windsor, C. R. 1990. A versatile grouted cable dowel reinforcing system for rock. The AusIMM Proceedings, 1:25–29.

Hutchinson, D., and Diederichs, M. 1996. Cablebolting in underground mines. Richmond: Bitech Publishers, 406.

Hyett, A. J., Bawden, W. F., and Coulson, A. L. 1992a. Physical and mechanical properties of normal Portland cement pertaining to fully grouted cable bolts. In: Kaiser and McCreath (eds) Rock support, Balkema, Rotterdam, 341–348.

Hyett, A. J., Bawden, W. F., and Reichardt, R. 1992b. The effect of rock mass confinement on the bond strength of fully grouted cable bolts. International Journal of Rock Mechanics and Mining Sciences, 29:503–524.

ISRM. 1978. Suggested methods for determining tensile strength of rock materials. International Journal of Rock Mechanics and Mining Sciences, 15:90–103.

Jager, A. J. 1992. Two new support units for the control of rockburst damage. Rock support in mining and underground construction. Rotterdam: Balkema, 621–631.

Jastrzebski, Z. D. 1976. The nature and properties of engineering materials (2nd ed.). New York: John Wiley & Sons, 633.

Jenkins, P. A, J. Mitchell, J., and Upton, B. 2004. Hydro scaling and in-cycle shotcrete at Waroonga mine, Western Australia, Ground Support in Mining and Underground Construction; Proc. 5th intern, symp, Balkema, Perth, Rotterdam.

Jolin, M., Beaupre, D., and Mindess, S. 1999. Tests to characterize properties of fresh dry shotcrete. Cement and Concrete Research, 29:753–760.

Jones, E., Ghabraieb, B., and Beck, D. 2018. A Method for determining field accuracy of mobile scanning devices for geomechanics. In: Proceedings of 10th Asian Rock Mechanics Symposium, Singapore.

Jung, W., Utagawa, M., Ogata, Y., Seto, M., Katsuyama, K., Miyake, A., and Ogava, T. 2001. Effects of rock pressure on crack generation during tunnel blasting. Japan Explosives Social 62:138–146.

Knight, B., Rasping, M., and Clegg, I. 2006, Wet-mix shotcrete as a material, process and ground control component of a 21st century underground mining operation. In: Morgan, D., and Parker, H. (eds) Proceedings of the 10th International Conference Shotcrete for Underground Support X, American Society of Civil Engineers, Whistler, BC, Canada, 298–306.

Kuchta, M. E. 2002. Quantifying the increase in adhesion strength of shotcrete applied to surfaces treated with high-pressure water. In: SME Preprint Number 02–35, SME Annual Meeting, Phoenix, AZ.

Kuijpers, J., Milev, A., Jager, A., and Acheampong, E. 2002. Performance of various types of containment support under quasi-static and dynamic loading conditions. Part I. Gap Report 810a. May. South Africa SIMRAC.

Kusui, A. 2015. Scaled down tunnel testing for comparison of surface support performance, PhD Thesis, WA School of Mines, Curtin University, Western Australia, Kalgoorlie, 216.

Kusui, A., and Villaescusa, E. 2016. Seismic Response Prior to Spalling Failure in Highly Stressed Underground Tunnels, Procc. Seventh International Conference and Exhibition on Mass Mining 2016, The AusIMM, Sydney, Australia.

Kusui, A., Villaescusa, E., and Funatsu, T. 2016. Mechanical behaviour of scaled-down unsupported tunnel walls in hard rock under high stress. Tunnelling and Underground Space Technology, 60:30–40.

Kutter, H. K., and Fairhurst, C. 1971. On the fracture process in blasting, International Journal of Rock Mechanics and Mining Sciences, 8:181–202.

Lajtai, E., and Dzik, E. 1996. Searching for the damage threshold in intact rock. Aubertin, H., and Mitri (eds) Rock mechanics: Tools and techniques, Balkema, 1:701–708.

Lane, R. O. 1983. Effects of fly ash on freshly mixed concrete. Concrete International, 10:50–52.

Lang, T. A. 1961. Theory and practice of rock bolting. Transactions of the American Institute of Mining, Metallurgical, and Petroleum Engineers, 220:333–348.

Lees, D. (Editor) 2009. The history of Australian tunnelling. West Ryde: Paragon Printers.

Levkovitch, V., Reusch, F., and Beck, D. 2010. Application of a non-linear confinement sensitive constitutive model to mine scale simulations subject to varying levels of confining stress. In: Zhao, J., Labiouse, V., Dudt, J., and Mathier, J. (ed) Proceedings of the European Rock Mechanics Symposium, Lausanne, 161–164.

Li, C. 1997. Swellex rock bolts in weak and soft rock (Research Report). Lulea: Lulea University of Technology, Department of Civil and Mining Engineering.

Li, C. 2010. A new energy-absorbing bolt for rock support in high stress rock masses. International Journal of Rock Mechanics and Mining Sciences, 47:396–404.

Li, J. 2004. Critical strain of intact rock and rock masses, PhD Thesis, Curtin University of Technology, Chile, 186.

Littlejohn, G. S. 1993. Overview of rock anchorages. In: Hudson, J. A. (ed) Comprehensive rock engineering, Pergamon Press, London, Volume 4, Chapter 15:413–450.

Littlejohn, G. S., and Bruce, D. A. 1977. Rock anchors- sate of the art. Brentwood: Foundation Publications.

Malmgren, L., Nordlund, E., and Rolund, S. 2005. Adhesion strength and shrinkage of shotcrete. Tunnelling and Underground Space Technology, 20(1):33–48. doi:10.1016/j.tust.2004.05.002.

Malvar, L. J., and Crawford, J. E. 1998. Dynamic increase factors for steel reinforcing bars. Department of Defence, Explosive Safety Board Seminar, Orlando, FL, August 1998.

Martin, C. D. 1993. The strength of massive Lac du Bonnet granite around underground openings, PhD Thesis, University of Manitoba, Manitoba.

Martin, C. D. 1997. Seventeenth Canadian geotechnical colloquium: The effect of cohesion loss and stress path on brittle rock strength. Canadian Geotechnical Journal, 34(5):698–725.

Mathews, K. E., Hoek, E., Wyllie, D. C., and Stewart, S. B. V. 1980. Prediction of Stable Excavation Spans for Mining at Depths Below 1,000 Meters in Hard Rock. Ottawa: Golder Associates Report to Canada Centre for Mining and Energy Technology (CANMET), Department of Energy and Resources.

Mathis, J. L. 1988. Development and verification of a three dimensional rock joint model, PhD Diss., Lulea University, 265.

Matthews, S. M. Thompson, A. G. Windsor, C. R., and O'Bryan, P. R. 1986. A novel reinforcing system for large rock caverns in blocky rock masses. In: Saari, K. H. O. (ed), International Conference on Large Rock Caverns, Pergamon Press, London, Volume 2, 1541–1552.

Matthews, S. M., Tillmann, V. H., and Worotnicki, G. 1983. A modified cable bolt system for the support of underground openings. Proc AusIMM Annual Conference, Broken Hill NSW, AusIMM, Melbourne, 243–255.

Mauldon, M. 1994. Intersection probabilities of impersistent joints. International Journal of Rock Mechanics and Mining Sciences, 31(2):107–115.

Melbye, T. A., and Garshol, K. F. 1997. Sprayed concrete for rock support (6th ed.). Zurich: MBT International Underground Construction Group, 198.

Mills, W. B. 2009. The Snowy men behind tunnel rock bolting. Cooma: Walter Mills.

Milne, D. P. Germain, D. G., and Noble, P. 1991. Systematic rock mass characterization for underground mine design. Procc. 7th Int Congress on Rock Mechanics, Aachen, AA Balkema, Volume 1, 293–298.

Morton, E. C. 2009. Static testing of large scale ground support panels, Master of Science Thesis, Western Australian School of Mines, Curtin University of Technology, 250.

Morton, E. C., Villaescusa, E., and Thompson, A. G. 2009. Determination of energy absorption capabilities of large scale shotcrete panels. In: Amberg, F., and Garshol, K. F. (eds) Shotcrete for Underground Support XI, Proc 2009 ECI Conf on Shotcrete for Underground Support, Paper, Davos, 6, 20.

MOSHAB. 1999. Surface rock support for underground mines. Perth: Code of Practice, The Government of Western Australia, Mines Occupational Safety and Health Advisory Board 17.

Mould, R. J., Campbell, R. N., and MacGregor, S. A. 2004. Extent and mechanisms of gloving and unmixed resin in fully encapsulated roof bolts and a review of recent developments. In: Villaescusa, E., and Potvin, Y. (eds) Ground support in mining & underground construction, Proc 5th int symp on ground support, Balkema, Leiden, Perth, 231–242.

Myrdal, R. 2007. State of the art report, accelerating admixtures for concrete. Trondheim: SINTEFF Building and Infrastructure, Concrete Innovation Centre (COIN), 23.

Nagi, M. A., Okamoto, P. A., Kozikowski, R. L., and Hover, K. 2007. Evaluating air-entraining admixtures for highway Concrete (NCHRP report 578). Washington, DC: Transportation Research Board, 49.

Neville, A. M. 1963. Properties of concrete. London: Pitman, 532.

Neville, A. M. 1995. Properties of concrete (4th ed.). Essex: Pearson Education Ltd, 844.

Nguyen, Q. D., and Boger, D. V. 1985. Direct yield stress measurements with the vane method. Journal of Rheology, 29(3):335–347.

O'Donnell, J. D. P. 1992. The use of distressing at Inco's Creighton Mine. In: Proceedings of MassMin 92, SAIMM, 71–74.

O'Donnell, J. D. P. 1999. The development and application of distressing techniques in the mines of INCO Limited, Sudbury, MSc Thesis, Laurentian University, Ontario.

O'Shea, W. M. 2005. In-Cycle fibrecreting at the Plutonic gold mine. B. Eng. Thesis, Western Australian School of Mines, Curtin University, WA, Australia, 44.

O'Toole, D., and Pope, S. 2006. Design, testing and implementation of in-cycle shotcrete in the northern 3500 orebody. In: Morgan, D., and Parker, H. (eds) Proceedings of the 10th international conference on shotcrete for underground support X, American Society of Civil Engineers, Whistler, BC, Canada, 316–327.

Ortlepp, W. 1983. Considerations in the design of support for deep hard rock tunnels. In: 5th international congress on rock mechanics. Vol. 2, Balkema, Rotterdam.

Ortlepp, W. D. 1993. High ground displacement velocities associated with rockburst damage. Rockbursts and Seismicity in Mines, 93:101–106.

Ortlepp, W. D., and Stacey, T. R. 1994. Rockburst mechanisms in tunnels and shafts. Tunnelling and Underground Space Technology, 9(1):59–65. http://dx.doi.org/10.1016/0886-7798(94)90010-8.

Paglia, C., Wombacher, F., and Böhni, H. 2001. The influence of alkali-free and alkaline shotcrete accelerators within cement systems: I. Characterization of the setting behavior, Cement and Concrete Research, 31(6), 913–918.

Pahl, P. J. 1981. Estimating the mean length of discontinuity traces. International Journal of Rock Mechanics & Mining Sciences, 18(3):221–228.

Pakalnis, V., and Ames, D. 1983. Load tests on mine screening. Underground support systems. Udd, J (ed), CIMM, Montreal, 79–83.

Peng, S. S., and Tang, D. H. Y. 1984. Roof bolting in underground mining: A state-of-the-art review. International Journal of Mining and Mineral Engineering, 2:1:42.

Player, J. R. 2012. Dynamic testing of rock reinforcement systems, PhD Thesis, Western Australian School of Mines, Curtin University of Technology, Perth, Western Australia, Australia, 501.

Player, J. R., Morton, E. C., Thompson, A. G., and Villaescusa, E. 2008. Static and dynamic testing of steel wire mesh for mining applications of subsurface support. In: Stacey, T. R., and Malan, D. F. (eds) Proc 6th Int Symp on Ground Support in Mining & Civil Engineering Construction, SAIMM, Cape Town, Johannesburg.

Player, J. R., Villaescusa, E., and Thompson, A. G. 2004. Dynamic testing of rock reinforcement using the momentum transfer concept. In: Villaescusa, E., and Potvin, Y. (eds) 5th international symposium of ground support in mining and construction, Balkema, Leiden, 327–339.

Player, J. R., Villaescusa, E., and Thompson, A. G. 2009a. Dynamic testing of friction rock stabilisers. In: Diederichs, M., and Grasselli, G. (eds) RockEng09, Rock Engineering in Difficult Conditions, Toronto, 9–15 May, Paper 4027, 12, Montreal CIM.

Player, J. R., Villaescusa, E., and Thompson, A. G. 2009b. Dynamic testing of threadbar used for rock reinforcement. In: In: Diederichs, M., and Grasselli, G. (eds) RockEng09, Rock Engineering in Difficult Conditions, Paper 4030, CIM: Montreal, 12.

Popovics, S. 1992. Concrete materials: Properties, specifications, and testing. New Jersey: Noyes Publications, 542.

Potvin, Y., and Giles, G. 2008. The development of a new high-energy absorption mesh. 10th Underground operators conference. Launceston Tasmania. Melbourne AusIMM.

Potvin, Y., Tyler, D. B., MacSporran, G., Robinson, J., Thin, I., Beck, D., and Hudyma, M. 1999. Development and implementation of new ground support standards at Mount Isa Mines Limited. In: Villaescusa, E., Windsor, C. R., and Thompson, A. G. (eds) Procc of rock support and reinforcement practice in mining, Balkema, Kalgoorlie, 367–371.

Powers, T. C. 1968. The properties of fresh concrete. New York: John Wiley & Sons, Inc, 664.

Priest, S. D. 1985. Hemispherical projection methods in rock mechanics. London: Allen & Unwin, 124.

Ramachandran, V. S. 1995. Concrete admixture handbook: Properties, science and technology (2nd ed.). Ottawa: National Research Council Canada, Noyes Publications, 1153.

Regourd, M. 1980. Characterization of thermal activation of slag cements. Proceeding, 7th International Congress on the Chemistry of Cements, Septima, Paris, Volume 2, 105–111.

Reny, S., and Jolin, M. 2011. Improve your shotcrete: Use coarse aggregates. *Shotcrete Magazine*, 26–28.

RILEM. 1986. The hydration of tricalcium aluminate and tetracalcium aluminoferrite in the presence of calcium sulphate, Mathematical Modelling of Hydration of Cement. RILEM Technical Committees 68-MMH Task Group 3 Report, RILEM publication, Bagneux, France, 137–147.

Rispin, M., Knight, B., and Dimmock, T. 2003. Early re-entry into working faces in mines through modern shotcrete technology—Part II. Canadian Institute of Mining—Mines Operations Centre, Saskatoon, SK, Canada.

Romualdi, J. P., and Batson, G. B. 1963. Mechanics of crack arrest in concrete. Journal of Engineering Mechanics, 89:147–168.

Ross, A., Barley, M., Brown, S., McNaughton, N., Ridley, J., and I. Fletcher 2004. Young porphyries, old zircons: new constraints on the timing of deformation and gold mineralisation in the Eastern Goldfields from SHRIMP U—Pb zircon dating at the Kanowna Belle Gold Mine, Western Australia. Precambrian Research, 128:105–142.

Roth, A., Windsor, C. R., Coxon, J., and de Vries, R. 2004. Performance assessment of high tensile wire mesh ground support under seismic conditions. In: Villaescusa and Potvin (ed), Ground support in mining and underground construction. Taylor and Francis Group, London, 589–594.

Roth, A.P., 2023. Dynamic loading of ground support schemes. PhD Thesis. Curtin University (in Preparation).

Roy, D. M., and Idorn, G. M. 1982. Hydration, structure and properties of blast furnace slag cement, mortars and concrete. ACI Journal, 79(6):445–457.

Sandvik. 2018. DS411 Rock Support Bolter—Technical Specification. Sandvik: Mining and Rock Technology.

Satola, I., and Aromaa, J. 2004. The corrosion of rock bolts and cable bolts. In: Villaescusa, E., and Potvin, Y. (eds) Ground support in mining and underground construction, Balkema, Leiden, 521–528.

Saw, H. A. 2015. Early strength of shotcrete, PhD Thesis, Curtin University of Technology, Chile, 277.

Saw, H. A., Villaescusa, E., Windsor, C. R., and Thompson, A. G. 2013. Laboratory testing of steel fibre reinforced shotcrete. International Journal of Rock Mechanics & Mining Sciences 57:167–171.

Saw, H. A., Villaescusa, E., Windsor, C. R., and Thompson, A. G. 2017. Surface support capabilities of freshly sprayed fibre reinforced concrete and safe re-entry time for underground excavations. Tunnelling and Underground Space Technology, 64:34–42. doi:10.1016/j.tust.2017.01.005.

Saw, H., Villaescusa, E., Windsor, C. R., and Thompson, A. G. 2009. Non-linear, elastic—plastic response of steel fibre reinforced shotcrete to uniaxial and triaxial compression testing. International Conference on Shotcrete for Underground Support XI, Davos, Switzerland. http://services.bepress.com/eci/shotcrete/.

Sellevold, E. J. 1987. The function of condensed silica fume in high strength concrete. In: Holand, I., Helland, S., Jakobsen, B., and Lenschow, R. (eds) Proceeding of Symposium on Utilisation of High Strength Concrete, Tapir Publishers, Trondheim, 39–49.

Sellevold, E. J., and Nilsen, T. 1987. Condensed silica fume in concrete: A world review, Supplementary cementing materials for concrete. Ottawa: CANMET, 165–243.

Simser, B., Andrieux, P., and Gaudreau, D., 2002. Rockburst support at Noranda's Brunswick Mine, Bathurst, New Brunswick. In R Hammah, W Bawden, J Curran & M Telesnicki (eds), Mining & Tunnelling Innovation & Opportunity, Proc 5th North American Rock Mech Symp & 17th Tunnelling Association of Canada Conf, Toronto, 7-10 July, 1: 805–813. Toronto: University of Toronto.

Slade, N., and Kuganathan, K. 2004. Mining through filled stopes using shotcrete linings at Xstrata Mount Isa Mines. Int. Conf. on Engineering Developments in Shotcrete, Cairns.

Smolczyk, H. G. 1978. The effect of the chemistry of slag on the strength of blast furnace cements. Zement KalGips, 31(6):294–296.

SS 13 72 43. 1987. Concrete Testing—Hardened Concrete—Adhesion Strength. Swedish Standard, in Swedish.

Stacey, T., and W. Ortlepp 2001. The capacities of various types of wires mesh under dynamic loading. Section 2. International seminar on mine surface support liners: Membrane, shotcrete and mesh. Australian Centre for Geomechanics, Perth.

Standards Australia, 1973. Ground anchorages. In Prestressed Concrete Code CA35, 50–53.

Standards Australia, 2007a. ASNZS 4672.1:2007 Steel prestressing materials Part 1: General requirements, 44, Sydney: Standards Australia.

Standards Australia, 2007b. ASNZS 4672.1:2007 Steel prestressing materials Part 2: Testing requirements, 16, Sydney: Standards Australia.

Stillborg, B., 1994. Professional users handbook for rock bolting. Series on Rock and Soil Mechanics, Vol. 18, Trans Tech Publications, 164.

Tannant, D. 1995. Load capacity and stiffness of welded wire mesh. 48th Canadian geotechnical conference. Vancouver.

Tannant, D. 2001. Load capacity and stiffness of welded wire, chain link and expanded metal mesh. Section 3. International seminar on mine surface support liners: Membrane, shotcrete and mesh. August. Australian Centre for Geomechanics, Perth.

Tannant, D., P. Kaiser and S. Maloney 1997. Load – displacement properties of welded wire, chain link and expanded metal mesh. International symposium on rock support – Applied solutions for underground structures. Lillehammer Norway. E. Broch, A. Myrvang and G. Stjern (Eds). June 22–25. 651–659.

Taylor, H. F. W, Barret, P., Brown, P. W., Double, D. D., Frohnsdorff, G., Johansen, V., Ménétrier-Sorrentino, D., Odler, I., Parrott, L. J., Pommersheim, J. M., Regourd, M., and Young, J. F. 1984. The hydration of tricalcium silicate. Journal of Mechanics of Materials and Structures 17:457–468.

Taylor, H. F. W. 1997. Cement chemistry (2nd ed.). London: Thomas Telford Services Ltd, 459.

Taylor, H. F. W. 1997. Cement. London: Thomas Telford, 459.

Tejchman, J., and Kozicki, J. 2010. Experimental and theoretical investigations of steel-fibrous concrete. Berlin: Springer, 280.

Thompson, A. G, Windsor, C. R., and Cadby, G. 1999. Performance assessment of mesh for ground control applications. In: Villaescusa, W. T (ed) Rock support and reinforcement practice in mining. Balkema, Rotterdam, 119–130.

Thompson, A. G. 1989. Analysis of a reinforced arbitrary shaped block. In: Hogarth, W. L., and Noye, B. J. (eds) Proc. International Conference on Computational Techniques and Applications, CTAC-89, Hemisphere Publishing, New York, 723–730.

Thompson, A. G. 1992. Tensioning reinforcing cables. Proc. Int. Symp. On Rock Support, Sudbury, Balkema, Rotterdam, 285–291.

Thompson, A. G. 2001. Rock support action of quantified by testing and analysis. Section 1. International seminar on mine surface support liners: Membrane, shotcrete and mesh. Perth: Australian Centre for Geomechanics.

Thompson, A. G. 2002. Stability assessment and reinforcement of block assemblies near underground excavations. In: Hammah, R., Bawden, W., Curran, J., and Telesnicki, M. (eds) Mining and Tunnelling Innovation and Opportunities, University of Toronto Press, Toronto, Volume 2, 1439–1446.

Thompson, A. G. 2004. Performance of cable bolt anchors—An update. In: Karzulovic, A., and Alfaro, M. A. (eds) MassMin2004, Instituto de Ingenerios de Chile, Santiago, 317–323.

Thompson, A. G. 2011. Reinforcement systems subjected to complex loadings. In: Proceedings of 14th Pan-American Conference on Soil Mechanics and Geotechnical Engineering, Paper 594, Canadian Geotechnical Society, Vancouver, 8.

Thompson, A. G., and Villaescusa, E. 2013. Case studies of rock reinforcement components and systems testing. In: Proceedings 47th US Rock Mechanics/Geomechanics Symposium, Paper ARMA 13–642, 10.

Thompson, A. G., and Villaescusa, E. 2014. Case Studies of Rock Reinforcement Components and Systems Testing, Rock Mechanics and Rock Engineering, Rock Mechanics and Rock Engineering, Special Issue, DOI 10.1007/s00603–014–0583-z.

Thompson, A. G., and Windsor, C. R. 1992. A classification system for reinforcement and its use in design. In: Szwedzicki, T., Baird, G. R., and Little, T. N. (eds) Proceedings of the Western Australian conference on mining geomechanics, Kalgoorlie, Western Australian School of Mines, 115–125.

Thompson, A. G., and Windsor, C. R. 1993. Theory and strategy for monitoring the performance of rock reinforcement. Proc. Australian Conference on Geotechnical Instrumentation and Monitoring in Open Pit and Underground Mining, Balkema, Rotterdam, Kalgoorlie, 473–482.

Thompson, A. G., and Windsor, C. R. 1995. Tensioned cable bolt reinforcement—An integrated case study. In: Proceedings of 8th International Society for Rock Mechanics and Rock Engineering, Tokyo, June, V2, 679–683, Rotterdam: Balkema.

Thompson, A. G., and Windsor, C. R. 1998. Cement grouts in theory and reinforcement practice. NARMS 98, Cancun, Paper Aus-330–2p.

Thompson, A. G., Player, J. R., and Villaescusa, E. 2004. Simulation and analysis of dynamically loaded reinforcement systems. In: Villaescusa, E., and Potvin, Y. (eds) Ground support in mining & underground construction, Proc 5th Int Symp on Ground Support, Perth, 28–30 September, 341–355. Leiden: Balkema.

Thompson, A. G., S. M. Matthews, C. R. Windsor, S. Bywater and V. H. Tillmann 1987. Innovations in rock reinforcement technology in the Australian mining industry. 6th ISRM International Congress on Rock Mechanics, Canada, September, V2:1275–1278.

Thompson, A. G., Villaescusa, E., and Windsor, C. R. 2012. Ground support terminology and classification: An update. Geotechnical and Geological Engineering, 30(3):553–580.

Thompson, A. G., Villaescusa, E., and Windsor, C. R. 2013. Planning and documenting reinforcement system test programs. In: Anagnostou, G., and Ehrbar, H (eds) World tunnel congress. Underground—the way to the future, Paper, Geneva, 1534, 8.

Thompson, A. G., and E. Villaescusa, 2022. Development, creation and application of a reinforcement systems database. Procc 56th US Rock Mechanics/Geomechanics Symposium, American Rock Mechanics Association, Paper Number 0299, Santa Fe.

Thompson, A. G., and S. M. Mathews 1989. Design packages for reinforced excavations in rock. Procc. Conf. Computer Systems in the Australian Mining Industry, The University of Wollongong.

Tinucci, J. P. 1992. A ground control computer program for support analysis of three-dimensional critical rock blocks. In: Kaiser, P. K., and McCreath, D. (eds) Rock support and underground construction, Balkema, Rotterdam, 49–56.

Tran, V. N. G., Beasley, A. J., and Bernard, E. S. 2001 Application of yield line theory to round determinate panels. In: Bernard, E. S. (ed) Shotcrete: Engineering developments, Swets and Zeitlinger, Lisse, 245–254.

Tran, V. N. G., Bernard, E. S., and Beasley, A. J. 2005. Constitutive modeling of fiber reinforced shotcrete panels. ASCE Journal of Engineering Mechanics, 512–521.

Van Sint Jan, M., and Cavieres, P. 2004. Large scale static laboratory test of different support systems. In: Villaescusa and Potvin (ed) Ground support in mining and underground construction, Taylor and Francis Group, London, 571–577.

Varden, R. 2005. A methodology for selection of resin grouted bolts. Master of Engineering Science Thesis, Western Australian School of Mines, Curtin University of Technology, Chile, 113.

Varden, R., and Esterhuizen, H. 2012. Kanowna Belle—evolution of seismicity with increasing depth in an ageing mine. In: Proceedings of Deep Mining, Australian Centre for Geomechanics, Perth.

Varden, R., Lachenicht, L., Player, J. R., Thompson, A. G., and Villaescusa, E. 2008. Development and implementation of the Garford dynamic bolt at the Kanowna Belle Mine. Proc 10th Underground Operators' Conf, Launceston, The AusIMM, 95–102.

Velasco, R. V., Filho, R. D., Fairbairn, E. M., and Silvoso, M. M. 2008. Basic creep of steel fibers reinforced composites. Proc. 8th International Conference on Creep, Shrinkage and Durability of Concrete and Concrete Structure, Ise-Shima, 1:735–739.

Villaescusa, E. 1991. A three-dimensional model of rock jointing, PhD Diss., University of Queensland, 252.

Villaescusa, E. 1999. Keynote lecture: The reinforcement process in underground mining. In: Villaescusa, E., Windsor, C. R., and Thompson, A. G. (eds) Rock support and reinforcement practice in mining, Proceedings of the international symposium on ground support, A.A. Balkema, Kalgoorlie, Rotterdam, 245–257.

Villaescusa, E. 1999. Laboratory testing of weld mesh for rock support. In E. Villaescusa, C. R. Windsor, and A. G. Thompson (eds.), Rock Support and Reinforcement Practice in Mining, Proceedings of the International Symposium on Ground Support, Kalgoorlie, Western Australia, Australia, 15–17 March, 155–159. Rotterdam, the Netherlands: A.A. Balkema.

Villaescusa, E. 2014. Geotechnical design for sublevel open stoping. Boca Raton, FL: CRC Press.

Villaescusa, E., A. de Zoysa, J. R. Player and A. G. Thompson 2016. Dynamic testing of combined rock bolt and mesh schemes. Procc. Seventh International Conference & Exhibition on Mass Mining, Paper 21, The AusIMM, Sydney, Australia.

Villaescusa, E., A. G. Thompson and J. R. Player 2005. Dynamic testing of ground support systems Phase I. The Minerals and Energy Research Institute of Western Australia. Report Number 249, 148.

Villaescusa, E., A. G. Thompson and J. R. Player 2015. Dynamic testing of ground support systems. The Minerals and Energy Research Institute of Western Australia. Report Number 312, 146.

Villaescusa, E., and Brown, E. T. 1992. Maximum likelihood estimation of joint size from trace length measurements. Rock Mechanics and Rock Engineering, 25:67–87.

Villaescusa, E., and Schubert, C. J. 1999. Monitoring the performance of rock reinforcement. Geotechnical and Geological Engineering, 17:1–13.

Villaescusa, E., and Wright, J. 1997. Permanent excavation reinforcement using cement grouted split set bolts. Proceedings of the AusIMM, 1:65–69.

Villaescusa, E., and Wright, J. 1999. Reinforcement of underground excavations using the CT bolt. In: Villaescusa, E., Windsor, C. R., and Thompson, A. G. (eds) Rock Support and Reinforcement Practice in Mining, Proceedings of the International Symposium on Ground Support, A.A. Balkema, Kalgoorlie, Rotterdam, 109–115.

Villaescusa, E., Hassell, R., and Thompson, A. G. 2008. Development of a corrosivity classification for cement grouted cable strand in underground hard rock mining excavations. In: Stacey, T. R., and Malan, D. F. (eds) Proc 6th int symp on ground support in mining & civil engineering construction, SAIMM, Cape Town, 461–476.

Villaescusa, E., Kusui, A., and Drover, C. 2016. Ground support design for sudden and violent failures in hard rock tunnels. Keynote Lecture, 9th Asian Rock Mechanics Symposium, Bali, Indonesia.

Villaescusa, E., M. Seto and G. Baird 2002. Stress measurements from oriented core. International Journal of Rock Mechanics and Mining Sciences, 39(5):603–615.

Villaescusa, E., Neindorf, L. B., and Cunningham, J. 1994. Bench stoping of lead/zinc orebodies at Mount Isa Mines Limited. In: New Horizons in Resource Handling and Geoengineering, Proceedings of the Joint MMIJ/AusIMM Symposium, Ube, Yamaguchi, 1–5 October, 351–359.

Villaescusa, E., Player, J. R., and Thompson, A. G. 2014. A reinforcement design methodology for highly stressed rock masses. 8th Asian Rock Mechanics Symposium, DOI: 10.13140/2.1.4423.8402.

Villaescusa, E., R. Varden and R. Hassell 2008. Quantifying the performance of resin anchored rock bolts in the Australian underground hard rock mining industry. International Journal of Rock Mechanics and Mining Sciences, 45(1):94–102.

Villaescusa, E., Sandy, M. P., and Bywater, S. 1992. Ground support investigations and practices at Mount Isa. In: Kaiser, P. K., McCreath, D. (eds) Rock support in mining and construction, Balkema, Rotterdam, 185–193.

Villaescusa, E., Thompson, A. G., Player, J. R., and Morton, E. C. 2010. Dynamic testing of ground support systems. Report No. 287 MERIWA Project M349, Phase 2-Final Report, 207.

Warburton, P. M. 1981. Vector stability analysis of an arbitrary polyhedral rock block with any number of free faces. International Journal of Rock Mechanics and Mining Sciences, 18(5):415–427.

Warburton, P. M. 1993. Some modern developments in block theory for rock engineering. In: Hudson, J. A. (ed) Comprehensive rock engineering, Pergamon, Oxford, Volumes 2, 293–315.

Wiles, T., Villaescusa, E., and Windsor, C. R. 2004. Rock reinforcement design for overstressed rock using three dimensional numerical modelling. In: Villaescusa, E., and Potvin, Y. (eds) Ground support in mining and underground construction, proceedings of the fifth international symposium on ground support, A.A. Balkema, Perth, Leiden, 483–489.

Windsor, C. R. 1992a. Block stability in jointed rock masses. In: Cook, N. G. E, Goodman, R. E, Myer, L. R., and Tsang, C. F (eds) Fractured and jointed rock masses, Balkema, Rotterdam, 59–66.

Windsor, C. R. 1992b. Cable bolting for underground and surface excavations. In: Kaiser, P. K., and McCreath, D. (eds) Rock support in mining and construction, Balkema, Rotterdam, 349–366.

Windsor, C. R. 1995. Improvements in rock reinforcement practice. Proc of The Australian Mining Summit, Conference 2 — Implementing and Managing New Technologies in Mining. Institute for International Research. IIR, Sydney, Australia. 1–17.

Windsor, C. R. 1995. Rock Mass Characterization—A course on structural characterization and structural analysis, Course notes for the Masters of Engineering Science in Mining Geomechanics at the Western Australian School of Mines, 462.

Windsor, C. R. 1996. Schlumberger Lecture—Rock Reinforcement Systems. International Journal of Rock Mechanics and Mining Sciences, 34(6):919–951.

Windsor, C. R. 1997. Rock reinforcement system. International Journal of Rock Mechanics and Mining Sciences, 5:919–951.

Windsor, C. R. 1999. Systematic design of reinforcement and support schemes for excavations in jointed rock. Keynote lecture, Procc. Int. Symp. Rock Support and Reinforcement Practice in Mining, Villaescusa, Windsor and Thompson (Eds), Balkema Rotterdam, 35–58.

Windsor, C. R. 2004. A review of long, high capacity reinforcing systems used in rock engineering. In: Villaescusa E. and Potvin Y. (eds) Ground support in mining and underground construction, Taylor and Francis Group, London, 17–41. 30.

Windsor, C. R. 2009. A Review of geotechnical practice at BHP Billiton Cannington Mine. CRC Mining – Western Australian School of Mines Confidential Report, Unpublished.

Windsor, C. R., and Thompson, A. G. 1992a. A new friction stabilizer assembly for rock and soil reinforcement applications. Proc. International Symposium on Rock Support, Sudbury, June, 523–529, Balkema, Rotterdam.

Windsor, C. R., and Thompson, A. G. 1992b. Reinforcement design for jointed rock masses. In: Tillerson, J. R., and Wawersik, W. R. (eds) Proceedings of the 33rd US symposium rock mechanics, A.A. Balkema, Santa Fe, NM, Rotterdam, 521–530.

Windsor, C. R., and Thompson, A. G. 1992c. SAFEX—A design and analysis package for rock reinforcement. In: Kaiser, P. K., and McCreath, D. (eds) Rock support and underground construction, Balkema, Rotterdam, 17–23.

Windsor, C. R., and Thompson, A. G. 1993. Rock reinforcement—Technology, testing, design and evaluation. In: Hudson, J. A., Brown, E. T., Fairhurst, C., and Hoek, E. (eds) Comprehensive rock engineering, Pergamon, Oxford, Volume 4, 451–484.

Windsor, C. R., Bywater, S., and Worotnicki, G. 1983. Instrumentation and observed behavior of the N663 trial stope Racecourse Area, Mount Isa Mine (Project Report No. 23). Melbourne: Geomechanics of Underground Metalliferous Mines, CSIRO Division of Geomechanics.

Windsor, C. R., and A. G. Thompson 1997. Reinforced rock system characteristics. Procc. Int. Symp. on Rock Support. E. Broch, A. Myrvang and G. Stjern (Eds), Lillehammer, 433–448.

Windsor, C. R., 1999. Structural design of shotcrete linings. Australian Shotcrete Conference and Exhibition, Sydney, IBC Conference, 34.

Wood, K. 1981. Twenty years of experiences with slag cement. Symposium on slag cement, University of Alabama, Birmingham.

Wu, Y. K., and Oldsen, J. 2010. Development of a new yielding rock bolt—Yield-Lok bolt. Proc. 44th U.S. Rock Mechanics Symposium, Document 10–197.

Zuo, Y. J., Li, X. B., and Zhou, Z. L. 2005. Determination of ejection velocity of rock fragments during rock bust in consideration of damage. Journal of Central South University, 12(5).

Index

Note: Page numbers in *italics* indicate a figure and page numbers in **bold** indicate a table on the corresponding page.

A

accelerator, 243
admixtures, 243–244, 253
aggregate, 239
air-entraining, 243
anchors
    barrel and wedge, 207–212
    barrel and wedge strand, 98–99, *98–99*
    expansion shell, 94–96
    slot and wedge, 93
areal support systems, 22, 25
assessment
    deterministic assessment of rock mass instabilities, 31–32
    probabilistic assessment of rock mass instabilities, 33–36
Australian threaded bar, 142
axial force-displacement responses, 335–336

B

barrel and wedge strand anchors, 98–99
bolts, *see* energy dissipation of rock bolts
boreholes
    grouting, 185–189
    installation, 132–133
    simulated, 111–112, 132–133
boundary conditions
    mesh testing, 220
brittle, 18

C

cable bolts, *see also* energy dissipation of cable bolts
    cable bolt pull tests, 104
capacity, 80–82, *81–82*, **81**
    performance indicators of, 19–20
case study (de-stress blasting), 366–378
cement
    hardened, 178–183
    hydration of, 236
    material properties, 236
    supplementary materials, 237–238
cement- and resin-encapsulated Cone bolt, 157–160
cement- and resin-encapsulated D bolt, 156
cement- and resin-encapsulated Garford Dynamic Bolt, 161–165, *162–165*
cement-encapsulated decoupled threaded bar, 148
cement-encapsulated Durabar bolt, 165
cement-encapsulated rebar and G80/4 mesh, 290–294
cement-encapsulated rebar and welded wire mesh, 295–297
cement-encapsulated threaded bar, 140–143
    Australian, 142
    Chilean, 142–143

cement-encapsulated Yield-Lok bolt, 166–169, *167–169*
cement grouts, 97, 175–188
    grouting reinforcement boreholes, 185–189
    physical and mechanical properties, 175–183
cement paste, fresh, 175–177
characterization, rock mass, 27–29, *28–29*, 61–62, *62*
Chilean threaded bar, 142–143
collar region
    load transfer within, 43–45
combined reinforcement and mesh schemes
    cement-encapsulated rebar and G80/4 mesh, 290–294
    cement-encapsulated rebar and welded wire mesh, 295–297
    data analysis, 288
    decoupled posimix and welded wire mesh, 298–305
    decoupled posimix and woven mesh, 306–320
    sample preparation and testing, 287
    summary of energy dissipation, 321
component testing, 87–89, *88*, **89**
construction and monitoring, 351
    deformation, 401
    de-stress blasting, 357–379
        case study, 366–378
        de-stress blasting patterns, 361
        explosive energy, 362–363
        geotechnical concerns, 365
        mechanics, 359–360
        micro seismic activity, 364
    drilling and blasting, 398–400
    excavation shape, 355–357, *356–357*
    high energy dissipation ground support scheme, 379–396
        clearing of temporary face support, 380
        final ground support scheme arrangement, 396
        installation of primary reinforcement and mesh, 384–391, *384–391*
        mechanical scaling, 380
        primary reinforcement mark-up, 383
        primary to secondary support installation sequence, 391–395, *392–395*
        shotcrete application, 381–382
        structural geological mapping with photogrammetry, 381
    induced stress, 351–353
continuous frictionally coupled, 16, 123–139
    expanded tube bolts, 126–132, *127–132*
    friction rock stabilizer, 123–125
    hybrid point anchored bar and split tube bolts, 132–140, *133–140*, **133**
continuous mechanically coupled, 13–15, 140–144, *141*, *143–144*
    cement-encapsulated threaded bar, 140–143
    resin-encapsulated threaded bar, 144–147, *145–147*

D

damage
    prior to violent failure, 66
data, 289, **290–291**
    analysis, 288
    cement-encapsulated rebar and G80/4 mesh, 290–294
    cement-encapsulated rebar and welded wire mesh, 295–297
    decoupled posimix and welded wire mesh, 298–305
    decoupled posimix and woven mesh, 306–320
    reinforcement databases, 328–339
        application of, 337
        generic reinforcement system, 332
        measurement of reinforcement system responses, 333
        reinforcement system responses, 333–336, *334–336*
        reinforcement system simulations, 333
decoupled posimix and welded wire mesh, 298–305
decoupled posimix and woven mesh, 306–320
decoupled strand, 201–203
deformation, 401
    mechanisms, 247
    prior to violent failure, 66
demand, ground support, 73–80, *74, 76–80*
design, ground support, 61
    ejection velocity, 67–72
    ground support capacity, 80–82
    ground support demand, 73–79
    rock mass characterization, 61–62
    tunnel instability, 62–66, *63–66*
de-stress blasting, 357–379, *358, 360–379*
    case study, 366–378
    de-stress blasting patterns, 361
    explosive energy, 362–363
    geotechnical concerns, 365
    mechanics, 359–360
    micro seismic activity, 364
deterministic assessment of rock mass instabilities, *31–32*, 31–32
discrete mechanically or frictionally coupled, 16, 147
    cement- and resin-encapsulated Cone bolt, 157–160
    cement- and resin-encapsulated D bolt, 156
    cement- and resin-encapsulated Garford Dynamic Bolt, 161–164
    cement-encapsulated decoupled threaded bar, 148
    cement-encapsulated Durabar bolt, 165
    cement-encapsulated Yield-Lok bolt, 166–169
    resin-encapsulated decoupled posimix, 149–156, *150–156*
    self-drilling anchor bolt, 170
displacement
    displacement controlled dynamic analysis methodology, 343
    force, 111
    mesh, 220–231
drilling and blasting, 398–401, *398–401*
ductile, 19
dynamic analysis methods, 343–345
dynamic block loading, 341–342, *341*
dynamic performance of ground support schemes, 283
    combined reinforcement and mesh schemes, 286–321
        cement-encapsulated rebar and G80/4 mesh, 290–294

cement-encapsulated rebar and welded wire mesh, 295–297
        data analysis, 288
        decoupled posimix and welded wire mesh, 298–305
        decoupled posimix and woven mesh, 306–320
        sample preparation and testing, 287
        summary of energy dissipation, 321
    free body diagrams, 285
    load transfer, 283–284, *284*
dynamic testing
    dissipated dynamic energy, 231
    expanded tube bolts, 129–131
    friction rock stabilizer, 124–125
    hybrid point anchored bar and split tube bolts, 134–139, **136**
    large scale, 270–280
    mesh and shotcrete layers, of, 119
    results, 113–118, 213
    welded wire mesh, 228
    woven mesh, 228–230

E

ejection velocity, 67–72, *68–70, 72–73*
elastic, 18
element
    load transfer along the, 46–47
    testing, 90–92, *90–93*
embedment length concept, 18
energy-based analysis, 343
energy dissipation of cable bolts, 173–174, *174*
    cable bolt plates, 206–213
    cable bolt types, 189–206, *189–206*
        decoupled strand, 201–203
        modified strand cable bolts, 189
        multiple dynamic impact testing, 204–205
        plain strand—15.2 mm diameter, 190–195
        plain strand—17.8 mm diameter, 196–200
    cement grout, 175–188, *176–178, 182–188*
        grouting reinforcement boreholes, 185–189
        physical and mechanical properties, 175–183
energy dissipation of mesh support, 215
    mesh force and displacement, 220–231
        dissipated dynamic energy, 231
        dissipated static energy, 227
        dynamic results for welded wire mesh, 228
        dynamic results for woven mesh, 228–230
        static results for welded wire mesh, 223
        static results for woven mesh, 224–226
    mesh load transfer, 215–218, *216–218*
    mesh testing, 218–220, *219*
energy dissipation of rock bolts, 121
    continuously frictionally coupled, 123–139
        expanded tube bolts, 126–131
        friction rock stabilizer, 123–125
        hybrid point anchored bar and split tube bolts, 132–139
    continuously mechanically coupled, 140–146, **141–142**
        cement-encapsulated threaded bar, 140–143
        resin-encapsulated threaded bar, 144–146
    discretely mechanically or frictionally coupled, 147–170

cement- and resin-encapsulated Cone bolt, 157–161, 158–161
cement- and resin-encapsulated D bolt, 156
cement- and resin-encapsulated Garford Dynamic Bolt, 161–164
cement-encapsulated decoupled threaded bar, 148
cement-encapsulated Durabar bolt, 165
cement-encapsulated Yield-Lok bolt, 166–169
resin-encapsulated decoupled posimix, 149–155
self-drilling anchor bolt, 170
momentum transfer, 121
reinforcement system load transfer, 121–122
energy dissipation of shotcrete support, 235
large scale dynamic testing, 270–280
large scale static testing, 267–269
material properties, 236–244, 237, 240–242
shotcrete failure mechanisms, 265–266
shotcrete mix design, 235
shotcrete support system, 244–247, 245–247
static performance of cured shotcrete, 259–264, 260–262, 264
static performance of freshly sprayed shotcrete, 248–258
excavation shape, 355–357
expanded tube bolts, 126–131
expansion shell anchors, 94–96
explosive energy, 362–363
external fixture testing, 97–100

F

failure
damage and deformation prior to violent failure, 66
geometry, 50–52, 51–52
shallow depth of, 75
shotcrete failure mechanisms, 265–266, 265–266, 273
spalling, 63–64, 75
structurally controlled, 65, 75–79
fibres, 240–242
fibre-reinforced shotcrete, 274–280, **274–275, 277–280**
fly ash, 238
force
mesh, 220–231, 221–230
transfer and displacement, 111
free body diagrams, 285
fresh cement paste, 175–177
frictionally coupled, *see* continuous frictionally coupled; discrete mechanically or frictionally coupled
friction rock stabilizer, 123–125

G

geotechnical concerns, 365
geotechnical mapping of underground exposures, 28–29
grouts, *see also* cement grouts
resin, 97

H

hardened cement, 178–183
high demand conditions, 82
high energy dissipation ground support scheme, 379–396
clearing of temporary face support, 380

final ground support scheme arrangement, 396
installation of primary reinforcement and mesh, 384–390
mechanical scaling, 380
primary reinforcement mark-up, 383
primary to secondary support installation sequence, 391–395
shotcrete application, 381–382, 382
structural geological mapping with photogrammetry, 381
history of ground support technology, 3–10
hybrid point anchored bar and split tube bolts, 132–139
hydration of cement, 236
stabilizer, 244

I

implemented theory, 344–345, 345
induced stress, 351–354, 352–354
instabilities
deterministic assessment of rock mass instabilities, 31–32
probabilistic assessment of rock mass instabilities, 33–37, 34–37, **40**
tunnel, 62–66
installation
anchor, 207–210
borehole, 132–133
sample, 111–112
internal fixture testing, 93–97, 94–97

L

laboratory testing, 87
component testing, 87–89
element testing, 90–92
external fixture testing, 97–100
internal fixture testing, 93–97
systems testing, 100–108, 101–108
WASM Dynamic Test Facility, 108–119, 109–118
large scale static testing, 267–269, 268–269
loading mechanisms, 22
load transfer
collar region, within the, 43–46, 44–46
concepts, 16–17, 17, 23–25
dynamic performance of ground support schemes, 283–284
element, along the, 46–47
mesh, 215–217
shotcrete, 266
toe anchor region, at the, 46–48, 47–48

M

mapping, *see* geotechnical mapping of underground exposures
materials behaviour terminology, 18
mechanically coupled, *see* continuous mechanically coupled; discrete mechanically or frictionally coupled
mechanics, ground support, 43
failure geometry, 50–51
ground support scheme, 56

reinforcement response, 43–47

surface layer toughness, 52–55, *53–56*

surface support response, 48–50, *49–50*

mechanism, anchor, 211–213, **213**

mechanized grouting, 188

mesh, *see also* energy dissipation of mesh support

combined reinforcement and mesh schemes, 286–321

cement-encapsulated rebar and G80/4 mesh, 290–295, *291–295*

cement-encapsulated rebar and welded wire mesh, 295–297

data analysis, 288

decoupled posimix and welded wire mesh, 298–305

decoupled posimix and woven mesh, 306–320

sample preparation and testing, 287

summary of energy dissipation, 321

dynamic testing of, 119

mesh-reinforced shotcrete, 280

micro seismic activity, 364

models

rock mass, 30

modified strand cable bolts, 189

momentum-based analysis, 342

momentum transfer, 121

monitoring, *see* construction and monitoring

multiple dynamic impact testing, 204–205

N

Newtonian mechanics-based analysis, 342

P

performance

anchor, 211–213

indicators of capacity, 19–20, *20*

plain strand cable bolts, 189

15.2 mm diameter, 190–195

17.8 mm diameter, 196–200

plastic, 18

plates, 100

cable bolt, 206–213, *207–211, 213*

point support systems, 21, 23–24

posimix

welded wire mesh, and, 298–305

woven mesh, and, 306–320

preparation, sample, 287

prestressing strand, 90

probabilistic simulation of rock mass demand, 27

deterministic assessment of rock mass instabilities, 31–32

probabilistic assessment of rock mass instabilities, 33–37

rock mass characterization, 27–29

rock mass model, 30

process of support, 3–9, *3–9*

R

re-entry time, 257–258

reinforced block analysis, 327

description of the problem, 327

displacement controlled dynamic analysis methodology, 343

dynamic block loading, 341–343

example of implemented theory, 344–345

reinforcement databases, 328–339

application of, 337

generic reinforcement system, 332

measurement of reinforcement system responses, 333

reinforcement system responses, 333–336

reinforcement system simulations, 333

reinforcement response at a block face, 327

static analysis of a reinforced arbitrarily-shaped block, 338–340, *339–340*

reinforcement, process of, 3–8

reinforcement boreholes

grouting, 185–189

reinforcements response, 13, 43–47

axial force-displacement responses, 335–336

combined reinforcement and mesh schemes, 286–321

cement-encapsulated rebar and G80/4 mesh, 290–294

cement-encapsulated rebar and welded wire mesh, 295–298, *296–298*

data analysis, 288

decoupled posimix and welded wire mesh, 298–306, *299–306*

decoupled posimix and woven mesh, 306–320, *307–320*, **313, 316, 318, 320**

sample preparation and testing, 287

summary of energy dissipation, 321

component properties, 333–334

interface properties, 335

load transfer within the collar region, 43–45

load transfer along the element and at the toe anchor region, 46–47

reinforcement system load transfer, 121–122, *122*

resilience, 19

resin-encapsulated decoupled posimix, 149–155

resin-encapsulated reinforcement, 105–107

resin-encapsulated threaded bar, 144–146

resin grouts, 97

rock bolts, *see* energy dissipation of rock bolts

rock mass demand, *see* probabilistic simulation of rock mass demands

S

sample

installation, 111–112

preparation and testing, 287

scheme, ground support, 56, *see also* dynamic performance of ground support schemes

self-drilling anchor bolt, 170

shallow depth of failure, 75

shotcrete, *see also* energy dissipation of shotcrete support

dynamic testing of, 119

silica fume, 237

simulation, *see also* probabilistic simulation of rock mass demands

reinforcement system, 333

simulated boreholes, 111–112, 132–133

slag cement, 238
slot and wedge anchors, 93
solid threaded bar, 90–92
spalling failure, 63–64, 75
split tube rings, 98
stabilizer
    friction rock stabilizer, 123–126, *123–126*
static analysis of a reinforced arbitrarily-shaped block,
    338–340
static performance of cured shotcrete, 259–264
    shear strength, 263
    tensile bond strength of rock—shotcrete interface,
        261–263, **263**
    tensile strength, 261
    toughness, 264
    uniaxial compressive strength (UCS),
        259–260
static performance of freshly sprayed shotcrete, 248–259,
    249–255, 257–259
    review of early strength, 249, 250
    safe re-entry time, 257–258
    shear strength of freshly sprayed shotcrete,
        251–253
    structural requirements for a freshly sprayed shotcrete
        layer, 253–256
static testing
    dissipated static energy, 227
    expanded tube bolts, 127–128
    friction rock stabilizer, 124
    large scale, 267–269
    welded wire mesh, 223
    woven mesh, 224–226
steel fibres, 242
stiffness, 19
strength, 62
strip support systems, 22, 25
structurally controlled failure, 65, 75–79
structure, 61
superplasticizer, 243
supplementary cementing materials, 237–238
surface layer toughness, 52–55
surface support response, 21–26, *21–26*, 48–49
    loading mechanisms, 22
    load transfer concepts, 23–25
    type of, 21–22
synthetic fibres, 242
systems testing, 100–107

T

terminology, 13
    continuous frictionally coupled, 16
    continuous mechanically coupled, 13–15, *14–16*, **14**
    discrete mechanically or frictionally coupled, 16
    embedment length concept, 18
    load transfer concept, 16–17
    materials behaviour terminology, 18–19
    performance indicators of capacity, 19–20
    reinforcement system response, 13
    surface support system response, 21–25
testing, *see also* laboratory testing
    dynamic, 124–125, 129–131, 134–139, 213
        large scale, 270–280, *270–280*
            multiple dynamic impact testing, 204–205
    mesh, 218–220
    sample, 287
    static, 124, 127–128, 267–269
toe anchor region
    load transfer at the, 46–47
toe-to-collar grouting, 186–188
toughness, 19
    surface layer, 52–55
tunnel instability, 62–66

U

underground exposures, *see* geotechnical mapping of
    underground exposures

V

velocity
    ejection velocity, 67–72

W

WASM Dynamic Test Facility, 108–119
water
    mixing water, 239
welded wire mesh, 223, 228, 295–297, 400–305
woven mesh, 224–226, 228–230, 306–320

Y

yield, 19